T0192620

Urban Soils

Advances in Soil Science
Series Editors: Rattan Lal and B.A. Stewart

Assessment Methods for Soil Carbon
R. Lal, J.M. Kimble, R.F. Follett, and B.A. Stewart

Food Security and Soil Quality
R. Lal and B.A. Stewart

Global Climate Change: Cold Regions Ecosystems
R. Lal, J.M. Kimble, and B.A. Stewart

Interacting Processes in Soil Science
R.J. Wagenet, P. Baveye, and B.A. Stewart

Methods for Assessment of Soil Degradation
R. Lal, W.H. Blum, C. Valentine, and B.A. Stewart

Principles of Sustainable Soil Management in Agroecosystems
R. Lal and B.A. Stewart

Soil Erosion and Carbon Dynamics
E.J. Roose, R. Lal, C. Feller, B. Barthès, and B.A. Stewart

Soil Management and Greenhouse Effect
R. Lal, J.M. Kimble, E. Levine, and B.A. Stewart

**Soil Management: Experimental Basis for Sustainability
and Environmental Quality**
R. Lal and B.A. Stewart

Soil Management of Smallholder Agriculture
R. Lal and B.A. Stewart

Soil Phosphorus
R. Lal and B.A. Stewart

Soil Processes and the Carbon Cycle
R. Lal, J.M. Kimble, R.F. Follett, and B.A. Stewart

Soil Quality and Biofuel Production
R. Lal and B.A. Stewart

**Soil-Specific Farming: Precision
Agriculture**
R. Lal and B.A. Stewart

**Soil Structure: Its Development
and Function**
B.A. Stewart and K.H. Hartge

Soil Water and Agronomic Productivity
R. Lal and B.A. Stewart

Soils and Global Change
R. Lal, J.M. Kimble, E. Levine, and B.A. Stewart

Structure and Organic Matter Storage in Agricultural Soils
M.R. Carter and B.A. Stewart

World Soil Resources and Food Security
R. Lal and B.A. Stewart

Urban Soils

Edited by
Rattan Lal and B.A. Stewart

CRC Press
Taylor & Francis Group
Boca Raton London New York

CRC Press is an imprint of the
Taylor & Francis Group, an **informa** business

CRC Press
Taylor & Francis Group
6000 Broken Sound Parkway NW, Suite 300
Boca Raton, FL 33487-2742

First issued in paperback 2021

ISBN 13: 978-1-03-209621-6 (pbk)
ISBN 13: 978-1-4987-7009-5 (hbk)

Library of Congress Cataloging-in-Publication Data

Names: Lal, R., author. | Stewart, B. A. (Bobby Alton), 1932- author.
Title: Urban soils / authors: Rattan Lal, and B.A. Stewart.
Other titles: Advances in soil science (Boca Raton, Fla.)
Description: Boca Raton : Taylor & Francis, 2017. | Series: Advances in soil science
Identifiers: LCCN 2017004393 | ISBN 9781498770095 (hardback : alk. paper)
Subjects: LCSH: Urban soils.
Classification: LCC S592.17.U73 L35 2017 | DDC 577.5/6--dc23
LC record available at https://lccn.loc.gov/2017004393

Visit the Taylor & Francis Web site at
http://www.taylorandfrancis.com

and the CRC Press Web site at
http://www.crcpress.com

Contents

Preface

The twenty-first century is the era of urbanization. Global urban population has increased from 746 million in 1950 to 3.9 billion in 2014 and is expected to add another 2.5 billion (with a total of 6.4 billion) by 2050. Globally, 54% of the world population lives in urban areas compared with 30% in 1950 and the projected 66% by 2050. At present, the most urbanized regions include North America (82%), Latin America (80%), and Europe (73%). In comparison, the urban population is 48% in Asia and 40% in Africa. However, these regions are also undergoing the urban revolution. About 90% of the expected increase of 2.5 billion will occur in Asia and Africa. India, China, and Nigeria will account for 37% of the projected growth with addition to urban population of 404 million, 292 million, and 212 million, respectively. At present, 2.7% of the world's land (excluding Antarctica) or 350 million ha (Mha) is under urban development. Similar to the world's area, ~2.6% of the land area of the United States is under urban centers (75.4 Mha). Estimates of rate of growth and land area under urban centers vary widely with estimated increase of land area between 2015 and 2030 of 34 Mha (2.25 Mha/yr) to 120 Mha. With removal of vegetation cover and use of asphalt and cement, urbanization strongly impacts numerous ecosystem properties and processes including soil biodiversity. Global hot spots of biodiversity threatened by urbanization, include the Eastern Afromontane, Guinea forests of West Africa, Western Ghats of India and the adjacent regions of Sri Lanka, and Central and South America. Urban encroachment also impacts the terrestrial carbon pool, including the soil carbon and the aboveground biomass carbon pool. Urbanization may enhance the flux of carbon from soil and vegetation into the atmosphere. There are also strong impacts on surface and subsurface hydrology. By drastically transforming the world's land sequence, urbanization has exacerbated ecological and environmental problems. Increase in artificial surfaces (build-up or impervious surface areas) has a strong impact on soil, soil carbon pool and flux, hydrology, and energy balance. Of the 2.7% of the global urban areas in 2010, 0.65% is build-up and 0.45% is impervious (sealed). Thus, 1.9% of the urban area is partly vegetated and can be managed. However, the manageable soil area is anthropic and comprised of drastically disturbed soils. Several abandoned urban areas can be restored for agricultural, forestry, and recreational uses. Judicious management of urban soils to improve their functionality and ecosystem services can lead to carbon sequestration, increased food production from urban agriculture, and restoration of the environment. The latter encompasses improvement of air quality, water quality, greenspace, and biodiversity, and use of urban soils for food production. Thus, sustainable management of urban soils is among the major challenges and opportunities of the twenty-first century. Urban ecology is an emerging science, and urban soils are an essential component of urban ecosystems.

The focus of this book is on "urban soils," which are composed of geological material that has been drastically disturbed by anthropogenic activities. Therefore, physical, chemical, biological, and ecological properties of urban soils have been severely altered. In general, properties of urban soils are not favorable to plant growth. Most soils are highly compacted, contaminated by heavy metals, and contain aromatic hydrocarbons. Thus, the quality of urban soils must be restored. This 18-chapter volume addresses these issues and presents innovative ideas on food production (e.g., sky farming) and ecological restoration.

The editors thank all the authors for their outstanding contributions and for sharing their knowledge and experiences. Despite busy schedules and numerous commitments, preparation of manuscripts in a timely manner by all authors is greatly appreciated. The editors also thank the editorial staff of Taylor & Francis for their help and support in publishing this volume. The office staff of the Carbon Management and Sequestration Center, who facilitated the flow of manuscripts between authors and editors and made valuable contributions, is greatly appreciated for their help and support. In this context, special thanks are due to Ms. Laura Conover who formatted the text and prepared the final submission. It is a challenging endeavor to thank by name all of those who contributed in one way or another to bringing this volume to fruition. Thus, it is important to build upon the outstanding contributions of numerous soil scientists, agricultural engineers, and technologists whose research is cited throughout the book.

Rattan Lal
B.A. Stewart

Editors

Rattan Lal, PhD, is a distinguished university professor of soil science and director of the Carbon Management and Sequestration Center, The Ohio State University, and an adjunct professor of the University of Iceland. His current research focus is on climate-resilient agriculture, soil carbon sequestration, sustainable intensification, enhancing the efficiency of agroecosystems, and sustainable management of soil resources of the tropics. He received the honorary degree of Doctor of Science from Punjab Agricultural University (2001), the Norwegian University of Life Sciences, Aas (2005), Alecu Russo Balti State University, Moldova (2010), Technical University of Dresden, Germany (2015), and Universitat de Lleida, Spain (2017). He was president of the World Association of the Soil and Water Conservation (1987–1990), the International Soil Tillage Research Organization (1988–1991), and the Soil Science Society of America (2005–2007), and is President of the International Union of Soil Science. He was a member of the Federal Advisory Committee on US National Assessment of Climate Change–NCADAC (2010–2013), and is currently a member of the SERDP Scientific Advisory Board of the US-DOE (2011–), Senior Science Advisor to the Global Soil Forum of the Institute for Advanced Sustainability Studies, Potsdam, Germany (2010–), member of the Advisory Board of the Joint Program Initiative of Agriculture, Food Security and Climate Change (FACCE-JPI) of the European Union (2013–), and Chair of the Advisory Board of the Institute for Integrated Management of Material Fluxes and Resources of the United Nations University (UNU-FLORES), Dresden, Germany (2014–2017). Professor Lal was a lead author of IPCC (1998–2000). He has mentored 106 graduate students and 54 postdoctoral researchers, and hosted 152 visiting scholars. He has authored/coauthored 818 refereed journal articles and has written 20 and edited/coedited 64 books. For 3 years (2014–2016), Reuter Thomson listed him among the world's most influential scientific minds, with citations of publications among the top 1% of scientists in agricultural sciences.

B.A. Stewart is director of the Dryland Agriculture Institute and a distinguished professor of agriculture at West Texas A&M University, Canyon, Texas. He is a former director of the USDA Conservation and Production Laboratory at Bushland, Texas; past president of the Soil Science Society of America; and member of the 1990–1993 Committee on Long-Range Soil and Water Policy, National Research Council, National Academy of Sciences. He is a fellow of the Soil Science Society of America, American Society of Agronomy, and Soil and Water Conservation Society. He has received the USDA Superior Service Award and the Hugh Hammond Bennett Award of the Soil and Water Conservation Society. He was named an honorary member of the International Union of Soil Sciences in 2008 and was inducted into the USDA Agriculture Research Service Science Hall of Fame in 2009. Dr. Stewart is very supportive of education and research on dryland agriculture. The B.A. and Jane Ann Stewart Dryland Agriculture Scholarship Fund was established at West Texas A&M University in 1994 to provide scholarships for undergraduate and graduate students with a demonstrated interest in dryland agriculture.

Contributors

N.D. Ananyeva
Institute of Geography
Institute of Physico-chemical and Biological
    Problems in Soil Science
Moscow Region, Russia

Nina L. Bassuk
School of Integrative Plant Sciences
Urban Horticulture Institute
Cornell University
Ithaca, New York

Michele Bigger
Department of Horticulture and
    Crop Science
The Ohio State University
Columbus, Ohio

Sally Brown
School of Environmental and
    Forest Sciences
University of Washington
Seattle, Washington

Joe Calus
USDA-NRCS
Flint, Michigan

Przemysław Charzyński
Department of Soil Science and Landscape
    Management
Nicolaus Copernicus University
Toruń, Poland

Silke Cram
Instituto de Geografía
Universidad Nacional Autónoma
    de México
Ciudad Universitaria
Ciudad de México, México

Susan Day
Department of Horticulture
Virginia Polytechnic Institute and State
    University
Blacksburg, Virginia

Stephen Dadio
CEDARVILLE Engineering Group, LLC
North Coventry, Pennsylvania

Dickson Despommier
Department of Public Health and
    Microbiology
Columbia University
New York, New York

Robert Dobos
USDA-NRCS
Lincoln, Nebraska

A.V. Dolgikh
Agrarian-Technological Institute
Peoples' Friendship University of Russia
Moscow, Russia
and
Institute of Geography
Moscow, Russia

Beth A. Egitto
Broome County Planning Department
Binghamton, New York

John Galbraith
Virginia Polytechnic Institute and State
    University
Blacksburg, Virginia

Yoshiki Harada
School of Integrative Plant Sciences
Urban Horticulture Institute
Cornell University
Ithaca, New York

Luis Hernandez
Soil Survey Region
USDA-NRCS
Amherst, Massachusetts

Ganga M. Hettiarachchi
Department of Agronomy
Kansas State University
Manhattan, Kansas

Piotr Hulisz
Department of Soil Science and Landscape
   Management
Nicolaus Copernicus University
Toruń, Poland

K.V. Ivashchenko
Agrarian-Technological Institute
Peoples' Friendship University of Russia
Moscow, Russia
and
Institute of Geography
Moscow, Russia

Dariusz Kamiński
Department of Geobotany and Landscape
   Planning
Nicolaus Copernicus University
Toruń, Poland

Arjun Kumar
Institute for Human Development
New Delhi, India

Maxine Levin
USDA-NRCS
Washington, DC

David Lindbo
Soil Science Division
USDA-NRCS
Washington, DC

Klaus Lorenz
Carbon Management and Sequestration
   Center
School of Environment and Natural Resources
The Ohio State University
Columbus, Ohio

Mary G. Lusk
Institute of Food and Agricultural Sciences
University of Florida Extension
Wimauma, Florida

Simi Mehta
School of International Studies
Jawaharlal Nehru University
New Delhi, India

Dorothy S. Menefee
Department of Agronomy
Kansas State University
Manhattan, Kansas

Edwin Muñiz
USDA-NRCS
Somerset, New Jersey

Lucy More Palomino
Instituto de Geología
Universidad Nacional Autónoma
   de México
Ciudad Universitaria
Ciudad de México, México

Steve Peaslee
USDA-NRCS
Lincoln, Nebraska

Anna Piotrowska-Długosz
Department of Soil Science and Soil Protection
University of Technology and Life Sciences
Bydgoszcz, Poland

Andrzej Plak
Department of Soil Science and Protection
Maria Curie-Skłodowska University
Lublin, Poland

Richard V. Pouyat
USDA Forest Service
Research and Development
Washington, DC

Randy Riddle
MLRA Soil Survey
USDA-NRCS
Oxnard, California

Carl Rosier
Department of Plant and Soil Sciences
University of Delaware
Newark, Delaware

Jonathan Russell-Anelli
Soil and Crop Sciences Section
School of Integrative Plant Science
Cornell University
Ithaca, New York

Kristine Ryan
USDA-NRCS
Aurora, Illinois

Bryant Scharenbroch
Department of Soil and Waste Resources
University of Wisconsin—Stevens Point
Stevens Point, Wisconsin

Richard K. Shaw
USDA-NRCS
Somerset, New Jersey

William D. Shuster
National Risk Management Research
    Laboratory
Office of Research and Development
Cincinnati, Ohio

Christina Siebe
Instituto de Geología
Universidad Nacional Autónoma de México
Ciudad Universitaria
Ciudad de México, México

Susan Southard
USDA-NRCS
Lincoln, Nebraska

J.J. Stoorvogel
Soil Geography and Landscape Group
Wageningen University
Wageningen, The Netherlands

Debbie Surabian
USDA-NRCS
Tolland, Connecticut

Gurpal S. Toor
Department of Environmental
    Science and Technology
University of Maryland
College Park, Maryland

Tara L.E. Trammell
Department of Plant and Soil Sciences
University of Delaware
Newark, Delaware

R. Valentini
Laboratory of Agroecological Monitoring,
    Ecosystem Modeling, and Prediction
Russian State Agricultural University
Moscow, Russia

V.I. Vasenev
Agrarian-Technological Institute
Peoples' Friendship University of Russia
Moscow, Russia
and
Soil Geography and Landscape Group and
    Environmental System Analysis Group
Wageningen University
Wageningen, The Netherlands
and
Laboratory of Agroecological Monitoring,
    Ecosystem Modeling and Prediction
Russian State Agricultural University
Moscow, Russia

Thomas H. Whitlow
School of Integrative Plant Sciences
Urban Horticulture Institute
Cornell University
Ithaca, New York

Ian D. Yesilonis
US Department of Agriculture
Forest Service
Baltimore, Maryland

Weixing Zhu
Department of Biological Sciences
State University of New York—Binghamton
Binghamton, New York

1 Urban Agriculture in the 21st Century

Rattan Lal

CONTENTS

1.1 INTRODUCTION

Urbanization may be defined on the basis of number of residents, population density, nonagricultural population, and the provision of basic amenities of life (PRB 2016). There is no standard criteria, and some use a threshold population of 2,500 and others of 20,000 for a community to be deemed "urban." However, urbanization and its strong impact on the environment (Shaw 2015) are integral ramifications of the Anthropocene (Crutzen 2002). Thus, urban soils are termed "Anthrosols" or "Technosols" (Dazzi et al. 2009; Scalenghe and Ferraris 2009; Rossiter 2007). Anthrosols are formed by man-made soil changes called metapedogensis (Yaalon and Yaron 1966), and man is an important factor affecting formation of these soils (Bidwell and Hole 1965; Amundson and Jenny 1991). With an unprecedented and a rapid urbanization, Anthrosols and Technosols are important soils of the twenty-first century. Worldwide, urban land area increased by 5.8 million ha (Mha) between 1970 and 2000 (Seto et al. 2011) and is projected to increase by an additional 122 Mha by 2030 (Biello 2012). Thus, it is important to understand its characteristics, changes over time, and management options (Shaw 2015).

Urbanization began about 5000 years ago, but the twenty-first century is the era of urbanization. However, only 3%–12% of the world population lived in urban centers even by 1800 (Eisinger 2006), 14% by 1900 with only 12 cities having population >1 million, 30% by 1950 with 83 cities, and 50% by 2008 with 400 cities (PRB 2016). In 2015, 54% of the population (4 billion) lived in urban centers. The urban population will increase to 5.3 billion (66% of the total) by 2030, to 6 billion (68% of the total) by 2045 (World Bank 2015), and to 6.8 billion (70% of the total) by 2050 (U.N. 2015). Europe (73%), South America (83%), and North America (82%) are already the most urbanized continents (Burdett 2015). Almost all of the future urban growth by 2050 and beyond will occur in developing countries. The global urban population is projected to grow by 1.84% per year between 2015 and 2020, 1.63% per year between 2020 and 2025, and 1.44% per year between 2025 and 2030 (WHO 2016). The United States has undergone an urban revolution since 1950 (Cox 2014). By 2050, 87% of the US population will live in urban centers (University of Michigan 2015). By 2030, the population of Asia will be 5 billion, and a majority will live in urban centers. The number of world cities with population >10 million was 1 (New York, 12.3 million) in 1950, 5 in 1975, 19 in 2000 (UN/UNCHS 2001), 34 in 2015 (Cox 2015b), and is projected to be 41 in 2030 (U.N. 2015). Trends in population and area of 10 major cities in the United States are given in Tables 1.1 and 1.2, respectively. Change in urban population of the world by continents between 1950 and 2050 will be in the order of Latin America > Asia > Africa > Europe > North America > Oceania (Table 1.3). The highest rate of future urban growth is expected to occur in West Africa, eastern Afromontaine, Western Ghats of India, and Sri Lanka.

TABLE 1.1
Trends in Population Growth (10^6) of 10 Cities in the United States

City	1950	1960	1970	1980	1990	2000	2010	2015	2020	2025
Atlanta	0.51	0.77	1.17	1.61	2.16	3.50	4.52	5.02	5.04	5.15
Boston	2.23	2.41	2.62	2.68	2.78	4.03	4.18	4.48	4.92	5.03
Chicago	4.92	5.96	6.72	6.78	6.79	8.31	8.60	9.16	9.76	9.93
Dallas–FW	0.86	1.44	2.02	2.45	3.20	4.15	5.12	5.16	5.30	5.42
Houston	0.70	1.14	1.68	2.41	2.90	3.82	4.94	5.76	–	–
Los Angeles	4.0	6.49	8.35	9.48	11.40	11.79	12.15	15.06	–	–
Miami	0.46	0.85	1.20	1.61	1.92	4.92	5.50	5.76	6.14	6.27
New York	12.30	14.11	16.21	15.59	16.04	17.80	18.35	19.90	20.37	20.63
Philadelphia	2.92	3.64	4.02	4.11	4.22	5.15	5.44	5.57	6.0	6.10
Washington, DC	1.29	1.81	2.48	2.76	3.36	3.93	4.46	4.78	4.78	4.89

Sources: Demographia. 2000. Urban areas in the United States: 1950 to 2010. Principal urban areas in metropolitan areas over 1,000,000 population in 2010. *Data tables.* http://www.demographia.com/db-uza2000.htm; U.N. 2015. The Sustainable Development Goals. https://sustainabledevelopment.un.org/?menu=1300; U.N. Habitat 2009.

TABLE 1.2
Land Area of 10 Large Cities in the United States

City	Area (10^3 ha)							
	1950	1960	1970	1980	1990	2000	2010	2015
Atlanta	27	64	113	234	294	508	685	685
Boston	89	134	172	222	231	450	485	533
Chicago	183	249	331	388	441	550	633	686
Dallas–FW	68	238	277	332	374	364	461	518
Houston	70	112	140	272	305	335	430	464
Los Angeles	556	335	407	473	509	432	450	630
Miami	30	47	67	88	91	289	321	321
New York	325	490	628	727	768	868	894	1164
Philadelphia	81	155	195	263	301	466	513	513
Washington, DC	46	88	144	209	209	245	342	342
Total of all cities in the United States	2373	4401	5878	7704	6202	10094	11811	–

Sources: Demographia. 2000. Urban areas in the United States: 1950 to 2010. Principal urban areas in metropolitan areas over 1,000,000 population in 2010. *Data tables.* http://www.demographia.com/db-uza2000.htm; U.N. Habitat 2009; U.N. 2015. The Sustainable Development Goals. https://sustainabledevelopment.un.org/?menu=1300.

TABLE 1.3
Change in Urban Population (%) by Continents between 1950 and 2050

Continent	% Urbanized		
	1950	2050	Increase over 100 years (%)
Africa	14.0	40.0	+26.0
Asia	17.0	47.5	+29.5
Europe	50.0	73.4	+23.4
Latin America and the Caribbean	42.0	79.5	+37.5
North America	63.6	81.5	+17.9
Oceania	61.5	70.8	+9.3

Source: U.N., *World Urbanization Prospects Highlights. 2014 Revision*, Department of Economic and Social Affairs, Population Division, New York, 2014.

1.2 GLOBAL LAND AREA UNDER URBAN CENTERS

There is a large range in available estimates of global area under urban lands. The discrepancy is caused by different definitions of urban lands. Meyer and Turner (1992) estimated urban land area in 1990 at 250 Mha. Yang and Zhang (2015) reported that <2% of the global land surface is used by cities. In comparison, Grimm et al. (2008) estimated that global urban centers cover ~3% of the terrestrial land area or 400 Mha on the basis of the ice-free land (13.4×10^9 ha). In a strong contrast to these estimates, Schneider et al. (2009) produced a new global map of urban area, and estimated that cities occupy less than 0.5% of the world's total land area.

Liu et al. (2014) proposed a hierarchy framework comprising of three spatially nested definitions: (1) "urban area" delineated by administrative boundaries, (2) "build-up area" that is dominated by artificial surfaces, and (3) "impervious surface area" that is devoid of life. In 2010, estimates of global area were 3% for urban area (400 Mha), 0.65% for build-up area (87 Mha), and 0.45% for impervious surface area (60 Mha) (Liu et al. 2014). The data in Table 1.4, based on 66 large cities in the world, show that it takes about on average 28,400 ha to house 1 million people, with a range of 2,100 ha for Dhaka to 137,000 ha for Atlanta. Indeed, the land area to house 1 million people varies widely among continents with a range of 11,600 ha for Asia to 70,900 ha for the United States/Canada.

1.3 URBANIZATION AND THE ENVIRONMENTAL IMPACT

The future growth of the world population will almost entirely occur in urban areas of developing countries (Cohen 2006; U.N. 2015). However, cities in developing countries are ill-prepared for explosion in urban population (Van Ginkel 2008). Food insecurity, in addition to lack of other amenities, is and will continue to be an important issue. Furthermore, land cover change by urban expansion also impacts the environment (e.g., biodiversity, hydrological balance, biogeochemical cycling of C, and other elements). Impacts of urbanization on the environment have been described by Lal (2012) and on climate change by Lal and Augustin (2012). The loss of the terrestrial C pool from deforestation for urbanization may be 0.05 PgC/yr with a total loss of 1.38 PgC or 5% of the total emissions from tropical deforestation (Seto et al. 2012). In addition to loss of vegetation cover and decline in the ecosystem C pool, urbanization impacts runoff and transport of pollutants into natural waters, heat balance, energy use, surface sealing, gaseous emissions, and food waste. Some of these ramifications are discussed in the following sections.

Estimates of area under urban lands vary widely among cities (Table 1.5) and continents (Table 1.6). On the basis of calculations made using the data in Tables 1.3 and 1.4, land area under urban use will increase by about 57 Mha between 2010 and 2035 (Table 1.5). However, some argue

TABLE 1.4
Population and Land Area of 20 Large Cities of the World in 2015 and 2030

City	Population (10^6)	Land Area (10^3 ha)	10^3 ha/10^6 People
Tokyo	37.84	854.7	22.6
Jakarta	30.54	322.5	10.6
New Delhi	25.70	207.2	8.1
Manila	24.12	158.0	6.6
Seoul-Incheon	23.48	226.6	9.7
Shanghai	23.41	382.0	16.3
Karachi	22.12	94.5	4.3
Beijing	21.00	382.0	18.2
New York	20.63	1164.2	56.4
Guangzhou	20.60	343.2	16.7
Sao Paulo	21.07	270.1	12.80
Mexico City	20.06	207.2	10.3
Mumbai	21.04	54.6	2.6
Osaka	20.24	312.2	15.9
Moscow	16.17	446.2	28.8
Dhaka	17.50	36.0	2.1
Cairo	18.77	176.1	9.4
Los Angeles	15.06	630.0	41.8
Bangkok	15.00	259.0	17.3
Kolkata	14.67	120.4	8.2
Buenos Aires	15.18	268.1	17.6
Tehran	13.53	148.9	11.0
Istanbul	13.29	136.0	10.2
Lagos	13.12	90.7	6.9
Shanzhen	12.08	174.8	14.4
Rio de Janerio	11.73	202.0	17.2
Kinshasa	11.50	58.3	5.1
Paris	10.89	284.5	26.1
Lima	10.75	91.9	8.5
London	10.24	173.8	17.0
Lahore	10.05	79.0	7.9
Bangalore	9.81	116.6	11.9
Chennai	9.71	97.1	10.0
Chicago	9.16	685.6	74.8
Bogota	8.99	49.2	5.5
Ho Chi Minh City	8.96	148.9	16.6
Johannesburg	8.43	259.0	30.7
Dusseldorf	6.68	265.5	39.7
Santiago	6.22	98.4	15.8
Toronto	6.46	228.7	35.4
Dallas–FW	6.17	517.5	83.9
Madrid	6.17	132.1	21.4
Nanning	6.16	126.9	20.0

(Continued)

TABLE 1.4 (*Continued*)
Population and Land Area of 20 Large Cities of the World in 2015 and 2030

City	Population (10^6)	Land Area (10^3 ha)	10^3 ha/10^6 People
Xian	6.0	93.2	15.5
San Francisco	5.9	279.7	47.4
Luanda	5.9	89.4	15.2
Houston	5.8	464.4	80.0
Miami	5.8	320.9	55.3
Bandung	5.7	46.6	8.2
Riyadh	5.7	150.2	8.8
Pune	5.6	47.9	8.6
Singapore	5.6	51.8	9.3
Philadelphia	5.5	513.1	91.3
Surat	5.5	23.3	4.2
Milan	5.3	189.1	35.7
Atlanta	5.0	685.1	137.0
Zhengzhou	4.9	82.9	16.9
Washington, DC	4.9	342.4	69.8
Surabaya	4.9	67.3	13.7
Harbin	4.8	57.0	11.9
Sydney	4.0	203.7	50.9
Brisbane	2.0	197.2	98.6
Melbourne	3.9	254.3	65.2
Perth	1.8	156.6	87.0
Auckland	1.4	54.4	38.9
Adelaide	1.1	85.2	77.5
Average			**28.4 (2.1–137.0)**

Sources: Demographia. 2015. Demographia World Urban Areas, 11th Annual Edition 2015:01, Belleville, IL; Allianz Risk Pulse. 2015. The megacity state: World's biggest cities shaping our future. 11th Annual Edition, Munich, Germany.

that if the current rate of urbanization continues, urban land cover will increase by an additional area of 120 Mha (Seto et al. 2012) to 153 Mha (Seto et al. 2012). These estimates by Seto et al. (2011, 2012) are 2.1–2.7 times more than the 57 Mha reported on the basis of data presented in Tables 1.3 through 1.5. Thus, there is a strong need to standardize the use of the term "urban land" and to provide credible estimates of the land area using satellite imagery and other innovative techniques.

1.4 ENVIRONMENTAL RAMIFICATIONS OF URBANIZATION

Urbanization has adverse environmental impacts with regards to climate, hydrological cycle, air quality, and soil health (Figure 1.1). Urban ecosystems are characterized by drastically disturbed lands, strongly denuded original vegetation cover, simplified new vegetation cover, and disrupted and highly fragmented habitat with a low biodiversity (Martin 2006). Human habitat uses 100–300 times more energy than natural ecosystems (Alberti 2005), and thus emission of greenhouse gases is high. Consequently, soils of urban ecosystems are also highly disturbed (Shaw 2015). These anthropogenic perturbations have important implications to food production under urban agriculture (UA) and must be addressed.

TABLE 1.5

Area under 10 Largest Cities in the World in 2015

Rank	City	Population (10^6)	Land Area (10^3 ha)	Area per 10 Million (10^3 ha/million)
1	Tokyo	37.84	854.7	22.6
2	Jakarta	30.54	322.5	10.6
3	Delhi	25.00	201.2	8.1
4	Manila	24.12	158.0	6.6
5	Seoul-Incheon	23.48	226.6	9.7
6	Shanghai	23.42	382.6	16.3
7	Karachi	22.12	94.5	4.3
8	Beijing	21.01	382.0	18.2
9	New York	20.63	1164.2	56.4
10	Guangzhou	20.60	343.2	16.7
	Average			**17.0 (6.6–56.4)**

Source: Cox, W. 2015a. The world's ten largest megacities. *The Huffington Post*, 19 April 2015. http://www.huffington-post.com/wendell-cox/the-worlds-ten-largest-me_b_6684694.html.

TABLE 1.6

Average Land Use under Urban Centers per 10 Million People

Region	Area (10^3 ha)
The United States/Canada	70.9
Australia/New Zealand	69.6
Europe	25.6
South America	12.5
Africa	13.5
Asia	11.6

The heat island effect refers to the warmer temperatures of 1–3°C in a city with 1 million people or more compared to rural areas, and differences in temperature can be even higher. This effect can exacerbate air pollution, water quality, greenhouse gas emissions, and increase summertime peak energy demands (USEPA 2016; Sachindra et al. 2016). Urban land cover can also significantly impact the precipitation variability. In Atlanta, Shem and Shepherd (2009) reported that rainfall amounts downwind of the city are higher by 10%–13% within a strip 20–50 km east of the city. Urbanization affects precipitation by enhancing convergence because of increase in surface roughness in the urban environment, and destabilization due to the urban heat island effect (Shepherd 2006). By impacting the natural drainage system (Booth 1991), urbanization impacts surface runoff. The latter is also exacerbated by increase in impervious surface coverage, thus generating floods of different recurrence intervals (Hollis 1975). The magnitude of impervious surface coverage (surface sealing) is a key environmental indicator (Arnold and Gibbons 1996). Croplands, forestlands, and grasslands are replaced by impervious surfaces, which intensify storm runoff, reduce groundwater recharge, and increase risks of stream erosion (Stone 2004). Urbanization also impacts regional and global biogeochemistry of nitrogen and other elements. In Shanghai, China, Gu et al. (2012) reported that total N input increased from 57.7 to 587.9 Gg N/yr during the period from 1952 to 2004. The per capita N input increased from 13.5 to 45.7 kg N/yr due to fossil fuel combustion and food/feed import. Urbanization also increases the risks of soil pollution through waste disposal and acid deposition due to air pollution (Chen 2007). Soil sealing and fragmented landscapes are a serious threat to soil resources (Li et al. 2015) because they are

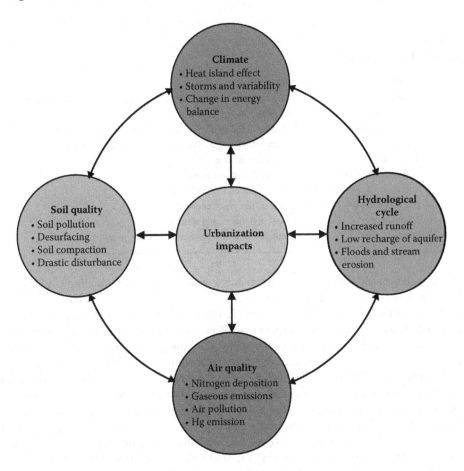

FIGURE 1.1 Environmental ramifications of urbanization with consequences on soil quality and food security.

detrimental to its ecological functions (Scalenghe and Marsan 2009) with strong environmental, social, and economic consequences (EC 2012). Gaseous emissions from urban lands include that of mercury (Eckley and Branfireum 2008), greenhouse gases (Kulchakova and Mozharova 2014), methane and nitrous oxide (Kaye et al. 2004), and a large amount of carbon dioxide.

1.5 URBAN ECOSYSTEMS, SOILS, AND THEIR MANAGEMENT FOR AGRICULTURE

Urban soils, Anthrosols or Technosols, are highly disturbed ecosystems because of human activities leading to alterations in natural landscape. Bockheim (1974) defined urban soil as "A soil material having a non-agricultural, man-made surface layer more than 50 cm thick, that has been produced by mixing, filling, or by contamination of land surface in urban and suburban areas." Craul (1985) defined urban soil as "A material that has been manipulated, disturbed or transported by man's activities in the urban environment and is used as a medium for plant growth." These two definitions are contradictory in terms of UA. the Bockheim excluded agricultural use while Craul focused only on agricultural use. In the context of UA, the definition by Craul is a good starting point and may be revised with the rapidly evolving field of UA.

Among several attributes of urban soils (Table 1.7) are high heterogeneity and large temporal and spatial variability and the presence of artifacts. These properties, especially soil structure and other physical attributes (e.g., bulk density or compactness, infiltration rate, plant available water capacity, and drought stress) and chemical properties (e.g., pH, electrical conductivity, nutrient/

TABLE 1.7
Attributes of Urban Soils

	Attribute	Description
1.	Variability	High temporal, spatial, and with depth because of mixing and filling
2.	Heterogeneity	High because of anthropogenic perturbations
3.	Physical properties	Shallow or no topsoil, stoniness, compact soil, and poor aeration
4.	Pollutant	High pollution, especially the presence of heavy metals
5.	Soil fertility	Nutrient imbalance
6.	Biogeochemical cycles	Highly disturbed and modified
7.	Artifacts	A large presence of anthropic artifacts
8.	Soil temperature	Higher than under undisturbed soils
9.	Infiltration rate	Low, high runoff and water logging
10.	Reaction, salts	Variable pH and accumulation of salts in dry regions

elemental imbalance, and soil C pool), must be duly considered for implementing UA using soil as a medium for growing crops and raising livestock.

Therefore, restoration of ecosystems and soil health are essential to improving the urban environments and increasing agronomic production of crops grown on urban soils. Martin (2006) outlined the basic concepts of restoring ecosystems: (1) maintaining large vegetation patches and corridors along with surrounding waterways for free movement of wild species, (2) creating an innovative urban habitat based on principles of landscape ecology, (3) enhancing the existing habitat for improving biodiversity, (4) maintaining and restoring vegetation cover, (5) using green roofs (Mentens et al. 2006), and (6) promoting UA on abandoned urban sites (Beniston et al. 2015). The problem of excessive runoff can be partly addressed by constructing green roofs (Peck 2003; Mentens et al. 2006). Similar to urban ecosystems, UA must also be developed following the basic concepts of restoration ecology and greening the landscape. Some innovative and emerging ideas of UA are discussed in Sections 1.6 through 1.8.

1.6   URBANIZATION AND FOOD SECURITY

Rapid urbanization in developing countries will impact all dimensions of food and nutritional security. Food security is impacted by the increasing trend from cities to megacities to gigacities. Some megacities of the world include Tokyo, Delhi, Shanghai, Sao Paulo, Mumbai, Mexico City, Beijing, Osaka, Cairo, New York, Paris, and London (Table 1.5). These mega and gigacities exacerbate the problems of food access, distribution, and safety along with that of sanitation, because of the complete lack and scarcity of clean and running water in developing countries. The problem may be exacerbated by lack of or weak infrastructure and poor institutional support. Preparation of food density maps of the growing number of megacities in developing countries may be a useful strategy to understand and address the complex issues (Matuschke 2009). The problems of these mega and gigacities will necessitate innovative approaches including vertical farms (Despommier 2011) and other forms of UA such as algal energy farms, and waste-to-power plants (Allianz Risk Pulse 2015), among others.

1.7   URBAN AGRICULTURE

Interest in UA is growing in many cities throughout the world (e.g., Havana). There are numerous factors that influence public involvement in UA. Important among these are society recognition, attitude, social impact, land tenure, economics and health impact, and knowledge (Rezai et al. 2014). Availability of space for UA, especially in densely populated countries of Asia (Table 1.5) is another issue (Schwarz et al. 2016). Thus, innovative ideas, such as vertical farms (Despommier 2011), are relevant to meeting the food needs for mega gigacities. The term "vertical farming" was initially proposed by Gilbert Bailey in 1915, but he used it to refer to stratification of soil properties according to their depth distribution within the soil profile (Bailey 1915). In contrast, the present concept by Despommier et al. refers to vertical glasshouses or skyscrapers using aeroponics, hydroponics, and other soil-less media (Despommier 2011). In addition to economics, there are also issues of energy use, gaseous emissions and pollution, etc., which need to be addressed. Yet, increase in local food production, weatherproofing of agriculture, efficient use and recycling of resources (e.g., nutrients, water), food safety, protection of soil, and biodiversity are among the merits of vertical farming.

1.8 NEXUS THINKING FOR WASTE AND WATER MANAGEMENT FOR UA

Understanding the waste–water–food nexus or interconnectivity is critical to promoting UA (Lal 2012, 2016). Water quality and waste management are intricately linked in urban ecosystems. The linkage can be used for enhancing human well-being through production of biofuel and food (Figure 1.2). Plant nutrients in gray and black water can be used to grow cyanobacteria (algae) for biofuel, and solid/liquid waste can be used in biodigesters to produce CH_4 for use in the household. Composting by-products of these energy-producing systems can be used as soil amendments or for producing synthetic soil as a plant growth media. Both water and nutrients available in these sources are important to grow vegetables (hydroponics and aeroponics in sky farming, home and community gardens) and in aquaculture. Thus, food and fuel can be produced from urban waste. Treated

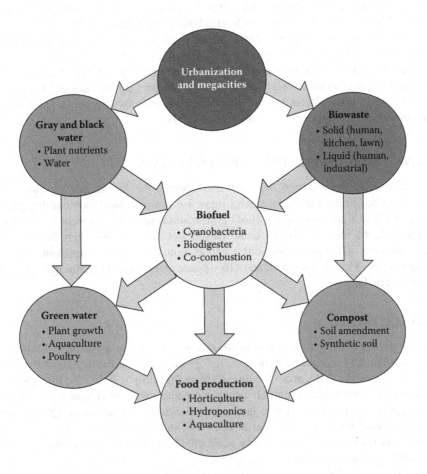

FIGURE 1.2 Recycling nutrients and water to produce biofuels and promote urban agriculture.

TABLE 1.8

Relevance of Urban Agriculture to Sustainable Development Goals of the United Nations

Goal Number	Objective	Relevance of UA
1	No poverty	Creating another income stream
2	Zero hunger	Enhancing access to fresh food
3	Good health	Improving nutrition
6	Clean water and sanitation	Converting gray and black water into green water
10	Reduced inequalities	Creating jobs
11	Sustainable cities	Producing food locally
12	Responsible consumption	Recycling biowastes
13	Climate action	Sequestering C, reducing transport
16	Peace and justice	Reducing poverty, hunger, and inequality

gray water, rich in N, P, and other nutrients, is a major cause of eutrophication and of algal blooms. Thus, these nutrients can be recycled to produce microalgae biomass for production of biofuels (Abdel-Raouf et al. 2012). Microalgae biomass, which can also remove heavy metals and other toxic organic compounds, is important to wastewater treatment. Because of high productivity, microalgae are an important feedstock for a renewable source for C-neutral energy production. Thus, coupling of microalgal cultivation with biofuel production and wastewater treatment is an important strategy (Bhatt et al. 2014) for producing energy and food (Metson and Bennett 2015).

Soils of urban ecosystems are drastically disturbed by construction and use of heavy equipment. Thus, soil quality must be restored to improve plant growth by using compost and soil amendments generated from bio-solids. Because of soil contamination, by heavy metals and organics/inorganic pollutants, food produced in urban areas may not be safe (Meharg 2016). Thus, UA has an important role in organic waste management but safety-oriented precautionary measures are essential. Safe use of human waste (e.g., night soil) is critical (Ling 1994).

1.9 UA AND SUSTAINABLE DEVELOPMENT GOALS

Judiciously and safely implemented, UA can also advance the Sustainable Development Goals (SDGs) of the U.N. (2015). Important among these are poverty alleviation and hunger elimination, for which UA is a pertinent option (FAO 2008). Sustainable development is one of the goals of UA (Mougot 2006). Some SDGs which are relevant to UA include #2—zero hunger, #6—clean water and sanitation, #11—sustainable cities and communities, #12—responsible consumption, #13—climate action, #15—life on land, and #16—peace and justice. The relevance of UA to these goals is outlined in Table 1.8.

1.10 CONCLUSIONS

- The twenty-first century is the era of urbanization. The urban population of 4 billion (54% of the total) in 2015 is projected to increase to 5.3 billion (66% of the total) by 2030, 6 billion (68% of the total) by 2045, and 6.8 billion (70% of the total) by 2050. The future growth of world population will almost entirely occur in urban areas of developing countries.
- Estimates of the area under urban land use vary widely ranging from 0.5% to 3.0% of the earth's total land area, and 250 Mha to 4 Mha of actual area. The wide range on estimates of land area is due to lack of a standard definition of the term "urban," and of the methodology to assess the area.
- The land area needed to accommodate a population of 1 million people ranges widely among cities from 2,100 ha in Dhaka to 137,000 ha in Atlanta, with a global average of

28,400 ha, and among continents from 11,600 ha for Asia to 70,900 ha for the United States and Canada.

- Urbanization has a drastic impact on the environment in areas such as climate and energy, hydrological balance, biogeochemical cycles, biodiversity, and vegetation cover. Notable among these are the heat island effect and increase in incidence of water runoff and inundation. Surface sealing and creation of impervious surfaces is an important ramification of urbanization.

- Urban soils, called Anthrosols or Technosols, are highly disturbed ecosystems created by mixing, filling, and use of artifacts. These soils are highly heterogeneous, compact, and characterized by a shallow or no surface soil and have low water infiltration rate.

- Rapid urbanization in developing countries does and will affect all dimensions of food and nutritional security, which are likely to be exacerbated by transformation of cities to megacities and gigacities. The problem is worsened by the lack of or at best the weak infrastructure and poor institutional support.

- Therefore, UA can play an important role in advancing food security, especially in densely populated cities of Asia and Africa. There are numerous types of UA ranging from a traditional backyard vegetation patch to green roofs, and glass skyscrapers called "vertical farming."

- The nexus thinking of waste and water management, based on the concept of recycling of nutrients and water and regenerating energy from waste, is an important concept in bringing innovation to UA.

REFERENCES

Abdel-Raouf, N., A.A. Al-Homaidan, and I.B.M. Ibrahim. 2012. Microalgae and waste water treatment. *Saudi Journal of Biological Sciences* 19:257–275.

Alberti, M. 2005. The effects of urban pattern on ecosystem function. *International Regional Science Review* 28 (2):168–192.

Allianz Risk Pulse. 2015. The megacity state: World's biggest cities shaping our future. 11th Annual Edition, Munich, Germany.

Amundson, R. and H. Jenny. 1991. The place of humans in the state factor theory of ecosystems and their soils. *Soil Science* 151 (1):99–109. doi: 10.1097/00010694-199101000-00012.

Arnold, C.L. and C.J. Gibbons. 1996. Impervious surface coverage—The emergence of a key environmental indicator. *Journal of the American Planning Association* 62 (2):243–258. doi: 10.1080/01944369608975688.

Bailey, G. 1915. *Vertical Farming*. Wilmington, DE: E. I. DuPont de Nemours Power Co. 84 pp.

Beniston, J., R. Lal, and K. Mercer. 2015. Assessing and managing soil quality for urban agriculture in a degraded vacant lot soil. *Land Degradation & Development*. 27(4):996–1006. doi: 10.1002/ldr.2342.

Bhatt, N.C., A. Panwar, T.A. Bisht, and S. Tamte. 2014. Coupling of algal biofuel production with wastewater. *Scientific World Journal* 26: 210504.

Bidwell, O.W. and F.D. Hole. 1965. Man as a factor in soil formation. *Soil Science* 99:65–72.

Biello, D. 2012. Gigalopolises: Urban land area may triple by 2030. *Scientific American*, 18 September 2012. http://www.scientificamerican.com/article/cities-may-triple-in-size-by-2030/.

Bockheim, J.G. 1974. Nature and properties of highly disturbed urban soils. Paper presented at Division S-5 Annual Conference. Soil Science Society of America, Chicago, IL.

Booth, D.B. 1991. Urbanization and the natural drainage system—Impacts, solutions, and prognoses. *Northwest Environmental Journal* 7 (1):93–118.

Burdett, R. 2015. Cities in numbers: How patterns of urban growth change the world. *The Guardian*, 23 November 2015. https://www.theguardian.com/cities/2015/nov/23/cities-in-numbers-how-patterns-of-urban-growth-change-the-world.

Chen, J. 2007. Rapid urbanization in China: A real challenge to soil protection and food security. *Catena* 69 (1):1–15. doi: 10.1016/j.catena.2006.04.019.

Cohen, B. 2006. Urbanization in developing countries: Current trend, future projections, and key challenges for sustainability. *Technology in Society* 28:63–80.

Cox, W. 2014. UN Projects 2030 US Urban Area Populations. *NewGeography*, 7 August 2014. http://www.newgeography.com/content/004464-un-projects-2030-us-urban-area-populations.

Cox, W. 2015a. The world's ten largest megacities. *The Huffington Post*, 19 April 2015. http://www.huffingtonpost.com/wendell-cox/the-worlds-ten-largest-me_b_6684694.html.

Cox, W. 2015b. Largest 1000 cities on earth: World urban areas, 2015 edition. *NewGeography*, 2 February 2015. http://www.newgeography.com/content/004841-largest-1000-cities-earth-world-urban-areas-2015-edition.

Craul, P.J. 1985. Urban soils. METRIA 5: Selecting and Preparing Sites for Urban Trees. Proceedings of the Fifth Conference of the Metropolitan Tree Improvement Alliance, The Pennsylvania State University, University Park, Pennsylvania, PA, May 23–24.

Crutzen, P.J. 2002. Geology of mankind. *Nature* 415 (6867):23–23. doi: 10.1038/415023a.

Dazzi, C., G. Lo Papa, and V. Palermo. 2009. Proposal for a new diagnostic horizon for WRB Anthrosols. *Geoderma* 151 (1–2):16–21. doi: 10.1016/j.geoderma.2009.03.013.

Demographia. 2000. Urban areas in the United States: 1950 to 2010. Principal urban areas in metropolitan areas over 1,000,000 population in 2010. *Data tables*. http://www.demographia.com/db-uza2000.htm.

Demographia. 2015. Demographia World Urban Areas, 11th Annual Edition 2015:01, Belleville, IL.

Despommier, D. 2011. *The Vertical Farm: Feeding the World in the 21st Century*. New York, NY: St. Martins Press, 311 pp.

EC. 2012. Soil Sealing. Science for environment policy: In depth report. European Commission, Brussels, Belgium.

Eckley, C.S. and B. Branfireun. 2008. Gaseous mercury emissions from urban surfaces: Controls and spatio-temporal trends. *Applied Geochemistry* 23 (3):369–383. doi: 10.1016/j.apgeochem.2007.12.008.

Eisinger, A. 2006. Urbanization. In Oswalt, P. and T. Rieniers (Eds.) *Atlas of Shrinking Cities*. Ostfildern: Hatje Cantz Verlag, pp. 28–29.

FAO. 2008. *Urban Agriculture: For Sustainable Poverty Alleviation and Food Security*. Rome: FAO, 84 pp.

GEOHIVE. 2016. Population of the entire world, 1950–2100. http://www.geohive.com/earth/his_history3.aspx.

Grimm, N.B., S.H. Faeth, N.E. Golubiewski, C.L. Redman, J.G. Wu, X.M. Bai, and J.M. Briggs. 2008. Global change and the ecology of cities. *Science* 319 (5864):756–760. doi: 10.1126/science.1150195.

Gu, B.J., X.L. Dong, C.H. Peng, W.D. Luo, J. Chang, and Y. Ge. 2012. The long-term impact of urbanization on nitrogen patterns and dynamics in Shanghai, China. *Environmental Pollution* 171:30–37. doi: 10.1016/j.envpol.2012.07.015.

Hollis, G.E. 1975. Effect of urbanization on floods of different recurrence interval. *Water Resources Research* 11 (3):431–435. doi: 10.1029/WR011i003p00431.

Kaye, J.P., I.C. Burke, A.R. Mosier, and J.P. Guerschman. 2004. Methane and nitrous oxide fluxes from urban soils to the atmosphere. *Ecological Applications* 14 (4):975–981. doi: 10.1890/03-5115.

Kulchakova, S. and S.V. Mozharova. 2014. Generation, sink and emission of greenhouse gases by urban soils or reclaimed filtration fields. *Journal of Soils and Sediments* 15:1753–1763.

Lal, R. 2012. Urban ecosystems and climate change. In Lal, R. and B. Augustin (Eds.) *Carbon Sequestration in Urban Ecosystems*. Dordrecht, Holland: Springer, pp. 3–19.

Lal, R. 2016. Global issues and nexus thinking. *Journal of Soil and Water Conservation* 71 (4):84A–90A

Li, J.D., J.S. Deng, Q. Gu, K. Wang, F.J. Ye, Z.H. Xu, and S.Q. Jin. 2015. The accelerated urbanization process: A threat to soil resources in Eastern China. *Sustainability* 7 (6):7137–7155. doi: 10.3390/su7067137.

Ling, B. 1994. Safe use of targeted night soil. *ILEIA Newsletter* 10 (3):10–11.

Liu, A., C. Ha, Y. Zhou, and J. Wu. 2014. How much of the world's land has been urbanized really? A hierarchical framework for avoiding confusion. *Landscape Ecology* 29 (5):763–771.

Martin, M. 2006. *Urban Ecosystems. Open Space Seattle 2100 Program*. Washington, DC: University of Washington, 1–7 pp.

Matuschke, I. 2009. Rapid urbanization and food security: Using food density maps to identify future food security hotspots. Paper presented at the International Association of Agricultural Economists Conference. Beijing, China, 16–22 August.

Meharg, A.A. 2016. Perspective: City farming needs monitoring. *Nature* 531:S60.

Mentens, J., D. Raes, and M. Hermy. 2006. Green roofs as a tool for solving the rainwater runoff problem in the urbanized 21st century? *Landscape and Urban Planning* 77 (3):217–226. doi: 10.1016/j.landurbplan.2005.02.010.

Metson, G.S. and Bennett, E.M. 2015. Phosphorus cycling in Montreal's food and urban agricultural systems. *PLOS ONE* 10 (3): e0120726.

Meyer, W.B. and B.L. Turner. 1992. Human-population growth and global land-use cover change. *Annual Review of Ecology and Systematics* 23:39–61.

Mougot, L.J.A. 2006. *Growing Better Cities: Urban Agriculture for Sustainable Development*. Ottawa, ON: IDRC.

Peck, S. 2003. Towards an integrated green roof infrastructure evaluation for Toronto. *Green Roofs Infrastructures Monitor* 5:4–5.

PRB. 2016. Human population urbanization. *Population Reference Bureau*. http://www.prb.org/Publications/Lesson-Plans/HumanPopulation/Urbanization.aspx.

Rezai, G., M.N. Shamsudin, Z. Mohamed, and J. Sharfuddin. 2014. Factor influencing public participation in urban agriculture in Malaysia. International Conference on Advances in Economics, Social Science and Human Behaviour Study (ESHB 2014), Hong Kong, China. 26–27 August, pp. 22–25.

Rossiter, D.G. 2007. Classification of urban and industrial soils in the world reference base for soil resources. *Journal of Soils and Sediments* 7 (2):96–100. doi: 10.1065/jss2007.02.208.

Sachindra, D.A., A.W.M. Ng, S. Muthukaumaran, and B.J.C. Pereira. 2016. Impact of climate change on urban heat island effect and extreme temperatures: A case-study. *Quarterly Journal of the Royal Meteorological Society* 142 (694):172–186.

Scalenghe, R. and S. Ferraris. 2009. The first forty years of a technosol. *Pedosphere* 19 (1):40–52.

Scalenghe, R. and F.A. Marsan. 2009. The anthropogenic sealing of soils in urban areas. *Landscape and Urban Planning* 90 (1–2):1–10. doi: 10.1016/j.landurbplan.2008.10.011.

Schneider, A., M.A. Friedl, and D. Potere. 2009. A new map of global urban extent from MODIS satellite data. *Environmental Research Letters* 4:1–11.

Schwarz, K., B.B. Cutts, J.K. London, and M.L. Cadenasso. 2016. Growing gardens in shrinking cities: A solution to the soil lead problem? *Sustainability* 8 (2):141.

Seto, K.C., M. Fragkias, B. Guneralp, and M.K. Reilly. 2011. A meta-analysis of global urban land expansion. *PLOS ONE* 6 (8):e23777.

Seto, K.C., B. Güneralp, and L.R. Hutyra. 2012. Global forecasts of urban expansion to 2030 and direct impacts on biodiversity and carbon pools. *PNAS* 109(40):16083–16088.

Shaw, R.K. 2015. Soils in urban areas: Characterization, management, challenges. *Soil Science* 180 (4–5):135.

Shem, W. and M. Shepherd. 2009. On the impact of urbanization on summertime thunderstorms in Atlanta: Two numerical model case studies. *Atmospheric Research* 92 (2):172–189. doi: 10.1016/j.atmosres.2008.09.013.

Shepherd, J.M. 2006. Evidence of urban-induced precipitation variability in arid climate regimes. *Journal of Arid Environments* 67 (4):607–628. doi: 10.1016/j.jaridenv.2006.03.022.

Stone, B. 2004. Paving over paradise: How land use regulations promote residential imperviousness. *Landscape and Urban Planning* 69 (1):101–113. doi: 10.1016/j.landurbplan.2003.10.028.

U.N. 2014. *World Urbanization Prospects Highlights. 2014 Revision*. New York, NY: Department of Economic and Social Affairs, Population Division.

U.N. 2015. The sustainable development goals. https://sustainabledevelopment.un.org/?menu=1300.

UN/UNCHS. 2001. Urbanization: Facts and figures. Special Session of the General Assembly for an Overall Review and Appraisal of the Implementation of the Habitat Agenda, New York, NY, 6–8 June.

U.N. Habitat. 2009. Annual Report 2009. *UN-Habitat*, p. 62, ISBN:978-92-1-132233-0.

University of Michigan. 2015. *U.S. Cities Factsheet*. Center for Sustainable Systems, University of Michigan.

USEPA. 2016. Heat Island Effect. https://www.epa.gov/heat-islands.

Van Ginkel, H. 2008. Urban future. *Nature* 456:32–33.

WHO. 2016. Urban population growth. *Global Health Observatory Data*, http://www.who.int/gho/urban_health/situation_trends/urban_population_growth_text/en/.

World Bank. 2015. Urban Development. http://www.worldbank.org/en/topic/urbandevelopment/overview.

Yaalon, D.H. and B. Yaron. 1966. Framework for man-made soil changes: An outline of metapedogenesis. *Soil Science* 102:272–277.

Yang, J.L. and G.I. Zhang. 2015. Formation, characteristics and eco-environmental implications of urban soils—A review. *Soil Science and Plant Nutrition* 61:30–46.

2 Urban Soil Mapping through the United States National Cooperative Soil Survey

Luis Hernandez, Maxine Levin, Joe Calus, John Galbraith, Edwin Muñiz, Kristine Ryan, Randy Riddle, Richard K. Shaw, Robert Dobos, Steve Peaslee, Susan Southard, Debbie Surabian, and David Lindbo

CONTENTS

2.1 INTRODUCTION

The US Soil Survey Program was established by the Agriculture Appropriation Act of 1896. Legislation passed in 1903, 1928, 1935, and 1966 clarified the purpose of the program and expanded the scope to further clarify the intent of the Soil Survey Program (Public Law 89–560, The Soil Survey for Resource Planning and Development Act). The US Congress directs the Secretary of Agriculture to (1) make an inventory of the soil resources of the United States, (2) keep the inventory current to meet contemporary needs, (3) interpret the information and make it available in a useful form, and (4) provide technical assistance and promote the use of soil survey for a wide range of community planning and resource development issues related to both nonfarm and farm uses.

Currently, the US Department of Agriculture-Natural Resources Conservation Service (USDA-NRCS), provides leadership, management, research, and coordination for the soil survey program, nationally known as the National Cooperative Soil Survey (NCSS).

The NRCS technically supports urban and community development and smart growth through the NCSS partnership, but most of the financial support for urban soil surveys comes directly from the urban communities themselves. Using mapping protocols, standards, and concepts from agricultural interpretations, urban soil surveys are pieced together with minimal recommendations for best management practices (BMPs) to develop storm water and sediment pollution controls. Over the years, urban areas have experienced more environmental pressures and public demands for improved quality of life so that cities have evolved from that time to use the data for green infrastructure projects—manipulating storm water, engineered structures, the soil resource baseline, and nature to move and conserve water ecologically in the urban system. Examples of these green infrastructure projects that involve analysis of soil function are rain gardens, innovative storm water systems, urban wetlands, and permeable pavement systems.

There has been growing demand for soil information to strengthen green economy initiatives and community climate change adaptation strategies. Increased interest in food security in the inner cities as well as green infrastructure projects to control surface water and temperature have required continuing collection of urban and nonfarm soil information and increased focus for these activities. Urban soil science has progressed and pushed pedology forward to meet the needs of the increasingly complex anthropogenic ecological network. It has developed new venues for the soil survey program that complement the most recent political demands for carbon (C) footprint analysis, public health issues, and emerging data needs for broad community climate change adaptation strategies.

The first urban soil survey, compiled in 1976, was of the District of Columbia. In 1998, the second urban soil survey in the United States was compiled of the City of Baltimore, Maryland, a heavily urbanized area. At the time, the USDA-NRCS had no specific field methodology, soil taxonomy, or interpretation classification standards to guide the field soil scientists in the urban environment, so USDA staff had to develop a whole new methodology for mapping and classifying urban soils while they mapped.

In the early 1990s, USDA enacted the Urban Conservation Initiative as an effort to better serve its clientele in major urban centers across the nation. The city of New York was selected as one of the urban centers for conservation activities, thus creating a local partnership for developing a soil survey program of New York City. Under this new partnership began a new direction for mapping, classification, and interpretation of soils developed in urban environments. Urban soil information users were interested in soil surveys containing interpretative data similar to the counterpart soil surveys completed in rural areas that mainly focused on agricultural uses.

The Soil Survey of South Latourette Park, in Staten Island, New York City, was completed in 1997 (Hernandez and Galbraith, 1997). This soil survey was the first one where human-modified soils (aka Anthrosols or anthropogenic soils) were characterized to the soil series level, thus providing the base for the development of interpretative data. In 2001, the Soil Survey of Gateway National Recreation Area in New York and New Jersey was completed and published by USDA-NRCS (Hernandez, 2001). This soil survey was instrumental for expanding urban soil mapping concepts to a larger area. In 2007, the Soil Survey of Bronx River Watershed, Bronx, New York, was completed and published by USDA-NRCS.

The objectives of this chapter are to provide and discuss information about urban soil surveys completed across the nation under the umbrella of the NCSS. Principal themes discussed in this chapter include: soil mapping techniques, soil survey products, and soil classification used for completing soil survey activities in Chicago, Los Angeles, Detroit, and New York City. Detailed discussion on the development of urban soil interpretations and urban soil classification is also covered in this chapter. The goal is to provide information to a wider audience on techniques and uses of urban soil mapping.

2.2 URBAN SOIL MAPPING

2.2.1 THE COOK COUNTY AND CHICAGO SOIL SURVEY PROGRAM

The Soil Survey of Cook County, including Chicago, is a subset of Major Land Resource Areas (MLRAs) 95B, Southern Wisconsin and Northern Illinois Drift Plain; 97, Southwestern Michigan Fruit and Truck Crop Belt; and 110, Northern Illinois and Indiana Heavy Till Plain. The survey area has a population of 5,194,675 (USDC Census Bureau, 2010), with Chicago being the county seat. This 600,000-acre (242,950 million ha) area contains a mix of urban, suburban, commercial, and industrial areas and 62,000 acres (25,101 ha) of forest preserves and natural areas. These open areas provide horse trails, bike trails, hiking trails, fishing, and other recreational opportunities for the residents of the county. Half of the county was soil mapped in the 1970s, with work concluding in 1976. The soil scientists did not investigate the more urbanized sections of the county, which resulted in a "hole."

The Chicago Soil Survey was completed (the remaining 298,761 acres or 120,956 ha) in 2012 in response to increasing demand for soil information in urban areas. Water is a major resource in the survey area and the ability to protect it from pollution and mitigate flooding is a concern due to low elevation and increased urbanization. Most of Cook County and Chicago are drained by the Des Plaines, Calumet, and Chicago Rivers. The flow of the Chicago and Calumet Rivers has been changed through engineering (Mapes, 1979). By the 1870s, there were visible signs of pollution and increased concerns about threats the rivers posed to public health. Between 1889 and 1910, the Metropolitan Sanitary District of Greater Chicago completed two major engineering projects to direct the flow of the Calumet and Chicago Rivers into the Des Plaines River to divert wastes away from Lake Michigan (Nilon, 2005). The Sanitary and Ship Canal was the first engineering project, completed in 1900. It diverted the flow of the three branches of the Chicago River (North, South, and Main Stem) into the Des Plaines River. The North Shore Channel was completed in 1910 and it diverted waste from the northern suburbs from Lake Michigan into the North Branch of the Chicago River. In the 1920s a project that connected the Calumet River to the Sanitary and Ship Canal was completed (Nilon, 2005). The flow of the three rivers is now westward (Mapes, 1979).

The Metropolitan Sanitary District of Greater Chicago is now called The Metropolitan Water Reclamation District (MWRD). The MWRD is the agency responsible for protecting the water supply source in the region. Despite the reversal of the Chicago River and the construction of the largest wastewater treatment plant in the world, contaminants continued to accumulate in the water supply, including Lake Michigan. A big reason was because Chicago and many of the older suburbs have combined sewers, where sanitary and storm flow are conveyed through the same pipes. The MWRD's Tunnel and Reservoir Project (TARP), also known as the Deep Tunnel Project, is one of the country's largest public works projects for controlling pollution and flooding. There are four tunnel systems that total 109 miles (175.4 km). These tunnels are 9–33 feet (2.74–10.0 m) in diameter and 150–300 feet (45.7–91.4 m) underground in the dolomitic limestone bedrock. These tunnels collect combined sanitary and storm sewer flows and route them to surface reservoirs for storage until the MWRD plants can treat and safely discharge the effluent (MWRD of Greater Chicago, 2010).

2.2.1.1 Mapping Techniques

Conducting a soil survey and creating a soil map in a major metropolitan area, such as Chicago, poses unique constraints due to scarce natural observation sites, multiple land use changes, anthropogenic soils, and a variety of customers with multiple needs and bureaucratic constraints. Natural soils are mixed with anthropogenic soils throughout the survey area and the variability of the anthropogenic soils can be high. This variability depends upon types of human-altered and human-transported (HAHT) material used, the location from which the HAHT was sourced, and the depth of the HAHT material. Human-altered (HA) material is defined as parent material for soil that has undergone anthroturbation (soil mixing or disturbance) by humans. It occurs in soils that have either been used for gardening, deeply mixed in place, excavated and replaced, or compacted in place for the artificial ponding of water (Soil Survey Staff, 2014). Human-transported (HT) material (HTM) is parent material for soil that has been moved horizontally onto a pedon from a source area outside of that pedon by purposeful human activity, usually with the aid of machinery or hand tools. This material often contains a lithologic discontinuity or a buried horizon just below an individual deposit (Soil Survey Staff, 2014). In the Chicago and Cook County Survey, anthropogenic soils became more extensive further east toward the city core.

Not all urban soil surveys pose the same challenges but there are many methods and procedures that can be shared from city to city. The soil–landscape paradigm is a concept understood by NRCS field soil scientists and one that is followed during soil mapping activities, whether in a traditional soil survey or an urban soil survey. Urban soil surveys are conducted differently from traditional nonurban soil surveys. The high density of impervious surfaces limits the amount of observation sites. In addition, the soil–landscape relationship cannot easily be determined—in some instances—because landforms have been developed (Calsyn and Ryan, 2012). When this occurs, other methods are employed.

Before survey work began, the Cook County Department of Geographic Information Systems (GIS) provided the Chicago soil survey crew with reference information such as raw slope data (that was later converted to 2-foot or 0.61-m contour data), raw digital elevation models (DEMs), cultural data (hospitals, libraries, schools), open space data (parks, empty lots, cemeteries, schools, forest preserves, etc.), and municipality data. All of the data provided by Cook County was in digital format that could be used in ArcGIS for spatial analysis and manipulation. Figures 2.1 through 2.3 show field work being conducted and the types of soils found in the survey area. An initial inventory and analysis was done on potential observation sites—parks, schools, cemeteries, golf courses, forest preserves, nature conservancy sites, and empty lots—and a contact list created for each site. Size, location, and ease of access to the observation sites were noted. This inventory and contact list was maintained in ArcGIS and updated throughout the survey. Additionally, historical aerial photographs, historical native vegetation maps (presettlement data), geology (Harlen, 1939, 1953) maps, historical "fill" maps, and impervious surface data layers were obtained and used in the analysis and development of the soil survey.

The geodatabase for the Chicago and Cook County survey was initially split into two separate databases—North Cook and South Cook—so it could be more efficiently managed and maintained. It was developed using a variety of reference materials as well as soil borings and soil pits documented in the field. The base map was 2005 aerial photography with the digital reference material used simultaneously for better analysis. The raw slope data and the raw DEM data were processed into 2-foot (0.61 m) contour slope maps and hill shade maps. The 2-foot (0.61 m) contour maps provided the subtleties in relief in a flat landscape whereas the hill shade maps helped identify natural and anthropogenic landforms throughout the survey area. Figure 2.4 shows the differences in elevation in Cook County. The information provided by these data helped identify and confirm the soil–landscape relationship; helped in the strategic location of sample points; and also helped in the map unit design. The J. Harlen Bretz Geology (Bretz) maps were provided in hard copy and digital formats. The Bretz maps were instrumental in confirming the surficial geology throughout the survey area and, subsequently, the parent material. The digital version was used in ArcMap for spatial analysis in conjunction with the 2005 imagery and the hill shade data. The digital presettlement vegetation data were overlain the Bretz maps and helped determine soil order in areas that were urbanized. This data layer delineated boundaries between native landscapes such as prairies, timber, scattering timber, marsh, river/creek, slough, swamp, and wet prairies. It helped the soil scientists piece together soil order and parent material. By using the Bretz maps, hill shade, and presettlement data simultaneously, the soil–landscape boundaries were delineated. By utilizing these tools, a solid map unit concept and design for the soil order, parent material, and landform was developed. Soil boundaries were defined and delineated using surficial geology maps and hill shade maps.

FIGURE 2.1   Field work being conducted along the Lake Michigan Shoreline. (Photo courtesy of Kristine Ryan, USDA-NRCS, 2016.)

FIGURE 2.2   A buried 'A' horizon found on a construction site where a new home is being built on a previously developed lot. This site was closer to the city and more disturbed and influenced by humans. (Photo courtesy of Kristine Ryan, USDA-NRCS, 2016.)

For Chicago, it was important to use historical records (i.e., photos, texts) and legacy maps to understand the survey area. The historical aerial imagery was georeferenced so it could be used for spatial analysis in ArcMap. The historical references provided clues about land use changes, age of neighborhoods, degree of disturbance, and taxonomic soil order. Historical aerial photos were compared with present-day aerial photos to determine the differences in land use

FIGURE 2.3 A natural soil (nonanthropogenic) in a cemetery in the western suburbs of Chicago. This soil was found on the Valparaiso Morainic System. These moraines are composed of relatively young geologic materials characterized by tills with a higher clay content. (Photo courtesy of Kristine Ryan, USDA-NRCS, 2016.)

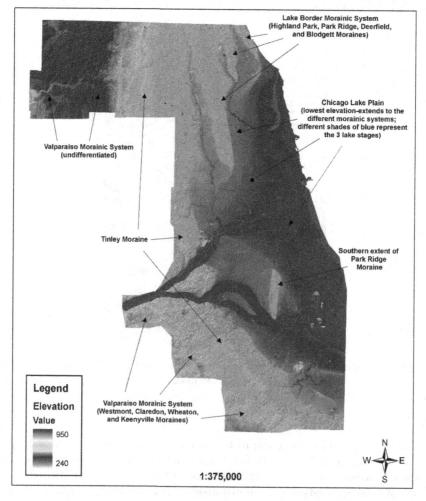

FIGURE 2.4 A Cook County elevation map showing surficial geology and boundaries of the geologic features of the survey area. (Map prepared by Kristine Ryan, USDA-NRCS, 2012.)

(agricultural to residential use). Knowing the timeline of when the land was converted gave clues about the degree and nature of disturbance as well as evidence of soil genesis. Neighborhoods that were built in the first half of the last century had a higher percentage of natural soils due to construction methods and types of development. Neighborhoods that were built after 1960 used more land leveling construction methods and, as a result, had a higher percentage of disturbed and anthropogenic soils. These differences helped in map unit design because some neighborhoods had more soil consociations whereas others had more soil complexes—natural soils with anthropogenic soils.

The impervious surface data were obtained from the National Land Cover Database. The impervious surface raster map was generated from three data sources: (1) high resolution imagery, (2) calibrating density prediction models with reference data, and (3) Landsat imagery. These digital layers helped with Chicago's "pavement problem" by offering a degree of imperviousness. The impervious surface raster was reclassified in ArcMap to reflect pavement percentages:

- 0%–15% paved
- 15%–24% paved
- 25%–49% paved
- 50%–85% paved
- 85%–100% paved

In neighborhoods across the survey areas, there are areas where runoff is high, but there are areas where precipitation will soak into the soil. It was important to capture the differences in areas that were high in pavement and buildings and those that had neighborhoods with large parks, lawns, and school yards. Map unit design took into account the percentage of pavement. For example, soils that had less than 15% pavement were named as soil consociations (natural soils with no anthropogenic components). These areas occurred in forest preserves, parks, and other natural areas. The same thought process went into delineating highway cloverleaves. Runoff versus infiltration is an important consideration for urban surveys.

Data collection in Chicago and Cook County was primarily in parks, empty lots, school yards, cemeteries, golf courses, forest preserves, construction sites, and other natural areas. HAHT soils were found throughout the survey area. Most of the residential neighborhoods, further west of the city core, had a thin layer of HAHT material over natural soils. This occurred in the neighborhoods that were older and were built when construction methods left a smaller impact on the soils. Many of these soils had an intact buried surface horizon over an intact subsoil. Soils that had less than 50 cm of disturbed or modified material over a natural soil were classified as the natural soil. The soil order, parent material, drainage, and other soil properties could be observed and the soil mapped. Soils that have HAHT materials to a depth greater than 50 cm and whose properties are dissimilar to the naturally occurring materials are considered anthropogenic soils (Soil Survey Staff, 2014). The Cook County and Chicago Soil Survey classified the anthropogenic soils down to the family level and not the series level. Due to time and staffing constraints, it was not possible to establish new series for the anthropogenic soils. Parks that were as small as 1 acre (0.40 ha) in size were investigated if they were located in highly urbanized areas closer to the city core. Densely urbanized areas had limited observation points so small parks offered glimpses into parent material and natural drainage. These small parcels of open space allowed the survey to be pieced together like a puzzle. Soil boundaries were delineated based on natural and anthropogenic landforms identified by the hill shade maps, surficial geology identified by the Bretz maps, presettlement data, the impervious surface raster map, and the historical aerial imagery (Figure 2.5). All of these tools were utilized to piece together a quality soil survey and provide useful interpretations for the urban user.

Chicago's history and physiographic characteristics presented an opportunity for the soil survey crew to act as "history detectives" to understand the landscape now and the soil capacity for the

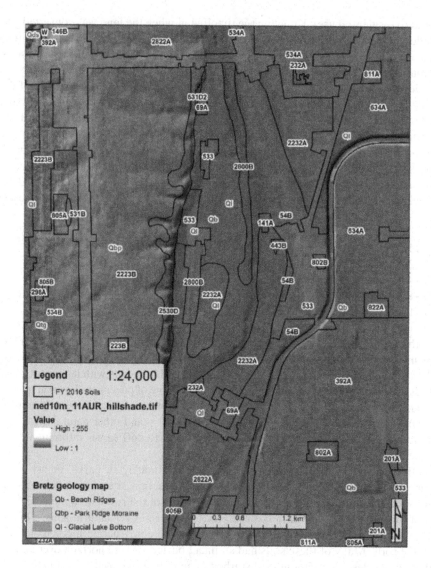

FIGURE 2.5 Chicago and Cook County soil survey–Blue Island Quarter Quad; 10-m DEM and surficial geology (Bretz) maps draped over 2005 aerial imagery. (Map prepared by Kristine Ryan, USDA-NRCS, 2016.)

future. With the maps and database complete it is hoped the urban soil team and city partners can explore applications of the soil database for new interpretations to express the specific needs of Chicago infrastructure and the surrounding area.

The Chicago Soil Survey was released digitally on the Web Soil Survey in 2012. There also is a supplemental manuscript available that provides users additional information about the survey.

2.2.2 THE LOS ANGELES SOIL SURVEY PROGRAM

The Soil Survey of Los Angeles County, Southeastern Part is located in the Southern California Coastal Plains and Mountains, Major Land Resource Area 19 and 20 (U.S. Department of Agriculture, Natural Resources Conservation Service, 2006). The survey area has a population of about 9.2 million people (USDC Census Bureau, 2010). This 681,000-acre (275,709 ha) area contains a mix of urban, suburban, commercial, industrial areas, and pockets of natural hills and mountains used for recreation or habitat areas. The past land use history has created a vast, sprawling urban landscape with both natural and anthropogenic soils. There are 80 incorporated municipalities including major cities such as Los Angeles, Long Beach, Burbank, Torrance, Pasadena, and Santa Monica.

The "Los Angeles Basin," as it is often referred, is an expansive sequence of coastal plain and coastal valleys that is surrounded by mountains and the ocean (Figure 2.6). It is not a true basin since it drains directly into the Pacific Ocean to the south and west. It is an area with a complex geomorphological history. The Los Angeles area is bounded by the east–west-oriented Transverse Range at the northern margin and north–south-oriented Peninsular Range to the southeast. Factoring a complex faulting history with rising and falling sea level, the complexity of the natural parent materials becomes evident. Natural parent materials range from igneous bedrock to fluvial sedimentary bedrock, marine sedimentary bedrock, volcanic bedrock, old alluvium on remnant alluvial fans and terraces from variable sources, young alluvium, old marine alluvium, palustrine deposits, and coastal eolian sand dunes. Surface modification from human activity often covers the natural soil properties and in other cases, partially or completely alters them.

The Los Angeles area has a variety of social and land management issues that can be more effectively managed with the availability of a modern soil survey. Water management is a major concern due to the high potential for urban runoff from impervious surfaces. The area also experienced devastating flooding in 1861 and 1938 in downtown Los Angeles (Bilodeau et al., 2007). As a result, the river and stream channels, with few exceptions, have been channelized and lined with concrete to alleviate flooding concerns. This has prevented mass flooding in high density commercial and residential areas, but has significantly reduced natural recharge into the local aquifer. As a result, in combination with groundwater pumping, the water table has been significantly lowered. The San Gabriel and Los Angeles River floodplain area is an extensive natural braided stream channel with a history of shifting channels. For example, until 1825 the Los Angeles River channel extended westerly from present-day downtown to the now mostly dry

FIGURE 2.6   A view of the Los Angeles Basin, looking northeast from the Palos Verdes Peninsula. This area is a broad coastal plain on the western and southern part, and a broad alluvial plain in the north and eastern part. It is surrounded by hills and mountains to the north and east with the Pacific Ocean to the southwest. Downtown Los Angeles is visible near the base of the mountains on the right side of the photo. (Photo courtesy of Randy Riddle, USDA-NRCS, 2016.)

Ballona Creek bed, which empties into the Pacific Ocean near Marina del Rey. The flood of 1825 caused the Los Angeles River to change channels to a southerly route, with its present mouth in what is now Long Beach Harbor on San Pedro Bay (Bilodeau et al., 2007). This shifting has reoccurred naturally over time. There are also a series of dams and debris basins located below major drainage ways at the base of the San Gabriel Mountains. These flood control structures and debris basins are used to capture high volume water flow and disperse debris flows from damaging infrastructure and neighborhoods below.

The hills in the Los Angeles area pose a different threat for developers and land use managers. The combination of faulting action, Mediterranean climate (cool and moist winters with warm and dry summers), and mineralogy of the marine sedimentary bedrock poses a threat for unstable slopes. This influenced the City of Los Angeles to be one of the first cities in the nation to adopt hillside-grading ordinances (Scullin, 1983). The marine sedimentary bedrock is mostly calcareous shales, fine-grained sandstones, siltstone, and mudstones (Yerkes and Campbell, 2005). They are often highly weathered, soft, and fractured. These rocks weather into soils with high shrink-swell (2:1) clays and other weakly developed soils due to the unstable nature of the landscape. The hills are generally rolling, but can have slopes in excess of 85% on the side slopes. Additionally, as the high shrink-swell soils erode from the hills, they are deposited in a sequence of broad alluvial fans that also possess similar soil properties. This creates a challenging environment for engineers and developers in an area with a strong demand for high-dollar residential neighborhoods on the hills.

2.2.2.1 Mapping Techniques

Mapping soils in an extensive urban environment, such as Los Angeles, can invite unique challenges that are not common to mapping in a natural setting. Natural soils are intermixed with variable depths of HTMs without regard to natural landscape position or landforms (Figure 2.7). This creates highly variable soil properties across an otherwise predictable soil–landform relationship. Anthropogenic soils make up additional components in map units found in the densely populated areas. HAHT

FIGURE 2.7 A middle school on an anthropogenic landform in northeast Los Angeles. (Photo courtesy of Randy Riddle, USDA-NRCS.)

materials that are recognized on identifiable, anthropogenic landforms are considered mappable. Soil components were also evaluated using this standard to determine similarity and limiting properties between soil types.

The geodatabase for this survey was developed using a variety of reference materials in addition to direct observations taken in the field. Using GIS, a 3-m light detection and ranging (LiDAR) dataset mostly was used for terrain analysis. This 3-m DEM was processed into various hillshade, contour, and slope maps. An older 10-m US Geological Survey (USGS) topo map, also processed using terrain analysis, was used to help identify and confirm anthropogenic surface modification. By comparing the two, a better understanding of the land use history is achieved. Anthropogenic soils can visually be predicted in many areas using high-resolution elevation data such as the 3-m DEM. There are surface signatures (shapes) that can be used to identify areas where HAHT materials are suspected.

Reference materials such as the USGS geology quad maps were georeferenced and draped over the terrain models (hillshade, etc.). Also available are a set of old reconnaissance soil surveys that was completed by the Bureau of Soils between 1916 and 1919. These historical documents provide vague soil descriptions, but were still helpful in identifying natural landforms and generalized descriptions of soil properties (observed in the predeveloped environment). Other resources included research papers, planning maps, and speaking with people who are familiar with the history of the area.

The landform mapping model was emphasized to capture the natural landscape units. This can be done even in dense urban environments. Every effort was made to achieve and maintain this standard. In places where mass grading has taken place and no historical evidence of the natural landform ends or begins, landform delineations were drawn at the extent of the current reshaped landform. All available reference materials are used to delineate the landforms properly.

After delineating the natural landforms, subset map unit polygons follow anthropogenic landforms, where they occur. This technique was used in areas where identifiable and repeatable anthropogenic landforms occur on the landscapes as independent landforms. The same rules apply in delineating anthropogenic landforms as natural areas. For example, hillsides that have been leveled and sub-terraced for building or recreational footprints are delineated as a new anthropogenic landform, the reasoning being that they differ from the adjacent naturally occurring landform (Figures 2.8 and 2.9). In these cases, the anthropogenic landforms (hillslope terraces) are identifiable and the soils are dissimilar to the natural soils. In rare cases, land use patterns are used for delineations where a highly disturbed or impervious situation is present. Dumps, mines, large oil refineries, tank farms, debris basins, seaports, and large commercial areas were delineated in this manner. These areas are exceptions to the landform mapping standard. Individual, small acreage, lots or parklands were not delineated, but were included in the greater map unit concepts. A strong map unit concept is key to ensure that areas too small to map have component data that sufficiently characterizes the soil properties therein.

Data collection in the Los Angeles area was primarily located in city parks, golf courses, airports, construction sites, and natural lands. After the soil–landscape model is properly developed, the soil type was projected into adjacent areas on common landforms where field data are not possible. Soil descriptions were taken opportunistically at locations where access was negotiated. This largely prevented traditional transects data from being collected due to unpredictable city park locations and other accessible lands.

Throughout the survey area, the distribution of HAHT soils was found to be irregular and unpredictably scattered. Impacted soils often exhibit hydrophobic soil surfaces, surface crust formation, and high bulk densities that restrict infiltration rates (Craul, 1992). However, a research study in Baltimore also suggests that although bulk densities are elevated, only 10% of sampled sites were root restrictive due to compaction (Pouyat et al., 2007). Most residential and commercial areas in the Los Angeles area have a thin layer of compacted HAHT overlying the natural soil. In many cases, the surface material was locally graded to smooth localized topography capped with a thin layer of topsoil or sod. These HAHT materials are often thin and may be similar to the natural material. Bulk densities in this material were documented as higher than those expected for natural material; however, most surface horizons were not root restrictive, which is consistent with Pouyats' findings in Baltimore. Areas with more substantial

FIGURE 2.8   The darker colored natural subsoil can be observed on the left side of the tarp. The lighter colored grayish surface fill can be observed on the right side of the tarp. This location was excavated on the uphill side by the retaining wall (cutbank) and filled on the tailing end of the locally raised landform. (Photo courtesy of Randy Riddle, USDA-NRCS, 2013.)

FIGURE 2.9   Mixed land use observed on marine terraces and risers in Palos Verdes Estates, CA. There is a clear difference between the natural landforms and the anthropogenic hillslope terraces (cut and filled slopes). Natural landforms in the left foreground and slopes on the right side of the road and in the neighborhood on the hillside in the background. (Photo courtesy of Randy Riddle, USDA-NRCS, 2015.)

modification are generally more isolated or are recognizable on an anthropogenic landform. For example, prior generation construction sites may not be visible today, but old foundations, cut and fill areas, or leveled areas leave the soil in a modified state.

Higher sloping lands tend to have more visible anthropogenic landform features. Large areas are graded to create level or more functional sites for construction and recreation. The HAHT materials in these area tend to be thicker and are more predictable to locate. Examples are large ballfields that are leveled on low elevation hills (Figure 2.10) or cut and filled hillslope terraces on steep side slopes in residential areas.

Soils that have HAHT materials to a depth greater than 50 cm and whose properties are dissimilar to the naturally occurring materials are considered anthropogenic soils (Soil Survey Staff, 2014). New soil series were created to incorporate components with HAHT materials possessing reasonably predictable soil properties (Figure 2.11). This is partly due to standardized building practices adopted by developers over the years. If soil has less than 50 cm of disturbed or modified material, or the HTM at the surface is similar to natural soil conditions, then the soil is considered a variation of the natural soil component.

Phillip Craul's 1999 book, *Urban Soils, Applications and Practices*, suggests there are four categories to consider when evaluating soils:

1. Not graded and not compacted
2. Not graded but compacted

FIGURE 2.10   Montebello Series found in La Verne, CA. Contact of the human-transported material with the natural substratum is observed at 110 cm. Evidence of transported materials includes a thin layer of construction gravels on the left sidewall at 70 cm. The natural substratum is an intact, truncated argillic horizon. (Photo courtesy of Randy Riddle, USDA-NRCS, 2014.)

FIGURE 2.11 A soil profile of a fine-loamy, spolic, mixed, semiactive, calcareous, mesic Anthroportic Udorthents, Midtown series (scale in centimeters). (Photo courtesy of Eric Gano, USDA-NRCS, 2012.)

3. Graded but not compacted
4. Graded and compacted

Graded soil is defined as soil that has had its topsoil disturbed, removed, or replaced. Compacted soil is defined as a soil that has been compressed and now has a bulk density >1.6 Mg/m³ (Craul, 1999). These separations are useful to consider when evaluating the intensity and impact of soil modification.

The Los Angeles County, Southeastern Part Soil Survey will be released digitally on the Web Soil Survey at its completion. There will also be a supplemental manuscript available that will provide users additional information about the information in the survey.

2.2.3 THE DETROIT SOIL SURVEY

Detroit, Michigan, is located in the far southeast corner of the state and is bordered by the Detroit River and Canada to the south and east. The Detroit metro area occupies several counties including Wayne, but the City of Detroit itself sits along the Detroit River and is completely contained within Wayne County.

In the early 1600s, Europeans (especially the French) began to explore the area. In the early 1700s, French settlers began to encourage farming (Detroit Historical Society, 2016) and by 1853 most of Wayne County was farmed (Farmer, 1890). During the Industrial Revolution of the late 1800s heavy industry developed along the bank of the Detroit River and farmland was gradually converted to urban and industrial land. During this time, the Detroit River provided a pathway for transportation, served as a source of fresh water, and was used as a dump for waste water (Howard et. al., 2013a; Figure 2.12). Detroit became a major urban area following the explosive growth of the automobile industry during the 1920s. By 1930, the city was the most industrialized area in the United States, and at its peak was home to over 2 million people (Hyde, 1980). Detroit's economy peaked in the 1950s. In the late 2000s, the population and economy shrank dramatically. In 2010, the USDC Census Bureau estimated that approximately 700,000 lived in Detroit itself. The loss of population has led to extensive demolition of abandoned homes and businesses. The creation of open spaces through demolition has stimulated growing interest in the use of vacant land for urban agriculture, urban forestry, and green infrastructure (Howard, 2015).

In 1974 Wayne County, Michigan correlated and published a soil survey; however, 23 metropolitan cities, including the City of Detroit (180,000 acres or 72,847.5 ha), were left unmapped due to urbanization. In the late 1990s, efforts were under way involving the Southeast Michigan Council of Governments, urban gardening groups, City of Detroit officials, Wayne County Conservation District (WCCD), and the NRCS to initiate an urban soil survey and map the remaining acres. In 2011, NRCS staff became available to start a soil survey in the Detroit area. The WCCD and NRCS were instrumental in setting up two informational meetings and a First Acre Ceremony on Belle Isle for perspective users of the soil survey. These perspective users included city governmental officials, NRCS, interested citizen groups, and environmental consultants. The informational meetings introduced the participants to soil surveys and allowed them a chance to list urban soil and water issues that affected them. Major

FIGURE 2.12 A view of downtown Detroit showing the densely spaced buildings and pavement (more than 80%) intermixed with HAHT soils (less than 15%). (Photo courtesy of William Baule, 2015.)

issues included basement flooding, trees to plant in urban areas, areas prone to flooding, areas of high storm water runoff, and areas most susceptible to urban sinkholes. These concerns gave the soil scientists some insight on what soil properties to collect that could ultimately lead to the development of local interpretations, if needed. Before field investigations began, the MLRA Soil Survey Leader met with all city officials to go over the logistics of soil mapping in their city. For example, most city officials were interested in how NRCS would avoid utilities and requested to be notified when survey crews were actively sampling. Memorandum of understandings between the cities and NRCS were then signed and the soil survey for Detroit and surrounding areas began in the spring of 2012. Cities granted NRCS access to public areas. NRCS expanded possible sample areas by enlisting the help of local newspapers to report on the soil survey and encouraged readers to participate in the survey.

The soil scientists began searching for reference material on soils and geomorphology. The following resources helped develop a premapping legend and a landform map:

- Current Wayne County, Oakland County, and Macomb County soil surveys adjoining survey area
- Detroit Folio (Sherzer, 1917) base map depicting surficial deposits, landforms, and relict glacial shorelines
- LiDAR derived 2 foot (0.61 m) bare earth DEMs for Wayne County
- Contact with private and public environmental firms for collection of historical soil boring logs
- Digital map of known brownfields
- Digital map of local roads
- Impervious surface dataset layer

Available aerial photography for the survey area included leaf-off black and white and color infrared imagery from the spring of 1999. The survey area has relatively small changes in elevation between landforms and constantly changing land use, making the black and white imagery unsuitable as a mapping base. Color infrared imagery would normally provide indications of soil moisture showing possible changes in drainage class or surface texture; however, artificial drainage throughout the survey area has masked these indicators especially in the extensive residential areas. True color imagery from the National Agriculture Imagery Program was also available; however, canopy cover from landscaping trees made it impossible to use this imagery as a mapping base.

The LiDAR-derived bare earth DEM proved to be the most useful mapping base for the survey area. LiDAR DEMs were used to create a hillshade image with a five times vertical exaggeration. This vertical exaggeration provided enough relief to identify most major landform changes. A color ramp with the lowest elevation in purple and the highest elevation in brown was applied to the DEM to give a visual indication of relative elevational changes on each mapping sheet and the DEM was overlain on the hillshade with a 40% transparency. Hand-digitized deposit feature boundaries and shorelines from the Detroit Folio (Sherzer, 1917) were overlain on top of the imagery as a guide to where soil properties may be changing. While this proved to be an excellent base for soil mapping, USGS topographic quadrangles and road maps printed on an aerial photograph base were still needed for navigation and identification of suitable locations for soil borings.

The area is highly urbanized with a loamy or sandy surface of HTM averaging 14 inches (35.6 cm) thick in residential areas. Industrial or commercial areas commonly have a thicker cover of HTM. The Detroit metro area is largely flat, with slopes only occasionally rising above 3%. The naturally subtle landform changes along with human alteration of natural surfaces confounds mapping based on geomorphic position. Instead, soil scientists premapped areas of the county using the methods described previously to plan field visits.

While performing field investigations, a grid system was used where possible. In areas where known brownfields were located, private environmental consulting and public engineering soil borings were used. Most of the survey area is considered residential and soil borings were made in front yards, side lots, backyards, and right-of-ways (Figure 2.13). Soil scientists worked in teams of two,

FIGURE 2.13 A soil boring taken in the back yard in a residential area in the City of Detroit. (Photo courtesy of Shawn Finn, USDA-NRCS, 2014.)

and standard equipment included a soil auger, tile spade, plastic tarp, GPS unit, and a five gallon (18.9 L) bucket. The bucket was used to remove any excess soil from the site.

Generally speaking, residential areas have notably less fill than industrial or commercial areas. However, fill depths ranged widely even in residential areas, and the search for patterns in residential fill properties led to a number of scientific publications by NRCS cooperators (see Research section). Examples of these map units include: Kibbie-Urban land complex, 0%–4% slopes and Avoca-Urban land-Blount complex, 0%–4% slopes. Minor components would include buried soils like the "Midtown" series, fine-loamy, spolic, mixed, semiactive, calcareous, mesic Anthroportic Udorthents, and the "Riverfront" series, fine-loamy, artifactic, mixed, semiactive, calcareous, mesic Anthroportic Udorthents (Figure 2.11). These buried soils contain more than 50 cm of fill and average fewer than 35% artifacts in the control section. Buried soils such as Midtown are commonly encountered on water-lain moraines. By contrast, Riverfront soils are formed mostly based on anthropogenic rather than natural landforms and include the deepest fill soils in the county.

Riverfront soils are found primarily along the Detroit River and include current and former industrial and commercial sites that have been converted to parks or green spaces. Anthropogenic landforms, such as railroads and major highways, were delineated from natural landforms by the use
of LiDAR and NAIP photography. Some map units covering federal and state highways included
Urban land sandy substratum, 0%–4% slopes and Urban land loamy substratum, 0%–4% slopes.
Railroads were covered by the map unit "Udorthents fragmental, 0%–8% slopes." It appears that
most urban soil surveys, depending on their geographical location, can have distinct soil mapping
techniques. No one method of soil mapping applies to all urban areas.

The urban soil survey in Detroit was supplemented by related research from NRCS cooperators,
and in turn the survey generated research ideas. Many research topics involve demolished buildings. Between 2000 and 2015, approximately 49,500 buildings were demolished in Wayne County,
about 43,400 of which were in Detroit (SEMCOG, 2015). As part of collaboration with the US
Environmental Protection Agency (EPA), NRCS soil scientists assisted the EPA with field investigations of demolition sites and their impact on runoff (W.D. Shuster, US EPA, personal communication,
2013). The EPA conducted a similar study in Cleveland, Ohio (United States Environmental
Protection Agency, 2011), and the study in Detroit will further advance understanding of urban soils.

Another resource used in soil mapping includes using research performed by Wayne State
University (WSU) in collaboration with NRCS. The Geology Department of WSU, located in downtown Detroit, is involved with numerous research projects involving urban soils, many of which
contributed to the Detroit survey. Some of the published research includes chemical and physical
weathering of artifactual fills (Howard et al., 2013a), geomorphology of the Detroit area (Howard
et al., 2013b), soil mapping of vacant lots in Detroit (Howard, 2015), and using geospatial tools to
quantify soil properties (Howard et al., 2016). These studies have helped soil scientists identify
anthropogenic surfaces from native soil surfaces, identify landforms in nearly level landscapes,
and help predict areas containing various types of artifacts and their effects on soil health. All the
aforementioned research has also been crucial in collecting data on chemical and physical properties of soils used in the development of interpretations and data population in national databases.

2.2.4 THE NEW YORK CITY SOIL SURVEY PROGRAM

New York City includes portions of two MLRAs: 144A, the New England and Eastern New York
Upland, Southern Part; and 149B, the Long Island-Cape Cod Coastal Lowland. The city's popula-
tion is 8.6 million (USDC Census Bureau, 2015); its area is 283,196 acres (114,654 ha), with 190,764
acres (77,234 ha) of land area. It is comprised of five boroughs (Manhattan, Bronx, Brooklyn,
Queens, and Staten Island), coextensive with five counties (New York, Bronx, Kings, Queens, and
Richmond), including four of the most densely populated counties in the United States (USDC
Census Bureau, 2010).

The city is surrounded by numerous waterways, and four of the five boroughs are situated on
islands. Extensive suburban areas border the city on the north, east, and west. Parts of three physiographic units are included: the New England Upland on the north and northwest, the Triassic
Lowland on the southwest, and the Atlantic Coastal Plain along the southeast.

New York City's complex geology includes layers of crystalline bedrock, sedimentary rocks
with associated igneous intrusive, coastal plain sediments, glacial deposits from several episodes,
and scattered postglacial materials. By nature of its geography, the city has had limited room for
expansion; transformation of the natural environment through extending the shoreline, draining and
filling of wetlands has been the norm since the late seventeenth century.

In the early 1990s, the USDA enacted the Urban Conservation Initiative in an effort to bring
together a group of federal and state agencies to better serve the urban clientele. As part of this
Initiative, the NRCS, in cooperation with the New York City Soil and Water Conservation District
(NYCSWCD), decided that New York City would benefit from a comprehensive effort to address
the sustainable use and management of urban soils. The agency entered into a formal agreement,

or memorandum of understanding, with the NYCSWCD and Cornell University to establish a soil survey program in New York City. Unlike other NCSS projects, this one sought to address the complexities of urban needs with not just one soil survey product, but with a multifaceted program, including a series of maps at variable scales; site evaluations; education, training and public outreach; and survey-related research. It proposed to produce and deliver soils data at multiple levels of scale and intensity commensurate with the needs of various users. The NYCSWCD role included the integration of soil survey elements with other city initiatives and projects, coordination of the soil survey program with local requests, university outreach and partnership development, marketing through electronic media, and providing cooperative program leadership and support. Cornell University provided technical guidance, research on important urban soil issues, and leadership in the description and classification of urban soils. A technical advisory committee (TAC) was established for the program with members from academia, government agencies, and environmental and community groups, to identify users' needs, review program progress, provide operational guidance, and disperse information. The TAC served to provide direction and identify priorities and goals for the program.

The first NCSS effort in New York City was the Soil Survey of Central Park, in cooperation with Cornell University and the Central Park Conservancy, completed in 1982 (Warner et al., 1982). Soils in the 843-acre (341.3 ha) park were classified to the great group and subgroup level and mapped at a scale of 1:4800. In general, soils formed in HTM with little profile development (Udorthents) were differentiated from those soils with little evidence of disturbance (Dystrochrepts). A small guide, the Tour of Soils of Central Park in New York City (Weber et al., 1984), which included a soil map, was published by Cornell Cooperative Extension.

The more recent soil survey program began with a 1:6000 scale map of the 320-acre (129.6 ha) South Latourette Park in Staten Island, which served as a pilot project (Hernandez and Galbraith, 1997). While this was not the first USDA survey to cover an urbanized or suburban area, nor was it the first to include "anthropogenic" soils at the soil series level, it did introduce five new series (Greenbelt, Central Park, Forest Hills, Canarsie, and Great Kills soil series) in HAHT materials, and focused on guidance for urban land use decisions (Figure 2.14), in a nontraditional format, including soil profile drawings and block diagrams that included these new soils. It was designed to be more user friendly, a model for future urban soil surveys, and part of the agency's ongoing urban initiative at the time.

The Soil Survey of Gateway National Recreation Area (Hernandez, 2001), one of the first urban parks in the National Park System, was a partnership effort with the US Department of the Interior's National Park Service (NPS). It was another high intensity, or Order 1, survey—8,300 acres (3360.3 ha) at 1:4800 scale, as requested by the NPS. This was commensurate with the mapping of other park resources, enabling its use for site suitability in project, facility, and resource planning; as a base layer for natural and cultural resource assessment; for wildlife and watershed management plans; and as a guide for research and restoration efforts.

The Bronx River Watershed Soil Survey (United States Department of Agriculture, Natural Resources Conservation Service, 2007), another Order 1 survey, covered about 7,000 acres (2834.0 ha) at a 1:6000 scale. As the agency's first urban watershed soil survey, it emphasized hydrologic implications. An accompanying infiltration study examined land use effects on Hydrologic Soil Group ratings (used in the estimation of runoff) and resulted in the creation of several land use-specific map units. The Bronx River Alliance, a local nonprofit working to protect, improve, and restore the river, was an important source of reference information and provided operational guidance and access throughout the survey area.

These surveys represented the high intensity phase of the survey program. Where selected, green space areas were mapped at large scales and soil data could be used to evaluate lot-sized areas. At the other end of the spectrum, a general or Reconnaissance Soil Survey (United States Department of Agriculture, Natural Resources Conservation Service, 2005), at 1:62,500 scale, provided an overview of soil patterns and the distribution of parent material types around the city, and served as a

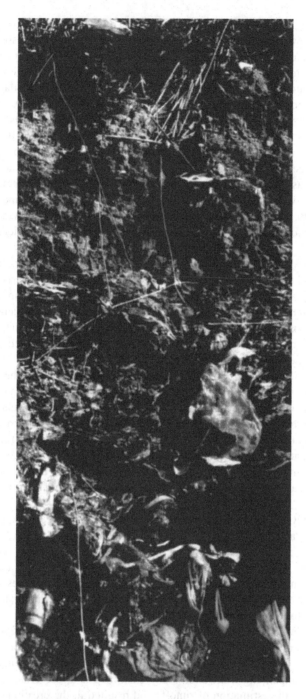

FIGURE 2.14 Soil profile of Freshkills series, anthropogenic soil, mapped in South Latourette Park Soil Survey. These soils formed in a thick mantle of human transported material that includes loamy soil material over a geomembrane over a mixture of household garbage. This particular soil profile does not have a geomembrane. It belongs to a landfill established when geomembrane technology did not exist and many landfills in NYC did not use geomembranes. But the main concept is captured with the picture. Also, the classification in the official soil series description (OSED) is coarse-loamy, mixed, active, hyperthermic oxyaquic eutrudepts—a temporary classification designated when soil series were established. Although the OSED has not been reclassified, this temporary classification is useful for this publication. (Photo courtesy of Luis A. Hernandez, USDA-NRCS.)

framework for more detailed survey work. Designed as an educational and outreach tool, it was distributed in poster format. The map was composed of 88 soil map units, with 20 soil series in HAHT materials and 24 soil series in naturally deposited materials. The map units in the Reconnaissance Survey are mostly all complexes, and the minimum size delineation is about 40 acres (16.2 ha).

The third phase of the soil survey program, a 1:12,000 scale city-wide soil survey, commensurate with the NCSS initial mapping scale and standards, was posted to the USDA-NRCS's Web Soil Survey in June 2014. In addition to providing a valuable resource inventory, information from the survey has useful applications for storm water management, restoration efforts, park and open space management, and urban soil education. As in previous NYC survey efforts, soils in HAHT materials, soils in "natural" materials, and miscellaneous areas (urban land, beaches, rock outcrop, etc.) were differentiated. A total of 66 soil series were included in the survey: 29 for HAHT soils and 37 in "natural" soils. Most of the HAHT soil series (27 of 29) were established in New York. According to the survey, the city's land area is comprised of 62.7% impervious surface (urban land), 27.6% HAHT soils, 8.6% natural soils, and 1.1% miscellaneous areas such as rock outcrop, beaches, and so on (Table 2.1). Staten Island remains the borough with the least amount of impervious surface and the greatest proportion of "undisturbed" soil. A significant amount of laboratory data have accompanied the various survey efforts; by the time of the 1:12,000 survey's completion, over 100 soil profiles in New York City had been sampled and analyzed for complete characterization by the USDA-NRCS Kellogg Soil Survey Laboratory. The soils of the city exhibit a wide range of physical, chemical, and mineralogical properties.

Soil survey rests on a strong scientific foundation, and soils in urban areas are among the last frontiers in soil science research. Investigations on properties of soils in urban areas are needed to allow for reliable ratings and interpretations for their use and management (Figure 2.15). Research under the leadership of Ray Bryant at Cornell in the 1990s examined land use effects on soil properties (Gossett, 1999; Singleton, 1998) and indicated that soil interpretations based on near surface properties are more closely related to land use types than soil types. Other projects have included soil temperature studies (Mount and Hernandez, 2000) that helped to establish the appropriate temperature regime for the city's landfill soils, and new applications for ground penetrating radar (GPR), identifying various substrate interfaces in the urban environment (Doolittle et al., 1997). Trace metal studies have included an examination of atmospheric additions to soils in wooded areas in the Bronx, Manhattan and Staten Island, the latter in cooperation with the College of Staten Island (Jovanovich, 2011), and a look at the correlation between soil metal levels and artifact contents (Shaw et al., 2010). To assist with the citywide survey efforts, the geochemical properties of serpentinite soils in Staten Island were examined by students from the College of Staten Island under Alan Benimoff (Ochieng, 2011; Ongoïba, 2012), and a project with Con-Edison to determine urban background levels of trace metals and PAHs in Manhattan helped in the characterization of soil properties in the city's inner core. An investigation of red parent material in Staten Island with regard to formation of redoximorphic features was conducted with assistance from the University of Maryland (Isleib et al., 2010). A sampling project designed to assess the city's carbon stocks,

TABLE 2.1

Percent of Land Area under Various Map Unit Component Types, 1:12,000 Scale Soil Survey of New York City

Area	Impervious	HAHT Soils	Natural Soils	Misc. Areas
Citywide	62.6	27.4	8.8	1.2
Manhattan	72.2	21.4	4.9	1.5
Brooklyn	67.0	29.6	2.8	0.6
Queens	63.7	28.9	6.2	1.2
Bronx	62.5	25.5	11.4	0.6
Staten Island	48.3	29.5	20.2	2.0

FIGURE 2.15 Soil compaction in urban area exposed by the tree leaning over. Root system is concentrated close to the surface and does not provide proper support (Photo courtesy of Richard K. Shaw, USDA-NRCS.).

including black carbon and, with Brooklyn College, a characterization of microbiological communities in important soil types and land uses. The survey staff in NYC also provided valuable input on the description and (soil taxonomic) classification of HA soils to the International Committee on Anthropogenic Soils (ICOMANTH), led by John Galbraith of Virginia Tech, who had helped establish the soil survey program in NYC while at Cornell in the 1990s.

With a significant extent of HAHT materials and a diversity of natural soil types in the survey area, an understanding of predevelopment conditions and land use history is invaluable. A comparison of old and new topographic maps was used to locate natural landforms and significant anthropogenic alterations. Aerial photography from multiple timeframes helped in identifying land use changes. Older surficial geology maps, in particular the USGS New York City Folio of Merrill (1902), were used to determine the nature of preexisting parent material (which can also serve as local fill) and to denote substratum conditions, and allow for some initial delineation of the survey area. A familiarity with the original parent materials and the properties of the associated soils in the survey area was important to determine whether a particular pedon is HT and HA or natural.

Soils in HAHT materials present a formidable challenge to soil survey; spatial and vertical variability can be complex and unpredictable, and soil conditions often change with little variation in landscape or vegetation. Small lot sizes, backyards in residential areas, narrow transportation corridors, and soils in small commercial zones, medians, and parking lots are likely better suited to onsite inspections. However, the larger open spaces (more than a hectare or so) in the city generally exhibited more uniformity in soil conditions: these areas were sometimes less disturbed, or, if HTM were involved, it was often as one big project, filled at the same time with similar materials. A general consistency was noted in the HTM in the larger open spaces. Eventually, certain HAHT soil types or HA landforms were associated with a particular surficial geology type, landscape position, or land cover type, allowing for some soil–landscape modeling. Such predictions need more field verification than in undisturbed areas.

A priority of the city's potential soil survey users is dependable information on the physical and chemical properties of the city's soils. Applications include revegetation purposes, such as recent efforts to increase the city's tree canopy; hydrologic applications for better storm water management; and a general open space inventory for wetland and other habitat types. The request for accurate soil property data was part of the justification to classify HAHT soils to the series level; this provided particle size and drainage classes, and degree of development. Additional series differentiation was made by HTM type and artifact content. Artifacts and their weathering byproducts can have significant effects on the physical and chemical properties of the soil. Among the most extensive HAHT series were low (<10%) artifact or "clean" fill, common in cemeteries and golf courses (Figure 2.16); and construction debris-enriched fill (Figure 2.17), used in the filling and "reclaiming" of wetlands and tidal marsh areas. Soil survey-related research in NYC established the predominant effect of land use on soil properties. Accordingly, the boundaries of soil map units reflect natural landforms, HA landforms, current land use or land cover patterns, or some combination of these.

For the citywide 1:12,000 survey, impervious surfaces were mapped as the miscellaneous area "Urban land." Urban land consociations were those areas with 90% or more impervious surface, and were differentiated by substratum type according to Merrill (1902). Urban land–soil complexes were set up to cover various lot sizes and land use types in residential, commercial, and industrial areas. To a large extent, the boundaries of *urban land* map units reflect preexisting surficial geology types.

NRCS and the NYCSWCD share a history of promoting soil science in the city for over twenty years. The multifaceted program has also included onsite investigations, research projects, and an

FIGURE 2.16   Soil formed in low artifact, human-altered material, Greenbelt series (tape in inches) (Photo courtesy of Richard K. Shaw, USDA-NRCS.).

FIGURE 2.17   Soil formed in human-transported materials enriched in construction debris, Laguardia series (tape in centimeters) (Photo courtesy of Richard K. Shaw, USDA-NRCS.).

education and outreach component including lectures, training sessions, workshops, as well as volunteer and internship opportunities. The NRCS and the NYCSWCD hosted two urban soils conferences in the city, a regional event in 1997 with the NPS; and an international one, the Fifth Soils of Urban, Industrial, Traffic and Mining Areas conference, with Queens College and The City University of New York. These allowed for the exchange of information both within the soil science community and with other environmental professionals. Field tours highlighted local soil conditions, applications of urban soil science, and partnerships.

Current interest in urban agriculture, revegetation and reforestation, and green infrastructure has brought about an increase in requests for analytical services, data and information, and soils training. The next step in the partnership, in 2015, was the creation of an Urban Soils Institute (USI), to advance the scientific understanding and promote the sustainable use of urban soils through education, conservation, and research, essentially to attend to all soil science needs in the city. Brooklyn College (the Environmental Sciences Analytical Center in the Department of Earth and Environmental Sciences) and the Gaia Institute, a nonprofit ecological engineering and restoration corporation from the Bronx with extensive experience in urban landscapes, also serve as partners in the Institute. The USI will serve as: (1) a data sharing depository and clearing house for all soil-related resources; (2) a provider of soil testing, interpretation, and other technical services; (3) the source for a wide range of education and outreach opportunities, and (4) a coordinator of soils-based research for the city. The USI partnership offers a model of how soil science can meet the needs of the changing world in the twenty-first century.

2.3 URBAN SOIL MAPPING TOOLS

Geophysical technology is becoming more available and experiencing an increase in usage over a wide range of applications and interpretations. The understanding of landscape processes and spatial variability of soil properties is essential (Corwin and Lesch, 2005). With the advantage of conducting noninvasive studies increasing the efficiency in data collection, less time is needed to collect the onsite information to make an accurate determination. Typically, this technology is implemented by NRCS soil scientists with the objective of increasing accuracy and quality in modern soil surveys, with an emphasis on agricultural applications. There are many publications available supporting or substantiating the use of the technology to collect information on various soil properties. These include depth to bedrock, bulk density, water table, salinity, compaction, and water content. However, even though this technology has been used widely in urban scenarios for applications like utilities and archeology, the acceptance in the urban environment for soil mapping is increasing at a slower rate. This could be due to limitations in space, the inability to find a representative landform, or the diversity in the anthropogenic soil material. Nevertheless, soil survey applications are possible with good study design, proper methodology, and the right adaptations to the urban environment (Pozdnyakova et al., 2001). The objective of this section is to explain the use, applicability, and limitations of portable X-ray fluorescence (p-XRF), ground-penetrating radar (GPR), and electromagnetic induction in urban soil survey.

p-XRF is a noninvasive analytical method developed for commercial use in the 1950s that determines elemental composition by measuring disperse energy or fluorescence (Innov-X Systems, 2010). According to Weindorf et al. (2012) and Carr et al. (2008) the major advantage of this technology is obtaining elemental information in a short time and the potential to capture on-site spatial variability. Some of the disadvantages include nonhomogenous samples and moisture content that could introduce a dilution effect. These limitations can be overcome by either using a correction factor for the moisture content or homogenizing the sample by sieving and grinding in the lab.

Environmental quality assessment is the most effective application for p-XRF for either urban agriculture (Weindorf et al., 2012), contaminant elements in anthropogenic material (Carr et al., 2008), or atmospheric additions of trace metals along the urban–rural gradient (Pouyat and McDonnell, 1991). HAHT soils in the urban environment often contain human artifacts, which can affect the physical and chemical properties of the soil. *The Keys to Soil Taxonomy*, Twelfth Edition (Soil Survey Staff, 2014) defines artifacts as material greater >2 mm of "nonnatural" origin, like asphalt, concrete, bricks, glass, metal, and others; twelve HAHT material classes are based on artifact types. Shaw et al. (2010) found that artifactual coarse fragments correlated to the trace metal content in anthropogenic soil material. Howard and Orlicki (2016) also found that even a small amount of micro artifacts in anthropogenic material could affect some geophysical processes. P-XRF technology allows for rapid *in situ* elemental analyses of artifacts and soil material.

In an investigation of atmospheric additions of trace metals to soils in the urban environment, examining woodland soils in western Staten Island, Jovanovich (2012) found elevated amounts of lead (Pb), copper (Cu), and zinc (Zn) concentrated in the highly or moderately decomposed fraction of the organic layer, with slight to negligible lixiviation of trace metals into the mineral soil fraction. It is also possible to observe where the Oi (slightly decomposed plant material) layer, or zone of more recent organic matter deposition, presented concentrations lower than those found in the underlying organic layers (Table 2.2). Lead levels in some horizons of these relatively "undisturbed" soils exceeded the New York State Department of Environmental Conservation Residential Cleanup Objective of 400 ppm (2006); such a study highlights the effects of the urban environment on "natural" soils. Ge et al. (2000) also found that organic matter content is an important soil property in the immobilization of trace metals in the soil profile, with the exception that in their study the pH levels were higher than in Jovanovich study. Ge et al. (2000) emphasized that little attention is paid to trace metal uptake by plants in urban soils (Table 2.2).

TABLE 2.2

Concentrations of Heavy Metals Obtained throughout Six Soil Profiles

Horizons	Saw Mill Creek		Deerfield	
	Pb	Cu	Zn	pH
Oi	156	126	146	5.1
Oe	326	253	155	5.2
A1	281	205	90	5.1
A2	57	81	32	5.2
Bw1	0	16	35	5.0
Bw2	0	9	30	5.4
South Ave		Deerfield		
	Pb	Cu	Zn	pH
Oi	31	34	93	4.8
Oa	992	806	179	
A	16	31	26	5.0
Bhs	6	10	27	5.1
Bw1	0	0	29	4.9
Bw2	0	0	0	5.0
Willowbrook Park		Boonton		
	Pb	Cu	Zn	pH
Oi	195	72	156	
Oa	335	149	114	5.2
BA	78	57	62	5.2
2Bx1	16	54	68	5.0
2Bx2	13	44	78	5.1
Forest Ave		Deerfield		
	Pb	Cu	Zn	pH
Oi	84	60	135	5.0
Oe	734	650	174	5.0
A	22	40	36	5.1
Bw1	4	13	39	5.1
Bw2	7	0	32	5.1
BC	8	16	52	5.1
2BCx	7	28	60	5.5
William T. Davis		Boonton		
	Pb	Cu	Zn	pH
Oe	212	44	73	4.8
Ap1	153	62	58	5.0
Ap2	135	49	52	5.0
Bw1	10	20	52	5.1
Bw2	10	13	60	5.6
2BC1	4	19	55	5.6
2BC2	N/A			
South Ave		Boonton		
	Pb	Cu	Zn	pH
Oi and Oe	408	260	162	5.1
A	54	42	49	5.1
BA	20	24	48	5.2
Bw1	9	16	43	5.3
Bw2	8	13	44	5.2
2Bx	10	20	58	5.5

Source: Modified from Jovanovich, V. (2011). Trace metal content of some forested soils on the west shore of Staten Island, New York. M.S. Thesis, College of Staten Island, Staten Island, NY.

In addition to environmental assessment, this technology provides potential application for soil survey mapping. Staten Island is the only location within the New York City Soil Survey where serpentine bedrock is either exposed or covered by a mantle of loamy soil material (Figure 2.18). In addition to a high magnesium (Mg) to calcium (Ca) ratio (U.S. Department of the Interior, 2015), serpentine-derived soils also contain higher concentrations of nickel (Ni) and chromium (Cr) than the surrounding soils (Rabenhorst et al., 1982). Ongoïba (2012) conducted a study in Staten Island with p-XRF examining the distribution of Ni and Cr in serpentine soils with an emphasis on the geochemical properties. Utilizing this kind of information, a soil and landscape model can be developed for correct and consistent mapping, presenting important geochemical characteristics.

Utilizing p-XRF, soil scientists from NRCS were able to identify and delineate shallow and moderately deep serpentinite till soils that presented elevated concentrations of Ni and Cr. In addition to producing a soil map showing this unique soil characteristic, it was possible to conduct an ecological site study to understand the ecological transformations and evolution in these areas and potentially design projects for the restoration of this unique and fragile ecosystem.

GPR is a system that consists of an antenna that transmits an electromagnetic pulse into the ground; the signal is reflected back between layers with contrasting dielectric permittivity (Soil Science Division Staff, 2017; Conyers, 2004; Huisman et al., 2003) to an antenna receiver in a time lapse generating a signal trace (Allred et al., 2016). A reduction in radar penetration is a direct impact of soil properties like moisture content, salinity (Doolittle and Butnor, 2008), clay content, and amount of iron oxide (Allred et al., 2016), dissipating or attenuating the signal energy. Antenna frequency is also a factor in signal penetration. Lower frequency antennas (10–300 MHz) provide greater penetration but lower resolution; higher frequency antennas (400–900 MHz) provide higher resolution but lower penetration (Doolittle et al., 2006). During the planning process, the frequency needs to be taken in consideration depending on the application (Figure 2.19).

FIGURE 2.18 Todthill (loamy skeletal, mixed, superactive, mesic Dystric Eutrudepts) well-drained soils formed in a loamy mantle of ablation till overlying serpentine bedrock (scale in centimeters). (Photo courtesy of Richard K. Shaw, USDA-NRCS.)

FIGURE 2.19 Ground-penetrating radar (GPR) unit with a 900-MHz antenna mounted in a utility cart with integrated survey wheel and GPR unit. (Photo courtesy of Edwin Muniz, USDA-NRCS.)

There are many applications for GPR in an agricultural setting used commonly NRCS soil scientist, and they are adaptable for urban setting, taking into consideration the differences in soil properties. Doolittle et al. (2006) conducted a study to investigate depth to water table and ground water flow patterns in coarse soil material. The study produced an extensive number of observations with a reasonable accuracy compared with 15 wells, giving a better understanding of the flow patterns and water table. Soil compaction and location of underground voids due to soil piping are some other potential uses for this technology in urban areas. A study was conducted to measure compaction distribution in soil; within the scope of the study soil series, depth to water table and land use were considered (Muñiz et al., 2016). GPR technology proved successful in providing information to generate a spatial distribution map showing depth to compacted layer. This kind of information is essential for the planning, management, and installation of conservation practices to reduce soil compaction and runoff. Soil rehabilitation is another suitable application for urban areas, in determining the thickness of construction debris, fly ash, coal ash, and dredge material deposits. The method is similar to that applied by Paterson and Laker in South Africa (1999), delineating the thickness of soil over spoil material. In the South Africa study, they found difficulties interpreting the results in areas with gradual boundaries and reduced penetration in clayey material. In addition to those limitations, it is possible to find limitations with increases in iron oxide material (Allred et al., 2016), and in the transition zone between anthropogenic material and tidal marsh. Splajt et al. (2003) utilized GPR for mapping landfills with the objective of identifying internal structure and geology. At the end of the project, they were able to compare the efficacy of multiple antenna frequencies that provided the necessary information for management and develop a leaching model. Orlando and Marchesi (2001) conducted another study with the intention of identifying and characterizing dump deposits. In their project, they found that the radar provided good resolution data to identify thickness, type, and shape of the waste deposits. Years earlier, a similar project was conducted in New York by Doolittle et al. (1997) mapping the fill cap thickness on the Brookfield landfill in Staten Island. This technique was successful in providing information about the fill cap thickness, which was variable and in some areas too thin for a proper reclamation, which included a proposed development for a recreational park. Because of the information gathered during the project, two soil series were proposed for mapping the landfill, Greatkills and Freshkills soils. GPR transect data indicated Greatkills was about 63% of the mapping area, Freshkills 16%, and 21% was other soils outside of the range for either soil. Presently, there are five established soil series with differences in type and thickness of cap, presence of liner, and depth to water table. This study provided an opportunity to gather a significant amount of quality information in a short time to determine the soil type distribution, propose new soil series as needed, and map the area consistently, providing a great deal of information for use and management. A landfill, in particular, is a prime example of an area where noninvasive technology lends itself to great advantage.

Electromagnetic induction, similar to GPR, rather than two antennas consists of two sets of coils located on opposite ends. One set of coils generates an electric current to the ground that generates a secondary magnetic field received by the other set of coils, indicating the electric conductivity (EC; GSSI, 2016). The distance between the two sets of coils, and the orientation, frequency, and soil conductivity determinate the penetration depth for the instrument (Tromp-van Meerveld and McDonnell, 2009). Salinity, water content (Corwin and Lesch, 2005), soil temperature, cation exchange capacity (CEC), and pore size, shape, number, and interconnectivity (Petersen et al., 2006) are dominant factors influencing variability in EC. Some of the applications utilizing the mentioned variabilities provide a stage to gather information on the spatial distribution of moisture content (Tromp-van Meerveld and McDonnell, 2009), magnetic properties (when used in combination with p-XRF) (Doolittle et al., 2013), compaction (Brevik and Fenton, 2004), and possibly the spatial distribution of soil organic carbon or SOC (Martinez et al., 2009). In correlating high EC levels with calcareous and ferruginous waste materials, Howard and Orlicki (2015) used EC to generate a better soil map in Detroit, in light of the increased interest in urban development and agriculture, green infrastructure, and other land uses. Concluding that EC and magnetic susceptibility are suitable methods to surveying urban soils.

Soil scientists from NRCS conducted a soil study in a community garden located in Rockaway, Queens, New York after Hurricane Sandy. The purpose of the study was to investigate trace metal deposition and salt intrusion from the floodwater. The study was conducted with multi-frequency electromagnetic induction equipment at 15MHz, 10MHz, and 3MHz in an average line spacing of 5 m. The area is mapped as Bigapple (dredgic, mixed, mesic anthroportic udipsamments), a well drained to excessively drained soil formed in a thick mantle of anthrotransported material from dredging activities in coastal waterways and rivers. It has a high or very high saturated hydraulic conductivity (K_{sat}) in the surface; it is very high in the subsoil and substratum. The data collected presented an increase in apparent EC with an increase in depth. This was expected due to the physical properties of the soils in the area, and indicated safe conditions within the root zone for proper crop establishment.

Geophysical technology has proven to be an invaluable and diverse tool in the soil scientist's toolbox with the potential to be used beyond environmental, archeological, and utility studies. Also, it is an instrumental tool for soil survey in urban areas with the potential for greater data collection and increased efficiency, reducing operational costs.

2.3.1   RECENT ADVANCES IN SOIL TAXONOMY TO CLASSIFY ANTHROPOGENIC SOILS

The International Committee for Anthropogenic Soils (ICOMANTH) is a soil taxonomy committee commissioned in 1988 to introduce differentiae and taxa for classification and survey of observed HAHT soils. Some of the HT soils are also known as "anthropogenic" soils or "technology-produced" soils if humans intentionally generate and emplace material designed and created to develop into soil. All HAHT soils form through intentional, profound, purposeful manufacture, alteration, or transportation of materials for a specific purpose, and do not include soils altered through standard agriculture practices such as shallow plowing, fertilization or liming, or collateral effects such as accelerated erosion. The committee introduced HAHT soils into the 12th edition of the Keys to Soils Taxonomy (Soil Survey Staff, 2014) to facilitate mapping of urban areas, allow meaningful interpretations for unique materials and soils, and ease establishment and correlation of new soil series. The HAHT soils are distinguished from other soils based on field properties, based on recommendations by Richard "Dick" Arnold, Hari Eswaran, Del Fanning, and John Sencindiver. Careful review of then new classification was provided by Craig Ditzler and Joseph Chiaretti of the USDA-NRCS.

Harmonization and consistency for HAHT soils are needed throughout soil taxonomy. Therefore, several changes were made between the 11th edition of the Keys to Soils Taxonomy (Soil Survey Staff, 2010) and the 12th edition (Soil Survey Staff, 2014). Confusion occurred in mapping urban areas because of the widespread alteration of soils, unusual parent materials encountered (Figure 2.20), and absence of suitable taxa.

Extensively modified areas with integrated land management are called Anthroscapes (Eswaran et al., 2005). Humans have built urban anthroscapes by extensively modifying the shape of the land over the last 10,000 years as they excavated areas or cut natural landforms to level the land for buildings and transportation, creating anthropogenic landforms and microfeatures (e.g., dikes and cutbanks). For complete definitions and lists of landforms and features, see the National Soil Survey Handbook, Exhibit 629-1, Part I.D. (U.S. Department of Agriculture, Natural Resources Conservation Service, 2013).

In urban areas, soil and rock excavated to build transportation corridors and tunnels (e.g., subways) and basements for buildings was used to fill low lying wetlands or submerged areas. The fill materials often contained debris from demolished buildings, construction projects, and manufactured products, as well as combustion by-products such as coal slag and ash. Many low areas were used as landfill sites and now contain waste material, along with extensive regulated landfills. Many coastal urban areas (e.g., Dubai, Venice, New York City) have extended their land area by filling shallow waters.

2.3.1.1   Classification

New subgroups were introduced and standardized so that all HAHT soils could be identified at the subgroup, family, and series levels. Formerly, HAHT soils were recognized at suborder, great group, and subgroup levels (Soil Survey Staff, 2010); all were deleted except select subgroups. Soil series of HAHT soils may be reclassified but were not deleted. Arents were used in agricultural or mined and dredged soils, and Anthrepts were not used because anthropic and plaggen epipedons were not recognized in urban areas. The anthropic and plaggen epipedons were therefore substantially revised, and the anthropic epipedon will now be widely used in urban areas.

In humid regions, almost all urban HAHT soils were classified as Typic Udorthents because of their young age and dearth of alternative taxonomic choices, despite enormous differences in materials, properties, and behavior. The Typic Udorthents is a heterogeneous subgroup that gives no information about the soil: it is the very last choice in the Keys if all other taxa are eliminated, with no qualifying properties. Because urban areas occur globally, it is logical that HAHT soils occur in every soil order, and in all moisture and temperature regimes. Soil variability is high in urban and other anthroscapes because of the heterogeneic deposition of transported material and property-altering amendments or pollutants. In New York City, urban soils were found in several great groups of Inceptisols and Entisols, but also retained properties from the buried Ultisols and Alfisols (Hernandez and Galbraith, 1997; Shaw et al., 2007).

FIGURE 2.20 Hassock (coarse-loamy, ashifactic, mixed, active, nonacid, mesic Anthroportic Udorthents) well-drained soil formed in human transported deposits of incinerator fly ash (scale in feet) (Photo courtesy of USDA-NRCS.).

Because of excavation, filling, and earth-moving, many HT fill soils have buried horizons and irregular decreases in SOC with depth. The criteria that led to allocation into the Fluvents suborder and Fluventic and Cumulic subgroups have been modified to prevent inclusion of HAHT soils with water-transported materials. A modification was also made to the definition of buried soils to make it easier to classify soils with buried horizons.

Finding HAHT soils for making digital maps or resource inventory would be difficult and impractical if they were classified only at the series level. There are tens of thousands of series and no method to consistently identify the HAHT soils besides individual inspection of the official description. However, it will be easy to query for and then correlate these soils if they occur in a few specific subgroups and in specific family classes because these criteria are listed in USDA databases.

At the family level, a new class was added for HAHT soils to provide specificity not differentiated at higher levels. The HAHT family classes are based on physical, chemical, and mineral properties that profoundly affect soil behavior or safety of use not identified using conventional family classes and not identified at the subgroup level.

Heterogeneity causes the need for many new soil series for mapping and interpretations, even if the extent of the series is low. Proposing new soil series by choosing among predetermined HAHT subgroup and family classes can be done easily and rapidly, and will result in meaningful ranges of properties within taxa and series. However, the ICOMANTH proposals must be fully vetted and tested in the field and laboratory and the taxa refined, deleted, or added to over time.

2.3.1.2   Identification

The following definitions were added to Chapter 3 of the 12th edition of the *Keys to Soil Taxonomy*, the *National Soil Survey Handbook*, Part 618.5, and to the *Field Book for Describing and Sampling Soils* (Schoeneberger et al., 2012). A chapter is in review for the revision and update of the *Soil Survey Manual*.

The anthropic epipedon was formerly defined with many properties similar to a mollic or umbric epipedon, but is now clearly differentiated and occurs in any soil moisture regime. The anthropic epipedon is not exclusive from other epipedons (e.g., an anthropic epipedon may also qualify as melanic). Anthropic now consists of mineral soil material that shows evidence of purposeful soil property alteration by human activity. The anthropic epipedon occurs on an anthropogenic landform or microfeature and either contains artifacts other than standard agricultural amendments or litter, kitchen midden material, or has anthraquic conditions. The requirement for high phosphorus (P) content was removed from anthropic epipedon due to conflicting lab data.

Artifacts were already mentioned in soil taxonomy in the description of the epipedons (Soil Survey Staff, 1999) but were not defined. Artifacts are ubiquitous in urban areas and should be described if they are cohesive enough to persist for decades. Artifacts should be first split into categories that relate to human safety concerns, and then categories that relate to their properties, behavior, and origin. These separations formed the basis for the HAHT family classes.

Manufactured layers are described with the new master layer letter "M" and are now defined. Manufactured layers include impervious materials such as geotextile liners and asphalt or concrete layers. The contact with a manufactured layer is now in the list of root-limiting layers.

HA (anthraltic) materials are soil materials that have either been intentionally mixed in place greater than 50 cm deep, excavated greater than 50 cm deep, excavated and replaced by humans, or compacted in place. HA material may contain artifacts but the soil material has no evidence that it was transported from outside of the pedon. HA material either occurs on excavated landforms, or contains evidence of intentional *in situ* alteration (e.g., soil in graves is anthraltic material).

HTM (anthroportic) has been excavated from a pedon and moved horizontally onto a different pedon by directed human activity with the aid of machinery or hand tools. HTM often contains an unconformity, lithologic discontinuity or buried horizon just below an individual deposit. HTM occurs on or above anthropogenic landforms or microfeatures and contains evidence of intentional transportation (e.g., the soils of Boston and Hong Kong airport runways are largely anthroportic).

Seven subgroup terms identify distinct groups of HAHT soils. All taxa that recognize HAHT soils above the soil series level use one of the following HAHT subgroups alone or in combination with other important subgroup modifiers (e.g., Anthraquic Sulfic). The HAHT subgroups are listed in order of interpretive importance: Anthraquic (ponded), Anthrodensic (compacted), Anthropic (epipedon), Plaggic (plaggen epipedon), Haploplaggic (thin Plaggen), Anthroportic (transported), and Anthraltic (altered).

In Chapter 17, ICOMANTH added a HAHT material family class between the particle-size and mineralogy class. The class is used for all soils that use one of the seven HAHT subgroups, or for soils that have at least 50 cm of HAHT material, or other soils that have HAHT materials from the surface to a root-limiting layer less than 50-cm deep. When no HA or HTM classes are used, the class is left blank. The control section is the zone from the mineral soil surface to a depth of 200 cm or to a paralithic or lithic contact, whichever is shallower. The following classes are listed in order of most importance to human health and safety: Methanogenic, Asphaltic, Concretic, Gypsifactic, Combustic (slag), Ashifactic (fly-ash), Pyrocarbonic (coke or biochar), Artifactic, Pauciartifactic (minimal amounts), Dredgic, Spolic, or Araric (deep-plowed).

2.4 INTERPRETATIONS IN URBAN SETTINGS

Since the late 1980s, the USDA-NRCS Soil Survey Program, International Union of Soil Science (IUSS), Soils of Urban, Industrial, Transportation and Military Areas (SUITMA) subgroup, and a few independent academic pedologists have been actively pursuing soil survey projects, pedological classification (Technosols, World Reference Base, IUSS, SUITMA) and soil taxonomy (ICOMATH, IUSS, J. Galbraith) in urban and nonfarm landscapes. Projects such as the Baltimore Ecosystem Study, NSF Longterm Ecological Research (Pouyat et al., 2010) applied the digital soil survey as a baseline to look more intensely at monitoring, transport, and network science questions in the urban and industrial landscape. Working with engineers, ecologists, and city and county planners, the USDA-NRCS soil survey projects have focused on developing field standards and interpretations of urban and nonfarm soil survey data and onsite sampling regimes to address community planning issues. In the new world of community interest in climate change adaptation and green infrastructure, related scientific disciplines are discovering the importance and need of soil survey investigations in the urban environment. There is a critical need for soil survey in planning for risk management and adapting to climate change and the emerging green economy (Levin, 2014).

Urban and nonfarm-use soil surveys have applications similar to traditional soil surveys, using tools of soil interpretation to transition data into information for use and planning. Issues of scale come into play in urban settings since the level of disturbance and predictability of the soil properties are difficult to describe in most areas of the country. Changes to the soil system and landscape are a dynamic feature that cannot always be described spatially or in a database. USDA-NRCS is transitioning to electronically interpreting soils onsite and point-to-point with a flexible Interpretations Generator. This will be especially useful in urban systems. With increased computing power and flexibility in data management, the hope is to be able to adjust soil properties from site-specific locations and generate on-the-fly interpretations from existing or create new interpretations for desktop or mobile applications. By using geospatial layers from many formats to integrate into the soils datum, it will be possible to develop more spatially precise interpretations. By being able to access open source statistical and analysis tools, it is possible to design and build engaging maps for our partners, both within and outside the USDA (Soil Science Division Staff, 2016).

2.4.1 WHAT IS A SOIL INTERPRETATION IN SOIL SURVEY?

The standard definition of "soil survey interpretations" are that they are algorithms from spatial and generalized soil data that predict soil behavior for specified soil uses and under specified soil management practices. Within the USDA-NRCS Soil Survey, there is a plan to develop a system that is citable, defensible, reproducible, and timely. With reasonable and effective use and management of soil, water, air, plant, and animal resources, soil interpretations in the United States are derived from data such as: (1) detailed soil survey maps, such as from the Soil Survey Geographic (SSURGO) database; (2) general soil association maps, such as from the General Soil Map of US (STATSGO2) database; and (3) more general soil maps, such as from the national major land resource area map (Soil Science Division Staff, 2017).

Soil interpretations (if described and recorded in maps and official databases) help implement laws, programs, and regulations at local, state, and national levels. They assist in the planning of broad categories of land use such as cropland, rangeland, pastureland, forestland, or urban development. They are also used to assist in pre- and postplanning activities for national emergencies, such as flooding from hurricanes, climate change sea level rise, and earthquake subsidence zones. Soil survey interpretations also help plan specific management practices that are applied to soils, such as storm water control planning in urban settings, irrigation of open space in urban areas (i.e., parks, golf courses) or equipment use for manipulation of landscape systems. The method by which soil interpretations are presented, such as tables, databases, interpretative sheets, thematic maps, and special reports provides easily understood soil limitations or potential for a specific use. Thematic maps effectively present soil limitations and potentials (USDA, 2013).

Whether in an urban setting or a native or agronomic setting, the soil properties and qualities that differentiate the soil on one site from another are the criteria for interpretation models. These properties and qualities may include: (1) site features, such as slope gradient; (2) individual horizon features, particularly physical characteristics, such as particle size, pH, bulk density, and plasticity; and (3) characteristics that pertain to soil as a whole, such as depth to a restrictive layer. Prediction of soil behavior results from the observation and recording of soil responses to specific uses and management practices or climatic factors such as seasonal wet soil moisture status and the resultant effect in a physical structure like a basement; storm water drainage tile or pipe; or rain garden systems for urban runoff from rooftops or parking lots. Recorded observations validate the predictive models of the soil interpretations. The models project the expected behavior of similar soils from the behavior of observed soils. Though standard interpretations in urban settings have been limited to construction and road building in the past. Interpretations are now extended to ecosystem services (biological, water quality, and structural); adaptation practices; green infrastructure; waste management; reclamation; and a myriad of other uses.

Soil interpretation criteria has changed some with technology. Onsite proximal sensing tools such as GPR and x-ray fluorescence have the potential to identify anomalies in the urban system that were observed more randomly before. Measurement data, especially in urban environments, was not available 10 years ago, which in turn left no potential to create maps for general planning. Laboratory and field measurements, models, and inferences from soil properties, morphology, and geomorphic characteristics provide the values used for estimating soil properties in the databases that can be interpreted spatially as well as using standard statistical regression of point data and tabular estimated information. Common sources of laboratory data are the NSSC Kellogg Soil Survey Laboratory, Agricultural Experiment Station laboratories, and State Highway Department testing laboratories. Pedon descriptions record field measurements, field observations, and descriptions of soil morphology (Soil Science Division Staff, 2017).

Published soil surveys include standard national soil interpretations which take account of many urban-related themes. These were developed in the 1970s as a reaction to Public Law 89–560, Soil Surveys for Resource Planning and Development, dated September 7, 1966. This law clarified the legal authority for the soil survey program of the USDA by specifying:

To provide that the Secretary of Agriculture shall conduct the soil survey program of the United States Department of Agriculture so as to make available soil surveys needed by States and other public agencies, including community development districts, for guidance in community planning and resource development, and for other purpose... That in recognition of the increasing need for soil surveys by the States and other public agencies in connection with community planning and resource development for protecting and improving the quality of the environment, meeting recreational needs, conserving land and water resources, providing for multiple uses of such resources, and controlling and reducing pollution from sediment and other pollutants in areas of rapidly changing uses, including farmlands being shifted to other uses, resulting from rapid expansions in the uses of land for industry, housing, transportation, recreation, and related services, it is the sense of Congress that the soil survey program of the USDA should be conducted so as to make available soil surveys to meet such needs of the States and other public agencies in connection with community planning and resource development. (U.S. Congress, 1966)

With this authority and change in paradigm, the soil surveys expanded their stakeholder base to community and urban planning.

2.4.1.1 Urban Interpretations-Protocol for Accessing Systems

The NCSS partnership with USDA-NRCS (formerly Soil Conservation Service) and use of related databases enhanced their capacity to make specialized applications for predictive urban-related planning tools. With this focus, engineering applications became a prominent part of the analysis. Single property maps were still useful but multifaceted spatial analysis became the norm for the categories of community suburban and urban planning; regulations; ecosystem services; reclamation potential; and onsite interpretive standards. With consideration particularly for regulations it is also critical that soil interpretations be developed from official or authoritative data and that protocols for development of interpretations be documented as well. The thousands of soil series recognized in the United States offer the possibility of acting as official reference states for the assessment, and to help monitor or reference conditions. See Table 2.3 for a list of properties of the National Soil Information System (NASIS) that affect soil capability for interpretations and act as reference states for assessment.

The hierarchy of property data used in models is at three levels. First, the characteristics of individual horizons are considered. These are properties such as pH in water or percentage of clay measured for a horizon. The second tier concerns the soil profile as an entity. These are items such as depth to a restrictive layer or root zone available water storage. The third level consists of site features like slope gradient or flooding frequency. Some features that are thought of as soil properties, such as erodibility or steel corrosion, are the function of more basic soil properties, such as SOC content, soil moisture status class, or EC. Some soil and site properties have strong temporal variation, such as temperature and depth to saturation. Site and climatic data are especially needed for ecologic soil capability assessments since a soil exists on a landscape within a climate (Soil Science Division Staff, 2017).

In developing new urban interpretations, USDA-NRCS has developed a protocol in the National Soil Survey Handbook that involves the user (U.S. Department of Agriculture, Natural Resources Conservation Service, 2015). See Table 2.4 for the generalized protocol for all USDA Official Soil

---

TABLE 2.3

Properties That Impact Soil Capability

Horizon Properties	Profile Properties	Site Properties
pH in water	Depth to saturation	Slope gradient
Content of sand, silt, and clay	Depth to bedrock	Slope shape
Saturated hydraulic conductivity	Available water storage	Slope aspect
Electrical conductivity	Depth to a restrictive layer	Precipitation
Rock fragment content		Temperature
Matrix color		Day length
Organic carbon content		Surface stones
Unified class		Elevation
AASHTO class		Frost-free days
Bulk density		Parent material
1/3 bar water holding capacity		
15 bar water holding capacity		
Cation exchange capacity		

Source: Levin, M., *Soil Horizons*, 2014.

TABLE 2.4

The Following Steps Lead to the Goals for All Interpretation Criteria

1. Define the Activity—Clearly and very specifically define the activity or use to be interpreted. Cite references that help to define the activity including performance, equipment for installation, resource conditions soil depths affected, specific geographic detail and map and interpretation reliability and uniformity. Literature citations, such as information from the State Health Department, bulletins, or soil performance research, support the decision made and help track the procedure.

2. Separate Aspects—Separate different aspects of the activity for separate interpretations. Aspects of interpretations are planning elements that require different criteria, such as installation, performance, maintenance, and effect. Proceed through the steps to develop criteria for each aspect. Each aspect is a unique interpretation that has separate criteria and users. Mention other aspects that may need interpretation but are not addressed.

3. Identify Site Features—Identify significant site features significant for the interpretation and any assumptions about them. Site features are not soil properties, but are instead features such as climate factors, landscape stability hazard, vegetation, and surface characteristics. Identify and record the implied effect of site features on each aspect of the interpretation. Although site features are not soil properties, they are commonly recorded on soil databases and are valuable for developing interpretations because they are geographically specific to soils.

4. List Soil Properties—Identify and list all specific soil properties that are significant to the interpretation. Use only basic properties, qualities, or observed properties. Do not make interpretations from previous interpretations or models. Generally, terms that refer to classes fit in this category. Only use derived soil qualities when they are derived within the criteria to ensure the integrity of the data and the resultant interpretation. Terms used as properties or qualities that have inconsistent entries or derivation pathways result in inconsistent interpretations. Concentrating on the basic influencing property that has the most consistent database entries provides for more consistent interpretations. For example, consider the soil moisture status during a construction period and not the drainage class. Minimize the list of properties by identifying only the basic properties. Review the list to ensure that the same property is not implied several times. For example, USDA texture, clay, and AASHTO do not need to appear on the same list.

5. Document Assumptions—Document assumptions about the significance of the property and established values for separating criteria.
   a. A record of the significance of the property helps to define the property and allows for future understanding and modification. It provides a basis for the criteria so that changes can be made if different equipment is used.
   b. Indicate why the feature is important and why the specific break was chosen, such as why 6% slope was used instead of 10% slope. If the limit is arbitrary or speculated, state that it is but also indicate the intent of the separation. The new interpretation generator recognizes the progressive effect of a property on the interpretation. The curve for approximate reasoning (fuzzy logic) reflects the increasing, decreasing, or constant effect that varying degrees of a property have on the interpretation. The evaluation phase of the interpretation generator uses the curve. Establish values that are significant to the interpretation and not to the mapping. The values should represent the significance to an activity.

6. Develop the Criteria Table—Assign feature and impact terms, and develop the criteria table. The following categories of column headings are recommended for use in the criteria table:
   a. Factor (this is the soil property);
   b. Degree of Limitation (such as slight, moderate, severe);
   c. Feature (the term to be displayed for soil property); and
   d. Impact (the dominant impact that the soil property has on the practice being rated).

7. Application, Presentation, and Testing—Provide a description of the calculation procedure. The calculation procedure is a set of instructions for the correct access to dataset entries. It is needed to sort criteria from a database without questioning the intention of the interpretation. The description should be specific to the database being used. Instructions for using high, low, or central values of data should be given in this description. Information on soil moisture status and freezing conditions are in the National Soil Information System. Include temporal, soil property and geographic reliability in the documentation. Interpretations should be tested against the actual effects on activities or practice performance.

Source: Soil Survey Staff. 2010. *Keys to Soil Taxonomy*, 11th ed. Washington, DC: USDA-Natural Resources Conservation Service. Retrieved from http://www.nrcs.usda.gov/Internet/FSE_DOCUMENTS/nrcs142p2_050915.pdf.

Interpretations from soil survey data. Interdisciplinary involvement particularly in urban systems assures that the needs of many potential stakeholders both technical and political are addressed as they are in city and community planning.

Soil interpretation is for planning purposes. Additional refinements or other resource information can be used for site selection particularly with onsite information. Many properties and criteria need further refinement before they can be used. Some terms, such as flooding, require clarifying statements such as for velocity, depth, or duration. Sources of information other than the NASIS soil interpretations may be available and should be considered at this stage of criteria development.

Urban soil interpretation provides a standard to compare options. Interpretive results are generated by applying interpretive criteria to soil data. Interpretive criteria are divided into three basic parts: properties, evaluations, and rules (which includes interpretation and subrules). A property is a specification used to extract soil property data from the database. This can be anything from a simple query statement to a complex calculation based on several queries. An evaluation is some test that will be applied to the data returned from a property to determine the truth of some proposition. Evaluations can be either fuzzy (returning any value between 0 and 1) or crisp (returning only 0 or 1).

2.4.2 Soil Criteria Considerations for Urban Interpretations

Soil survey interpretations for community planning use the standard national soil interpretations for construction as the foundation of urban interpretations. Soil survey data are critical for the analysis, and to spatially identify material can be helpful both in sourcing and construction and for deciding methods of stabilizing cuts and fills. Material that compacts readily and has high strength and low shrink-swell potential is preferred as base material under roads and foundations. Gravel and sand are used for concrete, road surfacing, filters in drainage fields, and other uses. Organic soil material is used widely as horticultural mulch, potting soil, and soil conditioner. Mineral soil material of good physical condition is generally rich in organic matter and is applied to lawns, gardens, and road banks. Soils can be rated as probable sources of these materials for fill or topsoil. The performance of local roads and streets, parking lots, and similar structures is often directly related to the performance of the underlying soil. Soil properties may affect the cost of installation and length of service of buried pipelines and conduits. Shallow bedrock, for instance, greatly increases the cost of installation. Rate of corrosion is related to wetness, electrical conductivity, acidity, and aeration. Differences in properties between adjacent horizons, including aeration, enhances corrosion in some soils. Soil properties affect the cathodic protection provided by sacrificial metal buried with pipes. Rock fragments can break protective coatings on pipes. Shrinking and swelling of some soils may preclude the use of certain kinds of utility pipe, or crack foundations and driveways. The amount of the clay in the soil at depth and its shrink-swell capacity is a major criteria in all construction considerations. Pipelines and conduits are commonly buried in soil at shallow depth. The properties of the soil may affect cost of installation and rate of corrosion. Soil material is used directly as topsoil, road fill, and aggregate for concrete. Soil information in conjunction with engineering testing can identify those soils that can be stabilized in place for a road base and establish where gravel or crushed stone will be needed. Usually, onsite evaluation is necessary (Soil Science Division Staff, 2017).

Hydrologic information and other data combined with interpretative soil properties, such as the soil hydrologic group, can be helpful for the estimation of potential runoff for design of culverts and bridges. The probability of bedrock and unstable soils that require removal or special treatment can be determined from soil surveys. Soil survey interpretations give general information about planning, planting, and maintaining grounds, parks, and similar areas. Particularly important is the suitability of the soil for turf and ornamental trees and shrubs; the ability to withstand trampling and traffic; the suitability for driveways and other surfaced areas; and the ability to resist erosion. A number of soil chemical properties may be critical, especially for new plantings. Interpretations for particular plants and the treatments for a specific site require other disciplines. Many lawn and ornamental plantings are made in leveled areas on exposed subsoil or substratum or on excavated material that has been spread over the ground. Interpretations can be made as to the suitability of such soil materials for lawns and other plantings, the amount of topsoil that is necessary, and other treatments required for satisfactory establishment of vegetation. Highway departments use soil interpretations to establish and maintain plantings on subsoil material in rights-of-way (Soil Survey Division Staff, 1993).

Waste treatment, dispersal, and disposal is divided on the basis of whether the practice places the waste in a relatively small area or distributes the waste at low rates over larger areas of soil. Waste in this context includes a wide range of material from household effluent, through solid waste, to industrial wastes of various kinds. Effluent from septic tanks is distributed in filter fields. Liquid wastes are stored and treated in lagoons constructed in soil material. Solid wastes are deposited in sanitary landfills and covered with soil material. Extremes in K_{sat} and free water at a shallow depth limit the use of soil for septic tank absorption fields. Sewage lagoons require a minimum K_{sat} to prevent rapid seepage of the water, a slope within certain limits, and slight or no possibility of inundation or the occurrence of free water at shallow depths. Soils are used to dispose of solid wastes in landfills, either in trenches or in successive layers on the ground surface. For trench disposal, properties that relate to the feasibility of digging the trench—depth to bedrock, slope—and factors that pertain to the likelihood of pollution of ground water—shallow zone of free water, inundation occurrence, and moderate and high K_{sat}—have particular importance. For disposal on the soil surface, K_{sat}, slope, and inundation occurrence are important. The rate at which wastes can be applied without contamination to ground water or surface water is called loading capacity. Low infiltration values limit the rate at which liquid wastes can be absorbed by the soil. Similarly, low K_{sat} through most of the upper meter limits the rate at which liquid wastes can be injected. Shallow depth of a hardpan or bedrock or coarse particle size reduces the amount of liquid waste that a soil can absorb in a given period. The time that wastes can be applied is reduced by the soil being frozen or having free water at shallow depths. Low soil temperatures reduce the rate at which the soil can degrade the material microbiologically (Soil Science Division Staff, 2017).

Soils differ in their capacity to retain pollutants until deactivated or used by plants. Though original interpretations were for agricultural land in this regard, now soils are evaluated for the same factors in terms of planning for water quality in an urban setting. Highly pervious soils may permit movement of nitrates to ground water. Similarly, saturated or frozen soils allow runoff to carry phosphates absorbed on soil particles or in waste deposited on the soil directly to streams without entering the soil. Soils that combine a limited capacity to retain water above slowly permeable layers and a seasonal water excess may allow water that is carrying pollutants to move laterally at shallow depths. Such water may enter streams directly. The first step in making interpretations of soils for disposal of wastes is usually to determine how disposal systems for each kind of waste have performed on specific kinds of soil in the area. Experience may have been acquired in practical operations or by research. Soil scientists and specialists in other disciplines determine what properties are critical and how to appraise the effects of the properties. Limiting values of critical properties can be determined through experience and may be used in making interpretations where data on soil performance are scarce or lacking (Soil Science Division Staff, 2017).

Water management in this context is concerned with the construction of relatively small or medium impoundments, control of waterways of moderate size, installation of drainage and irrigation systems, and control of surface runoff for erosion reduction. These activities may involve large capital expenditures. Onsite evaluation commonly should be conducted, particularly of soil properties at depth. The usual Order 2 or Order 3 soil survey can be helpful in the evaluation of alternative sites, but onsite investigations are required to design engineered projects (e.g., ponds and reservoirs). Soil information is used in predicting the suitability of soils for ponds and reservoir areas. Impoundments contained by earthen dikes and fed by surface water have somewhat different soil requirements than those that are excavated and fed by ground water. Separate interpretations are commonly made.

Soil survey interpretations are useful for comparing alternative sites, in planning onsite investigations and testing, and in land use planning. Soil maps can assist in selecting building sites that are near areas suitable for utilities, parks, and other needs. The preparation of building sites may alter soil properties markedly. Upper horizons may have been removed and locally translocated, which might either increase or decrease the depth to horizons important to behavior. The pattern of soil–water states may be changed. Areas may have been drained and, therefore, are not as wet as indicated. On the other hand, irrigation may be employed to establish and maintain vegetation leading to a more moist soil and possible deep movement of water. Pavements, roofs, and certain other aspects of construction increase runoff and may cause inundation at lower elevations where the soil survey does not indicate such a hazard (Soil Science Division Staff, 2017).

2.4.3 Overview of Green Infrastructure with Soil Interpretations

The USDA-NRCS has many examples of urban interpretation projects in the last 10 years that were facilitated by partnerships with NCSS, federal agencies, universities, state and local governments, and regional planning groups. Specific soil interpretations have been developed from these partnerships using the NASIS database in the context of Storm Water Control Systems; Waste and Public Health; Energy, Urbanization, Reclamation, Coastal Issues, Wetlands, and Urban Agriculture for regional assessment. The examples portrayed here are indicative of the potential for applications supporting nonfarm uses throughout the United States.

A few years ago, in 2011, in response to the EPA mandate for localities over 10,000 population to have storm water plans, the interpretations staff was contacted by USDA-NRCS and Agricultural Research Service (ARS) staff in West Virginia. A team was formed of university faculty (VA Tech, University of Akron) with ARS and USDA-NRCS staff to consider methods of storm water management applicable to the Eastern United States. The styles of systems include deep infiltration, shallow infiltration, and retention systems. These BMPs were modeled in the Soil Survey Database (NASIS) as a great example of urban soil interpretations and output was made publically available for states that export the data in 2015 based on criteria developed by multidisciplinary teams. Criteria for delineating soil rating for Storm Water Management Practices. The data can be custom made by anyone who uses the NASIS system. The most intractable problem is that of scaling. A soil survey at even 1:12,000 is still too small for planning in small city areas with a practice that occupies only a quarter acre or so. It still gives a good indication of potential in the area and can be used for planning, sizing of pipes and infrastructure, and cost estimates.

Another example of regional planning highly aligned with urban needs for water management would be the California Groundwater Banking Index (GWBI; O'Geen et al., 2015). The National Soil Survey Center has developed a NASIS-derived interpretation for this index. The interpretation is in review by the NRCS state soils staff in California and by the University of California Agriculture and Natural Resources Cooperative Extension (UCCE). The interpretation is modeled using criteria for the SSURGO-based Soil Agricultural Groundwater Banking Index (SAGBI), which was developed by UCCE. These indexes identify areas with potential for groundwater banking to restore aquifers deleted by drought on agricultural lands. Groundwater banking is a water management strategy that stores surface water in aquifers for future withdrawal. It is hoped water banking is achieved through application of surface water during the winter when crop trees and vines are dormant or when the fields are not planted with row crops. An additional benefit is the diversion of water away from and reduction of pressure on levees that protect urban areas such as Sacramento, CA. The interpretation has geographic applicability to the Central Valley of California, the California Delta, and California coastal plains and valleys. These areas are neither snow-covered nor frozen, and most have existing water management infrastructure.

The GWBI interpretation is based on five major soil factors that are critical to successful agricultural groundwater banking: deep percolation, root zone residence time, topography, chemical limitations, and soil surface condition. These five factors are weighted by a multiplier as originally modeled by UCCE. The factors are then summed to calculate an overall component-based rating. The major difference between the NASIS-based rule and the UCCE model is that the UCCE model uses an assigned component soil drainage class while the NRCS model uses soil aquic conditions which correlates to higher residence time of water in the soil and consequently a lower index rating. Research is currently being conducted by the University of California at Davis to determine whether permanent crops will survive this water management scenario and to see if yields are limited or if disease risks are increased, and whether nitrates and other pollutants will enter the groundwater. One potential source of water for recharge is river floodwaters. As mentioned earlier, using these floodwaters has the dual benefit of withdrawing large amounts of water from a river that is at or near flood stage and reducing downstream flood risks and levee breaches (O'Geen et al., 2015).

In the realm of public health, the staff at the National Soil Survey Center have developed a model for predicting the potential for various soils to provide favorable habitat for the growth of the soil-borne fungi *Coccidioides immitis* and *Coccidioides posadasii*. These are the organisms that cause the disease valley fever (coccidioidomycosis). This disease has been becoming more prevalent lately because of several factors, including increased soil disturbance and climatic change. Inhalation of the spores provides the environment for the infectious phase of the fungal life cycle. The soil-inhabiting phases of these organisms require a definite set of soil and site characteristics in order to out-compete other soil organisms. For example, the soil surface must be moist at some time during the year but then become dry and very hot. The soil surface must be saline and contain some organic matter. Since many soil and site attributes known to influence the fungi are available or can be derived from the soil survey database, the extent of soils amenable to the growth of the fungi and the degree to which conditions are favorable should be mappable. Typically, mapping of endemic areas is linked to observed cases of the disease and not the presence of the causative factor in soil; and the mapping scales are not amenable to planning. Also, in some years, such as during hot or dry spells, yellow areas are likely to be favorable habitat for the growth of the fungi. This information available at a planning scale should prove to be useful for proactive dust management in development and recreation.

In our energy-related toolbox the USDA-NRCS has built several regional nonfarm-related interpretations of interest, particularly for residential development—Soil Suitability for Closed-Loop Horizontal Residential Geothermal Heat Pumps; and Soil Limitations for Ground Mounted Solar Array Panels.

The Soil Suitability for Closed-Loop Horizontal Residential Geothermal Heat Pumps interpretation was developed by the NRCS Soil Interpretations staff and NRCS soils staff in CT-RI. This interpretation rates the suitability of soils for closed-loop horizontal residential geothermal heat pumps by rating the soil's relative thermal conductivity in the 0–200-cm zone. The thermal conductivity (K_T) of the soil directly impacts the size of the bore field needed for the closed-loop horizontal system. Soils that are good conductors (high thermal conductivity and low resistivity) are preferable in that they increase the heat loss from the geothermal heat exchanger plastic piping that is installed in the soil profile. K_T is defined as the amount of heat transferred through a unit area in unit time under a unit temperature gradient measured as W/m K (Watts per meter degree Kelvin). A soil is made of a matrix of minerals, organic matter, water/ice, and air. Soils vary widely in their K_T as shown in Table 2.5.

TABLE 2.5 Thermal Conductivity of the Soil (K_T)

Constituent	W/m K
Quartz	8.8 (poor insulator with higher heat loss)
Other minerals	2.9
Ice	2.2
Water	0.57
Organic matter	0.25
Air	0.025 (better insulator with less heat loss)

The relative K_T of a soil depends on its mineral composition and organic matter content as well as on the volume fractions of water and air. Soils with high air content and low water content with moderate to high organic matter content are the best insulators and are the most suitable for geothermal heat pumps because they minimize the heat loss from the heat exchange plastic piping installed underground. Soils with high quartz or other mineral content (high sand content, high coarse fragment content), low air content (high bulk density with less pore space) and/or high water content (poor drainage, seasonal high water table), low organic matter content are less suitable for geothermal heat pumps because they are poor insulators and can conduct heat away from the heat exchange plastic piping installed underground.

Since 2010, there has been a great deal of interest in developing nationwide reasonable sites on US public and private lands for solar array systems. Most recently in 2015 a site has been developed at the USDA headquarters, Washington Carver Center, in Beltsville, MD. Utility-scale solar energy environmental considerations include land disturbance and land use impacts; potential impacts to specially designated areas; impacts to soil, water, and air resources; impacts to vegetation, wildlife, wildlife habitat, and sensitive species; visual, cultural, paleontological, socioeconomic, and environmental justice impacts; and potential impacts from hazardous materials. There are several soil property considerations for successful installation and maintenance of ground-mounted solar array facilities that are integral into the design of a facility. A national soil interpretation from the official US NCSS for this land use has been developed using reasonable thresholds for optimum conditions of soil properties as a criteria table for high, medium, and low suitability.

Coastal concerns in the urban arena are a high priority in the NCSS interpretations program. Soil-based interpretations that are in development and could be applied directly to urban soil issues are Soil Potential for Coastal Acidification and Land Utilization of Dredge Materials. Coastal acidification is a problem that cuts across many disciplines and affects a diverse group of stakeholders. In an effort to map potentially vulnerable zones, the Soil Potential for Coastal Acidification interpretation will help users understand the spatial relationships between soils and coastal acidification as well as identify soils where coastal acidification may be the most problematic currently and in the immediate future. Research has shown that some subaqueous soils can induce physiological stress, shell dissolution, and mortality of recently set juvenile bivalves based on the dominant morphology and biogeochemistry operating within these soils. In developing this soil interpretation, NRCS is hoping to provide information that empowers stakeholders and enables decision makers to respond constructively to ocean acidification.

The Land Utilization of Dredge Materials soil interpretation displays coastal subaqueous soils that are at risk for developing acid sulfate soils if dredged. Disposal or use of dredged materials that have a high risk for developing acid sulfate conditions can result in land use problems. Acid sulfate soil conditions can result in limited plant growth (crops, landscaping, home gardens) and infrastructure corrosion. The interpretation looks for population of the reduced monosulfide indicator within 2 m depth for subaqueous soil components and is dependent on field tests with 3% hydrogen peroxide (H_2O_2) and oxidation pH measurements. The presence of reduced monosulfides can result in acid sulfate conditions when the soil or dredged material containing reduced monosulfides is exposed to air. If soil containing sulfidic materials is drained or otherwise exposed to air, the sulfides oxidize and form sulfuric acid. The pH value, which normally is near neutral before drainage or exposure, may drop below 3.

Sulfidic materials accumulate within a soil that is permanently saturated, generally within brackish water. The sulfates in the water are biologically reduced to sulfides as the materials accumulate. Sulfidic materials commonly accumulate in coastal marshes near the mouth of rivers that carry noncalcareous sediments, but they may occur in freshwater marshes if there is sulfur in the water. Acid sulfate soil formation occurs when sulfide minerals, such as pyrite and/or elemental sulfur in reduced sulfidic sediments, oxidize upon exposure to air through drainage or earth-moving operations. The oxidation products are jarosite and sulfuric acid. Jarosite undergoes hydrolysis in an oxidizing environment, releasing iron oxyhydrates and more sulfuric acid. This reaction is one of the most acid-producing reactions in soils. A sulfuric horizon is indicated if acid sulfate formation gives an end product pH of 3.5 or less.

The USDA has launched an Urban Agriculture tool book which is available on the internet as well as at USDA offices. Though focused more on the financial and legal aspects of setting up an urban farm, there are source information guides. USDA-NRCS has compiled a set of resources on its Urban Soils webpage. The site includes surveys, links, and guides including the Urban Soil Primer, an introduction to urban soils for homeowners and renters, local planning boards, property managers, students, and educators. NRCS's website on soil health is designed to help visitors understand the basics and benefits of soil health—and to learn about Soil Health Management Systems from farmers who are using those systems. The USDA-NRCS Urban Agriculture program has focused on funding and technical assistance for High Tunnel farming (U.S. Department of Agriculture, 2016).

The US-EPA Brownfields Program provides educational resources and guidelines for gardening on contaminated sites. It also encourages urban agriculture programs to work with local, state, or tribal brownfield programs to identify clean sites for food production and agriculture or to secure technical assistance to assess and clean proposed areas. Review their Interim Guidelines for Safe Gardening Practices for step-by-step recommendations on how to assess the risk of soil contamination on your land, how to conduct soil sampling and interpret the results, and how to conduct BMPs. The EPA maps past and active brownfield grants as well as Superfund and other land cleanup programs at: http://www2 .epa.gov/cleanups/cleanups-my-community. EPA's Superfund program also has established a website on Ecological Restoration that highlights the use of phytoremediation and soil amendments to address contaminated sites stabilization and exposure reduction (U.S. Department of Agriculture, 2016). Using the Web Soil Survey as a source for spatial data, the NRCS Soil Interpretations staff has been collaborating with local agencies to distribute this information for more specific recommendations.

2.5 CONCLUSION

This chapter has summarized and given specific examples of urban soil activity as part of the NCSS. Working with standards and techniques to address our heavily industrialized agricultural systems, the soil survey community has learned how to provide scalable soil information for food security in intensely farmed areas. With considering anthropogenic areas, the mapping paradigm has been transitioned to consider first the original landscape and genesis and then overlay recent history and management of the land to predict properties and soil condition as spatial mappable and scalable features for planning, use, and management. Working in urban areas has opened the door for soil science to inform public health and safety issues throughout the country and participate actively in the green infrastructure movement in cities and smaller communities. It is allowing us to come to the table to inform public policy and be the experts in the network of science and analysis that is driving development and reconstruction in nonfarm environments. By blending scientists' and practitioners' interconnected data and interpretations at all levels of the network, we are able to provide clearer messages as to how ecosystems will work adaptively in both farm and nonfarm systems. The current soil surveys are providing tailored interpretations to meet modern community planning needs and using digital processes to define, delineate, distribute, and convey soil information to new audiences with an eye toward establishing baselines for green infrastructure solutions and climate change adaptation strategies (Levin, 2014).

REFERENCES

Allred, B.J., Adamchuk, V.I., Viscarra Rossel, R.A., and Doolittle, J. 2016. Geophysical Methods. In *Encyclopedia of Soil Science*, 3rd ed, edited by Rattan Lal. Boca Raton, FL:. CRC Press by Taylor & Francis.

Bilodeau, W.L., Bilodeau, S.W., Gath, M.G., Oborne, M., and Proctor, R.J. 2007. Geology of Los Angeles, California, United States of America: The Geological Society of America. *Environmental and Engineering Geoscience*, 13(2):99–160.

Bretz, J.H. 1939. Geology of the Chicago region. *Illinois State Geological Survey Bulletin* 65, Part I—General.

Brevik, E.C. and Fenton, T.E. 2004. The effect of changes in bulk density on soil electrical conductivity as measured with the Geonics EM-38. *Soil Survey Horizon*, 45(3):96–102.

Calsyn, D. and Ryan, K. 2012. *Soil Survey of Cook County, Illinois*. Illinois: United States Department of Agriculture, Natural Resources Conservation Service.

Carr, R., Zhang, C., Moles, N., and Harder, M. 2008. Identification and mapping of heavy metal pollution in soils of a sport ground in Galway city, Ireland, using a portable XRF analyzer and GIS. *Environmental Geochemistry and Health*, 30:45–52.

Conyers, L.B. 2004. *Ground-Penetrating Radar for Archeology*. Ontario, Canada: Canadian Archaeological Assocation.

Corwin, D.L. and Lesch, S.M. 2005. Characterizing soil spatial variability with apparent soil electrical conductivity I. Survey protocols. *Computers and Electronics in Agriculture*, 46:103–133.

Craul, P.J. 1992. *Urban Soil in Landscape Design*. New York, NY: Wiley.

Craul, P.J. 1999. *Urban Soils, Applications and Practices*. New York, NY: John Wiley & Sons.

Detroit Historical Society. 2016. Timeline of Detroit. Retrieved from http://detroithistorical.org/

Doolittle, J.A. and Butnor, J.R. 2008. Soils, peatlands, and biomonitoring. In Jol, H.M. (Ed.), *Ground Penetrating Radar: Theory and Applications*, Chap. 6, pp. 179–202. Amsterdam, The Netherlands: Elsevier.

Doolittle, J., Chibirka, J., Muñiz, E., and Shaw, R. 2013. Using EMI and p-XRF to characterize the magnetic properties and the concentration of metals in soils formed over different lithologies. *Soil Horizons*, 54(3):1–10.

Doolittle, J., Hernandez, L.A., and Galbraith, J.M. 1997. Using ground penetrating radar to characterize a landfill site. *Soil Survey Horizons*, 38:60–67.

Doolittle, J.A., Jenkinson, B., Hopkins, D., Ulmer, M., and Tuttle, W. 2006. Hydropedological investigations with ground-penetrating radar (GPR): Estimating water-table depths and local ground-water flow pattern in areas of coarse-texture soils. *Geoderma*, 131:317–329.

Eswaran, H., Kapur, S., Akça, E., Reich, P., Mahmoodi, S., and Vearasilp, T. 2005. Anthroscapes: A landscape unit for assessment of human impact on land systems. In Yang, J.E., Sa, T.M., and Kim, J.J. (Eds.), *Application of the Emerging Soil Research to the Conservation of Agricultural Ecosystems*, pp. 175–192. Seoul, Korea: The Korean Society of Soil Science and Fertilizers.

Farmer, S. 1890. *History of Detroit and Wayne County and Early Michigan: A Chronological Cyclopedia of the Past and Present*. Detroit, MI, New York, NY: Silas Farmer & Cofor Munsell & Co.

Ge, Y., Murray, P., and Hendershot, W.H. 2000. Trace metal speciation and bioavailability in urban soils. *Environmental Pollution* 107:137–144.

Geophysical Survey Systems, Inc. (GSSI). 2016. Electromagnetic Induction Explained. Retrieved from http://www.geophysical.com/whatisem.htm. (accessed April 27, 2016).

Gossett, P.E. 1999. Urban land use effects on soil properties and their relevance to soil survey interpretations. M.S. Thesis, Cornell University, Ithaca, NY.

Harlen, B.J. 1939. *Geology of the Chicago Region. Illinois State Geological Survey Bulletin 65, Part I—General*. Urbana, IL: State Geological Survey.

Harlen, B.J. 1953. *Geology of the Chicago Region. Illinois State Geological Survey Bulletin 65, Part II: The Pleistocene*. Urbana, IL: State Geological Survey.

Hernandez, L.A. 2001. *Soil Survey of Gateway National Recreation Area, New York and New Jersey*. Collaborative Project Among USDA Natural Resources Conservation Service, USDI National Park Service Gateway National Recreation Area in Partnership with Cornell University Agricultural Experiment Station and New York City Soil and Water Conservation District.

Hernandez, L.A. and Galbraith, J.M. 1997. *Soil Survey of South LaTourette Park, Staten Island, New York City, NY*, p. 47. Ithaca, NY: Cornell Cooperative Extension and Syracuse, NY: USDA-Natural Resources Conservation Service.

Howard, J.L. 2015. Glaciolacustrine history of the Huron-Erie lowland in the southeastern Great Lakes region USA revisited. *Journal of Great Lakes Research*, 41(4):965–972.

Howard, J.L., Dubay, B.R. and Daniels, W.L. 2013a. Artifact weathering of waste building materials, and bioavailability of lead in a chronosequence at former demolition sites, Detroit, Michigan. *Environmental Pollution*, 179:1–12.

Howard, J.L., Putnam, S., Moorhead, S., and Schooch, R. 2013b. Preliminary quaternary geologic map of the Detroit, Michigan quadrangle. *Scale*, 1:24,000 (unpublished).

Howard, J.L. and Orlicki, K.M. 2015. Effects of anthropogenic particles on the chemical and geophysical properties of urban soils, Detroit, Michigan. *Soil Science*, 180(4/5):154–166.

Howard, J.L. and Orlicki, K.M. 2016. Composition, micromorphology and distribution of microartifacts in anthropogenic soils, Detroit, Michigan, USA. *Catena*, 138:103–116.

Howard, J.L. Orlicki, K.M., and LeTarte, S.M. (2016). Evaluation of geophysical methods for mapping soils in urbanized terrain, Detroit, Michigan. *Catena,* 143:145–158.

Huisman, J.A., Hubbard, S., Redman, J.D., and Annan, A.P. 2003. Measuring soil water content with ground penetrating radar: A review. *Vadose Zone Journal,* 2:476–491.

Hyde, C.K. 1980. *Detroit: An Industrial History Guide.* Detroit, MI: Detroit Historical Society.

Innov-X Systems. 2010. *DeltaTM Family: Handheld XRF Analyzers User Mannual,* p. 124. Canada: Innov-x Systems.

Isleib, J.I., Reinhardt, L.P., and Shaw, R.K. 2010. A study using IRIS tubes on problematic red parent material in Staten Island. *Soil Survey Horizons,* 51:17–21.

Jovanovich, V. 2011. Trace metal content of some forested soils on the west shore of Staten Island, New York. M.S. Thesis, College of Staten Island.

Levin, M. 2014. Green economy and infrastructure contributions of USDA urban and nonfarm soil projects in the U.S. *Soil Horizons.* 55(5):1–5. doi:10.2136/sh2013-54-5-gc

Levin, M. et al. 2016. Soil capability for the United States now and into the future. In *Global Soil Security,* Chap. 6, edited by D.J. Field et al. Switzerland: Springer International Publishing AG.

Mapes, D.R. 1979. *Soil Survey of Du Page and part of Cook Counties.* Illinois, IL: University of Illinois Agricultural Experiment Station Soil Report 108.

Martinez, G., Vanderlinden, K., Ordóñez, R., and Muriel, J.L. 2009. Can apparent electrical conductivity improve the spatial characterization of soil organic carbon? *Vadose Zone Journal* 8(3):586–593.

Merrill, F.J.H., Darton, N.H., Hollick, A., Salisbury, R.D., Dodge, R.E., Willis, B., and Pressey, H.A. 1902. *New York City folio No. 83: Paterson, Harlem, Staten Island and Brooklyn quadrangles, New York-New Jersey.* U.S. Geological Survey.

Metropolitan Water Reclamation District of Greater Chicago. 2010. Tunnel and Reservoir Plan. Retrieved from http://www.mwrd.org/irj/portal/anonymous/tarp.

Mount, H., and Hernandez, L.A. 2000. Temperature Signatures for Urban Soils of New York City. First International Conference on Soils of Urban, Industrial, Traffic and Mining Areas (SUITMA). University of Essen, Germany.

Muñiz, E., Shaw, R.K., Gimenez, D., Williams, C.A., and Kenny, L. 2016. Use of ground-penetrating radar to determine depth to compacted layer in soils under pasture. In Hartemink, A. and Minasny, B. (Eds.), *Progress in Soil Science: Digital Soil Morphometrics,* pp. 411–421. Switzerland: Springer International Publishing.

New York State Department of Environmental Conservation. 2006. Brownfield and Superfund Regulation, 6 NYCRR Part 375—Environmental Remediation Programs.

Nilon, C. 2005. Encyclopedia of Chicago: Cook County. Retrieved from http://www.encyclopedia.chicagohistory.org/pages/263.html.

Ochieng, M. 2011. Distribution of Mg to Ca ratio in the serpentine soil of the Greenbelt of Staten Island. M.S. Thesis, College of Staten Island, Staten Island, NY.

O'Geen, A.T., Saal, M.B.B., Dahlke, H.E., Doll, D.A., Elkins, R.B., Fulton, A., Fogg, G.E., Harter, T., Hopmans, J.W., Ingels, C., Niederholzer, F.J., Solis, S.S., Verdegaal, P.S., and Walkinshaw, M. 2015. Soil suitability index identifies potential areas for groundwater banking on agricultural lands. *California Agriculture,* 69 (2):75–84. doi:10.3733/ca.v069n02p75. April–June 2015.

Ongoïba, S. 2012. Distribution of Cr and Ni in the serpentine soils of the Greenbelt in Staten Island, New York. M.S. Thesis, College of Staten Island, Staten Island, NY.

Orlando, L. and Marchesi, E. 2001. Georadar as a tool to identify and characterize solid waste dump deposits. *Journal of Applied Geophysics,* 48:163–174.

Paterson, D.G. and Laker, M.C. 1999. Using ground penetrating radar to investigate spoil layers in rehabilitated minesoils. *South Africa Journal of Plant and Soil,* 16:3, 131–134.

Petersen, H., Fleige, H., Rabbel, W., and Horn, R. 2006. Geophysical methods for imaging soil compaction and variability of soil texture on farm land. *Advances in GeoEcology,* 38:261–272.

Pouyat, R.V. and McDonnell, M.J. 1991. Heavy metal accumulations in forest soils along an urban-rural gradient in southeastern New York, USA. *Water, Air, and Soil Pollution,* 57–58:797–807.

Pouyat, R.V., Szlavecz, K., Yesilonis, I.D., Groffman, P.M., and Schwarz, K. 2010. Chemical, physical and biological characteristics of urban soils. In Aitkenhead-Peterson, J., and Volder, A. (Eds.) *Urban Ecosystem Ecology. Agronomy Monograph 55,* Chap. 7, pp. 119–152. Madison, WI: American Society of Agronomy, Crop Science Society of America, Soil Science Society of America.

Pouyat, R.V., Yesilonis, I.D., Russell-Anelli, J., and Neerchal, N.K. 2007. Soil chemical and physical properties that differentiate urban land-use and cover types. *Soil Science Society of America Journal,* 71:1010–1019.

Pozdnyakova, L., Pozdnyakov, A., and Zhang, R. 2001. Application of geophysical methods to evaluate hydrology and soil properties in urban areas. *Urban Water*, 3:205–216.

Rabenhorst, M.C., Foss, J.E., and Fanning, D.S. 1982. Genesis of Maryland soils formed from serpentinite. *Soil Science Society of America Journal*, 46(3):607–616.

Schoeneberger, P.J., Wysocki, D.A., Benham, E.C., and Soil Survey Staff. 2012. *Field book for describing and sampling soils, Version 3.0*. Lincoln, NE: Natural Resources Conservation Service, National Soil Survey Center.

Scullin, C.M. 1983. *Excavation and Grading Code Administration, Inspection, and Enforcement*, p. 405. Englewood Cliffs, NJ: Prentice-Hall.

Shaw, R.K., Reinhardt, L., and Isleib, J. 2007. *Soil Survey of Bronx River Watershed, Bronx, NY*. U.S. Department of Agriculture, Natural Resources Conservation Service. Available online at http://www.nycswcd.net/soil_survey.cfm (accessed March 14, 2014).

Shaw, R.K., Wilson, M.A., Reinhardt, L., and Isleib, J. 2010. Geochemistry of artifactual coarse fragment types from selected New York City soils. In *Proceedings of the 19th World Congress Soil Science*, Brisbane, Australia, 25–27.

Sherzer, W.H. 1917. *Detroit Folio, Wayne, Detroit, Grosse Pointe, Romulus, and Wyandotte quadrangles, Michigan*. Michigan: U.S. Geological Survey, Geologic Atlas of the United States Folio GF-205. USGS Publications Warehouse.

Singleton, A.D. 1998. Impacts of Site Use on Soil Quality and Runoff Potential in a Degraded Watershed in Central Park, New York City. M.S. Thesis, Cornell University, Ithaca, NY.

Soil Survey Division Staff. 2017. *Soil Survey Manual*. Soil Conservation Service. U.S. Department of Agriculture Handbook 18.

Soil Survey Staff. 1999. *Soil Taxonomy: A Basic System of Soil Classification for Making and Interpreting Soil Surveys*. 2nd ed. Natural Resources Conservation Service. U.S. Department of Agriculture Handbook 436. Washington, D.C. Available at https://www.nrcs.usda.gov/Internet/FSE_DOCUMENTS/nrcs142p2_051232.pdf (accessed May 23, 2017).

Soil Survey Staff. 2010. *Keys to Soil Taxonomy*, 11th ed. Washington, DC: USDA-Natural Resources Conservation Service. Retrieved from http://www.nrcs.usda.gov/Internet/FSE_DOCUMENTS/nrcs142p2_050915.pdf (accessed March 14, 2014).

Soil Survey Staff. 2014. *Keys to Soil Taxonomy*, 12th ed. Washington, DC: USDA-Natural Resources Conservation Service.

Soil Science Division Staff. 2016. Interpretations Generator HLBR-Version 1.0. In house draft document (accessed May 10, 2016).

Southeast Michigan Council of Governments (SEMCOG). 2015. Community Profiles. Retrieved from http://semcog.org/Data-and-Maps/Community-Profiles.

Splajt, T., Ferrier, G., and Frostick, L.E. 2003. Application of ground penetrating radar in mapping and monitoring landfill sites. *Environmental Geology*. 44:963–967.

Tromp-van Meerveld, H.J. and McDonnell, J.J. 2009. Assessment of multi-frequency electromagnetic induction for determining soil moisture patterns at the hillslope scale. *Journal of Hydrology*, 368:56–67.

U.S. Congress. 1966. Public Law 89–560, Soil Surveys for Resource Planning and Development. Retrieved from http://uscode.house.gov/statutes/pl/89/560.pdf (accessed May 10, 2016).

USDC Census Bureau. 2010. Quick Facts: Detroit, Michigan. Retrieved from http://www.census.gov/quickfacts/table/PST045215/00.

USDC Census Bureau. 2015. Cities and Towns Totals: Vintage 2015. Retrieved from https://www.census.gov/popest/data/cities/totals/2015/.

U.S. Department of the Interior, U.S. Geological Survey. 2015. Geology of National Parks. Staten Island serpentinite. Retrieved from http://3dparks.wr.usgs.gov/nyc/parks/loc7.htm (accessed April 25, 2016).

U.S. Department of Agriculture (USDA). 2016. Urban Agriculture Toolkit. Retrieved from http://www.usda.gov/wps/portal/usda/knowyourfarmer?navid=kyf-urban-agric (accessed May 10, 2016).

U.S. Department of Agriculture, Natural Resources Conservation Service (USDA-NRCS). 2013. National Soil Survey Handbook, title 430-VI. Retrieved from http://www.nrcs.usda.gov/wps/portal/nrcs/detail/soils/survey/geo/?cid=nrcs142p2_054242 (accessed December 6, 2013).

U.S. Department of Agriculture, Natural Resources Conservation Service (USDA-NRCS). 2015. National Soil Survey Handbook, title 430-VI. Retrieved from http://www.nrcs.usda.gov/wps/portal/nrcs/detail/soils/ref/?cid=nrcs142p2_054242. (accessed May 10, 2016).

U.S. Department of Agriculture, Natural Resources Conservation Service (USDA-NRCS). 2006. *Land Resource Regions and Major Land Resource Areas of the United States, the Caribbean, and Pacific Basin*. Washington, DC: U.S. Department of Agriculture Handbook 296.

United States Department of Agriculture, Natural Resources Conservation Service (USDA-NRCS). 2005. *New York City Reconnaissance Soil Survey*. Staten Island, NY. Retrieved from http://www .soilandwater.nyc/urban-soils.html.

United States Department of Agriculture, Natural Resources Conservation Service (USDA-NRCS). 2007. Soil Survey of Bronx River Watershed, Bronx, NY. Retrieved from http://www.soilandwater.nyc/urban-soils .html.

United States Environmental Protection Agency. 2011. Land Revitalization Fact Sheet: Improving Demolition Practices. Retrieved from https://www.epa.gov/land-revitalization/fact-sheet-improving-demolition-practices.

Warner J.W., Jr, and Hanna, W.E. 1982. Soil Survey of Central Park, New York (preliminary report). Prepared for the Central Park Conservancy by the Soil Conservation Service, USDA, in cooperation with Cornell University Agricultural Experiment Station. 107 pp.

Weber, J., Olson, G.W., and Lopez, S.H. 1984. *Tour of soils of Central Park in New York City. Misc. Bulletin 132*. New York, NY: Cornell Cooperative Extension.

Weindorf, D.C., Zhu, Y., Chakraborty, S., Bakr, N., and Huang, B. 2012. Use of portable x-ray fluorescence spectrometry for environmental quality assessment of peri-urban agriculture. *Environmental Monitoring Assessment*, 184:217–227.

Yerkes, R.F. and Campbell, R.H. 2005. Preliminary Geologic Map of the Los Angeles 30' × 60' Quadrangle, Southern California: U.S. Geologic Survey Open-File Report 2005-1019, U.S. Geological Survey, Denver, CO, scale 1:100,000. Electronic document. Retrieved from http://pubs.usgs.gov/of/2005/1.

3 Changes in Soil Organic Carbon Stocks by Urbanization

V.I. Vasenev, J.J. Stoorvogel, A.V. Dolgikh,
N.D. Ananyeva, K.V. Ivashchenko, and R. Valentini

CONTENTS

3.1 GENERAL INTRODUCTION: URBANIZATION AS A THREAT AND AN OPPORTUNITY FOR SOIL CARBON

Soils accumulate about 1500–2000 Pg (10^{15} g) C, providing the largest stock in terrestrial ecosystems (Swift, 2001; Janzen, 2004). Historically, soil organic carbon (SOC) is a widely accepted indicator of soil quality. For example, SOC depletion is used as a basic indicator of soil degradation (Nortcliff, 2002; Bastida et al., 2008). The shift in recent decades from traditional agricultural attitudes of soil as a substrate for food production to its role in essential ecological processes and functions highlighted the importance of soil carbon stocks and fluxes (Bolin et al., 1979; Kovda and Rozanov, 1988). Carbon sequestration, for example, is an important process to mitigate climate change (IPCC, 2001; Lal, 2004; Janzen, 2004), whereas soil respiration is the largest biogeochemical carbon efflux into the atmosphere, contributing to climate change (Raich et al., 2002; Schulze, 2006). Soil microbial carbon indicates the soil's performance as a habitat for microorganisms. Soil microbial communities contribute to biodiversity and gene reservoirs (Andrews et al., 2004; Blum, 2005; Dobrovolsky and Nikitin, 2012). The relation between soil microbial carbon and microbial respiration defines the microbial metabolic coefficient, which is widely accepted as a relevant indicator of the state of microbial soil communities and ecosystem disturbance (Anderson and Domsch, 1985; Dilly et al., 2003; Bastida et al., 2006). Many studies classifying and assessing soil functions acknowledge the role of SOC (e.g., BBodSchG, 1998; Karlen et al., 2003; Andrews et al., 2004; Blum, 2005; Dobrovolsky and Nikitin, 2012; Table 3.1). Although reviewed approaches to classify soil functions differ in terms of definitions and labels of each function, their total number, and the classification's major purpose, they all consider SOC as an important parameter: up to two thirds of the soil functions are directly or indirectly related to SOC stocks. The recently emerged concept of ecosystem services (ESs; MA, 2003) expands the analysis of environmental properties, processes, and functions with human economic benefits (de Groot, 1992; Costanza et al., 1997). Although soil services are considered part of ESs (Breure et al., 2012), SOC directly or indirectly affects many specific ESs, including soil fertility maintenance, food production, and climate regulation (MA, 2003; TEEB, 2010). Currently, most of the carbon assessments focus on natural (forest/ meadows) and agricultural ecosystems (e.g., Islam and Weil, 2000; Valentini et al., 2000; Hamilton et al., 2002; Cruvinel et al., 2011; Fromin et al., 2012). Much less, however, is known about the effect of urbanization on soil carbon stocks and fluxes.

Urban soil is a specific phenomenon, exposed to the anthropogenic influence both directly (through, e.g., pollution, sealing, and overcompaction) and indirectly (through alteration of the soil

TABLE 3.1
Carbon-Related Soil Functions in Different Classifications

Blum (2005)	Nortcliff (2002)	Ritz et al. (2009)	BBodSchG (1998)	Andrews et al. (2004)	Dobrovolsky and Nikitin (2012)
• Biomass production • Participating in biogeochemical cycles, including gas exchange between soil and atmosphere • Source of raw materials	• Provision of physical, chemical, and biological settings for living organisms • Supporting biological activity and diversity for plant growth • Filtering, buffering, degrading organic and inorganic substances	• Food and fiber production • Environmental interactions, including carbon retention • Supporting habitats and biodiversity	• Participation in water and nutrient cycles • Decomposition • Basis for life of people, plants, animals, and soil organisms • Land for agriculture and sylviculture • An archive of natural and cultural history	• Nutrient cycling • Biodiversity and habitat • Resistance and resilience	• Influence on the gas content • Storage of nutrients • Habitat for terrestrial organisms • Source for minerals and fossils formation • Transformation and transfer of sun energy to the bowels of the earth • Storage of historical artifacts

We used the original names of the functions, given by the authors of each classification.

forming factors) (Stroganova et al., 1997). There is still no agreement in the research community on whether urban soils are carbon stocks or sources. Some comparative studies report higher SOC stocks in urban soils compared to nonurban soils (Lorenz and Lal, 2009; Vasenev et al., 2013a; Vodyanitskii, 2015). This is usually explained by adding of organic substrates (e.g., composts, turf, peat, and organic wastes) to the topsoil in greenery. Increased SOC stocks in urban subsoil refers to the so-called "cultural layers," resulting from residential activity over a long period of time (Alexandrovskaya and Alexandrovskiy, 2000). At the same time, substantial carbon outflow, resulting from high vulnerability of urban SOC stocks to anthropogenic influences and bioclimatic conditions, is also likely for urban soils (Kaye et al., 2005; Vasenev at al., 2015; Sarzhanov et al., 2015).

The assessment of carbon stocks in urban soils is complicated by high spatial-temporal variability, driven by a combination of bioclimatic and urban-specific factors (Vasenev et al., 2014a). Spatial variability of urban soils is a result of joined effect from zonal (regional) climatic conditions and urban-specific (local) land use and management. As a result, SOC stocks in urban soils differ between various bioclimatic regions (Stroganova et al., 1997; Pouyat et al., 2006); however, the difference in SOC stocks between the different functional zones can be even more substantial, although the climate is similar (Vasenev et al., 2012, 2013b). Combinations of different functional zones (e.g., industrial, residential, and recreational) and areas of historical development (e.g., historical core, central zone, and suburbs) contribute to this high heterogeneity of urban areas, generating a "patchy" spatial distribution. Spatial variability of the urban SOC stocks is further complicated by its temporal dynamics. Temporal changes in SOC stocks largely depend on a range of soil abiotic and biotic parameters that determine the balance between carbon uptake (through photosynthesis) and carbon efflux by soil respiration (Carlyle and Than, 1988; Gomes-Casanovas et al., 2012). Understanding spatial variability and temporal dynamics of urban soils' SOC stocks is critical to analyze and predict changes in soil SOC stocks by urbanization.

In this chapter, we review the factors and processes of the urbanization effect on SOC stocks, considering different space and time scales (Figure 3.1). The effect of urbanization on SOC stocks is complex and versatile. Building infrastructure (houses, roads, industries, etc.) has a direct anthropogenic influence on soils and can result in considerable SOC losses, whereas establishing urban green infrastructure (parks, lawns, etc.) likely increases soil carbon storage. In this chapter, projecting changes in SOC stocks by urbanization is addressed through analyzing different processes, driven by urbanization and effecting spatial variability and temporal dynamics of soil carbon.

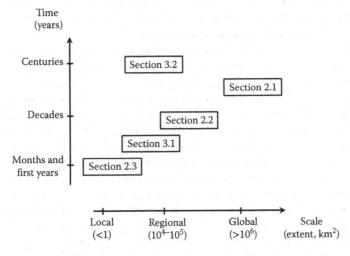

FIGURE 3.1 An outline of the chapter, representing different spatial and temporal scales of the analysis.

Section 3.2 presents urbanization consequences for SOC stocks at the global, regional, and local levels. At first, we explore relationships between global urbanization and dominating soil types and biomes (Section 3.2.1). Then, the net effect of urbanization on SOC stocks was modeled for the large Moscow region (Section 3.2.2). Finally, sustainability of SOC stocks in modeled urban soil constructions with contrast morphological features were analyzed at the experimental plot in the Moscow city (Section 3.2.3). The urbanization effect on SOC with regard to the response time is analyzed in Section 3.3. We discuss both the short-term urbanization effect on SOC, by comparison of the microbiological activity of urban and nonurban soils in the central chernozemic region of Russia (Section 3.3.1), and the potential of carbon storage in cultural layers of different age, origin, and climatic conditions (Section 3.3.2).

3.2 ANALYZING URBANIZATION EFFECTS ON SOIL CARBON STOCKS OVER A VARIETY OF SCALES

3.2.1 GLOBAL URBANIZATION AND POSSIBLE OUTCOMES FOR SOIL CARBON STOCKS

A range of different products characterizing the extent of global urbanization and its spatial distribution are available (Seto et al., 2012, 2015). The extent of the urbanized area varies between the different products typically because of the definition of the urban areas. For example, GlobCover (Bontemps et al., 2010) has a class defined as "Artificial surfaces and associated areas (Urban areas >50%), where the areas are defined on the basis of remote sensing," which is often interpreted as urban area. Other remote sensing' derived global maps on urbanization, include different definitions and different procedures. For example, the Global Rural-Urban Mapping Project (GRUMP; CIESIN et al., 2011; Balk et al., 2006) created a mask of urban areas on the basis of topographic information and the global Nightlight database. Differences between the different databases clearly occur based on the definition of urban areas (e.g., the GRUMP database explicitly excludes industrial areas as it is more focused on residential areas). In addition, the years of observation differ between the different products. As a result, it is difficult to combine the different maps for a more dynamic analysis of urbanization. However, the maps do coincide in the general patterns of urbanization. An overlay of the urban areas defined by the GlobCover database and the global biomes as defined by Olson et al. (2001) shows that the "Mediterranean forests, woodlands, and scrub" and the "temperate broadleaf and mixed forests" are the biomes that, relatively speaking, are the most urbanized from the world (Table 3.2). More than 10% of these biomes is urbanized. In absolute terms, however, the "temperate broadleaf and mixed forests" are really dominating, containing more than 40% of the urban areas of the world. The differences are not surprising if we see where the highly urbanized regions in the world are located. GlobCover is one of the few databases that provide data on land cover change for multiple years. The data in Table 3.3 show that, although expansion of the urban areas took place between 2005 and 2009, it occurred on the same soil types.

In a similar way, an overlay of GlobCover and the major soil groups derived from the S-World Database (Stoorvogel, 2017) can be performed. The data show that just like the major biomes, there are large discrepancies among the different soil types. This confirms the general belief that the urbanization process is highly concentrated on specific positions of the landscape. Most of the urban areas can be found on the Cambisols, Fluvisols, and Luvisols—each hosting more than 10% of the global cities. These soil types are typically found in lower and relatively young sedimentary areas. The areas are also suitable for agricultural practices. Other soil types that represent, for example, very poorly drained areas or mountainous regions are obviously less urbanized.

An overlay of the GlobCover map with a global map of topsoil SOC from S-World shows that the urbanized areas are located on soils with an average SOC concentration of 1.33% (0–50 cm) in contrast to the soils in areas that are not urbanized, which had an average SOC concentration of 2.31%. This is in strong contrast to the general belief that urbanization takes place in the fertile low lying areas like deltas. However, realizing that the soils in these areas are relatively young Cambisols, Fluvisols, and Luvisols in which SOC stocks have not been built up. The repercussions of the above

TABLE 3.2

The Global Distribution of the Urban Areas and Urbanization Patterns

	Area	GlobCover 2005		GlobCover 2009		
	10^6 km^2	% of all Urban Areas	% Urban	% of all Urban Areas	% Urban	Urbanization
Tropical and subtropical moist broadleaf forests	1989	11.3	1.82	11.4	1.91	2.1
Tropical and subtropical dry broadleaf forests	303	3.2	3.39	3.2	3.58	0.2
Tropical and subtropical coniferous forests	71	0.3	0.79	0.2	1.39	0.0
Temperate broadleaf and mixed forests	1284	43.2	11.14	44.7	11.39	68.9
Temperate coniferous forests	409	2.2	1.54	2.0	1.82	2.4
Boreal forests/Taiga	1508	1.5	0.40	1.9	0.33	7.6
Tropical and subtropical grasslands, savannas, and shrublands	2030	5.0	0.77	5.0	0.82	5.1
Temperate grasslands, savannas, and shrublands	1010	7.9	2.40	7.6	2.63	5.7
Flooded grasslands and savannas	110	1.5	4.44	1.5	4.85	0.2
Montane grasslands and shrublands	520	2.4	1.42	2.3	1.54	0.8
Tundra	1160	0.1	0.04	0.1	0.04	0.2
Mediterranean forests, woodlands, and scrub	323	12.7	11.97	12.2	13.22	3.0
Deserts and xeric shrublands	2798	7.8	0.81	7.2	0.93	3.6
Mangroves	35	0.7	7.04	0.7	8.12	0.3

Source: Based on the GlobCover Database (Bontemps et al., *ESA and Université Catholique de Louvain*, 53, doi:10.1594/PANGAEA.787668, 2010) and Global Biomes (Olson et al., *BioScience*, 51, 933–938, 2001.)

analysis for the effect of urbanization on SOC stocks is promising. The relatively low SOC contents result in a limited potential of the soils to lose carbon as a result of increased mineralization. At the same time, it is more likely that the soils can be enriched through, for example, greenery activities.

3.2.2 NET EFFECTS OF URBANIZATION ON CARBON STOCKS IN THE MOSCOW REGION

3.2.2.1 Introduction

Regional analysis of urbanization effect on SOC stocks may be even more challenging compared to the global scale, since the allocation of the urbanized areas and spatial variability of SOC carbon within and between the urban areas are complicated by multiple environmental and socio-economic factors.

TABLE 3.3

The Distribution of Urban and Urbanized Areas between the Different Soil Types

Soil Group	Area	% Urbanized	% of Global Cities
Anthrosols	55.54	22.61	4.25
Chernozems	210.40	4.50	3.21
Calcisols	750.05	1.17	2.97
Cambisols	1267.34	3.04	13.07
Fluvisols	412.64	8.63	12.07
Ferralsols	646.68	0.65	1.42
Gleysols	699.82	2.62	6.20
Greyzems	68.09	2.95	0.68
Gypsisols	133.76	0.48	0.22
Histosols	319.02	1.00	1.08
Kastanozems	340.43	1.22	1.41
Leptosols	1524.33	0.61	3.14
Luvisols	831.68	5.03	14.18
Lixisols	200.89	2.37	1.61
Nitisols	153.71	1.90	0.99
Podzoluvisols	231.71	2.75	2.16
Phaeozems	271.78	4.24	3.90
Planosols	134.15	6.69	3.04
Plinthosols	115.41	0.93	0.37
Podzols	580.48	1.66	3.26
Regosols	1100.01	1.06	3.96
Solonchaks	175.51	3.84	2.28
Solonetz	247.87	2.32	1.95
Vertisols	282.01	4.49	4.29

Urbanization coincides with different processes, having versatile effect on SOC stocks. Historically, investigation of urban ecosystems focused on the negative anthropogenic influences on soils. Some of them, like, for example, soil sealing or excavation of the fertile topsoil horizon for building construction, have a clear negative effect on SOC stocks in urban soils. Soil sealing is recognized as among the main threats for urban soils, constraining the performance of the soil functions. Ekranic Technosols (the WRB reference for sealed soils [WRB, 2014]) recently received increased attention regarding the amount of stored SOC and the availability of carbon stocks (Vasenev et al., 2013b; Lorenz and Lal, 2009; Piotrowska-Dlugosz and Charzynski, 2015). Although the values given by different authors to SOC stocks in Ekranic Technosols varies from negligible (Schaldach and Alcamo, 2007; Schulp and Verburg, 2009) to considerable (Wei et al., 2014), most of the research agrees on the substantial decrease of SOC stocks under the impervious areas compared to the open areas (Elvidge et al., 2004; Raciti et al., 2012). Although soil sealing in urban areas results in depletion of SOC stocks, establishing urban green infrastructures likely contributes to C storage. The juvenile age of trees used in greenery and the extended due to urban heat island effect vegetation season result in high net primary productivity and increased C stocks in urban vegetation (Nowak and Crane, 2001). Considerable SOC stocks are reported for urban lawns, which is explained by the addition of artificial substrates, such as turf, peat, and organic compost (Lorenz and Lal, 2009; Vasenev et al., 2014b). A remarkable increase of SOC stocks is also modeled for turf grasses after several decades of intensive management (e.g., cutting, irrigation, and fertilization), stimulating extra belowground and root biomass (Qian et al., 2002; Bandaranayake et al., 2003; Zircle et al., 2011). Both processes potentially increasing and depleting SOC stocks are likely brought by urbanization. The net effect of urbanization on the regional SOC stocks is determined by the environmental and socio-economic factors of the region.

In this study, we modeled the effect of urbanization on SOC stocks in the Moscow region, which is Russia's most urbanized region with high heterogeneity of environmental conditions. Urban expansion in the region has continued since the 1990s and recently the New Moscow project, announced by the Russian government, expands the legal borders of the city 2.4 times, resulting in rapid urbanization on approximately 2500 km². In our study, we analyzed the net effect of urbanization on the regional SOC stocks under different plausible urbanization scenarios.

3.2.2.2 Materials and Methods

The Moscow region is located in the central part of Russia (54–57°N; 35–40°E) and covers 46 700 km². The region has a temperate continental climate with mean annual temperatures between 3.5°C and 5.8°C. Average annual rainfall varies from 780 mm in the North to 520 mm in the South. Vegetation and soils of the region vary from Orthic Podzols to Eutric Podzoluvisols under south-taiga in the northern and central parts to Orthic Luvisols and Luvic Chernozems under deciduous forest and forest-steppes in the South, Dystric Histosols in the East, and Eutric Luvisols in the flood-plains of the Moskva and Oka rivers (Egorov et al., 1977; FAO, 1988; Shishov et al., 2004). More than half of the region's territory is covered by anthropogenic landscapes (e.g., settlements and croplands). Moscow city with a population of over 11.5 million people is the center of the region and the capital of Russia. It is an important political, economic, and cultural center, attracting intensive tourists, labor, and education migration (Kolossov et al., 2002, 2004; Gritsai, 2004).

Our analysis of the net urbanization effect on the SOC stocks in the region included two major steps: (1) modeling urbanization for the period 2014–2048 and development of the land-use map, considering urbanized areas; and (2) estimating SOC stocks for the new land-use map (for the year 2048) and comparing to the current situation (for the year 2014). Urbanization in the region was modeled by logistic regression, where urbanization probability was determined as a function of environmental, socio-economic, and neighborhood factors (Reilly et al., 2009; Dubovyk et al., 2011; Li et al., 2013). At first, we analyzed urbanization trends in the region for the period 1980–2014 by comparing the settlement boundaries of the 1980 1:200,000 topographic map to the open street map (OSM, www.openstreetmap.org) data for 2014. Only settlements larger than 2 km² were considered based on the accuracy of the topographic map.

Urbanization was related to several environmental, socio-economic, and neighborhood driving factors (Aspinall, 2004; Batisani and Yarnal, 2009; Li et al., 2013; Table 3.4). The probability of urbanization for the period 1980–2014 in the Moscow region was explained as a function of the driving factors using a logistic regression (Kleinbaum and Klein, 2010). The analyzed relationships between the binary urbanization and selected explanatory variables were used to model future urbanization in 2014–2048. Three alternative scenarios were analyzed: (1) the Full Urbanization Model (FUM), including the environmental, socio-economic, and neighborhood factors; (2) the Environmental Urbanization Model (EUM), excluding neighborhood factors; and (3) the New Moscow Urbanization Model (NMUM), combining FUM and the New Moscow zone. All the models resulted in the 771 m urbanization grid maps for 2048, used to identify the urbanized areas on the contemporary land-use map (Bartalev et al., 2003). SOC stocks for 2014 and 2048 were estimated using the methodology developed by Vasenev et al. (2014b), which modeled SOC stocks as a function of conventional (i.e., soil type, land-use type, relief, climate, and vegetation) and urban-specific (functional zoning, soil sealing, and settlement history) factors. Using the methodology, the 771 m grid map of topsoil (0–10 cm) and subsoil (10–150 cm) SOC stocks in the Moscow region was produced. The net effect of urbanization on SOC stocks was estimated as the difference between SOC stocks in 2014 and 2048 and analyzed for different bioclimatic zones and soil types at the region.

3.2.2.3 Results and Discussion

A substantial urbanization was obtained under all three analyzed scenarios, whereas the total extent and location of urbanized areas was different. EUM shows the lowest level of urbanization with 2834 km², of which the majority is located east from Moscow city, where the existing settlements

TABLE 3.4

Factors Driving Urbanization in the Moscow Region

Description of Variable	Source	Reason to Include in the Model
Environmental		
Mean annual temperature (°C)	www.worldclim.org (Hijmans et al., 2005)	
Total annual precipitation (mm)		
Elevation above sea level (m)	srtm.csi.cgiar.org/index.asp (Jarvis et al., 2008)	
Slope steepness (°)		
Leaf area index (July 12–19) for 1980 and 2014 (Liang and Xiao, 2012)	http://glcf.umd.edu/data/lai/	Conventional factors, influencing SOC distributions
Leaf area index (October 16–23) for 1980 and 2014		
Aggregated soil types (Soil 1 = podzols; Soil 2 = podzoluvisols; Soil 3 = luvisols; Soil 4 = chernozems; Soil 5 = alluvial and peat	Derived from Shishov and Voinovich (2002)	
Socio-economic		
Euclidian distance to the Moscow city polygon	Derived from the settlement layer, www.openstreetmap.org	
Euclidian distance to the federal roads	Derived from the highway layer, www.openstreetmap.org	Proximity factors, which determine the attractiveness of the location for settlement
Euclidian distance to the major rivers, lakes, and water basins	Derived from the hydrology layer, www.openstreetmap.org	
Neighborhood		
Proportion of urban areas within 100 m window	Derived from urban map 1980 and 2014	Likelihood of urbanization is much higher at locations surrounded by
Proportion of urban areas within 1000 m window	Derived from urban map 1980 and 2014	urban areas

are dense and several federal highways occur. The FUM yielded 3302 km² of urban lands in 2048. Urban expansion was the major trend under this scenario, since most of the new urbanized areas occurred close to the existing settlements. The highest urbanization was shown as expected by the NMUM scenario, yielding 4745 km² of urban area by 2048 (Figure 3.2). New settlements concentrate in the south-west direction within the legal New Moscow boundaries. Although this projection is likely an overestimation, it allows analyzing the potential land-use changes coinciding with the New Moscow project. For example, substantial loss of arable lands (i.e., 649 km²) was reported for the NMUM scenario, whereas under the two other scenarios urbanization mainly occurred on the forested land.

The reported land-use change resulted in 500 to almost 2000 km² of new urbanized areas and the corresponding loss of forest and arable lands in the Moscow region obviously should have an effect on SOC stocks. If only negative effects of urbanization are considered and zero SOC stocks are assigned to urban areas as often proposed by regional and global assessments (e.g., Schaldach and Alcamo, 2007; Schulp and Verburg, 2009), the projected urbanization would result in 9.0 TgC emissions (1.9 and 7.1 TgC from the topsoil and subsoil, respectively). This outcome is likely an underestimation, since the positive effects of, for example, urban green infrastructure, are not considered. The three other scenarios, considering this positive effect, resulted in a net increase of SOC stocks

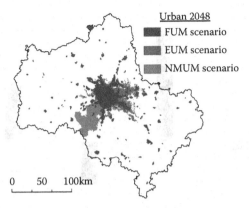

FIGURE 3.2 Urbanization in the Moscow region by 2048 as projected by FUM, EUM, and NMUM scenarios.

of 4, 10, and 11 TgC for the EUM, FUM, and NMUM, correspondingly. For all three scenarios, the changes in topsoil SOC stocks were minor, whereas subsoil stocks substantially increased. This is likely explained by the SOC stored in the cultural areas of urban soils. Cultural layers, containing plenty of organic compounds (e.g., coal, turf, wood remains, and buried soils horizons) can contain up to 3%–4% of SOC (Alexandrovskaya and Alexandrovskiy, 2000; He and Zhang, 2009; Dolgikh and Alexandrovskiy, 2010) and achieve several meters depth depending on the settlement history. The different net effect of urbanization of SOC stocks was also obtained for different bioclimatic zones and dominating soil types. For example, the largest increase was reported for the Eutric Podzoluvisols and Orthic Podzols, which contain relatively low amounts of carbon in natural conditions. However, the positive effects for more fertile Dystric Histosols and Eutric Luvisols were small (Figure 3.3).

3.2.2.4 Conclusion
Considerable urbanization is projected in the Moscow region in coming decades with up to 2000 km² of forest, meadow, and arable lands converted to urban areas by 2048. We showed that although the consequences of urbanization on SOC stock varied between different bioclimatic zones and soil types, the total positive net effect of urbanization on the regional SOC stocks was confirmed by all three scenarios. This outcome contradicts some studies, ignoring urban areas in carbon assessments and provides an optimistic message for scholars and urban planners, since we clearly showed that urban soils are capable of capturing and even enhancing SOC stocks.

3.2.3 Sustainability of SOC Stocks in Urban Soil Construction

3.2.3.1 Introduction
The local level of analysis brings a new set of factors influencing carbon stocks. Although the influence of more general natural and urban-specific factors discussed in Sections 3.2.1 and 3.2.2 remain relevant, spatial distribution and temporal dynamics in SOC stocks are mainly driven by more localized factors: the construction of urban soils, time passed from establishment, species and density of the soil cover, and management practices (clipping, irrigation, and fertilization). Combinations of these factors yield a substantial diversity of urban SOC stocks even at short distances. At the same time, the anthropogenic origin of most of the factors influencing urban soils at the local scale allows controlling the sustainability of SOC stocks via best management practices and sustainable construction of urban soils.

Straightforward universal guidelines on urban soil construction, including depth and substrate for organic and mineral layers, are still lacking in many countries, including Russia. This results in

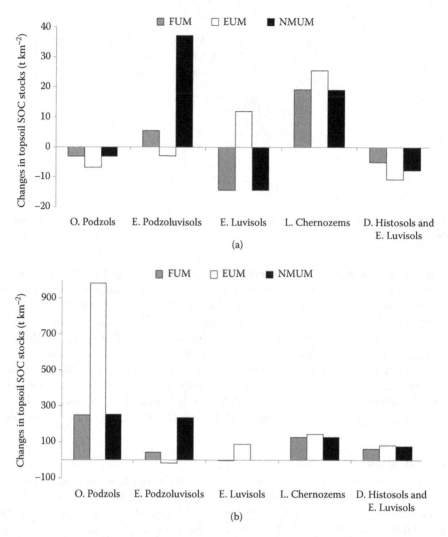

FIGURE 3.3 Urbanization effect on the topsoil (a) and subsoil (b) SOC stocks (t/km²) averaged per soil type in the Moscow region.

high heterogeneity in morphology of soil construction profiles. Analyzing the effect of contrasting morphological features of urban soil construction on environmental functions, including C sequestration, is necessary. This research aimed to study the effect of the morphological features of urban soils on the sustainability of their SOC stocks. We studied carbon emissions, temporal dynamics in soil, and biomass carbon stocks for urban soil models, difference in depths, and substrates of organic horizons.

3.2.3.2 Materials and Methods

Urban soil constructions were mounted at the experimental fields of Russian State Agrarian University (55°50′ N; 37°34′ E). The research area is located in the north administrative district in Moscow city. Model soil constructions were mounted at the open areas nearby a university golf course. Urban soil constructions were mounted in experimental containers. The experimental containers were produced from inert plastic with perforated wall, surface extent 80 cm². Soil constructions included a 20 cm layer of the zonal soddy-podzolic soil and an organic horizon of sandy basis of 30 cm depth in total (Figure 3.4).

Model urban soil constructions represented four different organic substrates: (1) valley peat (P), (2) peat–sand mixtures (PS) (valley peat to sand in weight proportion 30%–70%), (3) soil–peat mixture (SP) (valley peat and zonal soddy-podzolic soil (mixed A and AE horizons) in weight proportion 50%–50%), and (4) soil–sand mixture (SS) (sand and zonal soddy-podzolic soil in weight proportion 50%–50%). Two depth options (5 cm and 20 cm) were observed for each of the four substrates, giving 5P, 5PS, 5SP, 5SS, 20P, 20PS, 20SP, and 20SS samples, each in two replicas. Soil construction containing 10 cm of SP mixture and a 20 cm sand layer was considered as a control (C) for comparison. These parameters correspond to averaged parameters of urban soil constructions widely used in urban greenery. Control samples were analyzed in three replicas. In total, 19 containers were observed in the experiment. The experiment was set up in 2012 and monitoring took place during the vegetation season (June–November) in 2013. SOC concentrations and stocks, carbon stocks in aboveground biomass, CO_2 emissions, soil temperature and moisture, and projective cover were observed for the lawn ecosystems. SOC concentrations were measured using CN Elementar and carbon stocks in a 20 cm layer were estimated. *In situ* CO_2 emissions were measured by a LI-820 gas analyzer. Soil temperature was measured by Checktemp thermometer, and an SM300 moisture probe was used to measure soil moisture. Aboveground biomass was cut, air dried, and weighted afterward.

3.2.3.3   Results and Discussion

SOC concentrations in the 0–20 cm level of investigated soil constructions were very variable at the starting stage of experimental setup, since the organic substrates used for soil constructions were very different. The highest 36.5% SOC concentration was obtained for the 20P sample, whereas a minimal SOC concentration was 0.3% and was found for 5SS. SOC concentrations in all the samples were very unstable and highly dynamic. Average SOC concentrations decreased 30%–50% in the first three months after the experiment was started. The temopral dynamic of SOC concentration differed between the samples. The highest SOC loss was found for peat samples, where depletion in SOC concentrations was up to 26%, whereas SOC concentrations in SS and SP slightly increased (Figure 3.5).

The decline in topsoil SOC concentration is very typical for urban soils at the early stages of development. This is usually the result of low sustainability of peat-based soil mixtures in automorphic conditions. Organic substrates based on valley peat are easily mineralized during the first years after urban soil construction. This results in intensive CO_2 emissions. However, urban lawns can also store carbon through photosynthesis. The intensity of the latter process can be judged by net primary productivity and total biomass increase. We observed the highest CO_2 emissions for 20SP and 20P samples. Carbon efflux from these soil constructions was 23.5 ± 14 and 25.9 ± 14 mg CO_2 m^{-2}day^{-1}, respectively. The lowest CO_2 emissions were shown for SS mixtures, where

| 1 | 2 | 3 |

FIGURE 3.4   Model soil constructions at different stages of the experiment set up.

average carbon efflux for both observed depths was just 7.6 mg CO_2 m^{-2}day^{-1}. The lowest average soil moisture was also shown for the SS sample; however, we didn't find significant differences in soil temperature between observed soil mixtures.

Substantial dynamics in CO_2 emissions were found for all soil constructions during the growing season with the highest values obtained in July–August and the lowest in October (Figure 3.6a). This temporal trend was likely driven by seasonal changes in soil temperature and moisture (Figure 3.6b).

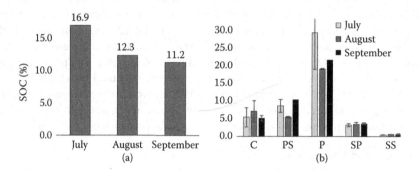

FIGURE 3.5 Temporal dynamics of SOC concentration in model soil constructions: averaged for all samples (a) and averaged for different substrates (b).

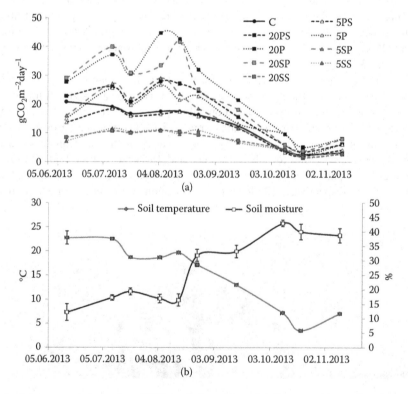

FIGURE 3.6 Interseasonal dynamics of CO_2 emissions (a), and soil temperature and moisture (b) in the model urban constructions

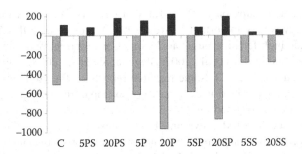

FIGURE 3.7 Total carbon emissions and carbon uptake in biomass in the model urban soil constructions for the vegetation season.

Total CO_2 emissions estimated for the vegetation period considering observed temporal dynamics allows us to judge the carbon losses by the urban soil constructions through autotrophic (root) and heterotrophic (microbial) respiration. Maximal total emission was obtained for the 20P and 20SP samples, where carbon losses were 640–700 g cm^{-2} during five months of observation. The minimal losses were shown for SS samples; however, initial carbon stocks in these SS mixtures were already very low. Carbon outflow through soil respiration was partly compensated by carbon uptake in biomass. The maximal biomass growth was shown for the same 20P and 20SP mixtures, where carbon emissions were also the highest. However, the carbon uptake in biomass doesn't compensate for the emitted carbon in these samples. Net carbon losses were 350–400 gCm^{-2}. Negative carbon balance was shown for all the samples (Figure 3.7).

3.2.3.4 Conclusion

Urban soil constructions based on valley peat are widely used in the urban greenery of Moscow city. We showed that this can have a negative environmental impact and decrease the green lawn's quality. Our research proved that peat-containing soil construction can be very vulnerable, especially during the first years after construction. Valley peat is rather unstable and is easily mineralized in automorphic conditions. The model soil constructions based on valley peat lost almost 25% of the initial carbon stocks during the first five months of the experiment. This was the largest carbon depletion compared to other soil constructions.

3.3 THE CONTRIBUTION OF URBANIZATION TO TEMPORAL DYNAMICS IN SOIL CARBON STOCKS

3.3.1 CHANGES IN URBAN SOIL CARBON STOCKS AND FLUXES BY ANTHROPOGENIC INFLUENCE IN THE URBAN ENVIRONMENT

3.3.1.1 Introduction

Analyzing the urbanization influences on SOC stocks based on spatial proxies is rather uncertain, if the temporal dynamics in SOC stocks are not considered. The urbanization effect on SOC stocks differs in the course of time. The most rapid changes in urban SOC stocks can be analyzed based on the alteration in soil microbiological activity responding to the anthropogenic effect. Anthropogenic disturbances resulting from urbanization cause instantaneous changes in soil respiration and microbial carbon content. The magnitude of changes depends on the zonal soil type; therefore, the urbanization effect on soils possessing high SOC contents and microbiological activity is substantial. One of the most evident examples is given by the urbanized Chernozems, widely recognized as the etalon soils.

Although Chernozems occupy only 6% of Russian territory, their contribution to the natural resources of the country is very important (Kovda, 1983; Avetov et al., 2011). The unique features of Chernozems include high fertility, large stocks of carbon, and nutrients over the whole profile (Rodionov et al., 2001; Stakhurlova et al., 2007; Blagodatskii et al., 2008; Polyanskaya et al., 2010; Shein et al., 2011; Kogut et al., 2012). Some research reports on the vulnerability of Chernozems to anthropogenic influence and land-use change. For example, the incorporation of Chernozems into intensive agriculture has resulted in 20%–30% loss of soil carbon over the last 150 years due to degradation as reported by Mikhailova and Post (2006). Recently the Chernozemic region is becoming more urbanized with the proportion of urban population increasing by 10% over the years 2010–2014 (Assessment of residential…, 2014), therefore substantial changes in soil microbiological activity and SOC stocks are likely. The research aimed to analyze the anthropogenic effect on microbiological activity and SOC stocks in Chernozems. This research problem was addressed by comparative research of Chernozems exposed to different land-use types, including urban.

3.3.1.2 Materials and Methods

Typical Chernozems (Luvic Chernozems; WRB, 2014) in the Kursk region (51°N / 36°E, Russia), under natural (virgin and mowing steppe) and anthropogenic disturbed (pasture, black fallow, arable, and urban) ecosystems, were studied (Table 3.5). Sampling points were selected randomly in each observed ecosystem and functional zone of Kursk city (recreational, residential, and industrial). In each observation point a mixed sample was taken from a 2×2 m plot (corners and center), from a 10 cm mineral layer (litter excluded) and along a soil profile (10–50, 50–100 and 100–150 cm). The profile distribution was analyzed in more detail for the virgin steppe, arable areas (cropland), and industrial zone, where five additional sampling points in each were located and samples from the top 50 cm were collected with the 10 cm interval.

Soil substrate-induced respiration (SIR) based on additional respiration response of soil microorganisms (initial maximum CO_2 production) enriched by available substrate (glucose) was measured (Anderson and Domsch, 1978; Ananyeva et al., 2011). Briefly, soil samples (2 g) were placed into a 15 mL vial, a solution of glucose was added (10 mg g^{-1} soil, 0.1 mL), the vial was closed hermetically, and time was recorded. The vial was incubated (3–5 h, 22°C) and an air sample was taken

TABLE 3.5
Land-Use Treatment and Chemical Properties of Typical Chernozems (0–10 cm) in Different Ecosystems and Functional Areas (Kursk Region)

Ecosystem/Zone (Number of Sites)	Land-Use History, years	Plants Dominated	C_{org} %	pH of Water
Steppe virgin (3)	74	Meadow grasses (*Stipa, Festúca*)	4.2 ± 0.6	6.1 ± 0.3
Steppe mowing (4)			4.5 ± 0.4	6.1 ± 0.4
Pasture (3)	79		4.3 ± 0.6	6.7 ± 0.1
Black fallow (2)	60	Bare	3.0 ± 0.4	6.2 ± 0.0
Arable (3)	60	Barley	2.7 ± 0.1	6.3 ± 0.5
Recreational (3)	40	Lawn grasses, *Betulapendula, Acerplatanoides, Aésculus hippocástanum*	3.0 ± 0.2	7.1 ± 0.4
Residential (2)	50	Lawn grasses, *Sorbusaucuparia, Tiliacordata Mill, Acerplatanoides*	1.7 ± 0.1	7.6 ± 0.2
Industrial (2)	68	Lawn grasses	2.3 ± 0.0	7.9 ± 0.1

and injected into a gas chromatograph (Kristallyuks 4000M, thermal conductivity) for measuring CO_2 production. Carbon of the microbial biomass was calculated according to: C_{mic} (μg C g^{-1}) = SIR (μL CO_2 g^{-1} h^{-1}) $\times$ 40.04 + 0.37 (Anderson and Domsch, 1978). Basal (microbial) respiration (BR) was measured as described for SIR, although glucose solution was substituted by distilled water (0.1 mL g^{-1} soil) and incubated (24 h, 22°C). The BR rate was expressed in μg CO_2-C g^{-1} soil h^{-1}. Specific respiration of the microbial biomass (the microbial metabolic quotient, qCO_2) was estimated as the ratio of BR/C_{mic} = qCO_2 (μg CO_2-C $mg^{-1}C_{mic}$ h^{-1}). The ratio between C_{mic} and SOC content (C_{org}) was calculated and expressed in %. The ratio qCO_2/C_{org} was calculated, illustrating the efficiency of soil organic matter consumption by soil microorganisms (Dilly et al., 2001; Dilly, 2005). Prior to the estimation of SIR and BR all soil samples (0.3–0.5 kg) were sieved (mesh, 2 mm), moistened up to 50%–60% water holding capacity, and pre-incubated in aerated bags at 22°C for 7 days to avoid excess CO_2 production after mixing, sieving, and moistening of soil sample (Ananyeva et al., 2008; Creamer et al., 2014). SOC (C_{org}) content in the collected soil samples was measured by the dichromate oxidation. The acidity of soil solution (soil: H_2O = 1:2.5) was determined.

The C_{mic} and BR measurements were performed in three replicas and results were expressed per dry weight of soil (105°C, 8 h) as mean $\pm$ standard deviation; a variation coefficient was also calculated (CV = mean/sd $\times$ 100%). Analysis of variance (ANOVA, post-hoc analysis by Tukey test) was carried out to determine the significant differences of soil indices (C_{mic}, BR, qCO_2, C_{mic}/C_{org}, qCO_2/C_{org}) for studied ecosystems. Data analysis and visualization (box plot: median, 1QR-3QR, min-max, outliers) were performed using the R software package (3.2.4, http://www.r-project.org/).

3.3.1.3 Results and Discussion

At first, the topsoil features were analyzed. Topsoil SOC ranged from 1.7% for the residential area to 4.5% in the mowed steppe (Table 3.5). Soil pH in urban areas was on average one unit higher than for the other ecosystems. The obtained C_{mic} values varied from 84 to 1954 μg C g^{-1} at the industrial area and the mowed steppe correspondingly. The highest C_{mic} was obtained for the undisturbed virgin steppe and pastures (1254 and 1088 μg C g^{-1}, respectively), whereas C_{mic} values for the disturbed areas were significantly lower (Figure 3.8). The C_{mic} at the industrial zone was on average half of that at the recreational area (308 and 630 μg C g^{-1} respectively). High spatial variability of C_{mic}, expressed via the coefficient of variance (CV), was obtained for the urban areas (41%–84%), which was almost twice that from the natural and arable areas (33%–37%). Although the spatial variability was high, the average C_{mic} in urban areas was significantly (two to four times, $p < .05$) less than in the natural steppe. This is clear evidence of the negative urbanization effect on microbial biomass. The BR in Chernozems ranged from 0.20 to 1.57 μg CO_2-C g^{-1} soil h^{-1} (surprisingly, both extreme

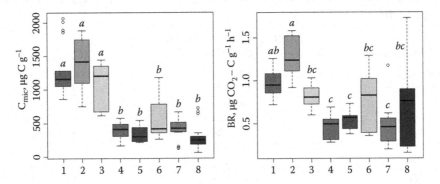

FIGURE 3.8 Distribution of soil microbial biomass carbon (C_{mic}) and basal respiration (BR) in typical Chernozems (0–10 cm) of different ecosystems and functional zones (1, steppe virgin ($n = 5$); 2, steppe mowing ($n = 4$); 3, pasture ($n = 3$); 4, black fallow ($n = 4$); 5, arable ($n = 3$); 6, recreational ($n = 3$); 7, residential ($n = 5$); 8, industrial ($n = 4$). The different letters indicate the significant difference, $p \leq .05$.

values were obtained at the industrial zone). The highest average BR was obtained for the moved steppe (1.24 µg CO_2-C g^{-1} soil h^{-1}) and the lowest for the arable and residential areas (0.54 and 0.56 µg CO_2-C g^{-1} soil h^{-1}, respectively). The highest CV in BE was estimated for urban soils (50, 65, and 90% for the recreational, residential, and industrial zones, respectively). High average BR in urban soils (up to 1.19 and 1.57 µg CO_2-C g^{-1} soil h^{-1}, respectively) provides an evidence of substantial potential CO_2 emission from urban soils.

The obtained qCO_2 values ranged from 0.50 to 6.83 µg CO_2-C $mg^{-1}C_{mic}$ h^{-1}. Although the difference in qCO_2 between the land-use types was not statistically significant, the average values increased from the virgin steppe to the industrial zone, indicating a more stressful condition for the microbial community at the disturbed sites. The highest spatial variability in qCO_2 was reported for urban soils and fallow lands, where CV reached 54%–91%. The C_{mic}/C_{org} ratio indicating the "quality" of SOC stocks ranged from 0.4 at the industrial zone to 4.1 at the mowed steppe (Table 3.6). Similarly to C_{mic}, BR, and qCO_2 the highest variability in C_{mic}/C_{org} was also shown for urban soils, where CV was double the natural ones (70% and 27%, respectively). The qCO_2/C_{org} ratio ranged from 12–140 µg CO_2-C mg^{-1} C_{mic} h^{-1}/g C_{org} g^{-1} at the virgin steppe and industrial zone correspondingly (Table 3.6). This ratio is a relevant indicator of the availability of SOC for microorganisms; when the ratio is high, the availability of SOC for microbes is limited (Dilly et al., 2001; Dilly, 2005). In our research the availability of SOC stocks in urban soils was almost 2.5 times less than for the undisturbed soils.

Based on the obtained results we conclude that the microbial biomass and the availability of SOC to microbes was substantially (from two to four times) less in the disturbed and urbanized Chernozems, compared to natural analogs. Apparently, continuous disturbance (i.e., introduction of plant species, withdrawal of plant residuals, overcompaction, alteration of soil structure, pollution, and perturbation of soil horizons) has a clear negative impact on the functioning of soil microorganisms and therefore decreases soil quality. We also found that topsoil CO_2 emission was higher in urban soils compared to agricultural and natural ones.

The outcomes obtained for the topsoil were later verified for the soil profile. The profile distribution of SOC and pH from the surface to 50 cm is presented in Table 3.7. The average subsoil (0–50 cm) urban SOC was lower compared to the virgin steppe, however pH in urban soil was

---

TABLE 3.6

Carbon-Use Efficiency and Quality of the Available Organic Matter in Soil (qCO_2/C_{org}) and the Portion of Carbon Microbial Biomass in Soil Organic Carbon (C_{mic}/C_{org}) in Typical Chernozems (0–10 cm) of Different Ecosystems and Functional Zones (Kursk Region) (Range/Mean, the Different Letters Indicate the Significant Difference, $p ≤ .05$, for Each Index Separately)

Ecosystem/Zone (Number of Sites)	qCO_2/C_{org}, µgCO_2-Cmg^{-1} $C_{mic}h^{-1}/$ g$C_{org}h^{-1}$ soil	C_{mic}/C_{org}, %
Steppe virgin (3)	12–25/19	2.9–4.1/3.3
Steppe mowing (4)	16–29/22	1.7–3.7/3.0
Pasture (3)	13–29/20	1.8–3.0/2.5
Natural (10)	*20 ± 6 b*	*2.9 ± 0.8 a*
Black fallow (2)	21–35/28	1.3–1.9/1.6
Arable (3)	46–73/61	0.8–1.7/1.2
Arable (5)	*48 ± 21 a*	*1.4 ± 0.4 b*
Recreational (3)	30–36/33	1.2–4.0/2.1
Residential (2)	29–54/41	2.5–3.6/3.0
Industrial (3)	37–140/61	0.4–1.0/0.7
Urban (8)	*51 ± 40 a*	*2.0 ± 1.4 ab*

one unit higher. The C_{mic} and BR decreased down the profile with the most abrupt decrease for the steppe soils (Figure 3.9). The 10 cm topsoil contributed respectively 27%–48% and 27%–34% to the average C_{mic} and BR estimated for the profile (0–50 cm). The qCO_2 didn't change significantly down the profile in the steppe and arable areas, whereas for the urban soils, the subsoil qCO_2 was from one third to three times higher that topsoil values (Figure 3.10). Urban subsoil (>20 cm) qCO_2 was 1.6–2.4 higher than those at the steppe sites. The qCO_2/C_{org} ratio also decreased down the profile, which is evidence of low effectiveness of subsoil SOC consumption by microorganisms. However, the qCO_2/C_{org} ratio in urban subsoil was on average from four to five times higher than in the steppe subsoil. At the same time the C_{mic}/C_{org} ratio in urban subsoil was on average three to five times lower than at the steppe subsoil. The highest C_{mic} stocks for the 0–50 cm layers were estimated for the steppe profile (206 g C_{mic} m^{-2}), whereas urban subsoil contained on average 82 g C_{mic} m^{-2}.

TABLE 3.7

Soil Organic Carbon Content (C_{org}) and Soil pH Value of Various Layers in Typical Chernozems of Different Ecosystems (Kursk Region)

Ecosystem (Number of Sites)	Layer, cm	Horizon	C_{org}, %	pH of Water
Steppe virgin (5)	0–10	A	5.10 ± 0.34	6.17 ± 0.53
	10–20	A	4.24 ± 0.42	6.13 ± 0.40
	20–30	A	3.62 ± 0.34	6.36 ± 0.54
	30–40	A	3.25 ± 0.23	6.45 ± 0.62
	40–50	A	2.96 ± 0.19	6.56 ± 0.49
Arable (3)	0–10	A$_{nax}$	2.67 ± 0.13	6.27 ± 0.46
	10–20	A$_{nax}$+ A	2.71 ± 0.23	6.03 ± 0.14
	20–30	A	2.45 ± 0.04	5.99 ± 0.24
	30–40	A	2.09 ± 0.25	6.19 ± 0.20
	40–50	A+AB	1.82 ± 0.17	6.30 ± 0.23
Urban (3)	0–10	A$_{ur}$/UR/A$_{ur}$	1.95 ± 0.55	7.93 ± 0.04
	10–20	A$_{ur}$/UR/UR	1.66 ± 0.36	8.08 ± 0.10
	20–30	UR/UR/UR	1.63 ± 0.28	8.11 ± 0.13
	30–40	UR/UR/UR	1.70 ± 0.27	8.16 ± 0.17
	40–50	AB$_{ur}$/UR/UR	1.74 ± 0.71	8.15 ± 0.15

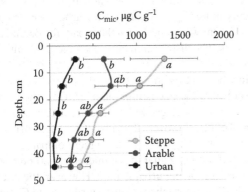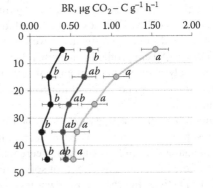

FIGURE 3.9 Distribution of soil microbial biomass carbon (C_{mic}) and basal respiration (BR) in various layers of typical Chernozems of steppe virgin, arable, and urban (Kursk region) ecosystems and functional zones. The different letters indicate the significant difference, $p \leq .05$, for each index and layer separately.

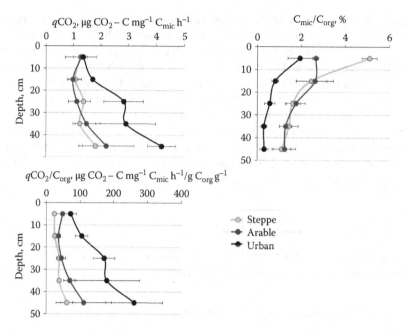

FIGURE 3.10 The specific microbial respiration (qCO_2) portion of microbial biomass carbon in soil organic carbon (C_{mic}/C_{org}), and C use efficiency and quality of the available organic matter in soil (qCO_2/C_{org}) in various layers of typical Chernozems of steppe virgin, arable, and urban (Kursk region) ecosystems.

The highest BR summarized for the 0–50 cm subsoil was estimated for the steppe and fallow lands (5.9 and 5.8 g CO_2-C m^{-2} d^{-1}), which was almost double that in urban soils (3.0 g CO_2-C m^{-2} d^{-1}). The contribution of the topsoil layer to the summarized C_{mic} and BR for the 0–50 cm profile were 66% and 52% for urban soils, which was substantially higher than for the other land-use types. The distribution of the average C_{mi},C and BR down the 0–50 cm profile is given in Table 3.8.

Further analysis of the profile distribution included depth down to 150 cm. The distribution of C_{mic} and BR down the profile is presented in Figure 3.11. Topsoil contributed 58%–79% and 51%–66% to the C_{mic} and BR summarized for the 0–150 cm profile. The largest difference between the topsoil and subsoil layers were shown for the virgin steppe, whereas C_{mic} and BR at the 100–150 cm layer of the urban soils was just 5–8 times less compared to the topsoil. This finding highlights the important role subsoil plays in the functioning of urban soils. The C_{mic} stocks summarized for the 0–150 cm profile ranged from 134 to 2618 µg C g^{-1} at the industrial zone and virgin steppe correspondingly. Undisturbed soil contained significantly more C_{mic} compared to disturbed ones (Figure 3.12). However, the BR values summarized for 0–150 cm were similar for the urban (residential and industrial) and natural (virgin and moved steppe) soils. The distribution of the average C_{mi},C and BR down the 0–150 cm profile is given in Table 3.9.

The analysis of the microbiological activity in Chernozems showed the substantial decrease in microbial biomass and BR, and the effectiveness of microbial consumption of SOC by the microorganisms in urbanized soils compared to natural ones. The negative influence of urbanization on soil microbiological activity was evident for the topsoil; however, the subsoil was less affected. Moreover, the contribution of urban subsoil to the profile C_{mic} stocks was higher compared to the natural analogs. Therefore, the rapid decrease in the quality and quantity of microbial carbon stocks is one of the instantaneous effects of soil disturbances following urbanization. However, the microbiological activity in urban subsoil was less disturbed, therefore the subsoil contribution to C stocks and fluxes in urban soil was higher, compared to the natural analogs. In contrast to the studied rapid influence of urban disturbance, this phenomenon may be the result of the urbanization effect over longer time periods, resulting in specific substrates, referred to as "cultural layers."

TABLE 3.8
Soil Bulk Density (SBD) of Various Layers of Typical Chernozems (0.5 m), Pool of Soil Microbial Biomass Carbon (C_{mic}), and Soil Microbial Production BR of Different Ecosystems (Kursk Region)

Ecosystem	Layer, cm	SBD, gcm^{-3}	C_{mic}, µg Cg^{-1}	BR, µg CO$_2$-C mg^{-1} h^{-1}	Pool (0.5 m) C_{mic}, gC$_{mic}$m^{-2}	BR, g CO$_2$-C m^{-2} d^{-1}
Steppe	0–10	1.05 ± 0.03	1024 ± 62	0.91 ± 0.08	108 ± 7	2.29 ± 0.20
	10–30	1.01 ± 0.07	327 ± 10	0.45 ± 0.13	66 ± 2	2.18 ± 0.63
	30–50	0.74 ± 0.04	215 ± 14	0.41 ± 0.04	32 ± 2	1.46 ± 0.14
Black fallow	0–10	0.98 ± 0.02	333 ± 46	0.60 ± 0.08	33 ± 5	1.41 ± 0.19
	10–30	1.01 ± 0.03	206 ± 15	0.49 ± 0.06	42 ± 3	2.38 ± 0.29
	30–50	1.00 ± 0.06	137 ± 3	0.43 ± 0.09	27 ± 1	2.06 ± 0.43
Urban	0–10	0.79 ± 0.01	681 ± 43	0.81 ± 0.04	54 ± 3	1.54± 0.08
	10–30	1.22 ± 0.42	81 ± 18	0.16 ± 0.02	20 ± 4	0.94 ± 0.12
	30–50	1.00 ± 0.06	39 ± 7	0.10 ± 0.02	8 ± 1	0.48 ± 0.10

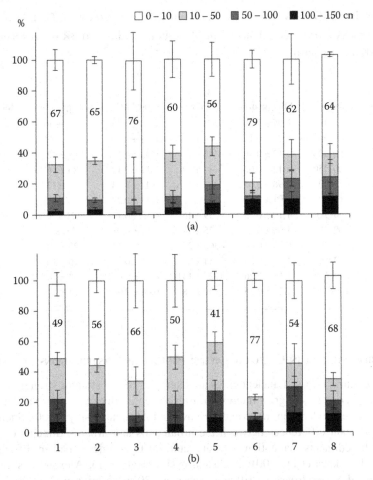

FIGURE 3.11 Contribution of various layers of typical Chernozems to total profile stocks of soil microbial biomass carbon (a) and basal respiration (b) for different ecosystems and functional zones: 1, steppe virgin ($n = 5$); 2, steppe mowing ($n = 4$); 3, pasture ($n = 3$); 4, black fallow ($n = 4$); 5, arable ($n = 3$); 6, recreational ($n = 3$).

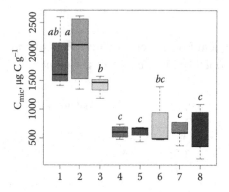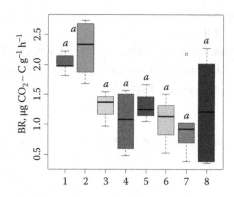

FIGURE 3.12 Distribution of soil microbial biomass carbon pool (C_{mic}) and basal respiration (BR) in typical Chernozems profile (1.5 m) for different ecosystems and functional zones: 1, steppe virgin ($n = 5$); 2, steppe mowing ($n = 4$); 3, pasture ($n = 3$); 4, black fallow ($n = 4$); 5, arable ($n = 3$); 6, recreational ($n = 3$); 7, residential ($n = 5$); 8, industrial ($n = 4$). The different letters indicate the significant difference, $p \le .05$.

TABLE 3.9

Soil Bulk Density (SBD) of Various Layers of Typical Chernozems (1.5 m), Pool of Soil Microbial Biomass Carbon (C_{mic}), and Soil Microbial Production BR of Different Ecosystems (Kursk Region)

Ecosystem (Number of Sites)	Layer, cm	SBD, gcm⁻³	C_{mic}, µg Cg⁻¹	BR, µg CO$_2$-C mg⁻¹ h⁻¹	Pool (1.5 m) C_{mic},gC$_{mic}$m⁻²	BR,g CO$_2$-C m⁻² d
Steppe virgin (5)	0–10	0.97 ± 0.20	1254 ± 413	0.95 ± 0.15	122 ± 40	2.21 ± 0.35
	10–50	0.88 ± 0.16	406 ± 152	0.53 ± 0.10	143 ± 54	4.48 ± 0.84
	50–100	1.12 ± 0.06	144 ± 21	0.35 ± 0.19	81 ± 12	4.70 ± 2.55
	100–150	1.12 ± 0.02	47 ± 29	0.12 ± 0.13	26 ± 16	1.61 ± 1.75
Black fallow (4)	0–10	0.98 ± 0.05	377 ± 136	0.47 ± 0.13	37 ± 13	1.11 ± 0.31
	10–50	1.01 ± 0.04	166 ± 4	0.33 ± 0.22	67 ± 2	3.20 ± 2.13
	50–100	1.10 ± 0.03	39 ± 15	0.17 ± 0.17	21 ± 8	2.24 ± 2.24
	100–150	1.06 ± 0.03	25 ± 14	0.11 ± 0.08	13 ± 7	1.40 ± 1.02
Urban (4)	0–10	0.76 ± 0.03	308 ± 258	0.71 ± 0.64	23 ± 20	1.30 ± 1.17
	10–50	1.19 ± 0.35	54 ± 22	0.12 ± 0.09	26 ± 10	1.37 ± 1.03
	50–100	1.11 ± 0.26	123 ± 141	0.13 ± 0.08	68 ± 78	1.73 ± 1.07
	100–150	1.20 ± 0.25	39 ± 18	0.08 ± 0.06	23 ± 11	1.15 ± 0.86

3.3.2 CARBON STORAGE IN CULTURAL LAYERS: HISTORICAL CONTRIBUTIONS OF URBANIZATION

The soils and cultural layers of ancient cities are examples of soil systems strongly transformed by anthropogenic activity. It is of great interest to study the entire thickness of urban deposits together with the soils buried under them (Alexandrovskaya and Alexandrovskiy, 2000). Such studies have shown the significant differences between the urban soils and urban sediments of ancient cities and towns (urbosediments, habitation deposits, cultural layers) developed in different landscape zones (Alexandrovskaya et al., 2001; Aleksandrovskii et al., 2015; Alexandrovskiy et al., 2012, 2013; Dolgikh and Alexandrovskiy, 2010; Golyeva et al., 2016; Mažeika et al., 2009; Mazurek et al., 2016; Sándor and Szabó, 2014).

Over the centuries, layers of habitation deposits (or archeological cultural layers) of average 3–5 m, in depressions up to 15–20 m in thickness have been formed in towns and cities of Europe. These specific deposits, partially transformed by pedogenesis and containing the profiles of weakly developed buried soils, as well as the underlying initial natural soils, may be referred to as urbosediments (Alexandrovskiy et al., 2012). They consist mainly of the remains of buildings constructed out of wood, stones, bricks, or other materials and may also include traces of manure and diverse municipal wastes. Their composition and morphology depend on the natural conditions of the territory. In the humid northern forest regions of Europe, the accumulation of wood remains, manure, and other organic substances is active. Thus, organic urbosediments are formed and the decomposition of organic matter in the deposits is hampered due to the high moisture content (Dolgikh and Alexandrovskiy, 2010). The formation of habitation deposits in the ancient cities in North and Central European Russia took place in two stages of predominant wooden construction (prior to the eighteenth century AD), followed by stone construction (Alexandrovskaya and Alexandrovskiy, 2000). The first stage is characterized by the accumulation of an urbo-organic layer composed of wooden chips, a peat-like mass from manure, and the remains of buildings made of wood (Figures 3.13 and 3.14). The stage of stone construction is characterized by the accumulation of an urban-mineral layer composed of the debris of bricks, stones, lime, sand, and clay with some inclusions of organic matter. Fragments of pottery and animal bones are present in both layers.

In the ancient and medieval towns in the semi-arid steppe zone of Europe, such deposits consist of remains of mudbrick or adobe made of loess. Antique cities in the south of European Russia are characterized by a considerable thickness of their cultural layers (urbosediments) accumulated as construction debris and household wastes. Under the impact of pedogenesis and weathering in the dry climate of the steppe zone, these sediments have acquired the features of loess-like low humus calcareous and alkaline deposits (Alexandrovskiy et al., 2015).

Over the course of time, these ancient urban deposits have been transformed as a result of the impact of pedogenetic processes developing under humid conditions in the North and under arid

FIGURE 3.13 Velikii Novgorod, the Troitsky-XIII, XIV archeological excavation (main historical area). The remains of the medieval woody pavement are seen in the foreground.

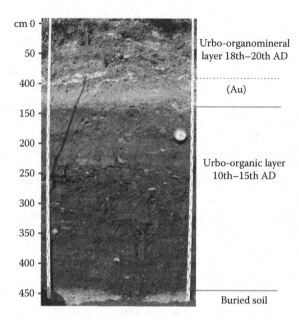

cm 0

50

400

(Au)

150

200

250

300

350

400

450

Urbo-organomineral
layer 18th–20th AD

Urbo-organic layer
10th–15th AD

Buried soil

FIGURE 3.14 Velikii Novgorod, the Troitsky-XIII archeological excavation (main historical area). Urbosediment profile.

conditions in the South. These deposits are formed under the impact of a set of processes of anthropogenic sedimentation combined with pedogenetic processes. The latter may differ from the pedogenetic processes that have shaped the initial (background) soils. As sedimentation processes are relatively fast, pedogenesis proper can only slightly transform the material of the cultural layer; later, it becomes buried at a considerable depth, where it is subjected to the action of deep soil and diagenetic processes differing from those in the upper part of the cultural layer. Thus, in ancient towns in the forest zone with humid climate and poor drainage conditions, thick urbo-organic layers, consisting of a peat-like mass saturated with woody remains, have formed, whereas in the steppe zone, urbo-mineral sediments have developed (Alexandrovskiy et al., 2012).

We observed two contrast cases: the urbosediments of Velikii Novgorod (since ninth century AD; 58.52500°N, 31.27500°E; forest humid zone, weak drainage, Northwest European Russia) and Phanagoria (sixth century BC to ninth century AD; 45.27694°N, 36.96611°E; semi-arid steppe zone; South of European Russia).

3.3.2.1   Organic Carbon in Urbosediments of a Humid Forest Zone

The morphology and property of urbosediments and the underlying buried soil in Velikii Novgorod were studied at two key sections (main area of the historical center—the Troitsky-XIII, XIV archeological excavation, 4–5 m thickness of urbosediments and periphery of the historical center—the Ilmensky excavation, 2–2.5 m of urbosediments) and are shown in Figures 3.15 and 3.16.

The soils buried under the habitation deposits in the historical center often include a plow horizon (Ap) with a thickness up to 12–15 cm. The urbo-organic sediments overlying the buried natural soil (OL, Figure 3.15) are saturated with water during a large part of the year. The raw organic material (peat-like mass) consists of the remains of herbs and straw (manure) and well-preserved wooden chips (Dolgikh and Alexandrovskiy, 2010). However, as the organic matter decomposes, an acidification takes place in this layer owing to the synthesis of organic acids. These urbosediments also contain inclusions of the medium loamy mineral matter with limestone debris, various artifacts, wooden pavements, and logs. In the central part of the city (Troitsky-XIII, XIV excavations),

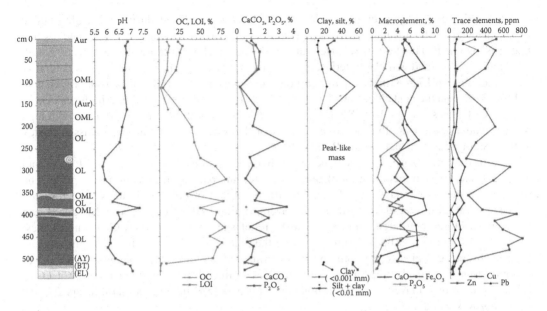

FIGURE 3.15 Velikii Novgorod, the Troitsky-XIIV excavation (main historical area). Structure and properties of urbosediments. LOI, loss on ignition; OC, organic carbon; OL, organic layer; OML, organo-mineral layer.

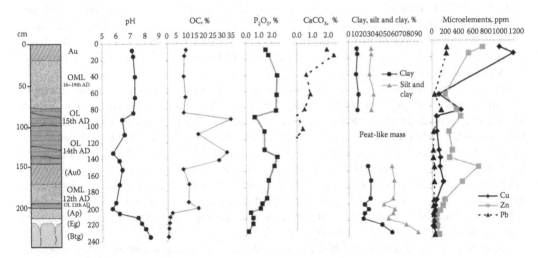

FIGURE 3.16 Velikii Novgorod, the Ilmensky excavation (periphery of historical center). Structure and properties of urbosediments.

the thickness of this waterlogging layer reaches 3–4 m, up to 5–6 m in paeodepressions. The portion of organic material in the organic layer varies from 50% to 90% (by volume). The organic horizons in the urban sediments prior to the eighteenth century AD have the most acid reaction in the entire profile (pH <5.8). The surface of aggregates is covered by films of kerchenite. These horizons are rich in phosphorus (1%–3%). Ash interlayers effervesce with HCl and are an important source of carbonates and carbon in the habitation deposits (Alexandrovskiy, 2007). The thickness of the organic layer in the periphery of historical center is 0.5–1.5 m and less than in the main area of historical center. Sometimes in areas with deep modern waterworks organic layers are not found. In these cases, they already have been transformed to an organo-mineral layer during modern times.

The overlying dark gray urbo-organic mineral layer accumulated during the eighteenth to twentieth century AD. It is much richer in mineral mass, although the content of humified organic matter is also high. This layer is relatively dry. The wooden chips in it are strongly decomposed; limestone, bricks, sand and clay interlayers, and lime are also present. The reaction is neutral to slightly alkaline (pH 7.5–8.0); the organic carbon content reaches 5%–15%. The phosphorus content is 1.5%–3.5%, and the content of $CaCO_3$ is 1.5%–8.0%. In the layers enriched in ash, the content of calcium carbonates reaches 40%. We observed similar properties in the urbosediments with high organic matter content in other old urban centers of the forest zone of European Russia, Rostov Velikii, Staraya Ladoga, and Staraya Russa.

Also in the forest zone, but under better drainage conditions (for instance, in Moscow and Yaroslavl), an active mineralization of organic matter and partial humification take place. The thickness of the organic layer is reduced, and the main part of urban sediments is represented by the mineral layer. Well-developed organic layers, similar to those described in Novgorod, can be found in Moscow in large depressions in the relief and in buried river valleys and ravines. They are also present on flat interfluves, where they alternate with plots in which the organic layer is absent. At higher topographic positions in the upland areas, the thickness of the initial organic layer is reduced, transformed into an organo-mineral or a humified mineral layer. In lower organo-mineral layers organic carbon content is 5%–15%, in upper mineral layers 2%–5% (Alexandrovskiy et al., 2012).

3.3.2.2 Carbon Stocks and Emission in an Ancient City with Thick Organic Urbosediments

Calculation of carbon stocks in urbosediments in the historical center of Velikii Novgorod was made on the basis of long-term geoarcheological research, geological drilling data (Kushnir, 1960), and archeological interpretations (Petrov and Tarabardina, 2012). The total carbon stocks in urbosediments of the historical center (4.5 km^2) were 1.95×10^6 t (4333 t C per ha). The obtained value is higher than SOC stocks estimated for the cultural layer of Moscow—6–7×10^6 t (Alexandrovskaya and Alexandrovskiy, 2000).

The comparative studies of soil carbon dioxide emissions were held in natural and anthropogenically transformed landscapes of the southern taiga (Novgorod region). Measurements of emission using the closed chambers method were carried out in the natural spruce forest, arable land, post-agrogenic meadow and forest, and residential areas of different ages in Veliky Novgorod. The soil CO_2 emissions in the city were influenced by the depth and SOC stocks in urban deposits. A 2.35 times reduction in soil emissions and decreasing carbon stocks were revealed from the historical center to modern residential areas. The maximum values of the soil CO_2 emissions in the southern taiga (Novgorod region) corresponded to the ancient residential areas of the city center (0.44–0.65 g C/(m^2 h)), whereas the minimum values were comparable to the modern agricultural arable lands (0.12 g C/(m^2 h)) (Dolgikh et al., 2015).

The hydromorphic organic layers of urbosediments (tenth to fifteenth century AD, raw organic material, organic carbon 25%–50%) in the historic center of the ancient cities can be important storage locations of adsorbed carbon dioxide. The results of the quantitative assessment and modeling of pulsed carbon dioxide emissions from hydromorphic organic urbosediments (Velikii Novgorod) showed that the thick urban habitation deposits accumulated significant amounts of carbon dioxide at the deep subsoil layers. At the first stages after the exposure of the cultural layer to the surface in archeological excavations, very high CO_2 emission values reaching 10–15 g C/(m^2 h^{-1}) have been determined (Figure 3.17). These values exceed the normal equilibrium emission from the soil surface by two orders of magnitude. However, they should not be interpreted as indications of the high biological activity of the buried urban sediments. A model based on physical processes shows that the measured emission values can be reliably explained by degassing of the soil water and desorption of gases from the urbosediments (Smagin et al., 2016).

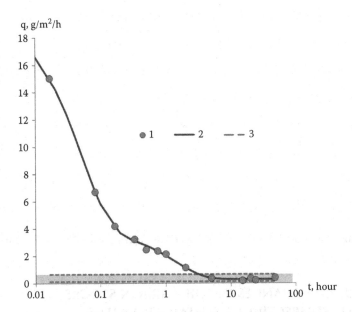

FIGURE 3.17 The dynamics of CO_2 emission from the exposed organic cultural layer (Velikii Novgorod): (1) real data, (2) model, and (3) background emission. (Adapted from Smagin, A.V. et al., *Eurasian Soil Science*, 49, 450–456, 2016.)

3.3.2.3 Organic Carbon in Urbosediments of a Semi-Arid Steppe Zone

Urbosediments of Pahanagoria (the largest ancient Greek city on the Taman peninsula, semi-arid steppe zone) were examined in the central part of this city. Their thickness reaches 5.5 m. After the city was abandoned in the ninth century AD, a well-developed chernozemic soil was formed in the surface part of the cultural layer. The soil buried under the cultural layer in the sixth century BC is a loamy sandy gray humus soil with ferruginous lamellae (Lamellic Arenosol). Urbosediments accumulated under the impact of anthropogenic activity differ from the natural sandy substrates; urbosediments are mainly derived from the debris of sun-dried bricks made of marine clay with some admixture of clay (Alexandrovskiy et al., 2012). The development of soils and urbosediments in ancient cities of European Russia was specified by the continuous accumulation of construction debris and other wastes, anthropogenic turbation of the material, arid climate, and the predominance of steppe vegetation. Under conditions of the dry climate of the Taman Peninsula coupled with good drainage and aeration of the cultural layer, the transformation of organic and mineral substances was very intense. Carbonates are present throughout, although their distribution in the profile is rather irregular. The organic carbon content is low (up to 1%), with a slight increase in the *Chernic* horizon (2.27%) developed from the urban sediments. The phosphorus content is higher than in the background natural soils (up to 1.2%) (Figure 3.18). The similar concentration of SOC is typical of urbosediments of other ancient cities of semi-arid steppe zone (Alexandrovskiy et al., 2015).

Thus, the urbosediments of ancient cities in the humid forest zone is a huge reservoir of organic carbon. Organic carbon and carbon dioxide from low water logging organic layers is available for the modern carbon cycle in the case of changing modern soil hydrological conditions in ancient cities. Organic carbon content in urbosediments of ancient cities in the semi-arid steppe zone is less than in topsoil and more than in subsoil horizons of background soils.

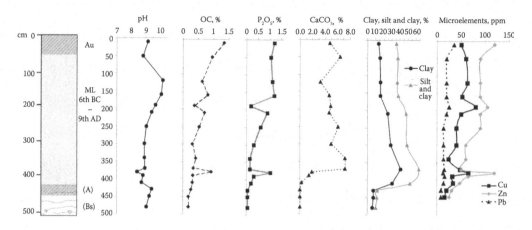

FIGURE 3.18   Properties of urbosediments in Phanagoria, the upper excavation. ML is the urbo-mineral layer.

3.4 CONCLUSION: CHANGES IN SOIL CARBON STOCKS FROM THE PERSPECTIVE OF SUSTAINABLE CITIES

The urban ecosystem is a relatively new ecological phenomenon. The first archeological settlements, recognized as towns, date back to 8000–10,000 years ago. The first megapolis with a population over 1 million, the city of Rome, achieved this threshold in the first century BC during the reign of Julius Caesar (Denisov et al., 2008). Only 16 cities in the world exceeded 1 million citizens by 1900. However, this number rapidly increased to more than 400 at the beginning of the twenty-first century (Berry, 2008). Urban ecosystems are more artificial than agricultural and natural systems. They are driven by humans and for humans, and anthropogenic factors have a predominant influence on such urban environments. These drivers allow the development of convenient and comfortable conditions for citizens, but as a drawback, sustainability of urban ecosystems and their components (air, water, vegetation, and soil) strongly depends on the decisions taken by humans, for example, through urban planning.

Historically, cities were planned mainly to support military defense, industry, and trade, which is vividly illustrated by the ring and sector structures of Moscow and London (Denisov et al., 2008) or block structure of New York (Pickett et al., 2011). With increasing industrialization this urban planning strategy resulted in multiple social and environmental problems, including soil, water, and air pollution, waste disposal, and traffic congestion. These problems constrained the quality of life in cities and created unfavorable conditions for residence. As a result, new ecological concepts of urban planning were developed, starting from the second half of the twentieth century (Jacobs, 1961; McHarg, 1969; Bacon, 1969). These new concepts highlighted the role of "green and blue spaces" (including parks, lawns, yards, gardens, rivers, streams, and ponds), and the functions and services they provide to people (Gómez-Baggethun and Barton, 2013). Major attention was given to planning small functional zones, following the ecological framework of patch dynamics in urban design (McGrath et al., 2008), rather than a traditional master plan of the settlement as a whole. This tendency in urban planning is defined by Pickett et al. (2011) as a shift "from sanitary city to sustainable city." A sustainable city is a social-ecological ecosystem, resilient to anthropogenic and environmental stresses (e.g., climate change) and providing ecosystem functions and services (Pickett et al., 2011).

Urban soil has a great but still poorly studied and rarely used potential to play a key role in the ecosystem of a sustainable city. Currently, urban soils face a paradox of being of the highest value regarding property and building issues, and being almost totally ignored with regard to the functions and ESs they can provide (Morel et al., 2014). According to a recent review by

Morel et al. (2014), urban soils provide from 9 to 15 ESs, depending on the surface features (sealed/open) and anthropogenic disturbance (pseudo-natural or engineered). The carbon stocks of urban soils directly or indirectly relate to at least five of these services: food production, nonfood biomass, global and local climate, and biodiversity. In this chapter, we focused on SOC stocks in urban soils and analyzed the factors and processes influencing their spatial variability and temporal dynamics.

Different driving factors were important for analyzing SOC stocks at different spatial scales. The global analysis (Section 3.2.1) highlighted the urbanization spots as a function of dominating soil type, biome, and climate. The regional mapping and modeling (Section 3.2.2) allowed analysis of the changes in SOC stocks under different urbanization scenarios. The positive net effect of urbanization on the regional C stocks highlights urban soils' potential to support C sequestration. To realize this potential, the best management practices should be implemented for urban soils at the local scale. The comparison of SOC stocks and CO_2 fluxes from different urban soil constructions (Section 3.2.3) is a promising background to develop such practices. Urbanization is a continuous process and the effects of urbanization change over time. We demonstrated that although the rapid disturbances coinciding with urbanization likely decrease microbial carbon and respiration (Section 3.3.1) the long-term residential activity may result in accumulation of substantial C stocks in the cultural layers as was shown for different climatic conditions in Section 3.3.2. Our findings can have a very clear addressee as, for example, in the Moscow region, where building and reconstruction processes take place on hundreds of square kilometers each day and the consequences of rapid urbanization on SOC stocks are still poorly investigated. They will be useful to many other cases worldwide, if future urbanization becomes more connected to sustainable urban planning.

ACKNOWLEDGMENT

The project was partly supported by Governmental Grant #11.G34.31.0079 and the Russian Foundation for Basic Research projects #15-34-70003 and 15-54-53117. The study of carbon storage in ancient cities was supported by the Russian Science Foundation, project no. 14-27-00133.

REFERENCES

Aleksandrovskii, A.L., Aleksandrovskaya, E.I., Dolgikh, A.V. et al. 2015. Soils and cultural layers of ancient cities in the south of European Russia. *Eurasian Soil Science.* 48(11): 1171–1181.

Alexandrovskaya, E.I., Alexandrovskiy, A.L., Gaidukov, P.G. et al. 2001. Woodland, meadow, field and town layout: Evidence from analyses of the earliest cultural deposits and buried soils in Novgorod. *Occasional Paper–British Museum.* 141: 15–21.

Alexandrovskaya, E.I. and Alexandrovskiy, A.L. 2000. History of the cultural layer in Moscow and accumulation of anthropogenic substances in it. *Catena.* 41: 249–259.

Alexandrovskiy, A.L. 2007. Pyrogenic origin of carbonates: Evidence from pedoarchaeological investigations. *Eurasian Soil Science.* 40: 471–477.

Alexandrovskiy, A.L., Alexandrovskaya, E.I., and Dolgikh, A.V. 2012. Pedogenetic features of habitation deposits in ancient towns of European Russia and their alteration under different natural conditions. *Bouletin de la Sociedad Geologica Mexicana.* 64(1): 71–77.

Alexandrovskiy, A.L., Krenke, N.A., Alexandrovskaya, E.I. et al. 2013. Early history of soils and landscapes of ancient Russian towns: To the 1150th anniversary of the Russian State. *Izvestiya of Russian Academy of Sciences, Geographical Series.* 5: 73–86.

Ananyeva, N.D., Susyan, E.A., Chernova, O.V. et al. 2008. Microbial respiration activities of soils from different climatic regions of European Russia. *European Journal of Soil Biology.* 44, 147–157.

Ananyeva, N.D., Susyan, E.A., and Gavrilenko, E.G., 2011. Determination of the soil microbial biomass carbon using the method of substrate-induced respiration. *Eurasian Soil Science.* 44: 1215–1221.

Anderson, J.P.E. and Domsch, K.H. 1978. A physiological method for the quantitative measurement of microbial biomass in soils. *Soil Biology Biochemistry.* 10: 215–221.

Anderson, T.H. and Domsch, K.H. 1985. Determination of ecophysiological maintenance requirements of soil micro-organisms in a dormant state. *Biology and Fertility of Soils.* 1: 81–89.

Andrews, S.S., Karlen, D.L., and Cambardella, C.A. 2004. The soil management assessment framework: A quantitative soil quality evaluation method. *Soil Science Society of America Journal*. 68(6): 1945–1962.

Aspinall, R. 2004. Modelling land use change with generalized linear models: Multi-model analysis of change between 1860 and 2000 in Gallatin Valley, Montana. *Journal of Environmental Management*. 72: 91–103.

Assessment of residential population on 1 January 2014. http://www.gks.ru.

Avetov, N.A., Alexandrovskii, A.L., Alyabina, I.O et al. 2011. *National Atlas of Russian Federation's Soils*. Dobrovolskii, G.V. and Shoba, S.A. (Eds.), p. 632. Moscow: Astrel.

Bacon, E. 1969. *Design of Cities*. New York, NY: Viking Penguin.

Balk, D.L., Deichmann, U., Yetman, G. et al. 2006. Determining global population distribution: Methods, applications and data. *Advances in Parasitology*. 62: 119–156. http://dx.doi.org/10.1016/S0065-308X(05)62004-0.

Bandaranayake, W., Qian, Y.L., Parton, W.J. et al. 2003. Estimation of soil organic carbon changes in turfgrass systems using the CENTURY Model. *Agronomy Journal*. 95: 558–563.

Bartalev, S.A., Belward, A.S., Erchov, D.V. et al. 2003. A new SPOT4-VEGETATION derived land cover map of Northern Eurasia. *International Journal of Remote Sensing*. 24(9): 1977–1982.

Bastida, F., Moreno Hernandez, J.M., and Garcia, C. 2006. Microbiological degradation index of soils in a semiarid climate. *Soil Biology and Biochemistry*. 38: 3463–3473.

Bastida, F., Zsolnay, A., Hernandez, T. et al. 2008. Past, present and future of soil quality indices: A biological perspective. *Geoderma*. 147: 159–171.

Batisani, N. and Yarnal, B. 2009. Urban expansion in Centre County, Pennsylvania: Spatial dynamics and landscape transformations. *Applied Geography*. 29: 235–249.

BBodSchG—Bundes-Bodenschutzgesetz—German Federal Soil Protection Act, published March 17, 1998.

Berry, B.J.L. 2008. Urbanization. In Marzluff, J., Shulenberger, E., Endlicher, W. et al. (Eds.), *Urban Ecology. An International Perspective between Humans and Nature*, pp. 25–48. New York, NY: Springer.

Blagodatskii, S.A., Bogomolova, I.N., and Blagodatskaya, E.V. 2008. Microbial biomass and growth kinetics of microorganisms in chernozem soils under different land use modes. *Microbiology*. 77: 99–106.

Blum, W.E.H. 2005. Functions of soil for society and environment. *Reviews in Environmental Science and Bio/Technology*. 4: 75–79.

Bolin, B., Degens, E. T., Kempe, S. et al. (Eds.). 1979. *The Global Carbon Cycle*. New York, NY: John Wiley.

Bontemps, S., Defourny, P., and van Bogaert, E. 2010. GLOBCOVER 2009, Products description and validation report. *ESA and Université Catholique de Louvain*. 53, doi:10.1594/PANGAEA.787668.

Breure, A.M., De Deyn, G.B., Dominati, E. et al. 2012. Ecosystem services: A useful concept for soil policy making! *Current Opinion in Environmental Sustainability*. 4(5): 578–585.

Carlyle, J.C. and Than, U.B. 1988. Abiotic controls of soil respiration beneath an eighteen-year old Pinus radiata stand in south-eastern Australia. *Journal of Ecology*. 76: 654–662.

CIESIN, Columbia University, IFPRI, the World Bank, and CIAT. 2011. Global Rural-Urban Mapping Project, Version 1 (GRUMPv1): Urban Extents Grid. Socioeconomic Data and Applications Center (SEDAC), Palisades, NY: Columbia University. http://sedac.ciesin.columbia.edu/data/dataset/grump-v1-urban-extents. Downloaded 23/2/2016.

Costanza, R., d'Are, R., de Groot, R. et al. 1997. The value of the world's ecosystem services and natural capital. *Nature*. 387: 253–260.

Creamer, R.E., Schulte, R.P.O., Stone, D. et al. 2014. Measuring basal soil respiration across Europe: Do incubation temperature and incubation period matter? *Ecological Indicators*. 36: 409–418.

Cruvinel, E.F., Bustamante, M.M.C., Kozovits, A. et al. 2011. Soil emission of NO, N_2O and CO_2 from croplands in the savanna region of central Brazil. *Agriculture Ecosystems and Environment*. 144: 29–40.

de Groot, R.S. 1992. *Functions of Nature: Evaluation of Nature in Environmental Planning, Management, and Decision Making*. Groningen: Wolters-Noordhoff.

Denisov, V.V., Kurbatova, A.S., Denisova, I.A. et al. 2008. *Ecology of a City*. Rostov on Don, Moscow (in Russian): MarT.

Dilly, O. 2005. Microbial energetics in soils. In Buscot, F. and Varma, A. (Eds.), *Microorganisms in Soils: Roles in Genesis and Functions*, pp. 123–138. Berlin: Springer.

Dilly, O., Blume, H.P., Sehy, U. et al. 2003. Variation of stabilized, microbial and biologically active carbon and nitrogen soil under contrasting land use and agricultural management practices. *Chemosphere*. 52: 557–569.

Dilly, O., Winter, K., Lang A. et al. 2001. Energetic eco-physiology of the soil microbiota in two landscapes of southern and northern Germany. *Journal of Plant Nutrition Soil Science*. 164: 407–413.

Dobrovolsky, G.V. and Nikitin, E.D. 2012. *Soil Ecology*. Moscow: Moscow University.

Dolgikh, A.V. and Alexandrovskiy, A.L. 2010. Soils and cultural layer in Velikii Novgorod. *Eurasian Soil Science*. 43: 477–487.

Dolgikh, A., Karelin, D., Kudikov, A. et al. 2015. Carbon stocks and fluxes in the ancient residential areas of the forest zone (European Russia). Geography, Culture and Society for Our Future Earth. Abstracts of International Conference of International Geographical Union. Moscow. 169.

Dubovyk, O., Sliuzas, R., and Flacke, J. 2011. Spatio-temporal modelling of informal settlement development in Sancaktepe district, Istanbul, Turkey. *ISPRS Journal of Photogrammetry and Remote Sensing*. 66: 235–246.

Egorov, V.V., Fridland, V.M., Ivanova, E.N. et al. (Eds.). 1977. *Classification and Diagnostics of Soils in USSR*. Moscow: Kolos.

Elvidge, C.D., Milesi, C., Dietz, J.B. et al. 2004. Paving paradise. *GeoSpatial Solutions*. 14: 58.

FAO. 1988. *Soils Map of the World: Revised Legend*. Rome: Food and Agriculture Organization of the United Nations.

Fromin, N., Porte, B., Lensi, R. et al. 2012. Spatial variability of the functional stability of microbial respiration process: A microcosm study using tropic forest soil. *Journal of Soils and Sediments*. 12: 1030–1039.

Golyeva, A., Zazovskaia, E., and Turova, I. 2016. Properties of ancient deeply transformed man-made soils (cultural layers) and their advances to classification by the example of Early Iron Age sites in Moscow Region. *Catena*. 137: 605–610.

Gomes-Casanovas, N., Matamala, R., Cook, D.R. et al. 2012. Net ecosystem exchange modifies the relationship between the autotrophic and heterotrophic components of soil respiration with abiotic factors in prairie grasslands. *Global Change Biology*. 18: 2532–2545.

Gómez-Baggethun, E. and Barton, D.N. 2013. Classifying and valuing ecosystem services for urban planning. *Ecological Economics*. 86: 235–245.

Gritsai, O. 2004. Global business services in Moscow: Patterns of involvement. *Urban Studies*. 41: 2001–2024.

Hamilton, J.G., DeLucia, E.H., George, K. et al. 2002. Forest carbon balance under elevated CO_2. *Oecologia*. 131: 250–260.

He, Y. and Zhang, G.-L. 2009. Historical record of black carbon in urban soils and its environmental implications. *Environmental Pollution*. 157: 2684–2688.

IPCC (Intergovernmental Panel on Climate Change). 2001. Climate change: The scientific basis. In Houghton, J.T., Ding, Y., Griggs, D.J. et al. (Eds.), *Contribution of Working Group I to the Third Assessment Report of the Intergovernmental Panel on Climate Change*. Cambridge, UK, New York, NY: Cambridge University Press.

Islam, K.R. and Weil, R.R. 2000. Land use effects on soil quality in a tropical forest ecosystem of Bangladesh. *Agriculture Ecosystems and Environment*. 79: 9–16.

Jacobs, J. 1961. *The Death and Life of Great American Cities: The Failure of Town Planning*. New York, NY: Random House.

Janzen, H.H. 2004. Carbon cycling in earth systems—A soil science perspective. *Agriculture, Ecosystems and Environment*. 104: 399–417.

Karlen, D.L., Ditzler, C.A., and Andrews, S.S. 2003. Soil quality: Why and how? *Geoderma*, 114(3–4): 145–156.

Kaye, J. P., R. L. McCulley and I. C. Burkez. 2005. Carbon fluxes, nitrogen cycling, and soil microbial communities in adjacent urban, native and agricultural ecosystems. *Global Change Biology*. 11: 575–587.

Kleinbaum, D.G. and Klein, M. 2010. *Logistic Regression: A Self-Learning Text*. New York, NY: Springer geocomputation. University of Southampton, UK.

Kogut, B.M., Sysuev, S.A., and Kholodov, V.A. 2012. Water stability and labile humic substances of typical chernozems under different land uses. *Eurasian Soil Science*. 45: 496–502.

Kolossov, V. and O'Loughlin, J. 2004. How Moscow is becoming a capitalist mega-city. *International Social Science Journal*. 56: 413–427.

Kolossov, V., Vendina, O., and O'Loughlin, J. 2002. Moscow as an Emergent World city: International links, business developments, and the entrepreneurial city. *Eurasian Geography and Economics*. 43: 170–196.

Kovda, V.A. 1983. Past and future of Chernozems. In Kovda, V.A. and Samoilova, E.V. (Eds.), *Russian Chernozems: 100 Years after Docuchaev*, pp. 253–280. Moscow: Nauka.

Kovda, V.A. and Rozanov, B.G. (Eds.). 1988. *Soil Science*. Moscow: V. Shkola.

Kushnir, I.I. 1960. Cultural layer of Novgrod. *Soviet Archaeology*. 3: 217–224 (in Russian).

Lal, R. 2004. Agricultural activities and the global carbon cycle. *Nutrient Cycling in Agroecosystems*. 70: 103–116.

Li, X., Zhou, W., and Ouyang, Z. 2013. Forty years of urban expansion in Beijing: What is the relative importance of physical, socio-economic, and neighborhood factors? *Applied Geography*. 38: 1–10.

Lorenz, K. and Lal, R. 2009. Biogeochemical C and N cycles in urban soils. *Environment International*. 35: 1–8.

MA—Millennium Ecosystem Assessment. 2003. *Ecosystems and Human Well-being: A Framework for Assessment*. Washington, DC: Island Press.

Mažeika, J., Blaževicius, P., Stancikaite, M. et al. 2009. Dating of the cultural layers from Vilnius Lower Castle, East Lithuania: Implications for chronological attribution and environmental history. *Radiocarbon*. 2: 515–528.

Mazurek, R., Kowalska, J., Gąsiorek, M. et al. 2016. Micromorphological and physico-chemical analyses of cultural layers in the urban soil of a medieval city—A case study from Krakow, Poland. *Catena*. 141: 73–84.

McGrath, B.P. 2008. *Digital Modelling for Urban Design*. London: Wiley.

McHarg, I. 1969. *Design with Nature*. Garden City, NY: Doubleday/Natural History Press.

Mikhailova, E.A. and Post, C.J. 2006. Organic carbon stocks in the Russian Chernozem. *European Journal of Soil Science*. 57: 330–336.

Morel, J.L., Chenu, C., and Lorenz, K. 2014. Ecosystem services provided by soils of urban, industrial, traffic, mining, and military areas (SUITMAs). *Journal of Soil and Sediments*. 15: 1659–1666.

Nortcliff, S. 2002. Standardization of soil quality attributes. *Agriculture, Ecosystems and Environment*, 88: 161–168.

Nowak, D.J. and Crane, D.E. 2001. Carbon storage and sequestration by urban trees in the USA. *Environmental Pollution*. 116: 381–389.

Olson, D.M., Dinerstein, E., Wikramanayake, E.D. et al. 2001. Terrestrial ecoregions of the world: A new map of life on earth (PDF, 1.1M). *BioScience*. 51: 933–938.

Petrov M.I. and Tarabardina O.A. 2012. Dynamics of changes of Novgorod territory in 10th-14th AD. *History and Archaeology of Pskov and Neighboring Territories*. 57: 139–147. (in Russian).

Pickett, S.T.A., Cadenasso, M.L., Grove, J.M. et al. 2011. Urban ecological systems: Scientific foundations and a decade of progress. *Journal of Environmental Management*. 92: 331–362.

Piotrowska-Dlugosz, A. and Charzynski, P. 2015. The impact of soil sealing degree on microbial biomass, enzymatic activity and physiochemical properties in the Ekranic Technosols of Torun (Poland). *Journal of Soils and Sediments*. 15:47–59.

Polyanskaya, L.M., Gorbacheva, M.A., Milanovskii, E. et al. 2010. Development of microorganisms in the chernozem under aerobic and anaerobic conditions. *Eurasian Soil Science*. 43: 328–332.

Pouyat, R.V., Yesilonis, I.D., and Nowak, D.J. 2006. Carbon storage by urban soils in the United States. *Journal of Environmental Quality*. 35: 1566–1575.

Qian, Y.L. and Follett, R.F. 2002. Assessing soil carbon sequestration in turfgrass systems using long-term soil testing data. *Agronomy Journal*. 94: 30–935.

Raciti, S.M., Hutyra, L.R., Rao, P. et al. 2012. Inconsistent definitions of "urban" result in different conclusions about the size of urban carbon and nitrogen stocks. *Ecological Application*. 22: 1015–1035.

Raich, J.W., Potter, C.S., and Bhagawati, D. 2002. Interannual variability in global respiration, 1980–1994. *Global Change Biology*. 8: 800–812.

Reilly, M.K., O'Mara, M.P., and Seto, K.C. 2009. From Bangalore to the Bay Area: Comparing transportation and activity accessibility as drivers of urban growth. *Landscape and Urban Planning*. 92: 24–33.

Rodionov, A., Amelung, W., Urusevskaja, I. et al. 2001. Origin of the enriched labile fraction (ELF) in Russian Chernozems with different site history. *Geoderma*. 102: 299–315.

Sándor, G. and Szabó, G. 2014. Influence of human activities on the soils of Debrecen, Hungary. *Soil Science Annual*. 65(1): 2–9.

Sarzhanov, D.A., Vasenev, V.I., Sotnikova, Y.L. et al. 2015. Short-term dynamics and spatial heterogeneity of CO_2 emission from the soils of natural and urban ecosystems in the Central Chernozemic Region. *Eurasian Soil Science*. 48: 416–424.

Schaldach, R. and Alcamo, J. 2007. Simulating the effects of urbanization, afforestation and cropland abandonment on a regional carbon balance: A case study for Central Germany. *Regional Environmental Change*. 7: 131–148.

Schulp, C.J.E. and Verburg, P.H. 2009. Effect of land use history and site factors on spatial variation of soil organic carbon across a physiographic region. *Agriculture, Ecosystems and Environment*. 133: 86–97.

Schulze, E.D. 2006. Biological control of the terrestrial carbon sink. *Biogeosciences*. 2: 147–166.

Seto, K., Güneralp, B., and Hutyra, L.R. 2012. Global forecasts of urban expansion to 2030 and direct impacts on biodiversity and carbon pools. *Proceedings of the National Academy of Sciences of the United States of America*. 109(40): 16083–16088. http://dx.doi.org/10.1073/pnas.1211658109.

Seto, K., Güneralp, B., and Hutyra, L.R. 2015. *Global Grid of Probabilities of Urban Expansion to 2030*. Palisades, NY: NASA Socioeconomic Data and Applications Center (SEDAC). http://dx.doi.org/10.7927/H4Z899CG. Accessed 23/2/2016.

Shein, E.V., Lazarev, V.I., Aidiev, A.Y. et al. 2011. Changes in the physical properties of typical chernozems of Kursk oblast under the conditions of a long-term stationary experiment. *Eurasian Soil Science*. 44: 1097–1103.

Shishov, L.L., Tonkonogov, V.D., Lebedeva, I.I. et al. 2004. *Classification and Diagnostics of Russian Soils*. Moscow (in Russian): Dokuchaev Soil Science Institute.

Smagin, A.V., Dolgikh, A.V., and Karelin D.V. 2016. Experimental studies and physically substantiated model of carbon dioxide emission from the exposed cultural layer. *Eurasian Soil Science*. 49(4): 450–456.

Stakhurlova, L.D., Svistova, I.D., and Shcheglov, D.I. 2007. Biological activity as an indicator of chernozem fertility in different biocenoses. *Eurasian Soil Science*. 40: 694–699.

Stoorvogel, J. J., Bakkenes, M., Temme, A. J. A. M., Batjes, N. H., and ten Brink, B. J. E. 2017. S-World: A global soil map for environmental modelling. *Land Degradation & Development*, 28: 22–33.

Stroganova, M.N., Myagkova, A.D., and Prokofieva, T.V. 1997. The role of soils in urban ecosystems. *Eurasian Soil Science*. 30: 82–86.

Swift, S. 2011. Sequestration of carbon by soil. *Soil Science*. 166(11): 858–871.

TEEB. 2010. *The Economics of Ecosystems and Biodiversity: Mainstreaming the Economics of Nature: A Synthesis of the Approach, Conclusions and Recommendations of TEEB*. Malta: Progress Press.

Valentini, R., Matteucci, G., Dolman, A.J. et al. 2000. Respiration as the main determinant of carbon balance in European forests. *Nature*. 404: 861–865.

Vasenev, V.I., Ananyeva, N.D., and Makarov, O.A. 2012. Specific features of the ecological functioning of urban soils in Moscow and Moscow oblast. *Eurasian Soil Science*. 45: 194–205.

Vasenev, V.I., Prokofieva, T.V., and Makarov, O.A. 2013b. Development of the approach to assess soil organic carbon stocks in megapolis and small settlement. *Eurasian Soil Science*. 6: 1–12.

Vasenev, V.I., Stoorvogel, J.J., Ananyeva, N.D. et al. 2015. Quantifying spatial-temporal variability of carbon stocks and fluxes in urban soils: From local monitoring to regional modelling. In Muthu, S.S. (Ed.), *The Carbon Footprint Handbook*, pp. 185–222. Boca Raton, FL: CRC Press.

Vasenev, V.I., Stoorvogel, J.J., and Vasenev, I.I. 2013a. Urban soil organic carbon and its spatial heterogeneity in comparison with natural and agricultural areas in the Moscow region. *Catena*. 107: 96–102.

Vasenev, V.I., Stoorvogel, J.J., Vasenev, I.I. et al. 2014a. How to map soil organic stocks in highly urbanized region? *Geoderma*. 226–227: 103–115.

Vasenev, V.I., Stoorvogel, J.J., Vasenev, I.I. et al. 2014b. How to map soil organic carbon stocks in highly urbanized regions? *Geoderma*. 226–227(1): 103–115.

Vodyanitskii, Y.N. 2015. Organic matter of urban soils: A review. *Eurasian Soil Science*, 48: 802–811.

Wei, Z., Wu, S., Yan, X. et al. 2014. Density and stability of organic carbon beneath impervious surfaces in urban areas. *PLOS ONE*. 9(10): e109380.

World Reference Base for Soil Resources. (2014). World soil resources reports, 106. Rome: FAO UNESCO.

Zircle, G., Lal, R. and Augustin, B. 2011 Modelling carbon sequestration in home lawns. *Hortscience*. 46: 808–814.

4 Drivers of Urban Soil Carbon Dynamics

Tara L.E. Trammell, Susan Day, Richard V. Pouyat, Carl Rosier, Bryant Scharenbroch, and Ian Yesilonis

CONTENTS

4.1 INTRODUCTION

Soils contain significant carbon (C) stocks in the form of soil organic carbon (SOC) and it appears that urban soils may provide an important C sink, at least for many regions (Scharenbroch et al. 2018). Consequently, there is an increasing interest in exploiting the potential of soil as a C sink, especially in urban areas where C volatilized during land-use change or another disturbance and has the potential to be recovered or even increased. While anthropogenic activities can increase variability in soil C stocks across urban landscapes (Scharenbroch et al. 2018), there is also evidence supporting convergence of urban soil C stocks across cities (Pouyat et al. 2015). Numerous anthropogenic factors drive SOC accumulation in urban landscapes. This chapter provides a conceptual model summarizing the major direct and indirect anthropogenic drivers of urban soil C dynamics and provides an in-depth discussion of how these drivers influence C stabilization and storage within the soil environment. A thorough assessment of the various mechanisms controlling SOC loss and accumulation within urban landscapes can both improve stewardship of urban soil C resources as well as increase global soil C model accuracy, which is especially important as urban landscapes continue to expand (Seto et al. 2011, 2012).

4.1.1 ANTHROPOGENIC DRIVERS OF SOC ACCUMULATION

While urban soils experience unique anthropogenic influences, proximate controls on soil organic matter (SOM) stabilization and cycling are universal. The contemporary view of SOM stabilization and long-term storage suggests that microbial ecology, environmental conditions, and the physical complexity of the soil matrix are controlling drivers. This new insight differs significantly from the traditional consensus that soil C storage is largely driven by the accumulation of complex recalcitrant compounds in soil over time (Lehmann and Kleber 2015; Stockmann et al. 2013). Furthermore, the activity of the microbial community has been identified as a critical factor driving long-term stabilization of SOM. Research findings suggest that long-term compound residence time within the soil environment is microbially derived; yet, chemical complexity of litter inputs also remains an important determinant of short-term decomposition (Stockmann et al. 2013).

The aim of this chapter is to introduce a conceptual model of urban SOC accumulation that illustrates this new paradigm shift in long-term soil C accumulation and stabilization, representing short-term decomposition as controlled by organic matter (OM) inputs to soil and long-term soil C stabilization as mediated by microbial community activity (Figure 4.1). Numerous direct and indirect anthropogenic drivers in urban and suburban areas affect the chemical composition of C inputs to soil, which influences rapid C cycling and microbial community structure, hence controlling slow C cycling. Some anthropogenic drivers, such as black carbon (BC) inputs and land-use and land-cover change (LULCC), directly influence the total soil C pool and are also

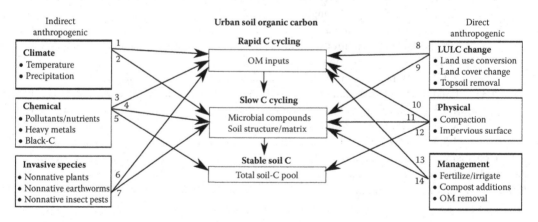

FIGURE 4.1 Conceptual model depicting the *indirect* and *direct* anthropogenic drivers of urban soil C dynamics. Anthropogenic activities indirectly alter both rapid and slow soil C cycling via changes in urban climatic conditions (temperature and moisture) and chemical inputs (pollutant and nutrient concentrations), as well as facilitate the spread of nonnative invasive species. Anthropogenic activities alter rapid and slow soil C cycling via direct modifications to the soil environment by land-use and land-cover change, transformation of physical soil properties, and management inputs to the soil.

included in the model. The discussion of direct and indirect drivers of SOC accumulation focuses on anthropogenic-derived controls within urban environments (e.g., the urban heat island [UHI]) and not on natural controls (e.g., rain storm events).

4.1.1.1 Indirect Anthropogenic Drivers

The conceptual model presented herein illustrates how indirect anthropogenic drivers alter urban soil C dynamics (Figure 4.1). Through increased fossil fuel consumption and land-use conversion, humans have altered urban climatic conditions (i.e., temperature and moisture) and chemical inputs (e.g., pollutant and nutrient concentrations), and facilitated the spread of nonnative invasive species, shifting urban soil conditions. Urban climates influence OM inputs to soil and microbial activity via UHIs and subsequent shifts in soil moisture and precipitation regimes (Oke 1973; Bonan 2002; Figure 4.1, arrows 1 and 2). Elevated nutrients (Lovett et al. 2000; Rao et al. 2014), heavy metals (Kumar and Hundal 2016; Yesilonis et al. 2008, and many others), and other pollutants (e.g., O_3) can indirectly alter the chemical composition of C inputs to soil and microbial activity (Figure 4.1, arrows 3 and 4). Anthropogenic activities can also directly alter soil C pools through BC inputs (Edmondson et al. 2015; Liu et al. 2011) and carbonate formation (Washbourne et al. 2012, 2015; Figure 4.1, arrow 5). The addition of nonnative species in urban landscapes (McKinney 2004) affects soil processes and C dynamics (Ehrenfeld 2010). Human-introduced invasive plants, soil microorganisms, and plant pests alter plant composition and subsequent plant chemical composition (Figure 4.1, arrow 6), in turn altering soil microbial community structure and activity (Figure 4.1, arrow 7).

4.1.1.2 Direct Anthropogenic Drivers

Anthropogenic activities alter urban soil conditions via LULC conversion (Figure 4.1, arrows 8 and 9). Urban and suburban land development create less permeable surfaces, reduce vegetation cover (Seto et al. 2012), and remove/move significant quantities of soil (e.g., Trammell et al. 2011), while human landscape decisions (e.g., xeriscaping) may convert land cover in urban landscapes. Anthropogenic activities directly influence soil microbial activity through transformed soil physical properties, such as soil compaction (Jim 1998; Yang and Zhang 2015) and impervious surface coverage (Figure 4.1, arrow 11). Management activities can vary from soil amendments (e.g., water, nutrients, and compost) to OM removal (i.e., litter/woody debris removal, lawn clippings), which directly affect the quantity and quality of SOM inputs as well as soil microbial activity (Figure 4.1, arrows 13 and 14).

The objective of this chapter is to discuss how anthropogenic drivers alter OM-C inputs and soil microbial activity, which control rapid and slow C cycling in urban soils, respectively. Since many anthropogenic drivers simultaneously impact soil C dynamics, these combined effects on soil C and the potential net effect on whether urban soil C is lost or stabilized are also discussed. Finally, a synopsis of the expected magnitude of impact (i.e., relative importance) and uncertainty (i.e., relative unknown impact) of these anthropogenic drivers on urban soil C dynamics is also provided.

4.2 INDIRECT ANTHROPOGENIC DRIVERS

Indirect anthropogenic drivers for this chapter are defined as the factors that control soil C dynamics via human actions that change the climatic (e.g., UHI) and chemical (e.g., heavy metals, BC) environment of urban areas, as well as the spread of nonnative species into urban ecosystems (e.g., nonnative plants and earthworms) (Yesilonis and Pouyat 2012).

4.2.1 URBAN CLIMATE

Urban landscapes experience altered climatic conditions, i.e., temperature and precipitation, due to human activities (Bonan 2002). It has long been recognized that urban environments are warmer on average than the surrounding rural areas, a phenomenon termed the "urban heat island" (Oke 1973). Cities are warmer due to increased heat-generating activities (e.g., fossil fuel burning) and reduced heat losses (less evaporative cooling surface) (Taha 1997; Botkin and Beveridge 1997; Zhao et al. 2014). While the UHI effect in cities has been well studied, the urban impact on precipitation is complex and is now a focus for atmospheric scientific research (Shepherd 2005; Song et al. 2016). Numerous studies have identified an "urban rainfall effect" (Shepherd 2005; Shem and Shepherd 2009; Lacke et al. 2009), where urban areas experience increased rainfall, snowfall, and convection storm events compared to nearby rural areas (Taha 1997; Shem and Shepherd 2009; Niyogi et al. 2011). Potential causes for increased precipitation have been linked to atmospheric aerosols, which act as cloud condensation nuclei (van den Heever and Cotton 2007; Lacke et al. 2009), as well as building density and heterogeneity and the UHI effect, which impact airflow and the urban boundary layer (Shepherd 2005; Song et al. 2016). These altered temperatures and precipitation regimes in urban environments affect plant productivity and thus OM inputs and microbial activity, all of which strongly influence urban soil C dynamics.

4.2.1.1 Temperature

The UHI effect is linked with changes in plant phenology, including timing and duration of canopy leaf out, leaf budburst, and flowering (Neil and Wu 2006; Jochner and Menzel 2015; Öztürk et al. 2015; Chen et al. 2016). As phenological response to temperature increase varies by species (Xu et al. 2016), the UHI effect can also affect species competition and influence the spread of invasive species. Longer growing seasons may increase aboveground vegetative growth, thereby increasing OM inputs to the soil environment. In addition to changes in phenology, the UHI effect has the potential to accelerate the rate of photosynthesis. However, it is difficult to separate the effects of the UHI on plant productivity in the urban environment where numerous factors, such as ozone, CO_2, and N deposition, influence plant growth (Gregg et al. 2003; Shen et al. 2008). Nonetheless, previous studies have documented increased productivity, i.e., aboveground biomass and plant height, in urbanized landscapes relative to surrounding rural areas (Ziska et al. 2003, 2004), and this increase in OM inputs may be an important control on rapid soil C cycling in urban soils (Figure 4.1, arrow 1). However, since the greatest differential temperature between urban and rural environments occurs at night when plant stomata are closed, the relatively greater impact of temperature alone on plants is most likely through changes in plant phenology than on plant productivity (Yesilonis and Pouyat 2012).

Increased temperatures alter the soil environment and soil microbial activity and processes, e.g., decomposition and N mineralization (Figure 4.1, arrow 2). Soil microbial activity responds directly to increased soil temperatures, enhancing decomposition rates of SOM (Kirschbaum 1995; Pietikäinen et al. 2005; Craine et al. 2010). Similarly, *in situ* net N mineralization and nitrification increased significantly under experimentally elevated temperatures (Verburg and van Breemen 2000; Xu et al. 2010), and net N mineralization remained elevated after 7 years of continuous experimental warming (Butler et al. 2012). These studies suggest that soil microbial activity and soil ecosystem processes in urban landscapes may be accelerated due to the UHI effect. In remnant forests along an urban–rural gradient, soil N mineralization rates may be greater in the urban compared to rural forest soils, which is partially attributed to higher temperatures in the urban forests (Pavao-Zuckerman and Coleman 2005; Chen et al. 2010; Enloe et al. 2015). Similarly, warmer temperatures may partially explain the faster litter decay rates in urban compared to rural forests along the NYC urban–rural gradient (Pouyat and Carreiro 2003). While it is expected that elevated temperatures will alter microbial activity in managed urban landscapes, such as lawns (Bijoor et al. 2008), it is difficult to separate the effects of management (e.g., fertilization) from warmer temperatures on soil C dynamics. In unmanaged urban ecosystems, the effect of warmer temperatures on increased microbial activity can result in greater decomposition and CO_2 flux potentially accelerating the rate of soil organic C loss. However, net soil C loss will depend on potential compensation by plant C gains and/or other anthropogenic factors that may restrict microbial activity in urban ecosystems.

4.2.1.2 Precipitation

While changes in precipitation could have substantial impacts on plant productivity and soil microbial activity in urban environments, few studies have investigated altered precipitation effects on urban ecosystems. A study conducted in a tropical forest fragment of São Paulo, Brazil found a strong positive correlation between precipitation and litter fall production and a positive correlation with decomposition rates (Ferreira et al. 2014). While this is typical of growing season conditions in the tropics, it does suggest that increased precipitation may have a greater influence on OM inputs relative to decomposition rates and would increase soil C over time within urban forests. However, separating the effects of temperature, precipitation, and other urbanization factors, such as nutrient deposition, on urban soil C dynamics is difficult (e.g., Chen et al. 2013b).

4.2.2 Urban Chemical Environment

Human activities in urban areas alter the chemical environment of soils by elevating nutrients (e.g., N deposition), increasing pollutant concentrations (e.g., O_3), intensifying heavy metal contaminants (e.g., Pb, Cd), and enhancing BC inputs, all of which can exert strong influences on the chemical composition and quantity of C inputs to soil and microbial community structure and function.

4.2.2.1 Urban Air Pollutants and Nutrients

Urban landscapes experience altered chemical inputs via gases (e.g., CO_2) and wet/dry deposition, which may function as pollutants and/or nutrients in urban ecosystems. Urban precipitation contains greater concentrations of inorganic nitrogen (N), calcium (Ca), and magnesium (Mg) that may have beneficial effects on plants and microbes via increased nutrient availability; however, the increase in hydrogen ions (acidity) in precipitation may be detrimental (Lovett et al. 2000; Carreiro et al. 2009). In addition to atmospheric deposition, urban environments experience elevated carbon dioxide (CO_2), nitrogen oxides (NO_x), ozone (O_3), and other air pollutants in the atmosphere. Nitrogen oxides, NO and NO_2, either singly or in combination with O_3 may depress plant growth, while some studies indicate NO_2 and NO_x can stimulate plant growth (Sparks 2009). CO_2 may enhance plant growth, yet in combination with O_3 in cities these gases may have a negative or neutral influence on plant growth. Volin et al. (1998) demonstrated that elevated CO_2 decreased stomatal conductance, which can protect leaves from O_3 damage. How altered chemical inputs act synergistically to affect urban plant growth is not fully understood (Calfapietra et al. 2015). In a study of potted cottonwood saplings placed in urban and rural locations throughout NYC, Gregg et al. (2003) found that saplings placed in the city grew more than saplings placed in rural areas. However, this increased urban sapling growth was not due to factors stimulating the urban trees (greater urban N deposition and CO_2), but due to detrimental effects of greater O_3 in rural areas downwind of the city (Gregg et al. 2003). The combined effects of urban air pollutants on aboveground vegetation will in turn alter belowground soil conditions and processes. If vegetation growth is stimulated, then increases in aboveground plant production will increase C inputs to the belowground environment (Figure 4.1, arrow 3). This may increase soil C pools by increasing soil OM quantity or may decrease soil C pools by stimulating decomposition. If air pollutants increase complex phenolic components of SOM, then rates of decomposition may be slowed and more soil C will be stabilized (Findlay et al. 1996). Atmospheric pollutants have short-term effects on urban soil C dynamics via impacts on plant productivity and litter quality.

While elevated gases (CO_2, O_3, and NO_x) and altered wet and dry deposition chemistry (e.g., N deposition) are expected to impact soil microbial activity in urban environments and subsequent long-term effects on soil C dynamics (Figure 4.1, arrow 4), few studies have directly investigated these chemical effects on soil processes in urban ecosystems. One study in tropical urban forests found that high soil N availability was associated with greater C and P acquiring enzyme activity, demonstrating a link between greater N availability and microbial activity in urban forests (Cusack

2013). The consensus suggests that N deposition increases soil N processes (N mineralization and nitrification) and N availability for microbial uptake (Pardo et al. 2011), which could increase C loss as CO_2 from soils. However, N-rich urban environments may enhance C accumulation due to enhanced plant biomass increasing soil C inputs. Furthermore, N additions were shown to depress ligninolytic enzyme activity, a potential mechanism explaining decreased litter decay rates in enriched N conditions (Carreiro et al. 2000). How N deposition will affect plant productivity and belowground processes will depend on responses to elevated CO_2 and O_3 in urban areas. A meta-analysis of 117 CO_2 enrichment studies demonstrated that elevated CO_2 stimulates microbial activity, yet soil C content increases due to greater relative increases in above- and belowground plant biomass and subsequent soil C inputs, and only with high N availability (De Graaff et al. 2006). Since urbanized regions are N-rich environments, it is possible that greater soil C content will occur in response to elevated CO_2 concentrations. However, while plant productivity is stimulated by N deposition and CO_2, increased O_3 typically dampens plant growth. Elevated O_3 concentrations can decrease soil microbial rhizosphere activity, suggesting microbes are impacted by detrimental effects of O_3 on plants rather than elevated O_3 directly influencing soil microbes (Chen et al. 2009). Furthermore, elevated O_3 shifts soil microbial structure and function and suppresses genes involved in soil N processes (He et al. 2014). The cumulative effects of altered wet and dry deposition as well as gases have the potential to stimulate or depress soil C pools over time, yet this will depend on how plants and microbes integrate these multiple factors.

4.2.2.2 Heavy Metals

Soil heavy metal deposition (i.e., Pb, Cd, Zn, Ni, and Cu) is toxic to humans, soil organisms, and vegetation (Adriano 1986). Industrial stacks (Govil et al. 2001; Kaminski and Landsberger 2000), vehicle emissions (Van Bohemen and Janssen van de Laak 2003; Zhang 2006), and paint (Trippler et al. 1988) are major sources of metal contamination. Even though metal concentration variability can be significant in urban soils, mean concentrations as elevated as 231 for Pb, 1.06 for Cd (Yesilonis et al. 2008), 156 for Zn (Bityukova et al. 2000), and 68 for Cu (Madrid et al. 2002) have been observed in urban soils. Metals interact with both abiotic and biotic components of soil. For example, abiotic processes complex metals to oxides, carbonates, OM, and clays. However, more importantly for soil C sequestration, metals in high concentrations may deleteriously affect the biotic communities and their function, such as vegetation and microbial community structure/ function (Kim et al. 2015).

The effect of metals on soil C sequestration is not entirely related to total metal concentration, but more importantly, the bioavailable concentration which is potentially toxic to biotic communities (Kim et al. 2015). For instance, bioavailable concentrations may harmfully compromise specific clade(s) of soil organisms, but at the same time not affect the overall functional capacity of groups, such as decomposers, shredders, predators, and microbial communities (Balsalobre et al. 1993; Berdicevsky et al. 1993; Romandini et al. 1992). Therefore, the soil system is less likely to collapse due to the functional redundancy (i.e., one species functionally substitutes for another) of the soil biotic community (Abu-Thiyab et al. 2012; Persiani et al. 2008; Rousk et al. 2009; Salminen et al. 2010). Long-term stress, however, can affect the tolerance of the soil biotic community to metals (Salminen et al. 2001). The chemical and physical properties of soil (e.g., pH, concentration of OM, iron oxides, forms of existing OM, soil texture, soil structure, redox status, and soil fertility) have an effect on the bioavailability of metals (Kim et al. 2015; Lorenz et al. 1994; Yesilonis et al. 2008). These factors influence the availability of resources and diversity of microorganisms that affect OM turnover as well as the molecular structure of OM (Giller et al. 1998). Toxicity thresholds for soil microflora are from 1,000 to 10,000 mg kg^{-1} metal cation (Smith 1990). Earthworms can persist in soils with 4,800 mg kg^{-1} metal cation (Bisessar 1982). Both of these values are above typical urban soil metal averages. Many studies using *in situ* contaminated soils found a poor relationship between the soil total metal concentration and earthworm accumulation due to factors such as pH,

OM content, and fine-sized particle content (Hopkin 1989; Ma 1982; Morgan and Morgan 1988; Spurgeion and Weeks 1998). In addition to influencing soil organisms, metals can also influence plant growth, reducing soil C inputs. Studies have shown that plant growth can be significantly reduced with high application rates of metals in soil experiments. For example, Athar and Ahmad (2002) found that the phytotoxic effect of heavy metals on the growth of wheat (*Triticum aestivum L.*) in a pot study for Cd at a concentration of 6.2 mg kg^{-1}. This concentration is much higher than what is found in a typical urban soils; Baltimore had a mean of 1 mg Cd kg^{-1} soil. Furthermore, the dry weight of shoot and root was decreased by 26% and 29%, respectively, at a Pb concentration of 97 mg kg^{-1}, which is lower than the mean Pb soil concentration found in many urban soils (e.g., Yesilonis et al. 2008; Madrid et al. 2002; Pouyat and McDonnell 1991).

Heavy metals affect both soil organism function and vegetation growth; however, the net effect on soil C sequestration is unknown due to the complex interactions between the soil matrix, bioavailable metal contamination levels, potential chemical reactions, and biotic communities. Studies examining extremely contaminated sites (e.g., close vicinity of smelters) observed increased OM on forest soil surfaces due to inhibition of soil microbial and fauna activity (e.g., Freedman and Hutchinson 1980). Aside from these toxic sites, there are studies that observed an increase in respiration rate possibly due to increases in microbial substrate utilization efficiency and maintenance energy requirement (Bååth 1989). Results of a meta-analysis of 160 studies suggest that within metal-polluted soils, SOC concentration decreased 5% relative to control soils, which may be related to a 50% decrease in plant biomass (Zhou et al. 2016). Furthermore, this meta-analysis revealed that metal contamination decreased SOC, yet varied with soil properties and climate. The analysis showed a pollution effect in soils high in initial SOC (>2%) compared to those with low initial SOC; soils that were acidic (pH < 6.5) and coarse-textured (sand) compared to fine-textured soils (i.e., clay). In other words, lower SOC percentages, higher pH values, and finer textures did not exhibit negative effects of metal concentrations on SOC relative to control treatments. Additionally, a climate condition effect had the greatest impact on SOC when mean annual temperatures (MATs) were between 10°C and 20°C (10.4%); however, when temperatures were less than 10°C or greater than 20°C, heavy metals had no effect on SOC. In addition, humid climates with mean annual precipitation (MAP) of greater than 1,000 mm showed the greatest response to metals compared to control treatments. Finally, the analysis suggested that lower respiration rates resulted in a buildup of soil C. However, this was offset by greater reduction in plant biomass causing an overall decrease in SOC.

Polluted urban soils with high bioavailable concentrations of heavy metals have a negative effect on soil C sequestration. Metals can influence the rate at which C is sequestered in the soil by affecting the communities that decompose and recycle SOC and through decreased plant biomass production. In general, plant biomass is more susceptible to soil metal concentration than microbial or microorganism populations at typical urban soil metal concentrations, which has the potential to decrease SOC. When the levels are extremely high, such as a Superfund site, the decomposer populations are affected more than the vegetation, which increases SOC.

4.2.2.3 Black Carbon

Globally, BC represents a significant slow cycling soil C pool, largely composed of soot and char resulting from the incomplete combustion of fossil fuels and vegetative biomass (Goldberg 1985). Current estimates suggest that BC contributes significantly to the total SOC pool (~20%–45% [depending on soil type]), despite BC's overall limited production rate (0.05–0.27 Pg yr^{-1}; Kuhlbusch and Crutzen 1995). However, a general understanding of the conditions that affect BC turnover is not fully realized. Preston and Schmidt (2006) suggested that BC decomposes extremely slowly, with turnover on millennial timescales (5–7 ky), whereas Bird et al. (1999) calculated BC's half-life < 100 years, and still others suggested that microbial usage of BC could further reduce BC residence time (10–100 years) (Steinbeiss et al. 2009; Zimmerman 2010). Due to BC's possible influence on soil C storage, largely based on BC's perceived long-term soil residence, ongoing research has focused on establishing connections between BC stability and soil environmental conditions (e.g., temperature, moisture, and texture), in turn identifying potential mechanisms influencing BC movement in soils. However, scientific understanding of BC soil dynamics remains incomplete. Furthermore, few studies have evaluated how BC accumulation and cycling in urban environments affect urban soil C sequestration and storage.

Direct inputs of BC represent a potentially important source of increased soil C storage (Figure 4.1, arrow 5), especially within urban landscapes, as urban soils are susceptible to large quantities of atmospheric BC deposition (Latha and Badarinath 2003; Mallet et al. 2004). BC has been identified as a significant proportion of total organic C stocks (28%–39%) in urban soils (<1 m) across a heavy industrial gradient, and the presence of trees may greatly increase the potential capture and accumulation of BC inputs (Edmondson et al. 2015). He and Zhang (2009) suggested that soil BC contents and the ratio of BC to SOC reflected different processes of human activities as well as pollution sources in urban areas; their results indicated that BC within urban soils was the result of traffic emissions while rural soils were more likely to accumulate BC from historic coal usage. Liu et al. (2011) found that fossil fuel combustion was the primary source of soil BC; however, the accumulation of BC in rural areas (i.e., sink) was largely controlled by regional-scale atmospheric deposition from urban areas. In contrast, Hamilton and Hartnett (2013), using soil BC concentrations combined with isotopic analysis, found that soot BC pools within urban desert soils were significantly enriched by both nonlocal (plant biomass) and local (fossil fuel) components.

Few if any studies have evaluated the influence of fossil fuel-derived BC (predominate form within urban setting) on either rapid or slow C cycling; however, a significant amount of research has been directed at understanding the effect of biomass-produced BC on soil C cycling. Several recent observations suggest that a portion of biomass BC is gradually (i.e., months–years) utilized biologically, contributing to the slow C-cycling pool (Baldock and Smernik 2002; Hilscher et al. 2009; Kuzyakov et al. 2009; Figure 4.1, arrow 4). Currently, two conceptual models have been proposed regarding the mineralization rate of biomass BC: (1) biomass BC is biphasic consisting of both a labile and stable C pool (Liang et al. 2008) and (2) biomass BC varies along a continuum of labile to extremely recalcitrant C pools and mineralization decreases as the labile pool is utilized leaving behind a physically inaccessible recalcitrant C source (Zimmerman 2010); either model would significantly influence the total soil C pool (Figure 4.1). Furthermore, the addition of biomass BC to soil may act as an immediate C source for microbial activity as well as stimulating the turnover of native C pools, i.e., a "priming" effect (Figure 4.1, arrow 3; Bell et al. 2003; Novak et al. 2010), or the biomass BC results in suppressing the turnover of labile C sources, i.e., a "negative priming" effect (Keith et al. 2011). The understanding that biomass BC results in negative priming is largely associated with the sequestration of labile C pools in biomass BC_rich Amazonian *terra preta* soils (Glaser 2007; Solomon et al. 2007). Zimmerman et al. (2011), through a series of laboratory incubation experiments < attempted to disentangle the complex interactions between biomass BC and non-BC sources by evaluating different soils and biomass BC types. Their results suggest that increasing soil C mineralization (i.e., "priming") or soil C sequestration (i.e., "negative priming") of non-BC soil C is dependent on the interactions between biomass BC, the native microbial community, and non-BC sources.

Several studies suggest that urban soils receive a significant portion of fossil fuel BC as a result of both industrial processes and traffic emissions (Bucheli et al. 2004; Liu et al. 2011). However, studies investigating the impact of fossil fuel BC on urban soil C cycling are nonexistent. Utilizing the results from biomass BC studies we expect industrial and traffic BC (due to increased recalcitrance as a result of combustion temperature) will lead to soil C sequestration via limited microbial usage of either industrial or fossil fuel BC (i.e., negative priming). Thus, greater BC and non-BC pools are anticipated in urban soils due to greater fossil fuel BC inputs and inhibited slow C cycling rates, with the caveat that plant biomass contributions to SOC remain unaffected by BC additions.

4.2.3 INVASIVE SPECIES

Invasive species are a known global phenomenon causing ecological and economic damage to ecosystems (Mack et al. 2000; Pimentel et al. 2005). Urban areas are often the epicenter for nonnative species introductions as urban landscapes serve as transportation hubs supporting global commerce and international trade. Nonnative invasive species commonly found in urban environments can be significant drivers of soil C dynamics (i.e., nonnative invasive plants, nonnative earthworms, and nonnative trees pests). In this chapter, nonnative species are defined as those from another country of origin.

4.2.3.1 Nonnative Invasive Plants

Urban and suburban areas are associated with a loss of native species and the spread of nonnative species. Nonnative plant populations are known to increase as cities and urban populations expand (Pyšek 1998; McKinney 2002, 2006). Nonnative plants increase across urbanization gradients, with greater populations and species richness occurring closer to city centers (Kowarik 2008). Plant invasion in urbanized landscapes results from a complex mixture of environmental (e.g., soil properties, adjacent land use; Ehrenfeld 2008) and social variables. Invasive plants alter ecosystem processes using a wide variety of mechanisms (i.e., altering forest structure or nutrient dynamics), resulting in a wide range of intensity of impacts (Ehrenfeld 2010). They exert similar impacts across ecosystem types (e.g., forests and grasslands) and life forms (e.g., woody shrubs and lianas), and while reported patterns are not necessarily universal, this chapter presents typical patterns observed across a range of studies establishing the potential impacts on soil C dynamics in urban landscapes.

Nonnative invasive plants exert pressure on ecosystems by altering plant community composition and altering soil processes. A meta-analysis of nonnative plant invasion effects in forests, grasslands, and wetlands found that aboveground primary productivity and litter decomposition was 50%–120% higher in invaded versus uninvaded ecosystems (Liao et al. 2008). In a common garden experiment, nonnative invasive plants had significantly faster leaf and root production and higher leaf N than native plants, resulting in nonnative plants having greater productivity than co-occurring native plants (Jo et al. 2015). Nonnative plants may also exhibit a longer growing season than many native species, initiating autumnal leaf senescence later than native plants (e.g., McEwan et al. 2009; Trammell et al. 2012), leading to an increase in C gain when native species are leafless (e.g., Harrington et al. 1989). Greater productivity in nonnative plants compared to congener natives suggests greater productivity in invaded ecosystems; however, this depends on the productivity of other native species encompassing different life forms. In urban forests invaded by honeysuckle shrubs (*Lonicera maackii*), honeysuckle leaf biomass was greater than in uninvaded sites as expected, but the leaf biomass of all other species combined was significantly less in invaded forests (Trammell et al. 2012). Another study in hardwood forests demonstrated reduced growth in tree basal area in forests invaded by honeysuckle shrubs (Hartman and McCarthy 2007). Invasion by nonnative plants into native ecosystems can alter productivity of congeners and other species causing cascading effects on the OM inputs to soil, hence altering soil C dynamics. Furthermore, several studies have demonstrated greater decay rates in invaded compared to uninvaded sites primarily due to increased leaf litter quality by nonnative plants; this pattern persisted in urban forests with variability in soil conditions (Liao et al. 2008; Trammell et al. 2012). The change in the OM quality entering soils will have large impacts on the rapid soil C cycle as decomposition rates increase with higher litter quality (Figure 4.1, arrow 6).

The above- and belowground plant C and N use strategies of nonnative invasive plants promote establishment and subsequent spread of invasive plants. However, belowground interactions with nonnative invasive plants and soil microbial community may significantly alter numerous soil processes within invaded ecosystems. The microbial community associated with a nonnative invasive plant was distinctly different than the native plant community suggesting that differences may originate prior to leaf senescence (Arthur et al. 2013). An experiment conducted on uninvaded soils demonstrated that the presence of nonnative shrubs and grasses altered the soil microbial community structure (i.e., phospholipid fatty acid [PLFA]) and significantly increased enzyme activity (Kourtev et al. 2003). Forests with two co-occurring nonnative invasive shrubs established nonadditive effects on extracellular enzyme activity (i.e., C-degrading enzymes), suggesting nonnative invasive species growing together may have greater impacts on ecosystem processes than when growing alone (Kuebbing et al. 2014). Soil processes, such as N mineralization and nitrification, tend to increase due to plant invasion (Liao et al. 2008), providing further evidence that invasive plants drive changes in microbial activity, hence microbial-derived soil compounds, i.e., slow soil C cycle (Figure 4.1, arrow 7).

The combined effects of altered productivity, leaf litter quality, decomposition rates, and microbial community structure and function can alter soil C dynamics depending on the relative magnitude of productivity, decomposition, and microbial activity within the soil environment. A review of 94 studies found plant invasion resulted in a 7% increase in soil C across a variety of plant life forms (i.e., grasses, shrubs, trees) and ecosystem types (Liao et al. 2008). In urban systems, the net effect on soil C storage will depend heavily on nonnative invasive species spread, decomposer community dynamics, and N deposition (e.g., Cusack et al. 2015). A study of forests along urban interstates found accelerated rates of litter decomposition in sites invaded by a nonnative shrub, thus indicating the strength of plant invasion to alter ecosystem function and soil C cycling even in locations with other urban stressors (Trammell et al. 2012, Trammell and Carreiro 2012). Plant invasion has the potential to increase soil C accumulation if OM inputs are relatively greater than decomposition rates.

4.2.3.2 Nonnative Earthworms

Earthworms facilitate several soil processes altering soil structure and biogeochemical cycles, such as N dynamics (Watmough and Meadows 2014), affecting plant growth and survival. In US northern regions, earthworm populations are nonnative (native earthworms were eliminated during the last glaciation) and their introduction has dramatically altered forest ecosystems and biogeochemical cycles. Nonnative earthworms have greater abundance in urban than in rural forests, and increased earthworm abundance is associated with enhanced N cycling in urban soils (Steinberg et al. 1997). Additionally, earthworms exert effects on soil structure and SOM content, which indirectly affects soil nutrient cycling, microbial community structure and function, and plant growth and survival. Indirect effects on plant N availability can feedback to reduced soil OM inputs. Furthermore, urban forests invaded with nonnative earthworms have faster litter decay rates, compensating for lower leaf litter quality from urban than rural trees (Pouyat and Carreiro 2003). Earthworm invasion in urban landscapes has the potential to reduce OM inputs and enhance litter decay rates, lowering SOM content, altering nutrient cycling and loss, and subsequently impacting soil microbial community composition and activity.

The relationship between nonnative earthworms and SOM is consistent across systems, i.e., lower OM is associated with greater earthworm abundance in forests and residential yards (Smetak et al. 2007; Sackett et al. 2013). Reduction in OM can have a cascading effect on soil C dynamics, yet the influences of earthworms on microbial composition and activity are less well known, especially in urban systems. A recent study investigating the spread of nonnative earthworms due to human activity in southern Québec forests found that as nonnative earthworm frequency increased, forest floor thickness decreased, while soil bacterial and fungal abundance increased (Drouin et al. 2016). Nonnative earthworms in temperate forests were associated with changes in soil microbial community composition and enzyme activity suggesting that the presence of earthworms promotes C metabolism through changes in both fungi and bacteria functional clades (Dempsey et al. 2013). Furthermore, nonnative earthworms increase C loss from mineral soils and decrease microbial biomass C (MBC); this may reflect a reduction in microbial community coupled to an increase in microbial activity (Sackett et al. 2013). In designed urban soils, the activities of *Lumbricus terrestris* enhanced microbial biomass, activity, and surface C efflux suggesting that nonnative earthworms can alter microbial community composition and activity (Scharenbroch and Johnston 2011).

The spread of nonnative earthworms into urban ecosystems alters rapid and slow soil C cycling. A complex relationship exists between earthworms, soil C, nutrient cycling, and soil structure in urban landscapes. In a small northern Idaho city, older residential yards had greater soil C and lower bulk density, which was associated with larger earthworm populations (Smetak et al. 2007). In urban forests with both native and nonnative earthworms, soil fungal abundance, enzyme activity, soil respiration, and tree seedling growth were modified by earthworm activity (Szlavecz et al. 2011). Furthermore, the combined influence of nonnative earthworm populations on SOM and microbial activity can indirectly alter plant community composition (Drouin et al. 2016), affecting OM inputs. Since earthworms have significant effects on soil dynamics and soil C cycling, more work is needed to fully understand the influence of these complex relationships and feedbacks in urban soils on soil C cycling.

4.2.3.3 Nonnative Insect Pests

Many nonnative pests that devastate native tree populations originate in cities. In the United States, the first detection of Asian long-horned beetle (ALB) (*Anoplophora glabripennis*) was in New York City, NY (1996) (Haack et al. 2010), and emerald ash borer (EAB) (*Agrilus planipennis*) was first discovered in Detroit, MI (2002) (Poland and McCullough 2006). Nonnative insects and pathogens have a long history of damaging native urban tree populations, such as Dutch elm disease (*Ophiostoma ulmi* and *O. novo-ulmi*), causing significant loss to the urban tree canopy (Poland and McCullough 2006). Many municipalities across the United States replaced lost elm trees with ash and maple species, which are currently threatened by the EAB and the ALB. Generalist insect pests feed on several tree species risking multiple urban trees populations (Raupp et al. 2006). For example, six popular urban genera (*Acer, Aesculus, Betula, Populus, Salix,* and *Ulmus*) are at risk due to ALB invasion (Haack et al. 2010) and the gypsy moth (*Lymantria dispar*) feeds on over 500 species resulting in major issues for urban tree species, especially oak species (Foss and Rieske 2003; Liebhold et al. 2005). Since many insect pests are known to destroy tree populations in infested cities, communities preemptively remove trees in order to reduce future risk (e.g., Heimlich et al. 2008), especially since early detection of some pests is difficult (e.g., EAB, Poland and McCullough 2006). This large loss of urban trees in a relatively short time period signifies a great loss of OM in urban ecosystems, particularly urban forests.

Nonnative pests and pathogens are a serious threat to many forests (e.g., Lovett et al. 2010); however, few studies have quantified the effects of urban tree loss and removal by nonnative pests on ecosystem processes, such as C cycling. In Minneapolis-St. Paul, Fissore et al. (2012) estimated the reduction in net primary production and C sequestration by the expected loss of ash trees due to EAB spread. A hypothetical assumption of total ash tree loss resulted in small effects on C cycling, approximately 4% of the total landscape C accumulation, which corresponded to the percent of total ash trees (Fissore et al. 2012). The overall effect of ash tree loss in this study was relatively small. However, since the ash tree loss effect on C paralleled the ash tree population, many cities with greater ash tree abundance may experience larger effects on C cycling. Thus, nonnative pests that defoliate and kill urban trees have the potential to greatly alter rapid C cycling via lower C inputs (Figure 4.1, arrow 6).

The effects of nonnative pests on microbial activity in urban environments are unknown, since there are few if any studies examining the effects of tree loss on urban soil C dynamics. However, a study modeling the potential of EAB to increase the potential of landslides in Pittsburgh, PA, demonstrated that a 75% loss of urban ash trees resulted in hundreds of potential landslide locations, which is dependent on soil compaction and species-specific root cohesion (Pfeil-McCullough et al. 2015). This suggests that tree loss or removal due to EAB could result in C exports due to soil erosion, potentially influencing soil C cycling. In nonurban ecosystems, beech bark disease, a nonnative pest/pathogen complex that causes decline in American beech trees (*Fagus grandifolia*), was shown to increase litter decomposition rates in forests, primarily due to shifts in tree species composition over time (Lovett et al. 2010). While these studies suggest altered soil processes and microbial community due to insect pests and pathogens, how tree defoliation, loss, or removal affects soil C cycling in urban systems is not well understood.

4.3   DIRECT ANTHROPOGENIC DRIVERS

The *direct* anthropogenic drivers are defined as the human activities in urban areas that control soil C dynamics via human transformations of soil, such as LULCC (e.g., topsoil removal), physical changes to soil (i.e., impervious surface), and management of soil additions and removals (e.g., compost; Figure 4.2).

(a)   (b)   (c)

FIGURE 4.2   Soil pits at the Morton Arboretum, Lisle, IL, demonstrating *direct* anthropogenic modification in urban soils. (a) Soil located within a compaction research plot. Transported material is present at 0–17 cm and compaction layers at 17–35 cm were created by removal of topsoil, compaction, and topsoil replacement in order to mimic common urban residential site preparation resulting in significant removal of topsoil organic C. (b) Soil located in a berm adjacent to I-88. Soil derived from transported material with subsoil low in organic C (~0.01%) and topsoil rich in organic C (~5%). (c) Highly disturbed soil containing buried road and subgrade material at 10–25 cm and transported topsoil rich with organic C (~5%) at 0–10 cm. (Courtesy of B. Scharenbroch.)

4.3.1 LAND-USE AND LAND-COVER CHANGE

LULCC affects soil C dynamics within urban ecosystems via disturbance events that operate along a fluctuating to predictable temporal and spatial scale continuum (Figure 4.1). Even in long urbanized lands, soils may be subject to disruption with regularity as utilities are replaced and landscapes reconfigured to accommodate changing needs. There are numerous examples of land-use conversion; however, this chapter is primarily focused on conversion from forested or agricultural lands to urban landscapes. Similarly, while there are many types of land-cover change in urban environments, the focus herein is primarily on changes that occur simultaneously with land-use change (i.e., the land development process; Figure 4.2a). In addition, land-cover changes also occur at a finer scale, in the absence of land-use change, and this can have consequences for C dynamics. Finally, topsoil removal represents a potentially significant loss of soil and soil C during land-use conversion and land-cover change, thus we highlight this consequence of LULCC.

4.3.1.1 Land-Use Conversion

Land-use conversion is strongly influenced by policy as well as a variety of social, economic, and environmental factors, both in terms of what land is converted and what practices will be employed during the land-use conversion process. Rapidly urbanizing megacities (e.g., Mumbai) struggle with unregulated land-use conversion, although this seemingly haphazard urbanization pattern appears less random when pre-existing spatial characteristics are considered, for example, the location of roads (Shafizadeh-Moghadam and Helbich 2015). In contrast, site practices employed during land-use conversion may be highly regulated in many regions. Most localities in the United States require review and approval of all development plans to ensure compliance with a variety of local ordinances. Additionally, since the implementation of the National Pollutant Discharge Elimination System Phase II (NPDES-Phase II) regulations in 2003, even small construction sites are regulated to ensure sediment and stormwater discharge are controlled, resulting in considerable scrutiny of soil grading and management (White and Boswell 2006). Unless soils are explicitly protected, site development processes commonly degrade soils via a range of modifications, including vegetation clearing, topsoil removal, grading, and compaction, influencing soil C dynamics (Figure 4.2). The current interest in green infrastructure practices (management of plant and soil resources for a broad array of specific ecosystem services) has generated new scrutiny of site-level practices (Windhager et al. 2010). The soil moved during land-use conversion can be considerable and increases with greater slopes (McGuire 2004).

4.3.1.2 Land-Cover Change

Land-cover change is typically a consequence of land-use conversion, but can also occur independently, in response to policy and cultural pressures (Cook et al. 2012). These land-use-associated shifts include forested land shifting to lawns or large-scale agriculture to home gardens. These land-cover changes have significant effects on ecological processes, as well as shifts in landscape design and management within a land use, which are increasingly the focus of environmental and development policy. For example, programs promoting xeriscape landscapes, where lawns are removed and replaced with rock mulches and widely spaced shrubs, can increase N pools in the soil and reduce C inputs greatly reducing NPP (Heavenrich 2015). Furthermore, altered land cover, from homogenous lawn to resource islands (i.e., shrubs), can greatly alter soil microbial processes. In remnant urban desert soils, plant resource islands were associated with greater microbial abundance and processes (McCrackin et al. 2008). In residential landscapes (constant land use), altered land cover has the potential to change the spatial arrangement and magnitude of soil C inputs and microbial activity (Figure 4.1, arrows 8 and 9). In addition to changes in land-cover type, alterations in soil and vegetation management practices and landscape design have the potential to alter soil C dynamics. For example, the Sustainable Sites Initiative (USGBC 2015) emphasizes protecting existing soils and plant communities and incorporating them into urban landscape design. Thus, design can influence whether particular maintenance practices are required or not. The ecological implications of landscape management practices are discussed throughout this chapter (e.g., see Section 4.3.3.1), but it is important to note that many management practices may initiate during the design process, which is driven by a variety of social, cultural, and economic factors. Land-cover changes within urban landscapes are increasingly important within the urban sustainability movement; consequently, opportunities to integrate ecological research, including studies of C dynamics, into the design process should be pursued where feasible (Felson et al. 2013).

4.3.1.3 Topsoil Removal

Topsoil removal is a consequence of both land-use change and infill development, where surface soil and vegetation are removed providing space for structural support of buildings and pavement. Topsoil removal constitutes an event directly removing stored soil C (Figure 4.1, arrow 8). The potential turnover of soil C is highly dependent upon handling and final destination. Erosion is the typical mechanism controlling topsoil loss in agricultural soils, yet the remaining soil has the potential to enhance C sequestration (via free mineral surfaces), accounting for an estimated 26% of soil C lost (Van Oost et al. 2007). However, this replacement effect is probably less significant in urban soil ecosystems. Accelerated C sequestration in eroded soils is due to reduced SOC available for decomposition combined with the continued C inputs from crops. In urban soils, C replacement may occur as a result of management practices. Several studies have observed increased SOC over time in some urban topsoil systems including managed lawns (Scharenbroch et al. 2005; Townsend-Small and Czimczik 2010). However, newly exposed soil may undergo sealing (i.e., paving or infrastructure development), thus reducing the potential for new C inputs (Raciti et al. 2012). Furthermore, vegetated urban soils are often exposed to repeated disturbance via redeveloped or disturbance for belowground infrastructure, establishing new opportunities for C loss.

During land development, removed topsoil may be exported or stockpiled. Stockpiled soil is rarely returned to the original removal site (Figure 4.2b). Few studies have examined SOC losses during scraping, stockpiling, and replacement although the length of the stockpiling period is known to influence C loss (Kundu and Ghose 1997). However, unlike stockpiling that occurs in mining operations or other large-scale disturbances, topsoil stockpiled at urban construction sites is stored for short durations (<1 year). Chen et al. (2013a) found scraping, stockpiling for approximately 1 month, and replacement resulted in a loss of 35% of topsoil SOC and 47% of mineral-bound C, suggesting that even short-term manipulations will result in significant C loss.

In recent years, manufactured or blended soils have been increasingly used as replacement soils when topsoil is removed (Sloan et al. 2012). Manufactured soils serve a variety of specialized purposes (e.g., green roofs, bioretention) or are used as replacement soils when topsoil is not available (Figure 4.2c). The majority of manufactured soils contain elevated sand and OM content coupled with relatively low silt and clay (Vidal-Beaudet et al. 2009; Paus et al. 2013). Since potential soil C accumulation is directly linked to abundance of clay fraction (O'Rourke et al. 2015), use of manufactured or blended soils may significantly decrease subsequent soil C storage potential. Net effects on soil C storage due to topsoil removal include (1) physical removal of soil and soil C (loss); (2) enhanced C sequestration potential in soil C-depleted soils (gain); (3) decomposition of soil C during handling and stockpiling (loss); (4) return of stockpiled or other soil and hence soil C (gain); and (5) creation of new conditions and management practices that potentially favor subsequent C sequestration (variable).

4.3.2 Urban Physical Conditions

Human activities in urban landscapes alter the soil physical environment directly by compacting soil and covering soil with an impervious surface, which potentially exerts strong influences on the soil C inputs and soil microbial community structure and function.

4.3.2.1 Soil Compaction

Soil compaction is widespread in urban settings due to grading, building construction practices, and heavy vehicular and pedestrian traffic (Figure 4.2a). Compacted soils restrict root growth and air/water movement limiting C inputs from vegetation (Figure 4.1, arrow 10). The effect of soil compaction on root distribution exerts considerable control on C inputs as roots are significant contributors to soil C compared to aboveground plant biomass (i.e., leaf litter) (Rasse et al. 2005; Kramer et al. 2010).

Soil compaction alters several physical properties, including reduction in micro- and macroaggregates, limiting water movement, and altering biological activity (Brevik et al. 2002; Deurer et al. 2012; Nawaz et al. 2012). The physical structure of soil, including pore structure (Ruamps et al. 2013) and aggregation, exerts considerable influence over microbial community access to soil C. There is evidence that microbial population structure shifts when plant biomass is nonlimiting (i.e., N additions; Janssens et al. 2010); however, the soil physical environment may play a greater role effecting microbial distributions and access to soil C (Ekschmitt et al. 2008). As a consequence, although extreme compaction consistently reduces microbial biomass C (MBC), the effects of soil compaction on MBC are otherwise inconsistent (Beylich et al. 2010). The ubiquity of compacted soils within urban landscapes significantly degrades soil structure, shifting microbial populations, and altering decomposition rates (Figure 4.1, arrow 11). This cascade of effects strongly suggests that soil compaction influences urban soil C dynamics, yet characterizing the nature and magnitude of this influence is difficult. While compaction in some examples limits C inputs from litter and roots, it may also increase microbial access to soil C.

4.3.2.2 Impervious Surface

Impervious surface area (ISA) formation (i.e., soil sealing) occurs when topsoil and vegetation are eliminated followed by the establishment of gray infrastructure construction (i.e., parking lots, buildings, and roads; Raciti et al. 2012). Current global estimates suggest that approximately 580,000 km^2 of the earth's surface is heavily influenced by constructed ISA (Elvidge et al. 2007), and land conversion to ISA is expected to increase as urbanized areas become denser and urban sprawl intensifies. Thus, ISA has been identified as one of the several significant threats to soil ecosystem health (Thematic Strategy for Soil Protection 2006) as all population growth is expected to occur within urban areas over the next half-century (United Nations 2011). Upon removal of native topsoil/vegetation and completion of soil sealing the remaining soils are sometimes considered biochemically inactive due to a reduction/elimination of C, N, nutrients, and water inputs (Zong-Qiang et al. 2014). However, specific studies aimed at evaluating soil C incorporation and sequestration into urban soils affected by ISA are not readily available. Due to our incomplete understanding of ISA on soil C cycling, it is difficult to estimate soil C storage at both regional and global scales (Pouyat et al. 2006).

The initial step in road bed construction involves the complete removal of the organic rich topsoil layer, depleting the remaining mineral soil layer of potential soil C inputs, and severely altering the microbial community structure. A few studies have measured the enzymatic activity (Raciti et al. 2012) or the respiratory capacity (Zong-Qiang et al. 2014) of the microbial community influencing slow C cycling pools beneath ISA soils (Figure 4.1, arrow 11). Raciti et al. (2012) measured the potential activity of three microbial enzymes under ISA; their results indicate that PEROX (lignin degrading) was the only enzyme comparable to non-ISA soils. Zong-Qiang et al. (2014) found that basal respiration rates in ISA soils were significantly reduced compared to non-ISA soils. The limited microbial activity associated with these studies supports the consensus that ISA soils are limited in soil organic C cycling.

The capacity of soils sealed by ISA to acquire and sequester SOC is significantly diminished as soil sealing limits new C inputs. However, ISA-impacted soils can become C sinks via absorption of inorganic C (IC; e.g., calcium carbonate) contributing to the total soil C pool (Figure 4.1, arrow 12).

In order for paved soils to sequester IC, a significant source of cations is required; paving materials often maintain high concentrations of either calcium (Ca^{2+}) or magnesium (Mg^{2+}; Kida and Kawahigashi 2015). Second, water must be available; several studies suggested that paved soils were not entirely impervious to water infiltration as a result of either pavement cracking (Morgenroth and Buchan 2009; Raciti et al. 2012; Zong-Qiang et al. 2014) or use of pervious paving materials (Li et al. 2013). In the presence of water, Ca^{2+} and Mg^{2+} minerals react with dissolved carbon dioxide (CO_2)-forming carbonates (Berner et al. 1983), in turn capturing and fixing atmospheric C. Engineered soils (Washbourne et al. 2012; Renforth et al. 2009), as well as soils beneath ISA (Kida and Kawahigashi 2015), have shown a greater capacity to sequester C as IC, suggesting that soils exposed to increased concentrations of cations will sequester greater amounts of IC compared to native topsoil. Future research investigating effects of soil sealing on C dynamics should consider the potential of both IC as well as organic sources to contribute to the total C pool. Additionally, greater characterization of the microbial community structure and functional diversity (e.g., quantitative polymerase chain reaction [PCR], molecular fingerprinting) would enable a greater understanding of the turnover potential of soil C sources beneath ISA.

4.3.3 Land Management

Urban landscapes are highly managed systems and primarily comprise residential, industrial, institutional, and commercial landscapes. Since the majority of urban green space area is residential land use and lawns, the following discussion of land management will focus on residential landscapes and urban lawns. Lawn management practices, e.g., fertilization, irrigation, and OM amendments, increase the potential for soil C accumulation.

4.3.3.1 Fertilization and Irrigation

Urban lawns, created and maintained by humans, are the dominant cover in residential landscapes, occupy the majority of open green space in urban landscapes, and are the largest irrigated crop in the United States (Milesi et al. 2005; Groffman et al. 2009; Ignatieva et al. 2015). Inputs of water and nutrients are used to establish and maintain lawns in many residential landscapes (Robbins et al. 2001; Polsky et al. 2014), and in the absence of water and fertilizer additions, turf grass species would not survive in many areas, especially arid regions (Milesi et al. 2005). Irrigation and fertilization are mechanisms driving soil C dynamics in lawns (e.g., Qian et al. 2003). Socioeconomic status (SES) of homeowners and age of lawn establishment influence homeowner management practices (i.e., higher SES and newer residential developments result in greater fertilization rates (Robbins et al. 2001; Law et al. 2004; Osmond and Hardy 2004), increasing soil C accumulation in these highly managed systems (Gough and Elliott 2012).

In general, greater soil N concentrations and mean annual precipitation (MAP) increases soil C pools (Selhorst and Lal 2012). Greater soil N availability *can* decrease litter and SOC decomposition rates, soil respiration, and mineralization, increasing total soil C pools (Hyvönen et al. 2007). Furthermore, increases in MAP typically improve plant growth and enhance soil C pools due to greater OM inputs from above- and belowground plant biomass (Figure 4.1, arrow 13). Thus, it is expected that fertilization and irrigation in urban systems will increase soil C pools due to enhanced plant productivity and reduced soil microbial activity. While soil properties can result in complex relationships influencing these dynamics (e.g., bulk density has a curvilinear interaction with soil C), soil N had the strongest consistent relationship (positive correlation) with soil C pools (Selhorst and Lal 2012).

The influence of irrigation or fertilizer alone on soil C dynamics in highly managed urban systems is not the focus of many research studies. However, a few studies have examined the effects of water and soil N availability on aboveground production in urban lawns, which affects OM inputs and rapid soil C cycling. Aboveground biomass in a Bermuda grass (*Cynodon dactylon*) lawn in Woodward, OK, was greater during years of high growing season precipitation suggesting that lawn production may increase with greater irrigation (Springer 2012). Leaf biomass production of four common turf species increased linearly with increase in fertilization rates in a growth chamber experiment demonstrating that greater N availability and leaf N concentrations increases aboveground production and OM inputs (Ericsson et al. 2012), which has been shown in other similar experimental studies (e.g., Wu et al. 2010). The relative importance of enhanced aboveground productivity due to fertilization and irrigation to increase soil C pools will depend on decomposition rates and microbial activity.

Soil characteristics and microbial activity are important determinants of soil C dynamics in highly managed urban systems. Previous studies on lawn C cycling found greater soil C in yards of older compared with newer housing developments (Golubiewski 2006; Gough and Elliott 2012), and lawn management practices, such as fertilization, are important mechanisms for soil C accumulation (Qian et al. 2003; Campbell et al. 2014). While most studies report increased soil C in highly managed lawns, soil C was not related to fertilization or irrigation in Auburn, AL, residential lawns suggesting increased decomposition rates in humid, subtropical regions may diminish the potential of soil C accumulation in these resource-rich environments (Huyler et al. 2014a). Other studies have demonstrated that urban soils have great potential to store and sequester C (e.g., Pouyat et al. 2009), yet the associated C costs of management (e.g., C emissions from mowing) should be considered when discussing the importance of soil C accumulation in highly managed systems (Selhorst and Lal 2013). Furthermore, there is a lack of understanding how increases in fertilization and irrigation practices effect microbial community structure and function in urban systems, an important control on slow soil C cycling and total soil C accumulation (Figure 4.1, arrow 14). While previous studies have not established direct links between fertilization and irrigation practices with soil microbial activity in urban landscapes, Raciti et al. (2011) found no relationship between net nitrification or mineralization and fertilization or irrigation practices in Baltimore, MD residential lawns. This implies high N inputs and/or irrigation may decrease microbial activity, diminishing slow soil C cycling rates and increasing soil C accumulation.

4.3.3.2 Compost Additions

Soil amendments including compost, leaf litter, or other OM can increase soil C storage (Chen et al. 2013a), increase water-holding capacity (Khaleel et al. 1981; Rawls et al. 2003), improve soil permeability (Brown and Cotton 2011; Chen et al. 2014; Martens and Frankenberger 1992), and enhance net primary productivity (De Lucia et al. 2013; Layman et al. 2016). These environmental shifts will influence soil C dynamics in urban soils (Figure 4.1, arrows 13 and 14) via microbial community and organic inputs from both above- and belowground plant parts. However, OM contributes binding agents required for aggregate formation (Six et al. 2004), slowing decomposition by physically protecting soil C stores from microbial access. Inputs of OM can result in a priming effect (Stockmann et al. 2013), where a small labile C input stimulates short-term mineralization of stable soil C. This could occur where compost or other OM is incorporated into the soil through tillage or a soil rehabilitation process (Cogger 2005; Layman et al. 2016) where existing soil C stores would be placed in proximity to fresh OM. However, unlike in agricultural settings, compost additions of this type in urban landscapes are likely to be discrete events, rather than an ongoing management practice.

4.3.3.3 OM Removal

In natural ecosystems, above- and belowground plant biomass is a vital input for SOM accumulation, and in unmanaged systems aboveground inputs naturally enter the soil environment. Management practices such as removing lawn clippings, clearing leaf and woody debris from planted shrubs and trees, and the removal of the autumn leaf pulse from deciduous trees result in removal of OM before it enters the soil environment. This OM removal represents a net loss of biomass from the system, altering soil C dynamics (Figure 4.1, arrow 13).

While it is expected that the removal of OM inputs will affect soils in highly managed urban ecosystems, like lawns, there is little research that has quantified how these management practices affect soil C accumulation (Guertal 2012). Using the CENTURY model, researchers estimated that the return of clippings to lawns would increase soil C sequestration by 11%–59% compared to removing lawn clippings over 10–50 years (under a low N fertilization regime) (Qian et al. 2003). The removal of litter fall from lawns during autumn leaf senescence represents a potentially substantial export of C from yards and other managed urban lawns where litter fall is not mulched and returned to the soil environment. The C loss associated with removal of yard waste was estimated at $52 \pm 17\%$ of residential C litter fall from Boston residential yards (Templer et al. 2015), which suggests a significant reduction of OM inputs and alteration of rapid soil C cycling. In residential yards and urban lawns, trees can increase soil C accumulation by greater OM inputs to soil. However, trees were found to play an important role in stabilizing soil C pools at lower depths (below 15 cm), yet play a minimal role at the soil surface (above 15 cm; Huyler et al. 2014b). These results may be due to increased fine root biomass and decreased OM inputs at the soil surface due to autumn leaf litter removal, though more research is needed to understand the influence of trees and management practices on soil C accumulation in highly managed urban areas.

The impact of OM removal from managed urban systems on the soil microbial structure and/or function is absent from the literature. However, as part of a long-term detritus input and litter removal (DIRT) experiment in deciduous forests, litter removal resulted in significant decreases in enzyme activity (β-glucosidase) after 10 years of treatment, which was attributed to changes in labile C availability (Veres et al. 2015). Furthermore, DIRT experiments in mixed deciduous forest and restored prairie grasslands exhibited litter removal treatments had significantly lower soil C concentrations, which varied by density fraction C content (i.e., light fraction C pools declined), age of C lost (i.e., younger C lost), and preferentially mineral C was lost (51% of mineral-associated C lost) (Lajtha et al. 2014). These results suggest that litter removal in highly managed urban soils can exert substantial impacts on the type of C stored or lost from the system, and hence microbial activity (Figure 4.1, arrow 14). Research is needed to understand how clippings removed and litter fall removal affects soil C dynamics, and what the long-term effects could be on soil C sequestration.

4.4 CONCLUSION

In this chapter, the challenge was to summarize the most important direct and indirect anthropogenic factors influencing urban soil C dynamics. The factors described are ubiquitous across urban landscapes, and thus exert significant effects on urban soil C. Scientific understanding of the relative impact of each factor is dependent on current research efforts, and the remaining gaps in our knowledge as to how these drivers influence soil C. Based on existing research findings, it is possible to assess the level of uncertainty (i.e., knowledge) and anticipate the magnitude of impact on the rapid versus slow C cycling via each factor (Table 4.1). However, it is difficult to generalize across all types of urban soils with respect to soil C dynamics since certain anthropogenic drivers may enhance C stabilization while others depress soil C stabilization. The net effect of these drivers will depend on the relative impact of factors that stimulate versus factors that depress soil C stabilization, resulting in urban soils accumulating more C than in the native soils they replace (Pouyat et al. 2010, 2015). Additionally, physical disturbances exert large impacts in urban landscapes, thus the age of the soil post-disturbance is an important consideration. Furthermore, it is critical to highlight that the mechanisms underlying changes in urban soil C are based on dynamic management decisions that may be driven by culture and/or policy and their interactions with climate. Human decisions and activities at small spatial scales can collectively translate into urbanization effects at larger scales. As urbanization and global change continue to advance, the impacts on urban ecosystems will remain complex and an important focus for ecological research. Understanding how multiple drivers in urban ecosystems alter urban soil C dynamics is a vital aspect of achieving sustainable cities.

TABLE 4.1

Indirect and Direct Anthropogenic Drivers of Urban Soil C Dynamics and Their Anticipated Magnitude of Influence and Level of Uncertainty (i.e., Knowledge) on Rapid and Slow Soil C Cycling

Anthropogenic Driver	Potential Magnitude		Current Level of Uncertainty	
	Rapid C Cycling	Slow C Cycling	Rapid C Cycling	Slow C Cycling
Indirect				
Climate				
Temperature	Moderate	High	Moderate	Moderate
Precipitation	Low	Low	High	High
Chemical				
Pollutants/nutrients	Moderate	Low	Low	Low
Heavy metals	Low	Low	Low	Low
Black carbon	High	Moderate	High	High
Invasive Species				
Nonnative plants	High	Moderate	Moderate	Moderate
Nonnative earthworms	Moderate	High	Moderate	Moderate
Nonnative insect pests	High	Moderate	High	High
Direct				
LULCC				
Land-use conversion	High	High	High	High
Land-cover change	Moderate	Very Low	High	High
Topsoil removal	High	High	High	High
Physical				
Soil compaction	Moderate	High	Moderate	High
Impervious surface	Moderate	High	High	High
Land Management				
Fertilization/irrigation	High	High	Moderate	Moderate
Compost additions	High	High	Moderate	High
Organic matter removals	Moderate	Moderate	Moderate	Moderate

Note: The potential magnitude and level of uncertainty for nonnative species are dependent on the species considered.

ACKNOWLEDGMENTS

The authors thank the University of Delaware Research Foundation for its support during the preparation of this chapter.

REFERENCES

Abu-Thiyab, H.M., Shair, O.H.M., Al-ssum, R.M., Aref, N.M., and Al-ssum, B.A. 2012. Functional redundancy diversity of gram positive bacteria as response to pesticide (malathion) exposure in soil. *J Pure Appl Microbio.* 6:201–207.

Adriano, D.C. 1986. *Trace Elements in the Terrestrial Environment.* New York, NY: Springer-Verlag.

Arthur, M.A., Bray, S.R., Kuchle, C.R., and McEwan, R.W. 2012. The influence of the invasive shrub, *Lonicera maackii*, on leaf decomposition and microbial community dynamics. *Plant Ecol.* 213:1571–1582.

Athar, R. and Ahmad, M. 2002. Heavy metal toxicity: Effect on plant growth and metal uptake by wheat, and on free living azotobacter. *Water Air Soil Pollut.* 138:165–180.

Bååth, E. 1989. Effects of heavy metals in soil on microbial processes and populations (a review). *Water Air Soil Pollut.* 47:335–379.

Baldock, J.A. and Smernik, R.J. 2002. Chemical composition and bioavailability of thermally, altered *Pinus resinosa* (Red Pine) wood. *Org Geochem.* 33:1093–1109.

Balsalobre, C., Calonge, J., Jimenez, E., Lafuente, R., Mourino, M., Munoz, M.T., Riquelme, M., and Mascastella, J. 1993. Using the metabolic capacity of rhodobacter-sphaeroides to assess heavy-metal toxicity. *Environ Toxic Water.* 8:437–450.

Bell, J.M., Smith, J.L., Bailey, V.L., and Bolton, H. 2003. Priming effect and C storage in semi-arid no-till spring crop rotations. *Bio Fert Soils.* 37:237–244.

Berner, R.A., Lasaga, A.C., and Garrels, R.M. 1983. The carbonate-silicate geochemical cycle and its effect on atmospheric carbon dioxide over the past 100 million years. *Amer J Sci.* 284:1183–1192.

Berdicevsky, I., Duek, L., Merzbach, D., and S. Yannai. 1993. Susceptibility of different yeast species to environmental toxic metals. *Environ Pollut.* 80:41–44.

Beylich, A., Oberholzer, H.R., Schrader, S., Hoper, H., and Wilke, B.M. 2010. Evaluation of soil compaction effects on soil biota and soil biological processes in soils. *Soil Till Res.* 109:133–143.

Bijoor, N.S., Czimzik, C.I., Pataki, D.E., and Billings, S.A. 2008. Effects of temperature and fertilization on nitrogen cycling and community composition of an urban lawn. *Global Change Biol.* 14:2119–2131.

Bird, M.I., Moyo, C., Veenendaal, E.M., Lloyd, J., and Frost, P. 1999. Stability of elemental carbon in a savanna soil. *Global Biogeochem Cycles.* 13:923–932.

Bisessar, S. 1982. Effects of heavy metals on microorganisms in soils near a secondary lead smelter. *Water Air Soil Poll.* 17:305–308.

Bityukova, L., Shogenova, A., and Birke, M. 2000. Urban geochemistry: A study of element distributions in the soils of Tallinn (Estonia). *Environ Geochem Health.* 22:173–193.

Bonan, G.B. 2002. *Ecological Climatology. Concepts and Applications.* Cambridge, UK: Cambridge University Press.

Botkin, D.B. and Beveridge, C.E. 1997. Cities as environments. *Urban Ecosyst.* 1:3–19.

Brevik, E., Fenton, T., and Moran, L. 2002. Effect of soil compaction on organic carbon amounts and distribution, South-Central Iowa. *Environ Poll.* 116:S137–S141.

Brown, S. and Cotton, M. 2011. Changes in soil properties and carbon content following compost application: Results of on-farm sampling. *Compost Sci Util.* 19:87–96.

Bucheli, T.D., Blum, F., Desaules, A., and Gustafsson, Ö. 2004. Polycyclic aromatic hydrocarbons, black carbon, and molecular markers in soils of Switzerland. *Chemosphere.* 56:1061–1076.

Butler, S.M., Melillo, J.M., Johnson, J.E., Mohan, J., Steudler, P.A., Lux, H., Burrows, E., Smith, R.M. et al. 2012. Soil warming alters nitrogen cycling in a New England forest: Implications for ecosystem function and structure. *Oecologia.* 168:819–828.

Calfapietra, C., Peñuelas, J., and Niinemets, U. 2015. Urban plant physiology: Adaptation-mitigation strategies under permanent stress. *Trends Plant Sci.* 20:72–75.

Campbell, C.D., Seiler, J.R., Wiseman, P.E., Straham, B.D., and Munsell, J.F. 2014. Soil carbon dynamics in residential lawns converted from Appalachian mixed oak stands. *Forests.* 5:425–438.

Carreiro, M.M., Pouyat, R.V., Tripler, C.E., and Zhu, W. 2009. Carbon and nitrogen cycling in soils of remnant forests along urban-rural gradients: Case studies in the New York metropolitan area and Louisville, Kentucky. In: *Ecology of Cities and Towns: A Comparative Approach.* eds. McDonnell, M.J., Hahs, A.K., and Breuste, J.H., 308–328. New York, NY: Cambridge University Press.

Carreiro, M.M., Sinsabaugh, R.L., Repert, D.A., and Parkhurst, D.F. 2000. Microbial enzyme shifts explain litter decay responses to simulated nitrogen deposition. *Ecology.* 81:2359–2365.

Chen, Y., Day, S.D., Wick, A.F., and McGuire, K.J. 2014. Influence of urban land development and subsequent soil rehabilitation on soil aggregates, carbon, and hydraulic conductivity. *Sci Total Environ.* 494:329–336.

Chen, Y., Day, S.D., Wick, A.F., Strahm, B.D., Wiseman, P.E., and Daniels, W.L. 2013a. Changes in soil carbon pools and microbial biomass from urban land development and subsequent post-development soil rehabilitation. *Soil Biol Biochem.* 66:38–44.

Chen, F., Fahey, T.J., Yu, M., and Gan L. 2010. Key nitrogen cycling processes in pine plantations along a short urban-rural gradient in Nanchang, China.

Chen, Z., Wang, X., Feng, Z., Xiao, Q., and Duan, X. 2009. Impact of elevated O_3 on soil microbial community function under wheat crop. *Water Air Soil Pollut.* 198:189–198.

Chen, Y., Wang, X., Jiang, B., Yang, N., and Li, L. 2016. Pavement induced soil warming accelerates leaf budburst of ash trees. *Urban For Urban Green.* 16:36–42.

Chen, H., Zhang, W., Gilliam, F., Liu, L., Huang, J., Zhang, T., Wang, W., and Mo, J. 2013b. Changes in soil carbon sequestration in *Pinus massoniana* forests along an urban-to-rural gradient of southern China. *Biogeoscience.* 10(10):6609–6616.

Cogger, C.G. 2005. Potential compost benefits for restoration of soils disturbed by urban development. *Compost Sci Util.* 13:243–251.

Cook, E.M., Hall, S.J., and Larson, K. L. 2012. Residential landscapes as social-ecological systems: A synthesis of multi-scalar interactions between people and their home environment. *Urban Ecosyst.* 15:19–52. doi:10.1007/s11252-011-0197-0.

Craine, J.M., Fierer, N., and McLauchlan, K.K. 2010. Widespread coupling between the rate and temperature sensitivity of organic matter decay. *Nature Geosci.* 3:854–857.

Cusack, D.F. 2013. Soil nitrogen levels are linked to decomposition enzyme activities along an urban-remote tropical forest gradient. *Soil Biol Biogeochem.* 57:192–203.

Cusack, D.F., Lee, J.K., McCleery, T.L., and Lecroy, C.S. 2015. Exotic grasses and nitrate enrichment altersoil carbon cycling along an urban-rural tropical forest gradient. *Global Change Biol.* 21:4481–4496.

De Graaff, M., Van Groenigen, K-J., Six, J., Hungate, B., and Van Kessel, C. 2006. Interactions between plant growth and soil nutrient cycling under elevated CO_2: A meta-analysis. *Globl Chng Biol.* 12:2077–2091.

De Lucia, B., Cristiano, G., Vecchietti, L., and Bruno, L. 2013. Effect of different rates of composted organic amendment on urban soil properties, growth and nutrient status of three Mediterranean native hedge species. *Urban For Urban Gree N.* 12:537–545.

Dempsey, M.A., Fisk, M.C., Yavitt, J.B., Fahey, T.J., and Balser, T.C. 2013. Exotic earthworms alter soil microbial community composition and function. *Soil Biol Biogeochem.* 67:263–270.

Deurer, M., Müller, K., Kim, I., Huh, K., Young, I., Jun, G., and Clothier, B. 2012. Can minor compaction increase soil carbon sequestration? A case study in a soil under a wheel-track in an orchard. *Geoderma.* 183:74–79.

Drouin, M., Bradley, R., and Lapointe, L. 2016. Linkage between exotic earthworms, understory vegetation and soil properties in sugar maple forests. *Forest Ecol Mgmt.* 364:113–121.

Edmondson, J.L., Stott, I., Potter, J., Lopez-Capel, E., Manning, D.A.C., Gaston, K.J., and Leake, J.R. 2015. Black carbon contribution to organic carbon stocks in urban soil. *Environ Sci Technol.* 49:8339–8346. doi:10.1021/acs.est.500313

Ehrenfeld, J.G. 2008. Exotic invasive species in urban wetlands: Environmental correlates and implications for wetland management. *J Appl Ecol.* 45:1160–1169.

Ehrenfeld, J.G. 2010. Ecosystem consequences of biological invasions. *Annu Rev Ecol Evol Syst.* 41:59–80.

Ekschmitt, K., Kandeler, E., Poll, C., Brune, A., Buscot, F., Friedrich, M., Gleixner, G., Hartmann, A. et al. 2008. Soil-carbon preservation through habitat constraints and biological limitations on decomposer activity. *J Plant Nutrition Soil Sci.* 171:27–35. doi:10.1002/jpln.200700051.

Elvidge, C.D., Tuttle, B.T., Sutton, P.C., Baugh, K.E., Howard, A.T., Milesi, C., Bhadura, B.L., and Nemani, R. 2007. Global distribution and density of constructed impervious surfaces. *Sensors.* 7:1962–1979.

Enloe, H.A., Lockaby, B.G., Zipperer, W.C., and Somers, G.L. 2015. Urbanization effects on soil nitrogen transformations and microbial biomass in the subtropics. *Urban Ecosyst.* 18:963–976.

Ericsson, T., Blombäck, K., and Neumann, A. 2012. Demand-driven fertilization. Part I: Nitrogen productivity in four high-maintenance turf grass species. *Acta Agri Scand B Soil Plant Sci.* 62(Suppl. 1):113–121.

Felson, A.J., Pavao-Zuckerman, M., Carter, T., Montalto, F., Shuster, B., Springer, N., Stander, E.K., and Starry, O. 2013. Mapping the design process for urban ecology researchers. *BioScience.* 63:854–865.

Ferreira, M.L., Silva, J.L., Pereira, E.E., and Lamano-Ferreira, A.P.N. 2014. Litter fall production and decomposition in a fragment of secondary Atlantic forest of São Paulo, SP, southeastern Brazil. *Revista Arvore.* 38(4):591–600.

Findlay, S., Carreiro, M., Krischik, V., and Jones, C. 1996. Effects of damage to living plants on leaf litter quality. *Ecol Apps.* 6:269–275.

Fissore, C., McFadden, J.P., Nelson, K.C., Peters, E.B., Hobbie, S.E., King, J.Y., Baker, L.A., and Jakobsdottir, I. 2012. Potential impacts of emerald ash borer invasion on biogeochemical and water cycling in residential landscapes across a metropolitan region. *Urban Ecosyst.* 15:1015–1030.

Foss, L.K. and Rieske, L.K. 2003. Species-specific differences in oak foliage affect preference and performance of gypsy moth caterpillars. *Entom Experi Appl.* 108:87–93.

Freedman, B. and Hutchinson, T.C. 1980. Effects of smelter pollutants on forest leaf litter decomposition near a nickel-copper smelter at Sudbury. *Ontario Can J Bot.* 58:1722–1736.

Giller, K.E., Witter, E., and McGrath, S.P. 1998. Toxicity of heavy metals to microorganisms and microbial processes in agricultural soils: A review. *Soil Biol Biochem.* 30:1389–1414.

Glaser, B. 2007. Prehistorically modified soils of central Amazonia: A model for sustainable agriculture in the twenty-first century. *Philos Trans Biol Sci.* 362:187–196.

Goldberg, E.D. 1985. *Black Carbon in the Environment.* New York, NY: John Wiley & Sons.

Golubiewski, N.E. 2006. Urbanization increases grassland carbon pools: Effects of landscaping in Colorado's Front Range. *Ecol Apps.* 16:555–571.

Gough, C.M. and Elliott, H.L. 2012. Lawn soil carbon storage in abandoned residential properties: An examination of ecosystem structure and function following partial human-natural decoupling. *J Environ Manage.* 98:155–162.

Govil, P.K., Reddy, G.L.N., and Krishna, A.K. 2001. Contamination of soil due to heavy metals in the Patancheru industrial development area, Andhra Pradesh, India. *Environ Geol.* 41:461–469.

Gregg, J.W., Jones, C.G., and Dawson, T.E. 2003. Urbanization effects on tree growth in the vicinity of New York City. *Nature.* 424:183–187.

Groffman, P.M., Williams, C.O., Pouyat, R.V., Band, L.E., and Yesilonis, I.D. 2009. Nitrate leaching and nitrous oxide flux in urban forests and grasslands. *J Environ Qual.* 38:1848–1860.

Guertal, E.A. 2012. Carbon sequestration in turfed landscapes: A review. In: *Carbon Sequestration in Urban Ecosystems.* eds. Lal, R. and Augustin, B., 197–213. Netherlands: Springer Science and Business Media.

Haack, R.A., Herard, F., Sun, J., and Turgeon, J.J. 2010. Managing invasive populations of Asian longhorned beetle and citrus longhorned beetle: A worldwide perspective. *Annu Rev Entomol.* 55:521–546.

Hamilton, G. A. and Hartnett, H.E. 2013. Soot black carbon concentration and isotopic composition in soils from an arid urban ecosystem. *Organ Geochem.* 59:87–94.

Harrington, R.A., Brown, B.J., and Reich, P.B. 1989. Ecophysiology of exotic and native shrubs in Southern Wisconsin: I. Relationship of leaf characteristics, resource availability, and phenology to seasonal patterns of carbon gain. *Oecologia.* 80:356–367.

Hartman, K.M. and McCarthy, B.C. 2007. A dendro-ecological study of forest overstorey productivity following the invasion of the non-indigenous shrub *Lonicera maackii. Appl Veg Sci.* 10:3–14.

He, Y. and Zhang, G.L. 2009. Historical record of black carbon in urban soils and its environmental implications. *Environ Pollut.* 157:2684–2688.

He, Z., Xiong, J., Kent, A.D., Deng, Y., Xue, K., Wang, G., Wu, L., Van Nostrand, J.D. et al. 2014. Distinct responses of soil microbial communities to elevated CO_2 and O_3 in a soybean agro-ecosystem. *ISME J.* 8:714–726.

Heavenrich, H. 2015. *Soil Biogeochemical Consequences of the Replacement of Residential Grasslands with Water-Efficient Landscapes.* Thesis. Arizona State University.

Heimlich, J., Sydnor, T.D., Bumgardner, M., and O'Brien, P. 2008. Attitudes of residents toward street trees on four streets in Toledo, Ohio, U.S., before removal of ash trees (*Fraxinus* spp.) from emerald ash borer (*Agrilus planipennis*). *Arbor Urban For.* 34:47–53.

Hilscher, A., Heister, K., Siewert, C., and Knicker, H. 2009. Mineralisation and structural changes during the initial phase of microbial degradation of pyrogenic plant residues in soil. *Org Geochem.* 40:332–342.

Hopkin, S.P. 1989. *Ecophysiology of Metals in Terrestrial Invertebrates.* London: Elsevier Applied Science.

Huyler, A., Chappelka, A.H., Prior, S.A., and Somers, G.L. 2014a. Drivers of soil carbon in residential 'purelawns' in Auburn, Alabama. *Urban Ecosyst.* 17:205–219.

Huyler, A., Chappelka, A.H., Prior, S.A., and Somers, G.L. 2014b. Influence of aboveground tree biomass, home age, and yard maintenance on soil carbon levels in residential yards. *Urban Ecosyst.* 17:787–805.

Hyvönen, R., Ågren, G., Linder, S., Persson, T., Cotrufo, M.F., Ekblad, A., Freeman, M., Grelle, A. et al. 2007. The likely impact of elevated (CO_2), nitrogen deposition, increased temperature and management on carbon sequestration in temperate and boreal forest ecosystems: A literature review. *New Phytol.* 173:463–480.

Ignatieva, M., Ahrné, K., Wissman, J., Eriksson, T., Tidáker, P., Hedblom, M., Kätterer, T., Marstorp, H., et al. 2015. Lawn as a cultural and ecological phenomenon: A conceptual framework for transdisciplinary research. *Urban For Urban Green.* 14:383–387.

Janssens, I.A., Dieleman, W., Luyssaert, S., Subke, J.A., Reichstein, M., Ceulemans, R., Ciais, P., Dolman, A.J. et al. 2010. Reduction of forest soil respiration in response to nitrogen deposition. *Nature Geosci.* 3:315–322. Available at: http://www.nature.com/ngeo/journal/v3/n5/suppinfo/ngeo844_S1.html.

Jim, C.Y. 1998. Soil characteristics and management in an urban park in Hong Kong. *Environ Mgmt.* 22:683–695.

Jo, I., Fridley, J.D., and Frank, D.A. 2015. Linking above- and belowground resource use strategies for native and invasive species of temperate deciduous forests. *Biol Invasions.* 17:1545–1554.

Jochner, S. and Menzel, A. 2015. Urban phenological studies—Past, present, future. *Environ Poll.* 203:250–261.

Kaminski, M.D. and Landsberger, S. 2000. Heavy metals in urban soils of East St. Louis, IL, part I: Total concentration of heavy metals in soils. *J Air Waste Mgmt Assoc.* 50:1667–1679.

Keith, A., Singh, B., and Singh, B.P. 2011. Interactive priming of biochar and labile organic matter mineralization in a smectite-rich soil. *Environ Sci Tech.* 45:9611–9618.

Khaleel, R., Reddy, K., and Overcash, M. 1981. Changes in soil physical properties due to organic waste applications: A review. *J Environ Qual.* 10:133–141.

Kida, K. and Kawahigashi, M. 2015. Influence of asphalt pavement construction processes on urban soil formation in Tokyo. *Soil Sci Plant Nutri.* 61:135–146.

Kim, R.Y., Yoon, J.K., Kim, T.S., Yang, J.E., Owens, G., and Kim, K.R. 2015. Bioavailability of heavy metals in soils: Definitions and practical implementation-a critical review. *Environ Geochem Health.* 37:1041–1061.

Kirschbaum, M.U.F. 1995. The temperature dependence of soil organic matter decomposition, and the effect of global warming on soil organic C storage. *Soil Biol Biochem.* 27(6):753–760.

Kourtev, P.S., Ehrenfeld, J.G., and Häggblom, M. 2003. Experimental analysis of the effect of exotic and native plant species on the structure and function of soil microbial communities. *Soil Bio Biochem.* 35:895–905.

Kowarik, I. 2008. On the role of alien species in urban flora and vegetation. In: *Urban ecology: An International Perspective on the Interaction between Humans and Nature.* eds. Marzluff, J.M., Shulenberger, E., Endlicher, W., Alberti, M., Bradley, G., Ryan, C.., ZumBrunnen, C., and Simon, U., 321–338. Springer-Verlag US.

Kramer, C., Trumbore, S., Fröberg, M., Cisneros Dozal, L.M., Zhang, D., Xu, X., Santos, G.M., and P.J. Hanson. 2010. Recent (4 year old) leaf litter is not a major source of microbial carbon in a temperate forest mineral soil. *Soil Biol Biochem.* 42:1028–1037. doi:10.1016/j.soilbio.2010.02.021

Kuebbing, S.E., Classen, A.T., and Simberloff, D. 2014. Two co-occurring invasive woody shrubs alter soil properties and promote subdominant invasive species. *J Appl Ecol.* 51:124–133.

Kuhlbusch, T. A. and Crutzen, P.J. 1995. Toward a global estimate of black carbon in residues of vegetation fires representing a sink of atmospheric CO2 and a source of O2. *Global Biogeoch Cycles.* 9:491–501.

Kumar, K. and Hundal, L.S. 2016. Soil in the city: Sustainably improving urban soils. *J Environ Qual.* 45:2–8. doi:10.2134/jeq2015.11.0589.

Kundu, N. and Ghose, M. 1997. Shelf life of stock-piled topsoil of an opencast coal mine. *Environ Conserv.* 24:24–30.

Kuzyakov, Y., Subbotina, L., Chen, H.Q., Bogomolova, I., and Xu, X.L. 2009. Black carbon decomposition and incorporation into soil microbial biomass estimated by C-14 labeling. *Soil Biol Bioch.* 41:210–219.

Lacke, M.C., Mote, T.L., and Shepherd, M. 2009. Aerosols and associated precipitation patterns in Atlanta. *Atmosph Environ.* 43:4359–4373.

Lajtha, K., Townsend, K.L., Kramer, M.G., Swanston, C., Bowden, R.D., and Nadelhoffer, K. 2014. Changes to particulate versus mineral-associated soil carbon after 50 years of litter manipulation in forest and prairie experimental ecosystems. *Biogeochemistry.* 119:341–360.

Latha, K. M. and Badarinath, K.V.S. 2003. Black carbon aerosols over tropical urban environment—A case study. *Atmos Res.* 69:125–133.

Law, N.L., Band, L.E., and Grove, J.M. 2004. Nitrogen input from residential lawn care practices in suburban watersheds in Baltimore County, MD. *J Environ Plan Mgmt.* 47(5):737–755.

Layman, R.M., Day, S.D., Mitchell, D.K., Chen, Y., Harris, J.R., and Daniels, W.L. 2016. Below ground matters: Urban soil rehabilitation increases tree canopy and speeds establishment. *Urban For Urban Greening.* (16:25–35).

Lehmann, J. and Kleber, M. 2015. The contentious nature of soil organic matter. *Nature.* 528:60–68.

Li, H., Harvey, J.T., Holland, T.J., and Kayhanian, M. 2013. The use of reflective and permeable pavements as a potential practice for heat island mitigation and stormwater management. *Environ Res Lett.* 8:015023.

Liang, B., Lehmann, J., Sohi, S.P., Thies, J.E., O'Neill, B., Trujillo, L., Gaunt, J., Solomon, D. et al. 2008. Black carbon affects the cycling of non-black carbon in soil. *Org Geochem.* 41:206–213.

Liao, C., Peng, R., Luo, Y., Zhou, X., Wu, X., Fang, C., Chen, J., and Li, B. 2008. Altered ecosystem carbon and nitrogen cycles by plant invasion: A meta-analysis. *New Phyto.* 177:706–714.

Liebhold, A.M., Raffa, K.F., and Diss, A.L. 2005. Forest type affects predation on gypsy moth pupae. *Agric For Entom.* 7:179–185.

Liu, S., Xia, X., Zhai, Y., Wang, R., Liu, T., and Zhang, S. 2011. Black carbon (BC) in urban and surrounding rural soils of Beijing, China: Spatial distribution and relationship with polycyclic aromatic hydrocarbons (PAHs). *Chemosphere.* 82:223–228.

Lorenz, S.E., Hamon, R.E., Mcgrath, S.P., Holm, P.E., and Christenson, T.H. 1994. Applications of fertilizer cations affect cadmium and zinc concentrations in soil solutions and uptake by plants. *Eur J Soil Sci.* 45:159–165.

Lovett, G.M., Arthur, M.A., Weathers, K.C., and Griffin, J.M. 2010. Long-term changes in forest carbon and nitrogen cycling caused by an introduced pest/pathogen complex. *Ecosystem.* 13:1188–1200.

Lovett, G.M., Traynor, M.M., Pouyat, R.V., Carreiro, M.M., Zhu, W., and Baxter, J.W. 2000. Atmospheric deposition to oak forests along an urban-rural gradient. *Environ Sci Technol.* 34:4294–4300.

Ma, W.C. 1982. The influence of soil properties and worm-related factors on the concentration of heavy metals in earthworms. *Pedobiologia.* 4:109–119.

Mack, R.N., Simberloff, D., Lonsdale, W.M., Evans, H., Clout, M., and Bazzaz, F.A. 2000. Biotic invasions: Causes, epidemiology, global consequences, and control. *Ecol Apps.* 10(3):689–710.

Madrid, L., Diaz-Barrientos, E., and Madrid, F. 2002. Distribution of heavy metal contents of urban soils in parks of Seville. *Chemosphere.* 49:1301–1308.

Mallet, M., Roger, J.C., Despiau, S., Putaud, J.P., and Dubovik, O. 2004. A study of the mixing state of black carbon in urban zone. *J Geophy Res Atmos.* 109(D4).

Martens, D. and Frankenberger, W. 1992. Modification of infiltration rates in an organic-amended irrigated. *Agronomy J.* 84:707–717.

McCrackin, M.L., Harms, T.K., Grimm, N.B., Hall, S.J., and Kaye, J.P. 2008. Responses of soil microorganisms to resource availability in urban, desert soils. *Biogeochemistry.* 87:143–155.

McEwan, R.W., Birchfield, M.K., Schoergendorfer, A., and Arthur, M.A. 2009. Leaf phenology and freeze tolerance of the invasive shrub Amur honeysuckle and potential native competitors. *J Torrey Bot Soc.* 136:212–220.

McGuire, M. 2004. Using DTM and LiDAR data to analyze human induced topographic change. In: Proceedings of ASPRS 2004 fall conference, September 12–16, 2004, Kansas City, MO. PDF Available at: http://eserv.asprs.org/eseries/source/Orders/

McKinney, M.L. 2002. Urbanization, biodiversity, and conservation. *BioScience.* 52:883–890.

McKinney, M.L. 2004. Citizens as propagules for exotic plants: Measurement and management implications. *Weed Technol.* 1:1480–1483.

McKinney, M.L. 2006. Urbanization as a major cause of biotic homogenization. *Biol Conserv.* 127:247–260.

Milesi, C., Running, S.W., Elvidge, C.D., Dietz, J.B., Tuttle, B.T., and Nemani, R.R. 2005. Mapping and modeling the biogeochemical cycling of turf grasses in the United States. *Environ Mgmt.* 36(3):426–438.

Morgan, J.E. and Morgan, A.J. 1988. Calcium-lead interactions involving earth worms. Part 1: The effect of exogenous calcium on lead accumulations by earthworms under field and laboratory conditions. *Environ Pollut.* 54:41–53.

Morgenroth, J. and Buchan, G.D. 2009. Soil moisture and aeration beneath pervious and impervious pavements. *J Arboriculture.* 35:135.

Nawaz, M.F., Bourrié, G., and Trolard, F. 2012. Soil compaction impact and modelling. A review. *Agron Sustain Dev.* 33(2): 291–309.

Neil, K. and Wu, J. 2006. Effects of urbanization on plant flowering phenology: A review. *Urban Ecosyst.* 9:243–257.

Niyogi, D., Pyle, P., Lei, M., Chen, F., and Wolfe, B. 2011. Urban modification of thunderstorms: An observational storm climatology and model case study for the Indianapolis urban region. *J Appl Meteor Climatol.* 50(5):1–13.

Novak, J. M., Busscher. W.J., Watts, D.W., Laird, D.A., Ahmedna, M.A., and Niandou, M.A.S. 2010. Short-term CO_2 mineralization after additions of biochar and switchgrass to a Typic Kandiudult. *Geoderma.* 154:281–288.

Oke, T.R. 1973. City size and the urban heat island. *Atmos Environ.* 7:769–779.

O'Rourke, S.M., Angers, D.A., Holden, N.M., and McBratney, A.B. 2015. Soil organic carbon across scales. *Global Change Biol.* 21:3561–3574. doi:10.1111/gcb.12959.

Osmond, D.L. and Hardy, D.H. 2004. Characterization of turf practices in five North Carolina communities. *J Environ Qual.* 33(2):565–575.

Öztürk, M., Bolat, İ., and Ergün, A. 2015. Influence of air-soil temperature on leaf expansion and LAI of *Carpinus betulus* trees in a temperate urban forest patch. *Agricult For Meteor.* 200:185–191.

Pardo, L.H., Fenn, M.E., Goodale, C.L., Geiser, L.H., Driscoll, C.T., Emmett, B., Gilliam, F.S., Greaver, T.L. et al. 2011. Effects of nitrogen deposition and empirical nitrogen critical loads for ecoregions of the United States. *Ecol Apps.* 21:3049–3082.

Paus, K.H., Morgan, J., Gulliver, J.S., Leiknes, T., and Hozalski, R.M. 2013. Assessment of the hydraulic and toxic metal removal capacities of bioretention cells after 2 to 8 years of service. *Water Air Soil Poll.* 225:1–12. doi:10.1007/s11270-013-1803-y.

Pavao-Zuckerman, M,A. and Coleman, D.C. 2005. Decomposition of chestnut oak (*Quercus prinus*) leaves and nitrogen mineralization in an urban environment. *Biol Fertil Soil.* 41:343–349.

Persiani, A.M., Maggi, O., Montalvo, J., Casado, M.A., and Pineda, F.D. 2008. Mediterranean grassland soil fungi: Patterns of biodiversity, functional redundancy and soil carbon storage. *Plant Biosyst.* 142:111–119.

Pfeil-McCullough, E., Bain, D.J., Bergman, J., and Crumrine, D. 2015. Emerald ash borer and the urban forest: Changes in landslide potential due to canopy loss scenarios in the city of Pittsburgh, PA. *Sci Total Environ.* 536:538–545.

Pietikäinen, J., Pettersson, M., and Bååth, E. 2005. Comparison of temperature effects on soil respiration and bacterial and fungal growth rates. *FEMS Microbiol Ecol.* 52:49–58.

Pimentel, D., Zuniga, R., and Morrison, D. 2005. Update on the environmental and economic costs associated with alien-invasive species in the United States. *Ecol Econ.* 52:273–288.

Poland, T.M. and McCullough, D.G. 2006. Emerald ash borer: Invasion of the urban forest and the threat to North America's ash resource. *J Forestry.* 104(3):118–124.

Polsky, C., Grove, J.M., Knudson, C., Groffman, P.M., Bettez, N., Cavender-Bares, J., Hall, S.J., Heffernan, J.B. et al. 2014. Assessing the homogenization of urban land management with an application to US residential lawn care. *PNAS.* 111(12): 4432–4437.

Pouyat, R.V. and Carreiro, M.M. 2003. Controls on mass loss and nitrogen dynamics of oak leaf litter along an urban-rural land-use gradient. *Oecologia.* 135:288–298.

Pouyat, R.V. and McDonnell, M.J. 1991. Heavy metal accumulation in forest soils along an urban-rural gradient in Southern New York, USA. *Water Air Soil Poll.* 57–58:797–807.

Pouyat, R.V., Szlavecz, K., Yesilonis, I.D., Groffman, P.M., and Schwarz, K. 2010. Chemical, physical, and biological characteristics of urban soils. In: *Urban Ecosystem Ecology.* eds. Aitkenhead-Peterson J. and Volder, A., 119–152. Madison, WI: Agronomy Monograph 55. ASA, CSA, SSSA.

Pouyat, R.V., Yesilonis, I.D., Dombos, M., Szlavecz, K., Setälä, H., Cilliers, S., Hornung, E., Kotze, D.J. et al. 2015. A global comparison of surface soil characteristics across five cities: A test of the urban ecosystem convergence hypothesis. *Soil Sci.* 180(4/5):1–10.

Pouyat, R.V., Yesilonis, I.D., and Golubiewski, N.E. 2009. A comparison of soil organic carbon stocks between residential turf grass and native soil. *Urban Ecosyst.* 12:45–62.

Pouyat, R.V., Yesilonis, I.D., and Nowak, D.J. 2006. Carbon storage by urban soils in United States. *JEQ.* 35:1566–1575.

Preston, C. M. and Schmidt, M.W.I. 2006. Black (pyrogenic) carbon: A synthesis of current knowledge and uncertainties with special consideration of boreal regions. *Biogeoscience.* 3:397–420.

Pyšek, P. 1998. Alien and native species in Central European urban floras: A quantitative comparison. *J Biogeo.* 25:155–163.

Qian, Y.L., Bandaranayake, W., Parton, W.J., Mecham, B., Harivandi, M.A., and Mosier, A.R. 2003. Long-term effects of clipping and nitrogen management in turfgrass on soil organic carbon and nitrogen dynamics: The CENTURY model simulation. *J Environ Qual.* 32:1694–1700.

Raciti, S.M., Groffman, P.M., Jenkins, J.C., Pouyat, R.V., Fahey, T.J., Pickett, S.T.A., and Cadenasso, M.L. 2011. Nitrate production and availability in residential soils. *Ecol Apps.* 21(7):2357–2366.

Raciti, S.M., Hutyra, L.R., and Finzi, A.C. 2012. Depleted soil carbon and nitrogen pools beneath impervious surfaces. *Environ Pollut.* 164:248–251.

Rao, P., Hutyra, L.R., Raciti, S.M., and Templer, P.H. 2014. Atmospheric nitrogen inputs and losses along an urbanization gradient from Boston to Harvard Forest, MA. *Biogeochemistry.* 121:229–245.

Rasse, D.P., Rumpel, C., and Dignac, M-F. 2005. Is soil carbon mostly root carbon? Mechanisms for a specific stabilisation. *Plant Soil.* 269:341–356. doi:10.1007/s11104-004-0907-y.

Raupp, M.J., Cumming, A.B., and Raupp, E.C. 2006. Street tree diversity in eastern North America and its potential for tree loss to exotic borers. *Arbor Urban For.* 32:297–304.

Rawls, W., Pachepsky, Y.A., Ritchie, J., Sobecki, T., and Bloodworth, H. 2003. Effect of soil organic carbon on soil water retention. *Geoderma.* 116:61–76.

Renforth, P., Manning, D.A.C., and Lopez-Capel, E. 2009. Carbonate precipitation in artificial soils as a sink for atmospheric carbon dioxide. *Appl Geochem.* 24:1757–1764.

Robbins, P., Polderman, A., and Birkenholtz, T. 2001. Lawns and toxins: An ecology of the city. *Cities.* 18(6):369–380.

Romandini, P., Tallandini, L., Beltramini, M., Salvato, B., Manzano, M., Debertoldi, M., and Rocco, G.P. 1992. Effects of copper and cadmium on growth, superoxide-dismutase and catalase activities in different yeast strains. *Comp Biochem Physio C-Pharmac Toxic Endocrin.* 103:255–262.

Rousk, J., Brookes, P.C., and Baath, E. 2009. Contrasting soil pH effects on fungal and bacterial growth suggest functional redundancy in carbon mineralization. *Appl Environ Microb.* 75:1589–1596.

Ruamps, L.S., Nunan, N., Pouteau, V., Leloup, J., Raynaud, X., Roy, V., and Chenu, C. 2013. Regulation of soil organic C mineralisation at the pore scale. *FEMS Microbiol Ecol.* 86:26–35. doi:10.1111/1574-6941.12078.

Sackett, T.E., Smith, S.M., and Basiliko, N. 2013. Indirect and direct effects of exotic earthworms on soil nutrient and carbon pools in North American temperate forests. *Soil Biol Biogeo.* 57:459–467.

Salminen, J., Hernesmaa, A., Karjalainen, H., Fritze, H., and Romantschuk, M. 2010. Activity of nematodes and enchytraeids, bacterial community composition, and functional redundancy in coniferous forest soil. *Biol Fert Soils.* 46:113–126.

Salminen, J., van Gestel, C.A.M., and Oksanen, J. 2001. Pollution-induced community tolerance and functional redundancy in a decomposer food web in metal-stressed soil. *Environ Toxic Chem.* 20:2287–2295.

Scharenbroch, B.C., Day, S., Trammell, T.L.E., and Pouyat, R.V. 2018. Urban soil carbon storage. In: *Urban Soils.* eds. Lal, R. and Stewart, B.A. Boca Raton, FL: Taylor and Francis Group.

Scharenbroch, B.C. and Johnston, D.P. 2011. A microcosm study of the common night crawler earthworm (*Lumbricus terrestris*) and physical, chemical and biological properties of a designed urban soil. *Urban Ecosyst.* 14:119–134.

Scharenbroch, B.C., Lloyd, J.E., and Johnson-Maynard, J.L. 2005. Distinguishing urban soils with physical, chemical, and biological properties. *Pedobiologia.* 49:283–296.

Selhorst, A. and Lal, R. 2012. Effects of climate and soil properties on U.S. home lawn soil organic carbon concentration and pool. *Environ Mgmt.* 50:1177–1192.

Selhorst, A. and Lal, R. 2013. Net carbon sequestration potential and emissions in home lawn turfgrasses of the United States. *Environ Mgmt.* 51:198–208.

Seto, K.C., Fragkias, M., Güneralp, B., and Reilly, M.K. 2011. A meta-analysis of global urban land expansion. *PLOS ONE.* 6(8):e23777. doi:10.1371/jounal.pne.0023777.

Seto, K.C., Güneralp, B., and Hutyra, L.R. 2012. Global forecasts of urban expansion to 2030 and direct impacts on biodiversity and carbon pools. *PNAS.* 109(40):16083–16088.

Shafizadeh-Moghadam, H. and Helbich, M. 2015. Spatiotemporal variability of urban growth factors: A global and local perspective on the megacity of Mumbai. *Int J Appl Earth Obs Geoinf.* 35(Pt. B):187–198.

Shem, W. and Shepherd, M. 2009. On the impact of urbanization on summertime thunderstorms in Atlanta: Two numerical model case studies. *Atmosph Res.* 92:172–189.

Shen, W., Wu, J., Grimm N.B., and Hope, D. 2008. Effects of urbanization-induced environmental changes on ecosystem functioning in the Phoenix metropolitan region, USA. *Ecosystems.* 11:138–155.

Shepherd, J.M. 2005. A review of current investigations of urban-induced rainfall and recommendations for the future. *Earth Interact.* 9(12):1–27.

Six, J., Bossuyt, H., Degryze, S., and Denef, K. 2004. A history of research on the link between (micro)aggregates, soil biota, and soil organic matter dynamics. *Soil Tillage Res.* 79:7–31.

Sloan, J.J., Ampim, P.A.Y., Basta, N.T., and Scott, R. 2012. Addressing the need for soil blends andamendments for the highly modified urban landscape. *SSSA J.* 76:1133–1141.

Smetak, K.M., Johnson-Maynard, J.L., and Lloyd, J.E. 2007. Earthworm population density and diversity in different-aged urban systems. *Appl Soil Ecol.* 37:161–168.

Smith, W.H. 1990. *Air Pollution and Forests: Interaction between Air Contaminants and Forest Ecosystems,* 2nd ed. New York, NY: Springer-Verlag.

Solomon, D., Lehmann, J., Kinyangi, J., Amelung, W., Lobe, I., Pell, A., Riha, S., Ngoze, S. et al. 2007. Long-term impacts of anthropogenic perturbations on dynamics and speciation of organic carbon in tropical forest and subtropical grassland ecosystems. *Global Change Biol.* 13:511–530.

Song, Y., Liu, H., Wang, X., Zhang, N., and Sun4, J. 2016. Numerical simulation of the impact of urban non-uniformity on precipitation. *Adv Atmos Sci.* 33:783–793.

Sparks, J.P. 2009. Ecological ramifications of the direct foliar uptake of nitrogen. *Oecologia.* 159:1–13.

Spurgeion, D.J. and Weeks, J.M. 1998. Evaluation of factors influencing results from laboratory toxicity tests with earthworms. In *Advances in Earthworm Ecotoxicology: Proceedings from the Second International Workshop on Earthworm Ecotoxicolgy.* eds. Sheppard, S.C., Bembridge, J.D., Holmstrup, M., and Posthuma, L., 15–25. Amsterdam: SETAC.

Steinbeiss, S., Gleixner, G., and Antonietti, M. 2009. Effect of biochar amendment on soil carbon balance and soil microbial activity. *Soil Biol Biochem.* 41:1301–1310.

Steinberg, D.A., Pouyat, R.V., Parmelee, R.W., and Groffman, P.M. 1997. Earthworm abundance and nitrogen mineralization rates along an urban-rural land use gradient. *Soil Biol Biochem.* 29:427–430.

Stockmann, U., Adams, M.A., Crawford, J.W., Field, D.J., Henakaarchchi, N., Jenkins, M., Minasny, B., McBratney, A.B. et al. 2013. The knowns, known unknowns and unknowns of sequestration of soil organic carbon. *Agricul Ecosys Environ.* 164:80–99.

Szlavecz, K., McCormick, M., Xia, L., Saunders, J., Morcol, T., Whigham, D., Filley, T., and Csuzdi, C. 2011. Ecosystem effects of non-native earthworms in mid-Atlantic deciduous forests. *Biol Invasions.* 13:1165–1182.

Taha, H. 1997 Urban climates and heat islands: Albedo, evapotranspiration, and anthropogenic heat. *Energy Build.* 25:99–103.

Templer, P.H., Toll, J.W., Hutyra, L.R., and Raciti, S.M. 2015. Nitrogen and carbon export from urban areas through removal and export of litterfall. *Environ Poll.* 197:256–261.

Thematic Strategy for Soil Protection. 2006. *Thematic Strategy for Soil Protection 231 final, 22.9.2006.* Brussels, EU: European Commission.

Townsend-Small, A. and Czimczik, C.I. 2010. Carbon sequestration and greenhouse gas emissions in urban turf. *Geophys. Res. Lett.* 37. doi:10.1029/2009GL041675.

Trammell, T.L.E. and Carreiro, M.M. 2012. Legacy effects of highway construction disturbance and vegetation management on carbon dynamics in forested urban verges. In: *Carbon Sequestration in Urban Ecosystems.* eds. Lal, R. and Augustin, B., 331–352. Netherlands: Springer Science and Business Media.

Trammell, T.L.E., Ralston, H.A., Scroggins, S.A., and Carreiro, M.M. 2012. Foliar production and decomposition rates in urban forests invaded by the exotic invasive shrub, *Lonicera maackii. Biological Invasions* 14:529–545. doi:10.1007/s10530-011-093-9

Trammell, T.L.E., Schneid, B.P., and Carreiro, M.M. 2011. Forest soils adjacent to urban interstates: Soil physical and chemical properties, heavy metals, disturbance legacies and relationships with woody vegetation. *Urban Ecosys.* 14:525–552.

Trippler, D.J., Schmitt, M.D.C., and Lund, G.V. 1988. Soil lead in Minnesota. In *Lead in Soil: Issues and Guidelines. Environmental Geochemistry and Health.* eds. Davies, B.E. and Wixson, B.G., 273–280. Northwood, U.K: Science Reviews Limited.

United Nations Department of Economic and Social Affairs, Population Division. 2011. *World Population Prospects: The 2010 Revision.* New York, NY: United Nations. Available at: http://www.un.org/esa/population/unpop.htm

USGBC. 2015. *SITES y2 Rating System for Sustainable Land Design and Development.* Washington, DC: U.S. Green Building Council.

Van Bohemen, H.D. and Janssen van de Laak, W.H. 2003. The influence of road infrastructure and traffic on soil, water, and air quality. *Environ Mgmt.* 31:0050–0068.

Van den Heever, S.C. and Cotton, W.R. 2007. Urban aerosol impacts on downwind convective storms. *J Appl Meteor Climato.* 46:828–850.

Van Oost, K., Quine, T.A., Govers, G., De Gryze, S., Six, J., Harden, J.W., Ritchie, J.C., McCarty, G.W. et al. 2007. The impact of agricultural soil erosion on the global carbon cycle. *Science.* 318:626–629. doi:10.1126/science.1145724.

Verburg, P.S.J. and van Breemen, N. 2000. Nitrogen transformations in a forested catchment in southern Norway subjected to elevated temperature and CO_2. *Forest Ecol Mgmt.* 129:31–39.

Veres, Z., Kotroczó, Z., Fekete, I., Tóth, J.A., Lajtha, K., Townsend, K., and Tóthmérész, B. 2015. Soil extracellular enzyme activities are sensitive indicators of detritus inputs and carbon availability. *Appl Soil Ecol.* 92:18–23.

Vidal-Beaudet, L., Caubel, V., and Olivier, R. 2009. Street tree root development in topsoil amended with high levels of compost. In: Second International Conference on Landscape and Urban Horticulture 881, June 9–13, pp. 511–516, Bologna, Italy.

Volin, J.C., Reich, P.B., and Givnish, T.J. 1998. Elevated carbon dioxide ameliorates the effects of ozone on photosynthesis and growth: Species respond similarly regardless of photosynthetic pathway or plant functional group. *New Phytol.* 138:315–325.

Washbourne, C-L., Lopez-Capel, E., Renforth, P., Ascough, P.L., and Manning, D.A.C. 2015. Rapid removal of atmospheric CO_2 by urban soils. *Environ Sci Technol.* 49:5434–5440.

Washbourne, C-L., Renforth, P., and Maning, D.A.C. 2012. Investigating carbonate formation in urban soils as a method for capture and storage of atmospheric carbon. *Sci Total Environ.* 431:166–175.

Watmough, S.A. and Meadows, M.J. 2014. Do earthworms have a greater influence on nitrogen dynamics than atmospheric nitrogen deposition? *Ecosystems.* 17:1257–1270.

White, S.S. and Boswell, M.R. 2006. Planning for water quality: Implementation of the NPDES phase II stormwater program in California and Kansas. *J Environ Plann Manag.* 49, 141–160. doi:10.1080/09640560500373386.

Windhager, S., Steiner, F., Simmons, M.T., and Heymann, D. 2010. Toward ecosystem services as a basis for design. *Landsc J.* 29:107–123.

Wu, L., Green, R., Klein, G., Hartin, J.S., and Burger, D.W. 2010. Nitrogen source and rate influence on tall fescue quality and nitrate leaching in a southern California lawn. *Agronomy J.* 102:31–38.

Xu, Z-F., Hu, R., Xiong, P., Wan, C., Cao, G., and Liu, Q. 2010. Initial soil responses to experimental warming in two contrasting forest ecosystems, Eastern Tibetan Plateau, China: Nutrient availabilities, microbial properties and enzyme activities. *Appl Soil Ecol.* 46:291–299.

Xu, S., Xu, W., Chen, W., He, X., Huang, Y., and Wen, H. 2016. Leaf phenological characters of main tree species in urban forest of Shenyang. *PLOS ONE*. 9(6):e99277. doi:10.1371/journal.pone.0099277.

Yang, J-L. and Zhang, G-L. 2015. Formation, characteristics and eco-environmental implications of urban soils—A review. *Soil Sci Plant Nutr*. 61:30–46.

Yesilonis, I.D., James, B.R., Pouyat, R.V., and Momen, B. 2008. Lead forms in urban turfgrass and forest soils as related to organic matter content and pH. *Environ Monit Assess*. 146:1–17.

Yesilonis, I.D. and Pouyat, R.V. 2012. Carbon stocks in urban forest remnants: Atlanta and Baltimore as case studies. In: *Carbon Sequestration in Urban Ecosystems*. eds. Lal, R. and Augustin, B., 103–120. Netherlands: Springer Science and Business Media.

Yesilonis, I.D., Pouyat, R.V., and Neerchal, N.K. 2008. Spatial distribution of metals in soils in Baltimore, Maryland: Role of native parent material, proximity to major roads, housing age and screening guidelines. *Environ Poll*. 156:723–731.

Zhang, C.S. 2006. Using multivariate analyses and GIS to identify pollutants and their spatial patterns in urban soils in Galway, Ireland. *Environ Poll*. 142:501–511.

Zhao, L., Lee, X., Smith, R.B., and Oleson, K. 2014. Strong contributions of local background climate to urban heat islands. *Nature*. 511:216–221.

Zhou, T., Li, L., Zhang, X., Zheng, J., Joseph, S., and Pan, G. 2016. Changes in organic carbon and nitrogen in soil with metal pollution by Cd, Cu, Pb and Zn: A meta-analysis. *Eur J Soil Sci*. 67:237–246.

Zimmerman, A.R. 2010. Abiotic and microbial oxidation of laboratory-produced black carbon (biochar). *Environ Sci Tech*. 44:1295–1301.

Zimmerman, A.R., Gao, B., and Ahn, M.Y. 2011. Positive and negative carbon mineralization priming effects among a variety of biochar amended soils. *Soil Biol Biochem*. 43:1169–1179.

Ziska, L.H., Bunce, J.A., and Goins, E.W. 2004. Characterization of an urban-rural CO_2/temperature gradient and associated changes in initial plant productivity during secondary succession. *Oecologia*. 139:454–458.

Ziska, L.H., Gebhard, D.E., Frenz, D.A., Faulkner, S., Singer, B.D., and Straka, J.G. 2003. Cities as harbingers of climate change: Common ragweed, urbanization, and public health. *J Allergy Clin Immunol*. 111:290–295.

Zong-Qiang, W.E.I., Shao-Hua, W.U., Sheng-Lu, Z.H.O.U., Jing-Tao, L.I., and Qi-Guo, Z.H.A.O. 2014. Soil organic carbon transformation and relate properties in urban soil under impervious surfaces. *Pedosphere*. 24:56–64.

5 Soil Carbon and Nitrogen Cycling and Ecosystem Service in Cities

Weixing Zhu, Beth A. Egitto, Ian D. Yesilonis, and Richard V. Pouyat

CONTENTS

5.1 INTRODUCTION

Soil is the foundation of life as well as all ecosystem processes. In urban ecosystems, soils are spatially heterogeneous and temporally dynamic. The spatial heterogeneities of urban soils are both horizontal and vertical (Craul 1985; Forman 2014); vertically, urban soils are often characterized by multiple buried layers, so-called lithologic discontinuity (Craul 1985; Short et al. 1986), while horizontally, they are associated with different land use and land cover, and topographic variations (Band et al. 2005; Pouyat et al. 2006a; Forman 2014). Urban soils are often disturbed and compacted, contain a variety of man-made materials as *fill*, and have lower levels of soil organic matter (SOM) and high levels of bulk density (BD) that could negatively affect water filtration, aeration, and plant growth (Craul 1985). Urban soils are often contaminated by heavy metals (Pouyat and McDonnell 1991; Yesilonis et al. 2008) and hydrocarbon pollutants and have elevated pH and sometimes very low nutrients (Forman 2014). Yet, dynamic life exists in urban soils and important ecosystem processes affect soil formation and function in cities. The objective of this chapter is to discuss the current understandings of soil carbon (C) dynamics and nitrogen (N) cycling in urban ecosystems, and ecosystem services provided by urban soils.

Urban soils, simply put, are soils found in urban areas. The pedologic definition of soil as "a collection of natural bodies on the earth's surface, in places modified or even made by man of earthy materials, containing living matter and *supporting or capable of supporting* plants out-of-doors" (Soil Survey Staff 1975; Effland and Pouyat 1997) seems to fit the purpose of studying urban soil well. The level of human disturbance varies greatly depending on the extent of urbanization, from the broad presence of relatively undisturbed soils in suburban and exurban areas, to the extensive addition of either earthy fill materials (excavated from either natural or other disturbed sites) or human-derived artifacts in urban centers (Effland and Pouyat 1997; Forman 2014). Although the definitions of urban ecosystems vary, generally speaking, urban ecosystems include cities and towns, as well as suburban areas, and are characterized by high human population densities, extensive built infrastructure, and impervious surface covers (ISCs), and have extremely high metabolisms and altered biogeochemical cycles associated with human activities (Pickett et al. 2001; Grimm et al. 2008). These characteristics have profound impacts on urban soils and C and N biogeochemistry.

5.2 ORGANIC C IN URBAN SOILS

Soil organic C (SOC) is a critical measurement of soil property. There has been interest recently in trying to quantify SOC storage and accumulation in urban ecosystems in the context of global climate and biogeochemical changes (Pataki et al. 2006; Pouyat et al. 2006b). SOC is the major component of SOM (~50% of the SOM mass) that controls an array of soil physical and chemical properties including BD, water filtration and aeration, cation exchange capacity (CEC), and soil structure (e.g., aggregates). SOM regulates microbial decomposition and C and N mineralization, supports soil fauna activities and diversity, and affects aboveground plant growth and subsequently, urban wildlife.

Early works, some in iconic places like Washington DC malls, Central Park in New York City (NYC), and Hong Kong Victoria Park, often reported urban soils were low in SOC and high in BD, and had low soil aeration and water drainage, and poor plant growth (Short et al. 1986; Jim 1998). For example, the Hong Kong study (Jim 1998) reported soils developed from mostly granite fill materials had high stone content, and high BD exceeding 1.75 Mg/m^3, but very low accumulation of SOM. The DC mall study reported similar soil compaction in a highly trampled public space with mean BD of 1.63 Mg/m^3 for the surface horizon and 1.74 Mg/m^3 at 0.3 m depth, and mean SOM of 19.7 g/kg in the surface horizon and 4.1–6.6 g/kg in the lower soil horizons (Short et al. 1986).

Not all urban land changes involve filling of man-made or transported materials with high BD and low SOC. In most North American cities, residential land development usually includes top soil being scraped away, stockpiled, and respread, followed by various lawn management activities including fertilization and irrigation (Craul 1985; Pouyat et al. 2009; Chen et al. 2013). Human intervention in urban soils can be both *direct* and *indirect* (Pouyat et al. 2006b, 2010; Yesilonis and Pouyat 2012). In the direct intervention, original earth material is excavated and moved, and the mixing and filling result in soil that is at least partially disturbed or a soil profile that is entirely remade. Depending on the nature of the fill material, SOC can be extremely low, resembling soil development following primary succession in natural ecosystems (i.e., soil developed after landslide, volcanic eruptions, etc.), or high due to buried formal A layer soil (Forman 2014).

The SOC concentration depends on the balance of C input through plant primary production and C loss through microbial decomposition; both are affected by human decisions and human-modified environments in urban ecosystems (Figure 5.1). The urban environment affects plant production and biomass accumulation (A) as well as decomposition (B) through changes in important drivers like CO_2 concentration, temperature, and nutrients (M). The primary production provides organic matter (OM) as litter, but, in particular, causes root turnover (C) that supports decomposer organisms in the soil (D). Most plant C and other OM are decomposed rapidly (H), but a small percentage can be retained for a very long time, centuries or even millenniums, as passive SOC (I). Recent advances in soil science point out the significance of organo-mineral associations, not the chemical recalcitrance of plant litter, as the main driver of long-term SOC stability (Schmidt et al. 2011). Thus in urban ecosystems, soil disturbance, or lack of it, often plays a dominant role in soil C accrual and stabilization.

Both irrigation and fertilizer input greatly stimulate lawn production in residential yards. Elevated CO_2 concentration in cities may also stimulate primary production and increase C flow into soils. In a study of the Denver–Boulder metropolitan ecosystem, Golubiewski (2006) reported aboveground grass productivity ranging from 80 to 1288 g/m^2/yr with a mean of 445 g/m^2/yr, significantly higher than the native prairie or agricultural fields. The intensity of lawn management was the main determinant of all grass production and biomass harvested, with intensive lawn care in the form of automatic sprinklers several times a week and multiple fertilizer applications throughout the year (Golubiewski 2006). However, the favorable water condition and in general a high

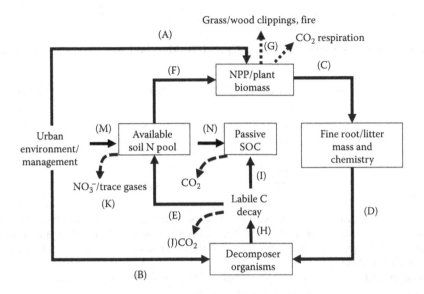

FIGURE 5.1 A conceptual diagram showing carbon (C) and nitrogen (N) cycling and their interactions in urban soils, and drivers affecting C and N processes in urban ecosystems. Processes are shown in capital letters and explained in the text. Solid lines represent internal flows of C and N within the ecosystem, dashed lines represent losses from the ecosystem. (Modified from Carreiro, M.M. et al., *Ecology of Cities and Towns—A Comparative Approach,* Cambridge University Press, 2009 and Yesilonis, I.D., and R.V Pouyat, *Carbon Sequestration in Urban Ecosystems,* Springer, 2012.)

urban temperature (i.e., urban heat island, or UHI effect) would also favor high decomposition and soil respiration that contributes to overall high urban CO_2 fluxes (Groffman et al. 2006; Decina et al. 2016). Nevertheless, age is often an important contributing factor to the residential SOC pool (Golubiewski 2006; Raciti et al. 2011), due to the lack of frequent soil disturbance, once the property is established. Golubiewski (2006) reported that established residential yards harbored more SOC, more than double in some cases, than the native prairie (limited by water and nutrients) and agricultural lands (regularly disturbed by plowing).

ISC represents a widely distributed type of direct soil disturbance in urban ecosystems, which prevents future C input into the soil and often involves surface soil removal, and with paved surface and traffic above, altered soil environment (temperature, moisture, and pH) (Figure 5.1). ISC varies greatly among cities. In the six major cities of the United States examined by Pouyat et al. (2006b), ISC ranges from 39.8% in Atlanta to 60.0% in Chicago (Table 5.1). In Phoenix, AZ, various urban land-use patches had ISC ranging from an average 26.7% in mixed land use to 65.8% in transportation land use, with large variation within each land-use category (Zhu et al. 2006). In Seattle, WA, ISC decreased significantly from the urban core (some places with >80% ISC) across a gradient of urbanization (Hutyra et al. 2011). Soil C is variable, but usually low under impervious surface. Pouyat et al. (2006b) assigned an SOC density of 3.3 ± 0.93 kg/m^2 under impervious cover (C density defined here as organic C content down to 1-m depth).

Residential land use occupies a large proportion of urban area, and in North America cities, often is the dominant land-use type (Table 5.1). Soil C dynamics in residential areas can be influenced indirectly by the site management associated with homeowner decisions and neighborhood locations (Raciti et al. 2011; Pickett et al. 2008), and by elevated atmospheric deposition and fertilizer input that change the pool of soil available N (Figure 5.1). Residential land use in the six cities examined by Pouyat et al. (2006b) ranged from 43.7% (Oakland, CA) to 54.6% (Atlanta, GA). In Phoenix, AZ, excluding the natural Sonoran Desert surrounding the city, urban residential area occupies 34.7% of the land, surpassing urban nonresidential (25.7%) and agriculture (22.3%) land use (Zhu et al.

TABLE 5.1

Percentage Impervious Surface Cover (ISC) and Residential Land Use in Seven Major US Cities

City	Total Area (km²)	% ISC	% Residential
Atlanta	341.4	39.8	54.6
Baltimore	209.2	50.4	48.1
Boston	142.8	53.9	44.1
Chicago	613.7	60.0	47.7
Oakland	132.4	47.8	43.7
Syracuse	65.0	46.5	50.0
Phoenix (CAP)[a]	3306.5	32.1	34.7

Note: Land-use and cover data of Atlanta to Syracuse were based on the UFORE Project, in which 200 (0.04 ha) plots were selected for each city (Pouyat et al. 2006). The Phoenix data were from the 203-plot survey of the Central Arizona Phoenix (CAP) project. Each plot was 30 m × 30 m (0.09 ha), stratified and randomly selected from the 6400 km² CAP area (Zhu et al. 2006).

[a] Data presented here excluded 3093.6 km² (73 plots) of Sonoran Desert surrounding the central urban area.

2006). Residential yards often contain a large proportion of turf grass cover (lawn), which during the initial stage of site building can experience direct disturbance of surface soil removal, stockpiling, and replacement, as well as subsoil compacting from heavy machinery (Chen et al. 2013), usually followed by management decisions involving fertilizer input, irrigation, and herbicide/pesticide control that could increase SOC accrual (Pouyat et al. 2006b, 2009; Raciti et al. 2011). Raciti et al. (2011) reported in Baltimore, residential soils had significantly higher C density than nearby forest soils of similar soil types. In Phoenix, soils from the urban residential and nonresidential sites had SOM 35% higher than the natural desert (Zhu et al. 2006). Soils at residential sites that were previously used for agriculture (Raciti et al. 2011) or natural lands with limited primary production (Golubiewski 2006) can rapidly accumulate organic C after residential development. Pouyat et al. (2006b) assigned an SOC density of 14.4 ± 1.2 kg/m² under residential grass cover, considering the nutrient and water subsidies in these systems, although that number likely varies across cities as well as within cities. But, ISC in residential areas can be as high as or even surpass the city average, which will bring down SOC storage (Pouyat et al. 2006b). On the other hand, alternative land development practices, including soil profile rebuilding with organic compost incorporated into subsoil, can increase SOC storage, reduce BD, and enhance microbial activities that are important for future soil development (Chen et al. 2013).

Natural remnant habitats along the urban–rural gradient provide a unique opportunity to examine the impact of urbanization on soil C (and N) dynamics (Pouyat and Carreiro 2003; Zhu and Carreiro 2004; Carreiro et al. 2009). In a 22-month litter decomposition study, Pouyat and Carreiro (2003) reported no difference of litter decay along a 130-km urban–rural transect in the NYC metropolitan area, when litter was left on the same site it was collected. In the transplant experiment, however, litter decomposition showed contrast controls of litter quality and decomposition environment. While leaf litter decomposed significantly faster in urban sites, regardless of whether the litter was collected from urban or rural forests, litter collected from the urban sites decomposed significantly slower, regardless of whether it was placed in urban or rural environments (Pouyat and Carreiro 2003). The fast decomposition in urban sites is likely due to warmer temperatures in cities and other environmental conditions (B in Figure 5.1). In the NYC urban–rural gradient study, high densities of exotic earthworms in the urban forests likely also contribute to the fast decomposition (Carreiro et al. 2009). The slower decomposition of urban litter, on the other hand, indicates lower litter quality of soil decomposers (i.e., fungi and bacteria), likely caused by various urban pollutants that affect plant growth and leaf senescence (Carreiro et al. 1999, A and C in Figure 5.1).

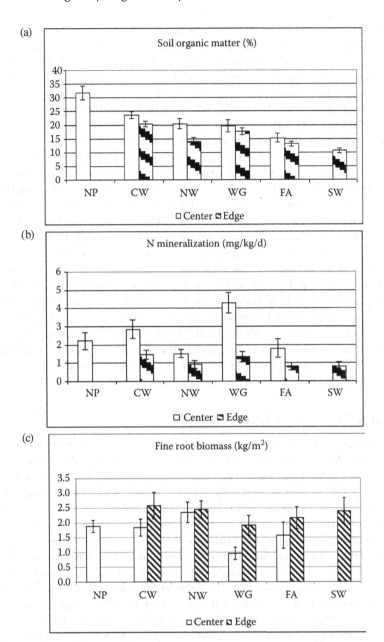

FIGURE 5.2 (a) Soil organic matter content (b), net nitrogen (N) mineralization rates, and (c) fine root biomass in forested patches from the largest nature preserve (NP) to the smallest science woods (SW), and the effect of location in the patch (Center or Edge of the patch) at the State University of New York, Binghamton campus. Only the center location was selected for the NP site and only edge habitat was available for the SW site. Means ± standard errors. (From Egitto, B.A., Nitrogen and root dynamics in urban forest patches. M.S. thesis, Binghamton University, State University of New York, 2005.)

Remnant forests in other urban regions sometimes show different patterns of C (and N) dynamics. In Louisville, KY, urban sites contained more SOM than rural plots, and litter decomposition in forests was not significantly different along the urban–rural gradient, despite the fact that leaf litter collected from urban and suburban sites had lignin:N ratios 15% higher than that in rural litter, a result similar to the NYC gradient study (Carreiro et al. 2009). In Baltimore, MD, comparisons of four urban forests and four rural forests showed significantly higher soil respiration (CO_2 flux) and lower CH_4 uptake in the urban sites, while greater productivity was more associated with natural soil conditions (urban sites were not necessarily nutrient rich, Groffman et al. 2006). In a study in Asheville, NC, Pavao-Zuckerman and Coleman (2005) reported slower litter decomposition toward the urban end of the gradient when a common reference litter was used, but higher *in situ* net N mineralization rates in urban soils. In Seattle, WA, Hutyra et al. (2011) found surprisingly high live aboveground plant biomass within forested and urban land covers (140 ± 40 and 18 ± 14 Mg C/ha, respectively), with a significant decrease of plant biomass as a function of distance from the urban core largely due to the change of land cover (more impervious land cover and less forests in the urban core), but with little statistically distinguishable differences within any land-cover type as a function of distance from the city (i.e., forests located in downtown areas had similar biomass and canopy cover as those 30 km away from the urban core). Such an aboveground trend of plant biomass and canopy cover distribution is expected to affect belowground soil C dynamics.

An important feature of urban ecosystems is landscape fragmentation (Luck and Wu 2002). Urban landscapes are increasingly compositionally diverse, structurally fragmented, and geometrically complex (Wu et al. 2011). These trends could have a major impact on soil C dynamics. Forest edges, for example, often have different temperature, moisture, and pH from the interior of the forest. At the State University of New York campus in Binghamton, remnant forests with different patch sizes showed a decrease in SOM from the largest patch to the smallest, and significantly lower SOM in the edges than in the centers of the forests (Figure 5.2a). The likely cause of this trend is the higher temperature or other environmental changes in the edges that accelerate SOM decomposition. The changes in SOM were also correlated to the changes of N mineralization rates and fine root biomass (Figure 5.2b and c).

5.3   NITROGEN CYCLING IN URBAN SOILS

Nitrogen cycling has profound impacts on C cycling because in most Earth ecosystems, available N (mostly ammonium, NH_4^+, and nitrate, NO_3^-) is scarce and often limit plant growth and microbial activities. In urban ecosystems, human activities, from importing food, fuel, and fertilizer to elevated atmospheric N deposition, have greatly changed N biogeochemistry in cities (Baker et al. 2001; Groffman et al. 2004; Zhu et al. 2006). While globally, anthropogenic N fixation due to fertilizer production, fossil fuel combustion, and legume crop expansion has more than doubled the annual reactive N input into the biosphere (Galloway et al. 2003), cities have increasingly become "hot spots" of N (and C) in regional landscapes (Grimm et al. 2008). Within cities, the removal, reduction, or accrual of SOC in urban ecosystems are interacting with changes of soil N since C and N are biogeochemically coupled (Figure 5.1). The decay of SOM not only releases CO_2 back to the atmosphere (J), but also replenishes available soil N pool (E), which then supports plant growth (F). The variation of SOM, which contains both organic C and organic N with its specific C:N ratio, often affects the transformation, transfer, and retention of inorganic N (NH_4^+ and NO_3^-).

Urban ecosystems can accumulate alarming levels of inorganic N. In Phoenix, soil inorganic N, mostly in the biogeochemical reactive form of NO_3^-, was on average 10 times higher in the transportation and mixed-use urban lands than in the natural Sonoran Desert soils, which themselves likely had N concentrations elevated several times compared to that of desert soils far away from urban impact (Zhu et al. 2006). The estimated increase of inorganic N in the 0–30 cm soil depth from the preurbanization level in Phoenix was on average 7.23 g N/m^2, or 46.3×10^9 g N in the 6400 km^2 Central Arizona–Phoenix urban ecosystem, which can account for several years of N accumulation calculated from the entire system input–output budget of the Phoenix urban ecosystem (Baker et al. 2001; Zhu et al. 2006).

Most N inputs into urban ecosystems, either as elevated atmospheric N deposition (Lovett et al. 2000), or direct fertilizer input, are likely retained as organic N and stored in soils along with SOC (M and N in Figure 5.1). Raciti et al. (2011) reported significantly higher soil organic N density in residential lawns that received direct fertilizer input than nearby forest soils of similar soil types (552 vs. 403 g N/m^2, down to 1 m depth). Among residential sites developed from previous agricultural lands, the rate of N accumulation (8.3 g/m^2/yr) was similar in magnitude to estimated fertilizer N inputs, confirming a high capacity for N retention in urban soils (Raciti et al. 2008, 2011). Pouyat et al. (2007) investigated multiple chemical and physical properties of soils collected from 122 plots in Baltimore, and found an average total N concentration of 1.6 g/kg in the surface 0–10-cm soils, with a wide range from 0.1 to 6.6 g/kg. However, no significant difference of N concentration was found among the six major land-use and cover types examined (commercial or transportation, industrial or urban open, unmanaged forest, park or golf course, residential, and institutional).

Large quantities of SOC found in urban ecosystems can affect the retention of inorganic N (N in Figure 5.1). Applying a pulse of ^{15}N labeled NO_3^-, Raciti et al. (2008) studied the retention of simulated inorganic N input in the Baltimore urban ecosystem. They found 99 ± 9% recovery of ^{15}N-NO_3^- in the lawn plots, and 83 ± 5% in the forest plots, 1 day after the N addition; ^{15}N label recovered declined steadily over time, but was significantly higher in lawns than in forests after 10, 70, and 365 days. The dominant N sink in lawns also changed over time: first in mineral soils (42% recovered N after 1 day), then plant biomass (10 days), followed by thatch and mineral soil pools at the inter-medium term (70 days), and long-term retention in mineral soils (1 year, 70% of recovered ^{15}N). Many studies in natural ecosystems reported high retention of inorganic N in soils containing high organic matter and organic C (Nadelhoffer et al. 1999; Perakis and Hedin 2001). In a laboratory study, Zhu and Wang (2011) collected forest soils with varying SOM associated with the dominant tree species of the sites (including coniferous species of Norway spruce and red pine, and deciduous species of sugar maple and red oak), and found the retention of $^{15}NH_4^+$ was positively correlated with the percentage of SOM, but the retention of $^{15}NO_3$- was not. In another greenhouse pot study, plants not only increased overall retention of $^{15}NH_4^+$ (including incorporation of ^{15}N into plant biomass), but also increased ^{15}N retained in the soil (Wang and Zhu 2012). Thus, urban land with healthy plants (trees or grasses) and high SOM content can effectively retain inorganic N from management fertilizer input and from atmospheric deposition (Groffman et al. 2004; Pickett et al. 2008). The interactions between reactive N compounds (NH_4^+, NO_3^-, NO_2^-, and dissolved organic N) and SOC (N in Figure 5.1) may contribute not only to the N retention, but also the long-term stability of soil C, especially at the organo-mineral surface of soil particles (Schmidt et al. 2011). However, much remains unknown about the biochemical and physical mechanisms involved. In cities, the potential to use urban soils and plants to overcome N pollution, especially from nonpoint sources of atmospheric deposition and fertilization, needs to be further studied.

The high SOC found in certain urban sites could also favor the removal of NO_3^- pollutants through microbial-driven denitrification. Denitrification is the process in which NO_3^- serves as an electron acceptor to oxidize organic C when O_2 level is low, and converts NO_3^- to chemically inactive N_2, although nitrous oxide (N_2O), a potent greenhouse gas, is often a by-product of denitrification. In urban soils, soil compaction can limit aeration (often due to deep soil compaction) and water stagnation can create localized anoxic conditions that favor denitrification. In the Phoenix urban ecosystem, Zhu et al. (2004) collected soils from green retention basins and city parks and found significantly higher SOM, extractable NO_3^- concentration, and nitrification rates than desert soils. These urban green spaces (UGSs) retain flood water during storms and serve as neighborhood playgrounds at other times. Denitrification rates, measured both as potential rates with substrate amendment, and as intact-core fluxes, were comparable to the highest rates reported in literature for natural ecosystems (Zhu et al. 2004). Urban ecosystems are hotspots of N contamination in regional landscapes due to intensive input of N compounds (Baker et al. 2001; Grimm et al. 2008). They also contain many hotspots of N removal due to heterotrophic distribution of NO_3^-, water, organic matter, and soil conditions favoring denitrification (Band et al. 2005; Grimm et al. 2008). These denitrification hotspots, both spatially and temporally highly variable, exist in riparian zones, urban wetlands, soils, and stream sediments (Groffman et al. 2003; Kearney et al. 2013). Budget analyses of urban watersheds support the notion of efficient N removal or retention in urban soils

(Groffman et al. 2004; Wollheim et al. 2005). While N losses from urban and suburban watersheds of Baltimore, MD, were unsurprisingly higher than forested watersheds, they were nevertheless much lower than the agricultural watershed, with the suburban watershed reaching a surprisingly high 75% retention of N inputs from fertilizer and atmospheric deposition (Groffman et al. 2004).

There has been a substantial amount of work examining the change of N cycling in remnant natural areas of urban ecosystems (Zhu and Carreiro 1999, 2004; Pavao-Zuckerman and Coleman 2005; Groffman et al. 2006; Hall et al. 2011). Nitrogen cycling is generally stimulated in urban ecosystems (Zhu and Carreiro 2004; Pavao-Zuckerman and Coleman 2005). Elevated atmospheric N deposition has been documented in multiple cities (Lovett et al. 2000; Carreiro et al. 2009; Hall et al. 2011). The change of N cycling could be shown both in terms of the increase of N mineralization rates (the conversion of soil organic N to inorganic NH_4^+), and probably more importantly, the shift from NH_4^+ production to NO_3^- production (nitrification). The induction and the continuous increase of nitrification, especially in regions that naturally had low rates of net nitrification (e.g., northeast temperate forests and boreal forests), is widely considered a sign of shifting from N-limited condition to N-saturation status (Aber et al. 1998).

In the study of remnant oak forests along an urban–rural gradient in the NYC metropolitan area, Zhu and Carreiro (2004) found substantially higher net N mineralization rates in urban forest soils than in rural forest soils. While net nitrification rates were mostly below zero in rural soils, in urban soils, they accounted for between 23.2% and 73.8% of the annual N mineralized. Furthermore, the seasonal increase of N mineralization from cold winter to warm summer tended to occur a month earlier in urban forests, which led to higher NO_3^- concentrations in urban soils in early spring when N losses, either through NO_3^- leaching, or denitrification, could occur before major plant uptake of N (Zhu and Carreiro 2004, K in Figure 5.1).

Urban natural areas can have elevated N availability and N turnover due to enhanced N input from atmospheric deposition, and/or due to the introduction of nonnative fauna (Steinberg et al. 1997; Zhu and Carreiro 1999, 2004) and flora (Ehrenfeld et al. 2001). Along the NYC urban–rural gradient, Lovett et al. (2000) reported that N entering urban forests was 50%–100% more than N flux into rural forests; furthermore, the difference was mainly due to dry N deposition (those associated with particulate N deposition or gaseous forms absorbed to leaf surface) that was 17 times greater in the urban end of the gradient. The N deposition was associated with the particulate distribution of Ca^{2+} and Mg^{2+}, which tended to increase the pH of the throughfall that entered the forest (Lovett et al. 2000). In cities, leaching from buildings and dusts from construction materials also contribute to the increase of alkalinity (and Ca and Mg availability) in urban soils (Pouyat et al. 2007; Forman 2014). The change of pH can be particularly important in terms of stimulating chemoautotrophic nitrification (bacteria gain energy by oxidizing NH_4^+ to NO_3^-), which is favored in less acidic soils (Zhu and Carreiro 1999). Interestingly, the replacement of fast turnover leaf litter from nonnative plants, often common in urban and suburban areas, could also raise soil pH and consequently lead to higher N mineralization and nitrification rates (Ehrenfeld et al. 2001; Ehrenfeld 2010).

The change of N cycling in urban ecosystems can affect multiple pathways in the C cycling. The stimulating effect of N on C fixation (i.e., higher net primary production [NPP]) in urban ecosystems remains largely inconclusive (Hall et al. 2011), possibly due to the confounding effect of other urban pollution (Forman 2014), or landscape positions and plant functional types, for example, downwind sites had higher precipitation than upwind sites and urban cores, which could affect NPP in arid regions (Hall et al. 2011). On the other hand, Groffman et al. (2006) found significantly less methane (CH_4, another potent greenhouse gas) consumption in urban forest soils than in rural forest soils. The microbial CH_4 oxidation shares a key enzyme with chemoautotrophic nitrification, with increased nitrification often leading to reduced CH_4 consumption in urban soils (Goldman et al. 1995). In a 4-year study in Baltimore, Groffman and Pouyat (2009) reported dramatic reduction of CH_4 uptake from rural forests (1.68 mg/m²/day) to urban forests (0.23 mg/m²/day) and almost complete elimination in urban lawns. Thus, N pollution in urban ecosystems can negatively affect an important ecosystem service provided by urban soils (CH_4 uptake) while having the potential to increase

NO_3^- leaching and N_2O emission. Highly fragmented forest patches in cities can have different C and N cycling than natural forests. In addition to exotic species, pH, human traffic, and disturbances often reported in urban literature, C and N cycling can engage in interesting interactions. For example, the decline of SOC in forest edges compared to forest centers can reduce soil N mineralization rates as microbial organisms decompose organic matter to release inorganic N (Figure 5.2b). Ironically, this actually reduces available N to plants in an urban environment, unless external N, for example, fertilizer, is applied. In a study on the State University of New York campus, higher fine root biomass was found in forest edges, not in forest centers, likely to compensate for lower N availabilities at the edges (Figure 5.2c). Soil C and N cycling in urban soils are affected by multiple, urban-specific drivers, and likely to continue challenging the conventional wisdom of soil processes in the cities (Figure 5.1).

5.4 ECOSYSTEM SERVICES OF URBAN SOILS AND UGSs

Soil is the foundation of vegetated urban habitat or UGS that provides key ecosystem services in cities. Widespread ISC affects urban hydrology, pollutant movement, and heat transfer, and contributes to the unique features of urban biogeochemistry (Kaye et al. 2006). Pervious soil and related vegetation are important components of an urban ecosystem and provide critically needed ecosystem services to urban dwellers. Urban soils provide important ecosystem services of N retention and C sequestration as discussed above.

Urban soils also vary greatly in their SOM contents. The quantity (determined by SOC) and characteristics (often by soil N, or the C:N ratio) of SOM affect the structure and function of urban soils, water infiltration, pollutant removal, vegetation growth, and evapotranspiration and consequently, both the hydrology and heat budget of urban ecosystems. Soils play important roles in supporting the structure and function of UGS, which include remnant natural areas, public parks, public and private landscaping, domestic gardens, sports and playing fields, allotments, cemeteries, nature reserves, and deserted/vacant lands (Gaston et al. 2013). In the past several decades, there have been extensive research documenting the critical importance of green space to the function of urban ecosystems, by providing essential socioeconomic, ecological, and planning amenities (Jim and Chen 2006; Baycan-Levent and Nijkamp 2009; Gaston et al. 2013). Numerous studies have shown that UGS plays a key role in urban sustainability, through mitigating heat stress, regulating hydrology, affecting C and N biogeochemistry, conserving biodiversity, and enhancing both the physical and mental health of urban dwellers (McKinney 2002; Alberti et al. 2003; Groffman et al. 2004; Fuller et al. 2007; Jenerette et al. 2011). Because the world is being urbanized at an accelerating rate, with more than 50% of the global population now living in cities and having pervasive impact on the global ecology in addition to economy (Grimm et al. 2008; Wu et al. 2011), urban sustainability is critically important as it affects and eventually determines global sustainability.

Bolund and Hunhammar (1999) analyzed seven within-city components including street trees, lawns/parks, urban forests, cultivated lands, wetlands, lakes/sea, and streams and discussed a range of ecosystem services they provided. Many of these ecosystem services depend on good quality soils that can support functionally diverse and sustainable green spaces. Urban soils can be highly compacted with high BD, although typically less than agricultural soils (Edmondson et al. 2011). Soils with higher organic matter and organic C, distributed along the soil profile, will have low BD and high porosity, and can greatly increase water filtration and minimize peak stormflows that are major problems in cities (Whitford et al. 2001; Pauleit et al. 2005). Urban plants, whether turf grass or trees, in residential plots or natural areas, can greatly influence C flow into soils, in particular through the turnover of roots (C and D in Figure 5.1), and through microbial decomposition, affect long-term SOC storage (H and I in Figure 5.1).

There are both challenges and great opportunities to improve urban soils for the sustainability of cities (Kumar and Hundal 2016). Urban soils can be amended with organic substances from local sources such as composts and biosolids to support better plant growth and improve ecosystem function related to greater stormwater infiltration, sequestrating C in deeper soil layers, and mitigating the UHI effect. Johnston et al. (2016) compared "prairie gardens" with traditional residential turf grass. Soil underneath prairie gardens (0–15 cm) had 25% greater SOM, 10% lower BD, 15% lower penetration resistance (PR), and 33% greater saturated hydraulic conductivity (Ksat) compared to soil underneath the adjacent lawns (Johnston et al. 2016).

However, such differences were not detected at deeper soil layers. Excavation and compaction during residential developments often affect lawn infiltration rates and stormwater runoff. In a comprehensive study of 108 residential sites and 18 agricultural sites in south-central Pennsylvania, Woltemade (2010) reported that measured infiltration rates were significantly higher for lots constructed pre-2000 (9.0 cm/h) than those post-2000 (2.8 cm/h), compared to the mean of agricultural sites of 10.2 cm/h, showing high soil compaction on recently developed residential lots and gradual improvement of infiltration with time, likely caused by the buildup of SOM. Using deep SOC amendment (compost incorporation), Chen et al. (2014) showed that rehabilitated soils had 48%–171% greater macroaggregate-associated C than the other treatments in a land development project in Montgomery County, VA, even though soil rehabilitation did not measurably enhance aggregate formation after 5 years. Deep incorporation of SOC did have about twice the Ksat of undisturbed soils (control) at 10–25 and 25–40 cm depths, and approximately 6–11 times that of soils subjected to typical land development practices (Chen et al. 2014). The effectiveness of UGS to infiltrate storms of different size and affordable methods to increase soil hydraulic conductivity, however, remain largely unknown.

Water infiltrated into soils can recharge groundwater and streams and provide the ecosystem service of supporting aquatic biodiversity (Paul and Meyer 2001). But, even when water infiltration into deeper soil layers is hampered due to soil compaction, water retained in surface soil layers still supports plant growth and productivity and affects urban microclimate (Livesley et al. 2016). NPP is positively related to the actual evapotranspiration rate, and in urban areas, evapotranspiration provides an essential ecosystem service of ameliorating the UHI effect by dissipating heat at critical times. Heat waves associated with global climate change and enhanced in urban areas (extreme heat events, chronically hot weather, or both) are a major concern of urban sustainability (Jenerette et al. 2011). In analyzing three decades (1970–2000) of land surface characteristics of residential neighborhoods in Phoenix, Jenerette et al. (2011) reported that vegetation provided nearly 25°C surface cooling on low-humidity summer days in this arid urban ecosystem. The ecosystem service of summer cooling comes at a cost, with an estimated 2.7 mm/d of water use on a typical summer day in Phoenix. Socioeconomically, neighborhood income was related to evaporative cooling and the availability of water resource use (Jenerette et al. 2011).

The need to reduce heat stress is not limited to arid cities with low precipitation (Whitford et al. 2001; Pauleit et al. 2005). There are often trade-offs between high vegetation cover, lower summer temperature, less runoff, more soil C storage, and the social and economic cost of maintaining green spaces, as well as equity implications (Jenerette et al. 2011; Whitford et al. 2001). On the other hand, there is a clear need to reconsider the policy of extensive engineering of underground pipes to remove water from cities as quickly as possible (Kaye et al. 2006), and balance that with water discharging into soils to support plant growth, C accrual in plants and soils, evapotranspirational cooling of cities, and maintaining baseflow for urban streams and wildlife. For example, water drained from the roof (a type of ISC) is usually channeled to yards where plants and soils are located or can be collected in rain barrels to irrigate vegetable gardens.

The spatial heterogeneity of urban landscape (Wu et al. 2011) greatly influences the interactions between urban soils and paved surface. Easton et al. (2007) modeled excess runoff from ISC infiltrated into the surrounding soils in a soil moisture distribution and routing (SMDR) model to simulate hydrologic processes in an upstate New York urban watershed. The SMDR captures the key concept of variable source areas (VSAs, e.g., paved surface generates runoff and affects moisture in the adjacent soil), and the spatial-temporal extent of VSAs in a watershed. Such high-resolution models are needed to accurately predict the hydrology and biogeochemical retention and movement of contaminants in urban watershed, as well as to analyze the interactions between ISC and soils in spatially explicit ways (Cadenasso et al. 2007). On the other hand, extensive storm sewer systems designed to move stormwater quickly from impervious surface via subsurface pipes in highly developed urban areas (Kaye et al. 2006) will greatly diminish such soil–ISC interactions and the ecosystem services provided by urban soils (Easton et al. 2007).

Riparian ecosystems occupy land–water junctions along natural waterways and are important habitats of urban landscape (Groffman et al. 2003). Riparian areas naturally flooded from overbank flows are unsuitable for property development and residential use, and yet are excellent places, along with streams and rivers, for parks and other natural green space. In Baltimore, MD, the Gwynns Falls Trail, a 14-mile stream–valley system, has been developed as a parkland trail running from the mouth of the Gwynns Falls to the city, much according to the original concept proposed by the legendary landscape architects, the Olmstead brothers (Groffman et al. 2003). In Scottsdale, AZ (part of the Phoenix metropolitan area), Indian Band Wash is a 11-mile long greenbelt that is characterized by a chain of shallow artificial lakes in a larger floodplain of irrigated turf, providing an oasis of parks, lakes, green paths, and even golf courses in a densely occupied urban landscape (Roach and Grimm 2011). Soil, plants, animals, and microbial organisms play important roles in the function of riparian ecosystems, yet all these can be affected directly and indirectly by human activities inside and outside the riparian zones. Hydrologic alterations have emerged as a major factor of riparian ecology in the urban landscape (Groffman et al. 2003; Stander and Ehrenfeld 2009). The common feature of incision or "downcutting" of urban stream channels leads to lower groundwater tables in riparian zones, and consequently aerobic soils instead of anerobic soil conditions (Groffman et al. 2003). In Gwynns Falls watershed, the change of riparian soil conditions has shifted denitrification of NO_3^- removal to nitrification of NO_3^- production, causing riparian soils to be sources rather than sinks of NO_3^- (Groffman et al. 2003). Similar hydrologic alteration was found in the urban wetlands of metropolitan New Jersey; when three wetland types, riverine, mineral flat, and mineral flat-riverine, were examined, N cycling function was found more associated with drier hydrology than with direct human disturbance of the site (Stander and Ehrenfeld 2009).

Most UGSs are small (<1000 m², Gaston et al. 2013) that support low biodiversity and are affected negatively by various urban pollutants which also limit soil development (Livesley et al. 2016). There is a debate between "land sharing"—extensive sprawling urbanization where built land and natural space are interspersed—and "land sparing"—intensive and extremely compact urbanization along separate, large, contiguous green space (Dallimer et al. 2011; Stott et al. 2015). From the fringe of the city to downtown urban center, patch size usually becomes smaller while patch richness, density, and landscape shape index all increase (Luck and Wu 2002; Wu et al. 2011). Contemporary urbanization often leapfrogs and embeds tracts of natural lands inside metropolitan areas surrounded by people and development (Ramalho and Hobbs 2012), which likely will be subjected to further fragmentation and other human impact.

The negative effect of small patch size and patch isolation on biodiversity is well recognized in urban literature (Forman 2014). They could also affect C and N cycling due to changes in temperature, moisture, and pollutant input (Figure 5.2, Egitto 2005), although these types of study are still very limited. On the other hand, the remnant natural areas play important roles in the function of a modern city. Besides ecosystem service provisions of water infiltration, microclimate amelioration, air pollutant removal, recreation, and esthetics (services also provided by man-made and managed green space), natural areas often occupy unique habitats that harbor biodiversity in cities, including riparian forests and stream corridors (Groffman et al. 2003).

Probably more importantly, they provide primary connections that link people living in cities to the natural world (Ramalho and Hobbs 2012; Forman 2014). The loss of natural experience has conservation implications well beyond the city boundary (Soga and Gaston 2016). Such ecosystem services, even from small green space patches, make the city more livable and are essential to the sustainability of a city, while soils are essential to the sustainability of these green spaces. The costs and benefits of UGSs are often poorly quantified (Pataki et al. 2011) but soil C and N usually enhance the value of green space and reduce its maintenance cost.

5.5 CONCLUSIONS

C and N cycling and their interactions differ greatly between urban soils and nonurban soils. In cities, they are affected by the urban environment and management, which regulate C fixation by plants and decomposition by microbial organisms, and increase N input via fertilizer application, fossil fuel combustion, and food importation (Figure 5.1). Other important drivers include urban microclimate (e.g., the UHI effect), higher CO_2 concentration, and nonnative plant and fauna species. The sequestration of C and retention of N lead to the accumulation of SOM that controls an array of soil physical and chemical properties ranging from BD, water filtration, and aeration to soil aggregate structure. Urban soils provide pivotal ecosystem services to the green space provision. In addition to human health benefits, green spaces benefit urban hydrology by increasing water infiltration and reducing stormwater runoff, and heat balance by providing evapotranspiration cooling and provide critical habitats for biodiversity conservation, features critical to urban sustainability.

Decades of scientific studies have demonstrated that urban soils are not all heavily disturbed and polluted. C and N contents, important features of soil, are spatially and temporarily variable, affected by management subsidies of irrigation and fertilizer input, warmer urban climate, pollution, nonnative species, and direct human disturbances. Urban soils, often isolated in small patches in the city, interact with landscape elements such as ISCs. The alteration of urban hydrology can have large impact on soil C and N cycling. Monitoring soil health and understanding the physical, chemical, and biological processes occurring in soils, and the social-economic drivers affecting these processes, are important for maintaining the functionality and the sustainability of urban soils, urban ecosystems, and the world.

REFERENCES

Aber, J., W. McDowell, K. Nadelhoffer, A. Magill, G. Berntson, M. Kamakea, S. McNulty, W. Currie, L. Rustad, and I. Fernandez. 1998. Nitrogen saturation in temperate forest ecosystems: Hypotheses revisited. *BioScience* 48: 921–934.

Alberti, M., J.M. Marzluff, E. Shulenberger, G. Bradley, C. Ryan, and C. Zumbrunnen. 2003. Integrating humans into ecology: Opportunities and challenges for studying urban ecosystems. *BioScience* 53: 1169–1179.

Baker, L.A., D. Hope, Y. Xu, J. Edmonds, and L. Lauver. 2001. Nitrogen balance for the central Arizona-Phoenix (CAP) ecosystem. *Ecosystems* 4: 582–602.

Band, L.E., M.L. Cadenasso, C.S. Grimmond, J.M. Grove, and S.T.A. Pickett. 2005. Heterogeneity in urban ecosystems: Patterns and process. In *Ecosystem Function in Heterogeneous Landscapes,* eds. G.M. Lovett, C.G. Jones, M.G. Turner, K.C. Weathers, pp. 257–278. New York, NY: Springer.

Baycan-Levent, T., and P. Nilkamp. 2009. Planning and management of urban green spaces in Europe: Comparative analysis. *Urban Plan. Develop.* 135: 1–12.

Bolund, P., and S. Hunhammar. 1999. Ecosystem services in urban areas. *Ecol. Econ.* 29: 293–301.

Cadenasso, M.L., S.T.A. Pickett, and K. Schwarz. 2007. Spatial heterogeneity in urban ecosystems: Reconceptualizing land cover and a framework for classification. *Front. Ecol. Environ.* 5: 80–88.

Carreiro, M.M., K. Howe, D.F. Parkhurst, and R.V. Pouyat. 1999. Variation in quality and decomposability of red oak leaf litter along an urban-rural gradient. *Biol. Fertil. Soils* 30: 258–268.

Carreiro, M.M., R.V. Pouyat, C.E. Tripler, and W-X. Zhu. 2009. Carbon and nitrogen cycling in soils of remnant forests along urban-rural gradients: Case studies in New York City and Louisville, Kentucky. In *Ecology of Cities and Towns—A Comparative Approach,* eds. M. McDonnell, A. Hahs, and J. Breuste, pp. 308–328. Cambridge: Cambridge University Press.

Chen, Y., S.D. Day, A.F. Wick, B.D. Strahm, P.E. Wiseman, and W.L. Daniels. 2013. Changes in soil carbon pools and microbial biomass from urban land development and subsequent post-development soil rehabilitation. *Soil Biol. Biochem.* 66: 38–44.

Chen, Y., S.D. Day, A.F. Wick, and K.J. McGuire. 2014. Influence of urban land development and subsequent soil rehabilitation on soil aggregates, carbon, and hydraulic conductivity. *Sci. Total Environ.* 494–495: 329–336.

Craul, P.J. 1985. A description of urban soils and their desired characteristics. *J. Arbori.*11: 330–339.

Dallimer, M., Z.Y. Tang, P.R. Bibby, P. Brindley, K.J. Gaston, and Z.G. Davies. 2011. Temporal changes in greenspace in a highly urbanized region. *Biol. Lett.* 7: 763–766.

Decina, S.M., L.R. Hutyra, C.K. Gately, J.M. Getson, A.B. Reinmann, A.G.S. Gianotti, and P.H. Templer. 2016. Soil respiration contributes substantially to urban carbon fluxes in the greater Boston area. *Environ. Pollut.* 212: 433–439.

Easton, Z.M., P. Gérand-Marchant, M.T. Walter, A.M. Petrovic, and T.S. Steenhuis. 2007. Hydrologic assessment of an urban variable source watershed in the northeast United States. *Water Resour. Res.* 43, W03413, doi:10.1029/2006WR005076.

Edmondson, J.L., Z.G. Davies, S.A. McCormack, K.J. Gaston, and J.R. Leake. 2011. Are soils in urban ecosystems compacted? A citywide analysis. *Biol. Lett.* 7: 771–774.

Effland, W.R., and R.V. Pouyat. 1997. The genesis, classification, and mapping of soils in urban areas. *Urban Ecosyst.* 1: 217–228.

Egitto, B.A. 2005. Nitrogen and root dynamics in urban forest patches. M.S. thesis, Binghamton University, State University of New York.

Ehrenfeld, J.G. 2010. Ecosystem consequences of biological invasions. *Annu. Rev. Ecol. Evol. Syst.* 41: 59–80.

Ehrenfeld, J.G., P. Kourtev, and W. Huang. 2001. Changes in soil functions following invasions of exotic understory plants in deciduous forests. *Ecol. Appl.*11: 1287–1300.

Forman, R.T.T. 2014. *Urban Ecology: Science of Cities*. Cambridge: Cambridge University Press.

Fuller, R.A., K.N. Irvine, P. Devine-Wright, P.H. Warren, and K.J. Gaston. 2007. Psychological benefits of greenspace increase with biodiversity. *Biol. Lett.* 3: 390–394.

Galloway, J.N., J.D. Aber, J.W. Erisman, S.P. Seitzinger, R.W. Howarth, E B. Cowling, and B.J. Cosby. 2003. The nitrogen cascade. *BioScience* 53: 341–356.

Gaston, K.J., M.L. Avila-Jiménez, and J.L. Edmondson. 2013. The UK National Ecosystem Assessment: Managing urban ecosystems for goods and services. *J. Appl. Ecol.* 50: 830–840, doi:10.1111/1635-2664.12087.

Goldman, M.B., P.M. Groffman, R.V. Pouyat, M.J. McDonnell, and S.T.A. Pickett. 1995. CH_4 uptake and N availability in forest soils along an urban to rural gradient. *Soil Biol. Biochem.* 27: 281–286.

Golubiewski, N.E. 2006. Urbanization increases grassland carbon pools: Effects of landscaping in Colorado's Front Range. *Ecol. Appl.*16: 555–571.

Grimm, N.B., S.H. Faeth, N.E. Golubiewski, C.L. Redman, J. Wu, X. Bai, and J.M. Briggs. 2008. Global change and the ecology of cities. *Science.* 319: 756–760.

Groffman, P.M., D.J. Bain, L.E. Band, K.T. Belt, G.S. Brush, J.M. Grove, R.V. Pouyat, I.C. Yesilonis, and W.C. Zipperer. 2003. Down by the riverside: Urban riparian ecology. *Front. Ecol. Environ.* 1: 315–321.

Groffman, P.M., N.L. Law, K.T. Belt, L.E. Band, and G.T. Fisher. 2004. Nitrogen fluxes and retention in urban watershed ecosystems. *Ecosystems.* 7: 393–403.

Groffman, P.M., R.V. Pouyat, M.L. Cadenasso, W.C. Zipperer, K. Szlavecz, I.D. Yesilonis, L.E. Band, and G.S. Brush. 2006. Land use context and natural soil controls on plant community composition and soil nitrogen and carbon dynamics in urban and rural forests. *For. Ecol. Manage.* 236: 177–192.

Groffman, P.M. and R.V. Pouyat. 2009. Methane uptake in urban forests and lawns. *Environ. Sci. Technol.* 43: 5229–5235.

Hall, S.J., R.A. Sponseller, N.B. Grimm, D. Huber, J.P. Kaye, C. Clark, and S.L. Collins. 2011. Ecosystem response to nutrient enrichment across an urban airshed in the Sonoran Desert. *Ecol. Appl.* 21: 640–660.

Hutyra, L.R., B. Yoon, and M. Alberti. 2011. Terrestrial carbon stocks across a gradient of urbanization: A study of the Seattle, WA region. *Glob. Change Biol.* 17: 783–797.

Jenerette, G.D., S.L. Harlan, W. Stefanov, and C.A. Martin. 2011. Ecosystem services and urban heat riskscape moderation: Water, green spaces, and social inequity in Phoenix, USA. *Ecol. Appl.* 21: 2637–2651.

Jim, C.Y. 1998. Soil characteristics and management in an urban park in Hong Kong. *Environ. Manage.* 22: 683–695.

Jim, C.Y., and W.Y. Chen. 2006. Recreation-amenity use and contingent valuation of urban greenspaces in Guangzhou, China. *Landsc. Urban Plan.* 75: 81–96.

Johnston, M.R., N.J. Balster, and J. Zhu. 2016. Impact of residential prairie gardens on the physical properties of urban soil in Madison, Wisconsin. *J. Environ. Qual.* 45: 45–52.

Kaye, J.P., P.M. Groffman, N.B. Grimm, L.A. Baker, and R.V. Pouyat. 2006. A distinct urban biogeochemistry? *Trends Ecol. Evol.* 21: 192–199.

Kearney, M.A., W. Zhu, and J. Graney. 2013. Inorganic nitrogen dynamics in an urban constructed wetland under base-flow and storm-flow conditions. *Ecol. Engineer.* 60: 183–191.

Kumar, K., and L.S. Hundal. 2016. Soil in the city: Sustainably improving urban soils. *J. Environ. Qual.* 45: 2–8.

Livesley, S.J., E.G. McPherson, and C. Calfapietra. 2016. The urban forest and ecosystem services: Impacts on urban water, heat, and pollution cycles at the tree, street, and city scale. *J. Environ. Qual.* 45: 119–124.

Lovett, G.M., M.M. Traynor, R.V. Pouyat, M.M. Carreiro, W.-X. Zhu, and J.W. Baxter. 2000. Atmospheric deposition to oak forests along an urban-rural gradient. *Environ. Sci. Technol.* 34: 4294–4300.

Luck, M.A., and J. Wu. 2002. A gradient analysis of urban landscape pattern: A case study from the Phoenix metropolitan region, Arizona, USA. *Landsc. Ecol.* 17: 327–339.

McKinney, M.L. 2002. Urbanization, biodiversity, and conservation. *BioScience* 52: 883–889.

Nadelhoffer, K.J., B.A. Emmett, P. Gunderson, O.J. Kjønaas, C.J. Koopmans, P. Schleppi, A. Tietema, and R.F. Wright. 1999. Nitrogen deposition makes a minor contribution to carbon sequestration in temperate forests. *Nature* 398: 145–148.

Pataki, D.E., R.J. Alig, A.S. Fung, N.E. Golubiewski, C.A. Kennedy, E.G. McPherson, D.J. Nowak, R.V. Pouyat, and P.R. Lankao. 2006. Urban ecosystems and the North American carbon cycle. *Glob. Change Biol.* 12: 1–11, doi:10.1111/j.1365-2486.2006.01242.x.

Pataki, D.E., M.M. Carreiro, J. Cherrier, N.E. Grulke, V. Jennings, S. Pincetl, R.V. Pouyat, T.H. Whitlow, and W.C. Zipperer. 2011. Coupling biogeochemical cycle in urban environments: Ecosystem services, green solutions, and misconceptions. *Front. Ecol. Environ.* 9: 27–36.

Paul, M.J., and J.L. Meyer. 2001. Streams in the urban landscape. *Annu. Rev. Ecol. Syst.* 32: 333–365.

Pauleit, S., R. Ennos, and Y. Golding. 2005. Modeling the environmental impacts of urban land use and land cover change—A study in Merseyside, UK. *Landsc. Urban Plan.* 71: 295–310.

Pavao-Zuckerman, M.A., and D.C. Coleman. 2005. Decomposition of chestnut oak (*Quercus prinus*) leaves and nitrogen mineralization in an urban environment. *Biol. Fertil. Soils* 41: 343–349.

Perakis, S.S., and L.O. Hedin. 2001. Fluxes and fates of nitrogen in soils of an unpolluted old-growth temperate forest, Southern Chile. *Ecology* 82: 2245–2260.

Pickett, S.T.A., M.L. Cadenasso, J.M. Grove, P.M. Groffman, L.E. Bend, C.G. Boone, W.R. Burch Jr., C.S.B. Grimmond, J. Hom, J.C. Jenkins, N.L. Law, C.H. Nilon, R.V. Pouyat, K. Szlavecz, P.S. Warren, and M.A. Wilson. 2008. Beyond urban legends: An emerging framework of urban ecology, as illustrated by the Baltimore ecosystem study. *BioScience* 58: 139–150.

Pickett, S.T.A., M.L. Cadenasso, J.M. Grove, C.H. Nilon, R.V. Pouyat, W.C. Zipperer, and R. Costanza. 2001. Urban ecological systems: Linking terrestrial ecological, physical, and socioeconomic components of Metropolitan areas. *Annu. Rev. Ecol. Syst.* 32: 127–157.

Pouyat, R.V., and M.M. Carreiro. 2003. Controls on mass loss and nitrogen dynamics of oak leaf litter along an urban-rural land-use gradient. *Oecologia* 135: 288–298.

Pouyat, R.V., and M.J. McDonnell. 1991. Heavy metal accumulations in forest soils along an urban-rural gradient in Southeastern New York, USA. *Water Air Soil Pollut.* 57–58: 797–807.

Pouyat, R.V., D.E. Pataki, K.T. Belt, P.M. Groffman, J. Hom, and L.E. Band. 2006a. Effects of urban land-use change on biogeochemical cycles. In *Terrestrial Ecosystems in a Changing World*, ed. P. Canadell, pp. 45–58. Canberra: Global Change and Terrestrial Ecosystem Synthesis Book.

Pouyat, R.V., K. Szlavecz, and I.D. Yesilonis. 2010. Chemical, physical, and biological characteristics of urban soils. In *Urban Ecosystem Ecology* (Agronomy Monograph 55), eds, J. Aitkenhead-Peterson, A. Volder, pp. 119–152. Madison, WI: American Society of Agronomy, Crop Science Society of America, Soil Science Society of America.

Pouyat, R.V., I.D. Yesilonis, and N.E. Golubiewaski. 2009. A comparison of soil organic carbon stocks between residential turf grass and native soil. *Urban Ecosyst.* 12: 45–62.

Pouyat, R.V., I.D. Yesilonis, and D.J. Nowak. 2006b. Carbon storage by urban soils in the United States. *J. Environ. Qual.* 35: 1566–1575.

Pouyat, R.V., I.D. Yesilonis, J. Russell-Anelli, and N.K. Neerchal. 2007. Soil chemical and physical properties that differentiate urban land-use and cover types. *Soil Sci. Soc. Am. J.* 71: 1010–1019.

Raciti, S.M., P.M. Groffman, and T.J. Fahey. 2008. Nitrogen retention in urban lawns and forests. *Ecol. Appl.* 18: 1615–1626.

Raciti, S.M., P.M. Groffman, J.C. Jenkins, R.V. Pouyat, T.J. Fahey, S.T.A. Pickett, and M.L. Cadenasso. 2011. Accumulation of carbon and nitrogen in residential soils with different land-use histories. *Ecosystems* 14: 287–297.

Ramalho, C.E., and R.J. Hobbs. 2012. Time for a change: Dynamic urban ecology. *Trends Ecol. Evol.* 27: 179–188.

Roach, W.J., and N.B. Grimm. 2011. Denitrification mitigates N flux through the stream—Floodplain complex of a desert city. *Ecol. Appl.* 21: 2618–2636.

Schmidt, M.W.I., M.S. Torn, S. Abiven, T. Dittmar, G. Guggenberger, I.A. Janssens, M. Kleber, I. Kögel-Knabner, J. Lehmann, D.A.C. Manning, P. Nannipieri, D.P. Rasse, S. Weiner, and S.E. Trumbore. 2011. Persistence of soil organic matter as an ecosystem property. *Nature* 478: 49–56.

Short, J.R., D.S. Fanning, M.S. McIntosh, J.E. Foss, and J.C. Patterson. 1986. Soils of the Mall in Washington, DC: I. Statistical summary of properties. *Soil Sci. Soc. Am. J.* 50: 699–705.

Soga, M., and K.J. Gaston. 2016. Extinction of experience: The loss of human-nature interactions. *Front. Ecol. Environ.* 14: 94–101.

Soil Survey Staff. 1975. *Soil Taxonomy—A Basic System of Soil Classification for Making and Interpreting Soil Surveys.* Washington, DC: U.S. Department of Agriculture, Agriculture Handbook 436.

Stander, E.K., and J.G. Ehrenfeld. 2009. Rapid assessment of urban wetlands: Do hydrogeomorphic classification and reference criteria work? *Environ. Manage.* 43: 725–742.

Steinberg, D.A., R.V. Pouyat, R.W. Parmelee, and P.M. Groffman. 1997. Earthworm abundance and nitrogen mineralization rates along an urban-rural land use gradient. *Soil Biol. Biochem.* 29: 427–430.

Stott, I., M. Soga, R. Inger, and K.J. Gaston. 2015. Land sparing is crucial for urban ecosystem services. *Front. Ecol. Environ.* 13: 387–393.

Wang, W. and W. Zhu. 2012. Soil retention of ^{15}N in a simulated N deposition study: Effects of live plant and soil organic matter content. *Plant Soil* 351: 61–72. doi:10.1007/s11104-011-0929-1.

Whitford, V., A.R. Ennos, and J.F. Handley. 2001. "City form and natural process"—Indicators for the ecological performance of urban areas and their application to Merseyside, UK. *Landsc. Urban Plan.* 57: 91–103.

Wollheim, W.M., B.A. Pellerin, C.J. Vörösmarty, and C.S. Hopkinson. 2005. N retention in urbanizing headwater catchments. *Ecosystems* 8: 871–884.

Woltemade, C.J. 2010. Impact of residential soil disturbance on infiltration rate and stormwater runoff. *Journal of the American Water Resour. Assoc. (JAWRA)* 46: 700–711.

Wu, J.G., A. Buyantuyev, G.D. Jenerette, J. Litteral, K. Neil, and W. Shen. 2011. Quantifying spatiotemporal patterns and ecological effects of urbanization: A multiscale landscape approach. In *Applied Urban Ecology: A Global Framework*, eds, M. Richte, U. Weiland, pp. 35–53. Oxford: Blackwell.

Yesilonis, I.D., and R.V. Pouyat. 2012. Carbon stocks in urban forest remnants: Atlanta and Baltimore as case studies. In *Carbon Sequestration in Urban Ecosystems*, eds. R. Lal and R. Augustin, pp. 103–120. Dordrecht, Netherlands: Springer.

Yesilonis, I.D., R.V. Pouyat, and N.K. Neerchal. 2008. Spatial distribution of metals in soils in Baltimore, Maryland: Role of native parent material, proximity to major roads, housing age and screening guidelines. *Environ. Pollut.* 156: 723–731.

Zhu, W.-X., and M.M. Carreiro. 1999. Chemoautotrophic nitrification in acidic forest soils along an urban-to-rural transect. *Soil Biol. Biochem.* 31: 1091–1100.

Zhu, W.-X., and M.M. Carreiro. 2004. Temporal and spatial variations in nitrogen cycling in deciduous forest ecosystems along an urban-rural gradient. *Soil Biol. Biochem.* 36: 279–288.

Zhu, W.-X., N.D. Dillard, and N.B. Grimm. 2004. Urban nitrogen biogeochemistry: Status and processes in green retention basins. *Biogeochemistry* 71: 177–196.

Zhu, W.-X., N. Hope, C. Gries, and N.B. Grimm. 2006. Soil characteristics and the accumulation of inorganic nitrogen in an arid urban ecosystem. *Ecosystems* 9: 711–724.

Zhu, W.-X., and W. Wang. 2011. Does soil organic matter variation affect the retention of ^{15}NH$_4^+$ and ^{15}NO$_3^-$ in forest ecosystems? *For. Ecol. Manage.* 261: 675–682.

6 Urban Soil Carbon Storage

Bryant Scharenbroch, Susan Day,
Tara Trammell, and Richard Pouyat

CONTENTS

6.1 INTRODUCTION

6.1.1 URBANIZATION AND URBAN LANDS

Global population in 2015 was 7.4 billion people and this number is projected to exceed 11 billion before the end of the twenty-first century (United Nations 2015). In addition to rising human population, more people are living within urban areas. Since 2010, more people live in cities compared to rural areas and the ratio of urban to rural dwellers is projected to increase in the future (United Nations 2015). Populations within cities are also increasing. The number of mega- (10 million people or more), large- (5–10 million people), and medium-sized (1–5 million people) cities will have increased from 10, 21, and 239 in 1990 to 41, 63, and 588 in 2030, respectively (United Nations 2015).

Increases in the global human population and the proportion of the population residing within cities are leading to the expansion of urban lands across the globe. However, there is no global consensus on the definition of urban land. In the United States, urban lands are defined by population densities of at least 386 people km^2 (1,000 people mi^2) (US Census Bureau 2016). Surrounding areas (exurban) have 50% of those urban densities (193 km^2). European countries often define urban lands on the basis of land use or by the proportion of area that is gray infrastructure (buildings, roads, and other sealed surfaces). Urban lands by this approach would not contain large (typically less than 200 linear m or ha by area) nonurban gaps. Urban areas have also been defined by their physical attributes and composition of the land cover. Schneider et al. (2009) defined urban areas as built-up lands, which were areas inclusive of all nonvegetative, human-constructed elements, such as roads, buildings, and impervious surfaces. Additionally, to be an urban area, there must be >50% built-up land and the areas of built-up land must be continuous patches of greater than 1 km².

Partly because of this lack of consensus, the extent of the global urban land cover is not clearly understood. Estimates of global urban land range from 308,007 km² (0.24% of total global land) (Bartholomé and Belward 2005) to 3,524,109 km² (2.74% of total global land) (CIESIN 2011). Schneider et al. (2009), using the Moderate Resolution Imaging Spectroradiometer (MODIS), concluded that the land footprint of urban areas is 657,000–727,000 km² (0.51% and 0.57% of total global land). Although increasing, urban lands are not evenly distributed across the globe. In Europe, for example, cities and towns tend to densify rather than sprawl. European urban lands are estimated to cover 9% of the continent (Scalenghe and Marsan 2009) and approximately 3% of the land surface is covered by impervious surfaces and buildings (Elvidge et al. 2007). These estimates of urban land and gray infrastructure far exceed proportional amounts in other industrialized countries, such as the United States (Elvidge et al. 2007). Because of its focus on physical transformations of the landscape rather than political boundaries or regionally relevant population densities, this chapter uses the definition of urban land utilized by Schneider et al. (2009) and deliberates on the assumption that urban land may range from 0.24% to 2.74% of the total global land surface. This chapter utilizes the MODIS findings of Schneider et al. (2009) as conservative estimates of global urban land (0.51% and 0.57%).

6.1.2 Carbon in the Urban Ecosystem

Concerns of climate change and increasing atmospheric carbon (C) concentrations have created interest in utilizing soils for long-term C storage, particularly in an attempt to offset C emissions associated with anthropogenic activities (DeLuca and Boisvenue 2012). Urbanization is generally viewed as a depletion of C stores and thus regaining some of that lost C is viewed as a potential strategy for climate mitigation. A slight change in the size of soil organic carbon (SOC) reservoir can potentially alter the atmospheric CO_2 concentration and ultimately the global climate (Guo and Gifford 2002). Assessments of ecosystem C stocks are crucial to understand anthropogenic changes to the global C cycle (Jobbágy and Jackson 2000).

Human activity produces about 8.78 GtC yr^{-1} through fossil fuel emissions (7.89 GtC) and land-use change (0.89 GtC) (Le Quéré et al. 2014; Ballantyne et al. 2015). These anthropogenic C emissions end up in the atmosphere, oceans, and land. Over the past 50 years, the oceans and land have absorbed more than 50% of the total CO_2 that has been emitted to the atmosphere (Ballantyne et al. 2015). Terrestrial C uptake has increased since the 1960s (1.39 GtC yr^{-1}) and has remained relatively constant over the last few decades (1990s, 2000s, and 2010s), ranging from 2.3 to 2.5 GtC yr^{-1} (Pan et al. 2011; Ballantyne et al. 2015).

Soils are the major stock of the terrestrial global C. Most estimates would agree that the amount of C in soil across the globe is nearly twice the amount in the atmosphere (816 GtC) and vegetation (497 GtC) combined (Scharlemann et al. 2014). In a review of global C storage, Scharlemann et al. (2014) found estimates of total global SOC stock to range from 504 to 3,000, with a median value of 1,461 GtC (Scharlemann et al. 2014). In a more recent study, Batjes (2016) utilized approximately 21,000 soil profiles and estimated the global C storage in soil to a 2 m depth to be 2,060 ± 215 GtC. It is unknown how much of this global C is contained within urban ecosystems.

Despite increasing rates of urbanization and potential C storage within urban ecosystems, the organic C storage in urban areas has not been well quantified. Churkina et al. (2010) reported that C stored in urban (23–42 kg m^{-2}) and exurban areas (7–16 kg m^{-2}) may be as high as C stored in tropical forests (4–25 kg m^{-2}). They suggest that the total C storage in urban lands is 10% of the total US conterminous land C storage. The SOC stock is a major component of the urban ecosystem C, although not well quantified. Jo and McPherson (1995) found SOC to represent 79%–85% of the total C stock in urban neighborhoods of Chicago, IL. Golubiewski (2006) found that in suburban areas around Denver, CO, 70% of total urban C is contained as SOC. Similarly, Edmondson et al.

(2012) reported that 82% of total urban C in Leicester, England is found in the SOC pool. Churkina et al. (2010) estimated SOC to contain 64% of the total urban C stock. From the limited body of research on the topic, it is apparent that soils represent the major pool (64%–85%) of C within urban ecosystems.

6.1.3 Objectives

Although research on urban SOC is rapidly growing, the field is still relatively new compared to other disciplines in soil science. It is critical that the scientific understanding of urban ecosystem C storage and in particular C storage within urban soils be improved and strengthened. An improved understanding of the urban SOC stocks will provide better regional estimates and a more accurate assessment of the global storage of C in urban ecosystems. The objectives of this chapter are to: (1) conduct and present a literature review on the topic of urban SOC stocks; (2) identify correlates for urban SOC and discuss potential drivers of the spatial distribution of urban SOC; and (3) estimate global urban SOC storage and calculate the global C storage of urban soils and ecosystems.

Across the globe, differences in soil properties and processes are driven by soil forming factors of climate, parent material, organisms, relief, and time (Jenny 1941). However, urban soils are different because their properties are also strongly determined by an anthropogenic factor (Pouyat et al. 2010). This chapter provides a review of the literature and is a compilation of the data to test the hypothesis that urban SOC is related to Jenny's soil forming factors. To do so, the available data are utilized to explore relationships of SOC and climate, parent material, vegetation, and age of the urban soils. An attempt is made to explore relationships of the anthropogenic soil forming factor and urban SOC. Relationships of urban SOC and the relief factor are not examined in this chapter because the effects of relief on soil tend to be local (i.e., within landscapes). The intent is to examine potential drivers of urban SOC on regional and global scales. This chapter also explores relationships among urban SOC and other soil properties that may influence SOC or might be used as predictors of urban SOC. Lastly, a summary of the literature is provided to extrapolate an estimate of global urban SOC storage.

6.2 METHODS

A literature search was conducted to identify studies and gather data on SOC in urban soils. The Academic Search Complete (EBSCO), JSTOR, Web of Knowledge, and Google Scholar search engines were utilized to find relevant articles. Articles were scanned and selected for review if they included data on SOC density ($kg\ m^{-2}$). Studies which reported SOC concentration data (%), soil bulk density ($g\ cm^{-3}$), and depth were also used and the data converted to C density ($kg\ m^{-2}$) prior to analysis. If studies reported total C, inorganic C, and organic C, only SOC data were used. Mean SOC was computed for each study included in the analysis as well as the standard deviation of the SOC data reported by each individual study. The maximum soil sampling depth and total number of soil data points used to compile the SOC estimate for each study were recorded. In total, 63 data points were compiled for the urban SOC assessment (Appendix).

In addition to SOC, data on soil texture (sand and clay %), bulk density, and pH were compiled for examining relationships among soil properties and urban SOC. Means were computed for these soil properties across all data reported and all depths for each study that included the data. If soil classification was provided by the study, it was also recorded. The global soil order map (US Soil Taxonomy) was used in cases where soil classification was not provided to determine the main soil order for the city and surrounding region. Mean annual temperature (MAT) and mean annual precipitation (MAP) were recorded for each city. Climate data were gathered from the individual studies or were attained from NOAA (2016). Climate data were partitioned in classes of MAT (<5°C, 5°C–10°C, 10.1°C–15°C, 15.1°C–20°C, and >20°C) and MAP (<500, 500–1,000, 1,001–1,500, and >1,500 mm yr^{-1}). The effects of aboveground organisms on urban SOC were examined via correlations with vegetation. The dominant vegetative cover reported in the individual studies as well as the ecoregions classification (Olson et al. 2001) was used as vegetation correlates for urban SOC. The ecoregion classification defines relatively large units of land with distinct assemblages of natural communities. The boundaries are intended to approximate the original extent of the natural communities prior to major land-use change. The reported vegetation on the study sites for each paper was generalized to three different community types: trees, grasses, or mixed vegetation. Most studies did not report local soil parent materials. The OneGeology Portal was used to access the World CGMW 1:50M Geological Units Onshore map to identify the main type of rock (lithology) and the age of the rock (stratigraphy) for each city. The relative ages of the urban soils were compiled from the studies when provided. Many of the studies reported approximate ages since last disturbance. The urban forest studies often reported this as the age of the trees in a forest stand. Urban factors compiled from the studies or gathered included land use and population density. The main land use was compiled from each individual study. The population and population densities were computed from online databases (United Nations 2015; US Census Bureau 2016).

Statistical analyses were conducted using SAS JMP 11.0 software (SAS Institute Inc., Cary, NC). Differences in urban SOC by sample depth, MAP and MAT class, parent material, time, ecoregion, vegetation type, soil order, country, population density, and land use were assessed using analysis of variance (ANOVA) and Wilcoxon/Kruskal–Wallis nonparametric tests. Numeric variables (clay, sand, bulk density, pH, age, MAT, MAP, sample depth) were analyzed for correlations with urban SOC with a fit y by x and the nonparametric Spearman's ρ statistic. Significant correlations were identified at the 95% confidence level ($P \leq .05$).

6.3   RESULTS AND DISCUSSION

The mean values of urban SOC in the studies ranged from 0.3 to 134.8 kg m^{-2}, with an overall mean of 12.7 kg m^{-2} (Table 6.1). The standard deviation of means reported in each study ranged from 0 to 113 kg m^{-2}, with an overall mean of 6.2 kg m^{-2}. The number of sample points in the study ranged from 3 to 220, with an overall mean of 40. The maximum depths sampled in the studies ranged from 10 to 235, with an overall mean of 50 cm. The mean SOC and SOC standard deviation within each study were significantly and positively correlated ($P < .0001$; $r = 0.5717$). Studies that found higher SOC contents also found greater variation of SOC in the urban soils. This is important information to consider when characterizing urban SOC. The number of sample points was not significantly correlated with mean SOC ($P = .3770$; $r = -0.1132$). Consequently, it is difficult to draw conclusions from this data set to suggest sample point requirements for specific ranges of SOC content or targeted variance tolerances.

TABLE 6.1

Summary Statistics (Mean, Standard Deviation, Sample Points, and Maximum Depths Sampled) of Urban SOC Studies ($n = 63$)

Statistic	Mean SOC (kg m^{-2})	SD of SOC (kg m^{-2})	Sample Points (#)	Maximum Depth Sampled (cm)
Maximum	135	113	220	235
75th quartile	9.90	1.60	50.0	90.0
Median	6.20	0.90	14.0	30.0
25th quartile	4.30	0.40	12.0	15.0
Minimum	0.30	0.00	3.00	10.0
Mean	12.7	6.23	40.0	50.4
SE of mean	2.84	2.43	6.50	6.49
SD of mean	22.5	19.3	51.7	51.6

6.3.1 Urban SOC by Climate

Climate is a major soil forming factor and has strong influence on primary production and decomposition and consequently SOC. On a global scale, Jobbágy and Jackson (2000) found SOC to increase with precipitation and decrease with temperature. Consequently, this chapter discusses the climate hypothesis that urban SOC will be greater in cities with higher precipitation and lower temperatures. Whether SOC stocks in urban areas are related to climate is not well established. Across village landscapes of China, Jia-Guo et al. (2010) found SOC to not vary with climate. However, many other studies, listed below, have found climate to have an effect on urban SOC. The following section explores relationships of temperature and precipitation and urban SOC in the 63 studies including in the analysis.

Across all cities, MAT ranged from 1.7 to 24.1, with a mean of 13.7°C. Further, the SOC in the 5°C–10°C temperature class was significantly ($P = .0179$) greater than that in the higher temperature classes (Figure 6.1). Furthermore, there existed a significant negative relationship ($P = .0179$; $r = -0.3974$) between SOC and temperature (Figure 6.1). Selhorst and Lal (2012) also found SOC to decrease with increasing temperatures across 16 US cities. At the lower temperatures (<5°C), primary production is likely limited and SOC is lower due to reduced organic inputs from plants. Conversely, at higher temperatures (>10°C), decomposition rates exceed primary production, thereby decreasing SOC. Green and Oleksyszyn (2002) found microbial activity to be positively related to soil temperature in urban soils in Phoenix, AZ, suggesting increased decomposition with higher temperatures. Overall, results presented herein are in agreement with the hypothesis that SOC tends to be lower in cities with higher MAT.

MAP ranged from 110 to 2,150 mm yr^{-1}, with a mean MAP of 884 mm yr^{-1} across the 63 studies. Cities with MAP of 500–1,000 mm yr^{-1} had significantly ($P = .0001$) greater SOC compared to cities with <500, 1,001–1,500, and >1,500 mm yr^{-1} (Figure 6.2). A significant linear relationship of MAP and SOC ($P = .0825$; $r = -0.2205$) was not observed with these data (Figure 6.2). Selhorst

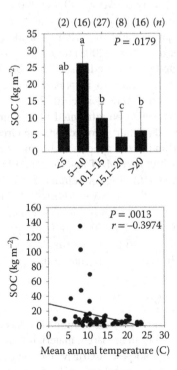

FIGURE 6.1 Urban SOC and mean annual soil temperature.

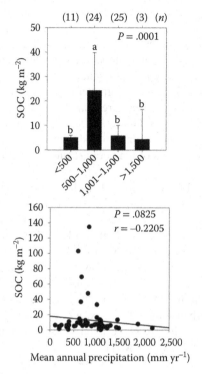

FIGURE 6.2 Urban SOC and mean annual precipitation.

and Lal (2012) also reported a nonlinear relationship with MAP and SOC in urban soils and found the greatest SOC in US cities receiving between 600 and 700 mm yr^{-1} of MAP. Increasing SOC with increasing MAP is likely due to greater plant growth and subsequent SOC sequestration (Shen et al. 2008). This mechanism is particularly important in arid environments where vegetation is limited by moisture. For example, Kaye et al. (2008) found SOC to be significantly greater in urban mesic compared to urban xeric and desert conditions. The observed decrease in SOC above the 1,000 mm yr^{-1} threshold may be attributed to increased soil microbial biomass and increasing decomposition with increasing soil moisture (Lin et al. 2010). Accelerated decomposition under increased moisture conditions might also be attributed to a priming effect on the soil microbial population associated with increased primary production (van Groenigen et al. 2014).

The interaction between MAT and MAP appears to be important for SOC in urban areas. The highest SOC was found in cities with lower MAT and moderate MAP (500–1,000 mm yr^{-1}). These observations are in agreement with the findings of Selhorst and Lal (2013) who also found the highest SOC in cities with MAT of 6°C–9°C and MAP of 600–900 mm yr^{-1}. The hypothesis stated that urban SOC would be positively correlated with soil moisture, but this was not the case because the highest SOC was found in areas with MAP of 500–1,000 mm yr^{-1}. Overall, the climate hypothesis was partially supported, and climate conditions favoring higher production and lower decomposition appear to be correlated with higher SOC in urban soils.

6.3.2 URBAN SOC BY PARENT MATERIAL

The effects of parent materials on urban SOC were examined with correlations among the SOC and the local lithology and stratigraphy. The four main parent materials found in the 63 studies were sedimentary, endogenous (plutonic and/or metamorphic), extrusive (volcanic rocks), and undifferentiated facies. Sedimentary rocks or undifferentiated facies (49) were the most common types of rocks found in the cities. Endogenous (8), extrusive (4), and undifferentiated (2) comprised only 22% of the cities. Ages of the parent materials ranged from the Precambrian (9) to the Cenozoic (13). The Mesozoic comprised the majority of the cities (23) and the Paleozoic the second most, split into upper (6) and lower (13).

The parent material hypothesis of this study was not supported because SOC did not differ by type or age of rocks. No significant differences were observed for SOC when the cities were grouped by the type ($P = .1736$) and age ($P = .2034$) of most common rocks. However, the data indicate that parent materials are not an important driver of differences in urban SOC. In Baltimore, MD, surface geology was more important than urban factors for explaining the distribution of surface soil characteristics (Pouyat et al. 2007). Parent materials are strong soil forming factors and the major driver of inherent soil properties, like soil texture. The latter strongly influences water and nutrient supply; by extension it impacts primary production and decomposition hence SOC. The data further indicate the linkages of soil texture and other soil properties with urban SOC below. Regarding parent material effects on urban SOC, the level of precision in this analysis for the parent materials was too coarse for detecting relationships with SOC across these cities. If each study reported parent materials for their sites, then a more precise analysis might have been performed. Unfortunately, very few of the studies reported soil parent materials.

6.3.3 Urban SOC by Time

The relative ages of soils in the 63 studies were examined for differences in urban SOC. The ages ranged from 4 to 275 years, with a mean age of 42 years. The oldest urban sites appeared to have the greatest urban SOC, but these differences were not significant ($P = .1694$) due to the high variation and low sample size of the oldest age class (Figure 6.3). A significant and positive relationship was observed with urban SOC and age ($P = .0255$; $r = 0.3455$). However, upon closer examination of the data, two studies reporting the highest SOC: 134.8 kg C m^{-2} in Eckernförde, Germany (Schleuß et al. 1998) and 34.8 kg C m^{-2} in Moscow, Russia (Vasenev et al. 2013) were from older urban soils (>100 years), but they also examined SOC to 150 cm. The third study with soils >100-years-old was conducted in Washington, DC, at 0–75 cm depth, finding a mean SOC of 5.3 kg m^{-2}.

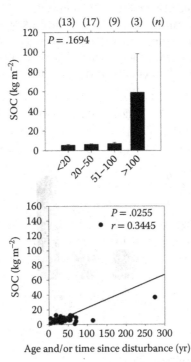

FIGURE 6.3 Urban SOC and age of urban landscape or time since last disturbance.

A few of the individual studies have found age or time since last disturbance to be an important driver of urban SOC. Scharenbroch et al. (2005) found a 35% increase in SOC comparing older (>50 years) and younger (<10 years) urban soils in Moscow, ID and Pullman, WA. Urban SOC in Auburn, AL, was found to be significantly and positively correlated with age (Huyler et al. 2014). Schmitt-Harsh et al. (2013) found the highest SOC stocks in parcels developed in the early 1970s and mid-1990s and they reported a significant, negative, and quadratic fit between SOC and urban soil age in Bloomington, IN. At this time, we are unable to confirm or deny the proposed hypothesis that urban SOC is greater with age of landscape or time since disturbance. Given the likely interaction with depth and age in the present analysis and the divergence in the literature on the effects of age on urban SOC, it is not possible to know if time is or is not an important factor for urban SOC. Furthermore, the relationship of age and urban SOC is likely confounded by many other factors, such as type and degree of disturbance.

6.3.4 Urban SOC by Organisms (Vegetation)

The majority of the mixed vegetation communities included trees, shrubs, grasses, and other herbaceous vegetation. The 63 studies included in the analysis spanned seven ecoregions. The majority of cities were located in temperate ecoregions: temperate broadleaf and mixed forests (TeBMF) (34), temperate coniferous forests (TeCF) (8), and temperate grasslands, savannas, and shrublands (TeSGSS) (4). Other ecoregions included tropical and subtropical moist broadleaf forests (TrSMBF) (8), deserts and xeric shrublands (DrXS) (5), Mediterranean vegetation (MedV) (3), and tropical and subtropical grasslands, savannas, and shrublands (TrSGSS) (1). There were no significant differences in urban SOC among the seven ecoregions ($P = .1883$); however, two temperate ecoregions appeared to have greater urban SOC than the others. Consequently, urban SOC did not appear to be distinguished by the original natural ecosystem community. As lands are urbanized for human dwellings, they may not retain their original community composition. Furthermore, little is known about whether original ecosystem communities may return in cities over time. Temperate ecoregions having greater urban SOC would support the previous climate finding of higher SOC in urban areas of lower MAT and moderate MAP (500–1,000 mm yr^{-1}).

Across all cities, there were significant differences in urban SOC according to the vegetation on site (Figure 6.4). Urban SOC was significantly ($P = .0444$) greater in mixed vegetation cover types compared to grass vegetation alone. Many studies included in this analysis found greater SOC under trees compared to grasses. Edmondson et al. (2014) found SOC to be 22% higher under trees and

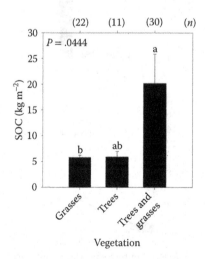

FIGURE 6.4   Urban SOC and type of vegetation.

shrubs compared to other urban vegetation. Mulches are often applied to urban surfaces with woody materials. These mulches help buffer climate, promote production, slow decomposition, and thus, have the effect of working to increase SOC. Livesley et al. (2010) found SOC contents under woody mulches to be 56% greater than those under turf grass lawns. Livesley et al. (2016) also found the C:N ratio to be greater under trees compared to that under grasses, which could work to slow decomposition and increase SOC under trees compared to grasses. Edmondson et al. (2015) found that the biochar rich fraction increased under trees compared to grasslands in cities of northeast England. Biochar or black C is considered relatively recalcitrant and may represent a significant portion of the SOC in urban soils. However, few studies have examined its distribution in soils of urban areas.

Although SOC under urban grasses appear to be lower than that under trees, this is not always the case. In Montgomery and Roanoke counties, VA, Campbell et al. (2014) found SOC to be significantly greater under lawn compared to that under forest environments. Golubiewski (2006) found that urban lawns contained significantly greater SOC compared to local grassland soils in Denver, CO. Root growth is rapid in turf grass systems and the constant and heavy input of fine roots may also work to increase SOC. Townsend-Small and Czimczik (2010) found SOC to increase 392% over 31 years under urban turf. Increased SOC with turf might even exceed local "natural" SOC. Urban SOC under turf grass was 200% greater compared to rural SOC in Baltimore, MD and Denver, CO (Pouyat et al. 2009). However, high intensity maintenance required for many turf grass systems means that these systems might not be perpetual C sinks. Kong et al. (2014) found that after 5–24 years the C sink capacity of turf grass is reached, then turf grass systems are net C sources.

Although the data presented herein suggest urban SOC is higher under mixed vegetation compared to grasses alone, it is also pertinent to consider confounding factors like sampling depths of the studies. The data show that the mean sampling depth (75 cm) of the studies of mixed vegetation was significantly ($P = .0005$) greater than the studies of grasses alone (22 cm). The mean sampling depth of the tree studies (41 cm) was intermediate between the two. This observation is reasonable considering sampling depth is often driven by rooting depth and turf grass roots are often thought to be confined to the upper 25 cm.

The data support the conclusion that vegetation type is likely an important determinant of urban SOC. While the data do suggest an effect of vegetation on urban SOC, it is pertinent to preface the findings by recognizing the likely interaction of sampling depth with vegetation. These results suggest that the natural vegetation of the region surrounding the urban area is not a determining factor for urban SOC, but vegetation on site appears to be a driver of urban SOC. It seems that higher urban SOC may occur on mixed and woody vegetation, which rejects the proposed vegetation hypothesis. Further, the hypothesis of expected greater SOC content under grassland compared to woody vegetation was derived from the association of higher SOC contents in prairie ecosystems. Prairie plants tend to have deep root systems with high fine-root turnover rates. Most grassland vegetation in these urban areas was turf grass. Although the root systems of turf grass are actively contributing organic matter to soil, the root systems are rarely as deep as natural prairie communities.

6.3.5 Urban SOC by Anthropogenic

The influence of humans as a soil forming factor and impact on SOC is likely strong in urban areas. To examine the anthropogenic effects, an attempt was made to assess urban SOC by country, population density, and land use across the 63 studies.

The vast majority of research on urban SOC in the literature appears to have been conducted in the United States. Forty-two of the 63 studies included in the analysis were from the United States (Figure 6.5). Other countries included China (7), England (4), Germany (4), Australia (2), Russia (2), Canada (1), and South Korea (1). Mean urban SOC reported in the German papers was significantly ($P = .0012$) greater than that in all other countries, and no other significant differences were observed. It is postulated that SOC in the German cities was greater primarily as a result of

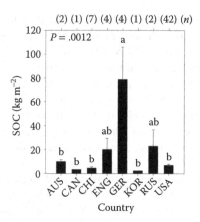

FIGURE 6.5 Urban SOC and country.

the sampling depths in those studies. On average, the German studies sampled to 148 cm, which was significantly ($P = .0049$) greater than depths in England (59 cm), China (43 cm), the United States (41 cm), Australia (28 cm), and Canada (25 cm). Mean sampling depths in Russia (100 cm) and South Korea (60 cm) were similar to German cities.

Total human population in these cities ranged from 2,231 to 394,072,000 people, with a mean of 985,285 people. Population densities ranged from 64 to 15,046 people km^{-2}, with a mean of 2,641 people km^{-2} (Figure 6.6). Although population density is not likely a direct driver of SOC contents in urban areas, the hypothesis was that there would be a positive correlation between population density and SOC. However, there were no significant ($P = .1919$) differences in SOC in relation to population densities across these cities, and there was not a significant positive relationship between population density and urban SOC ($P = .8196$; $r = -0.0293$), which rejected the proposed hypothesis.

There were five types of land use in the 63 studies: mixed (24), parks and openlands (5), forest (4), residential (27), and transportation (3) (Figure 6.7). The SOC appeared to be greater in mixed (unspecified or a heterogeneity of uses) and openlands compared to that in the forest, residential, and transportation land uses, but variances were high and these differences were not significant ($P = .3371$). With the current data set, it is not logical to reject the hypothesis that urban SOC differs by urban land use. Although the data did not show significant differences in urban SOC across land uses, a large body of literature suggests anthropogenic factors are important drivers of urban SOC.

Many studies have found significantly higher SOC in urban compared to rural lands. Sun et al. (2010) found urban SOC to be 72% greater than suburban lands near Kaifeng, China. Pouyat et al. (2009) found that SOC in urban lands was 70%–200% greater than that in rural areas near Baltimore, MD. In New York City, NY, SOC was significantly greater in urban forests compared to rural and suburban ones (Pouyat et al. 2002). Vasenev et al. (2013) found that urban SOC was 22% greater than nonurban SOC near Moscow, Russia.

The reasons reported for increased SOC in urban lands are many and quite variable. Belowground C allocation and respiration was found to be greater in urban compared to nonurban soils, and urbanization was proposed to enhance C cycling in Fort Collins, CO (Kaye et al. 2005). Beyer et al. (1995, 1996, 2001) found that urban soils in Kiel and Rostock, Germany, had higher aromatic compounds which slowed decomposition and increased SOC with urbanization. Gough and Elliot (2012) found urban SOC to increase 66% in 80 years and this increase was associated with neighborhood affluence, assessed by factors like land value and years since landscaping violation. Presumably, management actions in highly maintained, more affluent urban landscapes contribute to observed urban SOC increases. Vasenev et al. (2013) found increased urban SOC in Moscow, Russia attributable to the cultural layer in these urban soils, which is an organic-enriched layer from

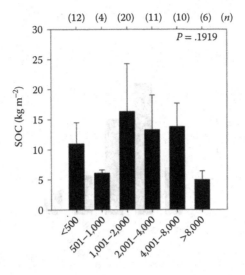

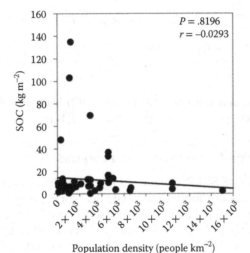

FIGURE 6.6 Urban SOC and population density.

residential and settlement history. Beesley (2012) found compost additions responsible for increased SOC in Liverpool, England, urban soils.

Within urban lands, land use may also have significant effects on SOC. Raciti et al. (2011) observed a 28% increase in residential compared to forest land uses in Baltimore, MD. Low-density residential and institutional lands were found to contain 38%–44% more SOC compared to commercial land uses in New York City, NY (Pouyat et al. 2002).

Although most studies report greater SOC in urban compared to nonurban soils, this is not always the case. In Montgomery County, VA, Chen et al. (2013) found urban and suburban SOC to be significantly less than rural SOC due to lower fine root biomass and greater decomposition. Jo (2002) also found 27% less SOC in urban compared to nonurban lands in Chuncheon, Kangleung, and Seoul, Korea. One possible explanation for decreased SOC in urban areas might be the relative increase in impervious to pervious surfaces in urban areas compared to nonurban areas. Impervious surfaces are likely to support less vegetation and therefore have lower organic matter inputs to the soils. However, pervious surfaces may also have slower decomposition rates due to changes in

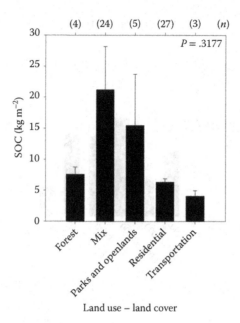

FIGURE 6.7   Urban SOC and land use–land cover.

microclimate, such as increased moisture. Raciti et al. (2012) found that SOC was 66% less under pervious compared to impervious surfaces in New York City, NY.

6.3.6   URBAN SOC BY SOIL CLASSIFICATION AND PROPERTIES

Soil classification to order in US Soil Taxonomy was investigated for a relationship with urban SOC. The proposed hypothesis was that urban SOC would not be related to this classification because many urban soils are highly altered and the current soil maps do not accurately represent soils in urban environments. A total of eight soil orders were included in the analysis: Ultisols (15), Alfisols (12), Inceptisols (12), Mollisols (9), Entisols (9), Aridisols (4), Spodosols (3), and Vertisols (1) (Figure 6.8). Entisols had the highest mean urban SOC, but due to the large variance, significant differences ($P = .2436$) in urban SOC by soil order were not detected. This finding supports the hypothesis that the current soil taxonomic system and soil maps do not accurately relate information of the current SOC in cities.

Concentration and stock of SOC may be impacted by other soil properties through effects on primary production and decomposition. The proposed hypothesis was that soil texture, density, and pH would be correlated to urban SOC. Specifically, it was expected to see the highest SOC in urban soils with higher clay and lower sand contents. Clay is important because it may help stimulate production through increased nutrient and water supply. Clay may also help to lower decomposition through protection of organic matter in aggregates. Rapid decomposition is common in sandy soils due to less climate buffering and less organic protection from decomposition. Decomposition and primary production are both limited in compacted soils. Likewise, primary production and decomposition tend to be limited under both alkaline and acidic soil conditions. Thus, it was expected that extreme soil pHs and elevated bulk density would reduce primary production more than they would reduce decomposition, and the net effect would be decreased urban SOC.

Mean sand and clay, from the 16 studies that reported these data, were 49% (range = 8%–81%) and 22% (range = 4%–75%), respectively. Mean bulk density from the 26 studies reporting was 1.3 g cm^{-3} and it ranged from 0.9–1.7 g cm^{-3}. Soil pH averaged 6.6 and ranged from 4.5–8.7 from the

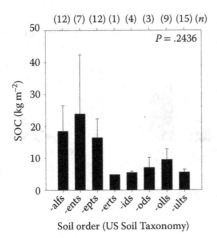

FIGURE 6.8 Urban SOC and soil order in US Soil Taxonomy.

21 studies reporting pH. However, there were no significant relationships among sand ($P = .6798$; $r = 0.1119$), clay ($P = .6228$; $r = -0.1012$), bulk density ($P = .8885$; $r = -0.0326$), or pH ($P = .3735$; $r = -0.2386$) with urban SOC. However, many of the studies included in the analysis did not include data on soil texture, bulk density, and/or pH, so the data set used to examine these hypotheses was limited. With this limited data set, it is not logical to accept or reject the hypothesis that soil properties might impart influence on SOC in urban soils.

A few of the studies in the analysis identified significant effects of soil properties on urban SOC. In Auburn, AL, Huyler et al. (2014) found SOC to be positively and significantly correlated with silt content. In 16 US cities, Selhorst and Lal (2012) found SOC to be positively correlated with soil N and bulk density, but not correlated with soil texture. They reported a curvilinear relationship between SOC and bulk density with maximum SOC at bulk density values of 1.4–1.5 g cm^{-3}. In Phoenix, AZ, Kaye et al. (2008) found SOC to be positively correlated with soil N and P contents. In Revda, Russia, Meshcheryakov et al. (2005) found SOC to be positively correlated with Ca, percent base saturation, and Fe oxides. Reasons for correlations with other soil properties and SOC are variable, but most exist because those other soil properties influence either primary production or decomposition. For example, Jim (1998) found compaction, low nutrients, and high pH to limit tree growth and thus potential supply of organic matter restitution in soils of Hong Kong, China.

6.3.6.1 Urban SOC by Sampling Depth

Soil sampling depth appears to help explain SOC differences among these studies (Figure 6.9). The highest SOC values were found with studies that sampled deeper soil profiles. Studies that sampled only topsoil (0–30 cm) found significantly ($P < .0001$) less SOC than those studies that sampled deeper (30+ to 235 cm). Furthermore, a significant positive ($P < .0001$; $r = 0.5577$) relationship was observed with sampling depth and SOC content. Sampling depth appears to be the confounding factor in the present analyses examining SOC relationships with some soil forming factors, particularly time and vegetation. Initially, the data indicated that urban SOC might have been influenced by time and vegetation. But, after closer examination of sampling depth, the data were found to have potential confounding by sampling depth of those main factor effects.

Sampling deeper is always challenging in any environment, but urban landscapes pose their unique set of difficulties. Pavements, buried roads, confined spaces, inaccessibility, and costs are a few of the obstacles that might prevent deep sampling in urban soils. However, the present data show the importance of sampling deeper in order to fully characterize urban SOC. The mean SOC for the 40 studies sampling the upper 30 cm of soil was 5.4 kg m^{-2}, compared to a mean 25.5 kg m^{-2}

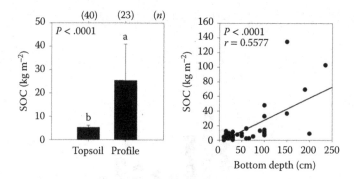

FIGURE 6.9 Urban SOC and sampling depth.

for the 23 studies that sampled up to 235 cm depth. Confining soil sampling to the upper 30 cm of soil might underestimate the urban SOC stock by 20.1 kg m^{-2} or an underestimate of 79%.

Unlike forests and agricultural lands, it does not appear that urban SOC drops sharply from topsoil to subsoil. In Chicago, IL, Scharenbroch et al. (unpublished) found that SOC contents actually increased with depth, with the maximum SOC found at the depths of 20–80 cm. They found that the 0–20 cm depth accounted for only 23% of the total SOC in the entire 0–100 cm profile, that is, sampling topsoil underestimates SOC stock by 77%. Many of the soil profiles in the Chicago sampling had multiple and deep (A1–A2, etc.), buried A horizons (Ab), human-altered and human-transported horizons that were enriched in organic material (^A, or ^O), and subsoil horizons enriched in illuvial organic matter (Bh).

6.3.7 GLOBAL ESTIMATE OF URBAN SOC

In comparison to other soils across the world, urban SOC is relatively high. The mean urban SOC across these 63 studies was 12.7 kg C m^{-2}. The three German studies in this analysis had extremely high values for SOC: 69.5 (Lorenz and Kandeler 2005), 103.0 (Beyer et al. 2001), and 134.8 kg C m^{-2} (Schleuß et al. 1998). A more conservative estimate of global urban SOC of 8.2 kg C m^{-2} may be used by removing the German studies. The present estimate of 8.2 kg C m^{-2} is in exact agreement with the 8.2 kg C m^{-2} estimate by Pouyat et al. (2002) and very close to the Pouyat et al. (2006) estimate of 7.7 kg C m^{-2} for urban SOC storage.

In comparison to other soil orders, the present estimate of urban SOC is greater than all soil orders but Histosols (organic soils), which are estimated to contain 22.8 kg C m^{-2} (Guo et al. 2006). The 8.2 kg C m^{-2} estimate of urban SOC is 332% higher than the global estimate for Aridisols (desert soils) and 90% higher than the global estimate for Mollisols (prairie soils).

Urban lands occupy 0.54% of the total global land surface (Schneider et al. 2009) with a total global urban land area is 800,820 km^2. Assuming an urban SOC density of 8.2 kg m^{-2}, the total urban SOC stock is estimated at 6.57 GtC. This urban SOC estimate is a minor fraction (0.32%–0.45%) of the total global soil C estimates of 1,461 GtC (Scharlemann et al. 2014) and 2,060 GtC (Batjes 2016). Assuming less conservative estimates of global urban land (4,004,100 km^2) (CIESIN 2011) and urban SOC densities (12.7 kg C m^{-2}), the global urban SOC is estimated at 50.9 GtC (2.5%–3.5% of global SOC). However, this may still be an underestimate of the urban C stock, since 40 of the 63 studies used to create this mean SOC value did not sample past 30 cm depth. Assuming deep urban SOC density (up to 235 cm) of 25.5 kg C m^{-2} and global urban land of 4,004,100 km^2 (CIESIN 2011), the global urban SOC might be estimated at 102.1 GtC, which is 5.0%–7.0% of global SOC.

6.4 CONCLUSION

The percentage of global SOC in urban areas may range from 0.32% to 7% depending on the estimation of total urban land area, SOC density of urban soil, and the estimate of total global SOC. This literature review and analysis suggest that urban SOC is quite high. The mean values of urban SOC ranged from 0.3 to 134.8, with an overall mean of 12.7 kg m^{-2}. The reasons for relatively high SOC in urban soils are varied and are discussed in Chapter 4. The data presented herein lead to the conclusion that urban SOC is partially related to climate, with higher SOC in areas with lower temperatures and moderate precipitation. The data also show that the vegetation on site may be an important correlate for urban SOC, with greater SOC under mixed compared to grasses. Conversely, there were no significant effects or associations between urban SOC and parent material, time, anthropogenic factors, or soil properties. However, it should be noted that others have found associations of urban SOC with these other soil forming factors. The results of this literature review and analysis must be interpreted with caution given the wide variability of sampling procedures in the individual studies and relatively limited number of studies that have been performed on the topic of urban soil C storage. For example, sampling depth was found to be an important determinant of C storage within urban soils, yet a consistent or standard sampling depth was not utilized by the individual studies. Going forward, more studies need to be conducted to better quantify urban SOC. It is imperative that these studies consider deep (>100 cm) soil C storage in urban ecosystems since sampling depth is a critical factor for properly detailing the storage of C in urban soils. Furthermore, to help understand drivers of urban SOC, studies must provide specific details on the soil forming factors that might be important drivers of urban SOC.

ACKNOWLEDGMENTS

The authors gratefully acknowledge and thank all the efforts of the researchers who have conducted studies on urban soil C storage. This publication would not have been possible without these publications and data. This research was supported by funding from The Morton Arboretum—Center for Tree Science.

REFERENCES

Ballantyne, A. P., R. Andres, R. Houghton, et al. 2015. Audit of the global carbon budget: Estimate errors and their impact on uptake uncertainty. *Biogeosciences* 12 (8):2565–2584.

Bartholomé, E., and A. S. Belward. 2005. GLC2000: A new approach to global land cover mapping from Earth observation data. *International Journal of Remote Sensing* 26 (9): 1959–1977.

Batjes, N. H. 2016. Harmonized soil property values for broad-scale modelling (WISE30 sec) with estimates of global soil carbon stocks. *Geoderma* 269: 61–68.

Beesley, L. 2012. Carbon storage and fluxes in existing and newly created urban soils. *Journal of Environmental Management* 104: 158–165.

Beyer, L., H-P. Blume, D-C. Elsner, et al. 1995. Soil organic matter composition and microbial activity in urban soils. *Science of the Total Environment* 168 (3): 267–278.

Beyer, L., E. Cordsen, H-P. Blume, et al. 1996. Soil organic matter composition in urbic anthrosols in the city of Kiel, NW-Germany, as revealed by wet chemistry and CPMAS 13 C-NMR spectroscopy of whole soil samples. *Soil Technology* 9 (3): 121–132.

Beyer, L., P. Kahle, H. Kretschmer, et al. 2001. Soil organic matter composition of man-impacted urban sites in North Germany. *Journal of Plant Nutrition and Soil Science* 164 (4): 359–364.

Campbell, C. D., J. R. Seiler, P. E. Wiseman, et al. 2014. Soil carbon dynamics in residential lawns converted from Appalachian mixed oak stands. *Forests* 5 (3): 425–438.

Center for International Earth Science Information Network—CIESIN—Columbia University, International Food Policy Research Institute—IFPRI, The World Bank, and Centro Internacional de Agricultura Tropical—CIAT. 2011. Global Rural-Urban Mapping Project, Version 1 (GRUMPv1): Urban Extents Grid. Palisades, NY: NASA Socioeconomic Data and Applications Center (SEDAC). http://dx.doi.org/10.7927/H4GH9FVG (accessed 22 June 2016).

Chen, Y., S. D. Day, A. F. Wick, et al. 2013. Changes in soil carbon pools and microbial biomass from urban land development and subsequent post-development soil rehabilitation. *Soil Biology and Biochemistry* 66: 38–44.

Churkina, G., D. G. Brown, and G. Keoleian. 2010. Carbon stored in human settlements: The conterminous United States. *Global Change Biology* 16 (1): 135–143.

Deluca, T. H., and C. Boisvenue. 2012. Boreal forest soil carbon: Distribution, function and modelling. *Forestry* 85 (2): 161–184.

Edmondson, J. L., Z. G. Davies, S. A. McCormack, et al. 2014. Land-cover effects on soil organic carbon stocks in a European city. *Science of the Total Environment* 472: 444–453.

Edmondson, J. L., Z. G. Davies, N. McHugh, et al. 2012. Organic carbon hidden in urban ecosystems. *Scientific Reports* 2: 963.

Edmondson, J. L., I. S. Jonathan Potter, E. Lopez-Capel, et al. 2015. Black carbon contribution to organic carbon stocks in urban soil. *Environmental Science & Technology* 49 (14): 8339–8346.

Elvidge, C. D., B. T. Tuttle, P. C. Sutton, et al. 2007. Global distribution and density of constructed impervious surfaces. *Sensors* 7 (9): 1962–1979.

Golubiewski, N. E. 2006. Urbanization increases grassland carbon pools: Effects of landscaping in Colorado's front range. *Ecological Applications* 16 (2): 555–571.

Gough, C. M., and H. L. Elliott. 2012. Lawn soil carbon storage in abandoned residential properties: An examination of ecosystem structure and function following partial human-natural decoupling. *Journal of Environmental Management* 98: 155–162.

Green, D. M., and M. Oleksyszyn. 2002. Enzyme activities and carbon dioxide flux in a Sonoran Desert urban ecosystem. *Soil Science Society of America Journal* 66 (6): 2002–2008.

Guo, L. B., and R. M. Gifford. 2002. Soil carbon stocks and land use change: A meta-analysis. *Global Change Biology* 8 (4): 345–360.

Guo, Y., R. Amundson, P. Gong, et al. 2006. Quantity and spatial variability of soil carbon in the conterminous United States. *Soil Science Society of America Journal* 70 (2): 590–600.

Huyler, A., A. H. Chappelka, S. A. Prior, et al. 2014. Drivers of soil carbon in residential 'pure lawns' in Auburn, Alabama. *Urban Ecosystems* 17 (1): 205–219.

Jenny, H. 1941. *Factors of Soil Formation: A System of Quantitative Pedology*. Chelmsford: McGraw-Hill.

Jia-Guo, J., Y. Lin-Zhuang, W. Jun-Xi, et al. 2010. Land use and soil organic carbon in China's village landscapes. *Pedosphere* 20 (1): 1–14.

Jim, C. Y. 1998. Urban soil characteristics and limitations for landscape planting in Hong Kong. *Landscape and Urban Planning* 40 (4): 235–249.

Jo, H-K. 2002. Impacts of urban greenspace on offsetting carbon emissions for middle Korea. *Journal of Environmental Management* 64 (2): 115–126.

Jo, H-K., and G. E. McPherson. 1995. Carbon storage and flux in urban residential greenspace. *Journal of Environmental Management* 45 (2): 109–133.

Jobbágy, E. G., and R. B. Jackson. 2000. The vertical distribution of soil organic carbon and its relation to climate and vegetation. *Ecological Applications* 10 (2): 423–436.

Kaye, J. P., A. Majumdar, C. Gries, et al. 2008. Hierarchical Bayesian scaling of soil properties across urban, agricultural, and desert ecosystems. *Ecological Applications* 18 (1): 132–145.

Kaye, J. P., R. L. McCulley, and I. C. Burke. 2005. Carbon fluxes, nitrogen cycling, and soil microbial communities in adjacent urban, native and agricultural ecosystems. *Global Change Biology* 11 (4): 575–587.

Kong, L., Z. Shi, and L. M. Chu. 2014. Carbon emission and sequestration of urban turfgrass systems in Hong Kong. *Science of the Total Environment* 473: 132–138.

Le Quéré, C., R. Moriarty, R. M. Andrew, et al. 2015. Global carbon budget 2014. *Earth Systems Science Data* 7: 47–85.

Lin, H. A. N., Y-L. Zhang, J. I. N. Shuo, et al. 2010. Effect of different irrigation methods on dissolved organic carbon and microbial biomass carbon in the greenhouse soil. *Agricultural Sciences in China* 9 (8): 1175–1182.

Liu, Y., C. Wang, W. Yue, et al. 2013. Storage and density of soil organic carbon in urban topsoil of hilly cities: A case study of Chongqing Municipality of China. *Chinese Geographical Science* 23 (1): 26–34.

Livesley, S. J., B. J. Dougherty, A. J. Smith, et al. 2010. Soil-atmosphere exchange of carbon dioxide, methane and nitrous oxide in urban garden systems: Impact of irrigation, fertiliser and mulch. *Urban Ecosystems* 13 (3): 273–293.

Livesley, S. J., A. Ossola, C. G. Threlfall, et al. 2016. Soil carbon and carbon/nitrogen ratio change under tree canopy, tall grass, and turf grass areas of urban green space. *Journal of Environmental Quality* 45 (1): 215–223.

Lorenz, K., and E. Kandeler. 2005. Biochemical characterization of urban soil profiles from Stuttgart, Germany. *Soil Biology and Biochemistry* 37 (7): 1373–1385.

Meshcheryakov, P. V., E. V. Prokopovich, and I. N. Korkina. 2005. Transformation of ecological conditions of soil and humus substance formation in the urban environment. *Russian Journal of Ecology* 36 (1): 8–15.

National Oceanic and Atmospheric Administration. 2016. USA Department of Commerce. http://www.noaa.gov/ (accessed 15 May 2016).

Olson, D. M., E. Dinerstein, E. D. Wikramanayake, et al. 2001. Terrestrial ecoregions of the world: A new map of life on earth a new global map of terrestrial ecoregions provides an innovative tool for conserving biodiversity. *Bioscience* 51 (11): 933–938.

OneGeology Portal. World CGMW 1:50M Geological Units Onshore. http://www.onegeology.org/portal/ (accessed 27 May 2016).

Pan, Y., R. A. Birdsey, J. Fang, et al. 2011. A large and persistent carbon sink in the world's forests. *Science* 333 (6045): 988–993.

Pouyat, R. V., P. M. Groffman, I. D. Yesilonis, et al. 2002. Soil carbon pools and fluxes in urban ecosystems. *Environmental Pollution* 116: S107–S118.

Pouyat, R. V., K. Szlavecz, I. D. Yesilonis, et al. 2010. Chemical, physical, and biological characteristics of urban soils. In *Urban Ecosystem Ecology*, eds. J. Aitkenhead-Peterson and A. Volder, pp. 119–152. Madison, WI: Soil Science Society of America.

Pouyat, R. V., I. D. Yesilonis, and N. E. Golubiewski. 2009. A comparison of soil organic carbon stocks between residential turf grass and native soil. *Urban Ecosystems* 12 (1): 45–62.

Pouyat, R. V., I. D. Yesilonis, and D. J. Nowak. 2006. Carbon storage by urban soils in the United States. *Journal of Environmental Quality* 35 (4): 1566–1575.

Pouyat, R. V., I. D. Yesilonis, J. Russell-Anelli, et al. 2007. Soil chemical and physical properties that differentiate urban land-use and cover types. *Soil Science Society of America Journal* 71 (3): 1010–1019.

Raciti, S. M., L. R. Hutyra, and A. C. Finzi. 2012. Depleted soil carbon and nitrogen pools beneath impervious surfaces. *Environmental Pollution* 164: 248–251.

Raciti, S. M., P. M. Groffman, J. C. Jenkins, et al. 2011. Accumulation of carbon and nitrogen in residential soils with different land-use histories. *Ecosystems* 14 (2): 287–297.

Scalenghe, R., and F. A. Marsan. 2009. The anthropogenic sealing of soils in urban areas. *Landscape and Urban Planning* 90 (1): 1–10.

Scharenbroch, B. C., R. T. Fahey, and M. Bialecki. (unpublished). Soil organic carbon storage in the Chicagoland ecosystem.

Scharenbroch, B. C., J. E. Lloyd, and J. L. Johnson-Maynard. 2005. Distinguishing urban soils with physical, chemical, and biological properties. *Pedobiologia* 49 (4): 283–296.

Scharlemann, J. P. W, Tanner, E. V. J., Hiederer, R., and Kapos, V. 2014. Global soil carbon: Understanding and managing the largest terrestrial carbon pool. *Carbon Management* 5 (1): 81–91.

Schleuß, U., Q. Wu, and H-P. Blume. 1998. Variability of soils in urban and periurban areas in Northern Germany. *Catena* 33 (3): 255–270.

Schmitt-Harsh, M., S. K. Mincey, M. Patterson, B. C. Fischer, and T. P. Evans. 2013. Private residential urban forest structure and carbon storage in a moderate-sized urban area in the Midwest, United States. *Urban Forestry and Urban Greening* 12 (4): 454–463.

Schneider, A., M. A. Friedl, and D. Potere. 2009. A new map of global urban extent from MODIS satellite data. *Environmental Research Letters* 4 (4): 044003.

Selhorst, A., and R. Lal. 2012. Effects of climate and soil properties on US home lawn soil organic carbon concentration and pool. *Environmental Management* 50 (6): 1177–1192.

Selhorst, A. and R. Lal. 2013. Net carbon sequestration potential and emissions in home lawn turfgrasses of the United States. *Environmental Management* 51(1): 198–208.

Shen, W., G. Darrel Jenerette, D. Hui, et al. 2008. Effects of changing precipitation regimes on dryland soil respiration and C pool dynamics at rainfall event, seasonal and interannual scales. *Journal of Geophysical Research: Biogeosciences* 113: G03024.

Sun, Y., J. Ma, and C. Li. 2010. Content and densities of soil organic carbon in urban soil in different function districts of Kaifeng. *Journal of Geographical Sciences* 20 (1): 148–156.

Townsend-Small, A., and C. I. Czimczik, 2010. Carbon sequestration and greenhouse gas emissions in urban turf. *Geophysical Research Letters* 37(2).

United Nations, Department of Economic and Social Affairs, Population Division. 2015. *World Population Prospects: The 2015 Revision, Key Findings and Advance Tables*. Working Paper No. ESA/P/WP.241.

United States Census Bureau (2016). http://www.census.gov/en.html (accessed 2 June 2016).

Van Groenigen, K. J., X. Qi, C. W. Osenberg, et al. 2014. Faster decomposition under increased atmospheric CO_2 limits soil carbon storage. *Science* 344 (6183): 508–509.

Vasenev, V. I., J. J. Stoorvogel, and I. I. Vasenev. 2013. Urban soil organic carbon and its spatial heterogeneity in comparison with natural and agricultural areas in the Moscow region. *Catena* 107: 96–102.

7 Sealing Effects on Properties of Urban Soils

Przemysław Charzyński, Piotr Hulisz, Anna Piotrowska-
-Długosz, Dariusz Kamiński, and Andrzej Plak

CONTENTS

7.1 SOIL SEALING AS A GLOBAL PROBLEM

Surface sealing is one of the main causes of soil degradation in the European Union, together with erosion, decrease in the soil organic matter (SOM) content, compaction, and so on (European Commission 2006). Soils are degraded by sealing in all developed countries (e.g., the United States, Japan, and China), and this phenomenon also has an increasing impact on soils in developing countries. Relatively large areas in the cities are sealed. In some municipalities, the percentage of such areas may even exceed 70%, such as in Tirana and Bucharest (Figure 7.1). For example, in the region of the urban agglomeration around Hangzhou Bay, one of the largest metropolitan areas in China, the percentage of sealed soils increased from 2.7 in 1994 to 8.7 in 2009 (Xiao et al. 2013).

Soil sealing is one of the most damaging and destructive processes to the soil environment. The process is basically irreversible and can be defined as a destruction of soil cover by total or partial application of an impermeable layer on the soil surface (Wesolek 2008; Scalenghe and Ajmone Marsan 2009); it has a significant impact on the functioning of soils. This problem often affects the fertile agricultural lands by increasing flood risks and water deficiency, and contributes to global warming (Couch et al. 2007; Scalenghe and Ajmone Marsan 2009; EU Technical Report 2011). Two degrees of soil sealing can be distinguished (Nestroy 2006): (1) total impermeability caused by solid concrete or asphalt and (2) partial impermeability caused by sealing with semipervious surface (e.g., cobblestones, concrete paving setts, or openwork concrete structures), which allow some penetration of moisture and air. In the course of construction work during sealing, the soils are either completely destroyed, especially under asphalt-concrete road surfaces, or are substantially truncated in the case of pavements.

Soil sealing interrupts or greatly reduces the exchange of matter and energy between the pedosphere and lithosphere, biosphere, hydrosphere, and atmosphere. It also affects the processes occurring in the water cycle, biogeochemical cycles, and energy transfer. As a result, this leads to many negative consequences. Depending on the degree of sealing, natural soil functions and ecosystem services provided by such an area are reduced or completely prevented in many cases. Properties of such soils are completely different in relation to morphologically similar but unsealed soil. Changes in temperature, water, and

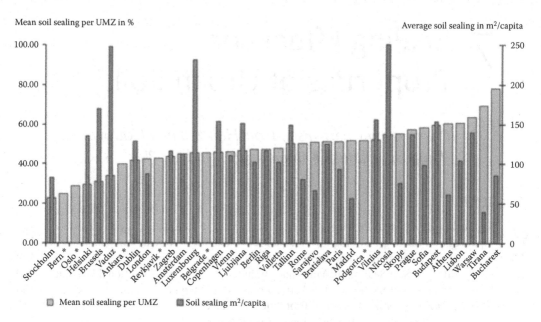

Mean soil sealing per UMZ in %   Average soil sealing in m²/capita

☐ Mean soil sealing per UMZ   ■ Soil sealing m²/capita

FIGURE 7.1   Mean soil sealing in European capitals (Urban MZ) and soil sealing per inhabitant. (From European Environment Agency, *Urban Soil Sealing in Europe*. http://www.eea.europa.eu/articles/urban-soil-sealing-in-europe. 2011.)

air regimes in such soils induce changes in their functioning (which is still insufficiently studied) and the subsequent alteration of the properties of these soils (Prokofyeva et al. 2011).

Sealed soils were named ekranozems by Russian scientists (Stroganova and Prokofieva 1995; Stroganova et al. 1998). Soils covered with asphalt are called *Ekranosols*. The name comes from French, *écran*—screen, and from Russian, *zemla*—soil, earth. The systematics of urban surface deposits developed by the aforementioned research team defines Ekranosols as one of the urban soil types under the road foundation, asphalt, concrete, and others. Burghardt (2001) classifies the covered soils into the group of lithosols as ekranolits and defines them as young urban soils where diagnostic horizons have not been developed yet. Greinert (2003) also included a unit of sealed soils (in Polish: *gleby przykryte*) in his concept of urban soil classification.

Sealed soils debuted in the world classification system in 2006 in the second edition of the World Reference Base (WRB) classification (IUSS Working Group WRB 2007). Sealed soils were included in the newly introduced Reference Soil Group, Technosols. Such soils can be identified on the lower level of classification with the qualifier Ekranic (ek). This qualifier can be used when technic hard material starts within 5 cm of the soil surface, covering at least 95% of the horizontal extent of the pedon (IUSS Working Group WRB 2015). Since that time "ekranic" soils have been included in some national classification systems, for example, the Polish Soil Classification (2011), Romanian (Florea and Munteanu 2012), and Slovak (Societas Pedologica Slovaca 2014). When compared with WRB, the criteria for horizontal coverage were reduced to 90% in the Romanian classification and even up to 80% in the Slovakian one. Sealed soils as a classification unit are still not present in the soil taxonomy (Soil Survey Staff 2014), and Russian (Shishov et al. 2004) and German (Ad-Hoc-Arbeitsgruppe Boden 2005) systems for soil classification.

7.2   IMPACT OF SOIL SEALING ON SOME PHYSICAL AND CHEMICAL SOIL PROPERTIES

Urban sealed soils are often characterized by a large horizontal and vertical heterogeneity. Soil properties are significantly related to the human impact intensity and the form of land use (Scalenghe and Ajmone Marsan 2009; Xiao et al. 2013; Greinert 2015). Sealed soils are characterized by specific morphology resulting from the presence of so-called *technic hard material* (IUSS Working Group WRB 2015), usually introduced in the form of a solid, bituminous, concrete layer, concrete slabs, setts (paving stones), and so on, whose presence affects the water resource management and gas exchange in the soil. Furthermore, the preparatory work related to the leveling and compressing of the bedding material results in a considerable compaction of the topsoil. Sealed soils usually lack the humus horizon as it is destroyed during surfacing (Stroganova et al. 1998; Greinert 2003; Prokofyeva et al. 2011). A sequence of horizons of primary, natural soil usually is disturbed or even completely destroyed and soil material undergoes substantial transformations, for example, mixing *in situ* (Figure 7.2). Similarly to other urban soils, sealed soils are usually alkaline (Table 7.1). The additional input of carbonates comes from technogenic materials used to cover the construction. Thus, the highest pH values and $CaCO_3$ concentration are usually recorded in the uppermost soil horizons. Stroganova et al. (1998) revealed that due to the specific water regime (very limited water penetration below the cover and impeded evaporation), carbonates may accumulate at a depth of 10–15 cm. Moreover, the content of carbonates and their distribution in the soil profile may also be affected by the presence of artifacts of other origin, as well as the primary soil properties.

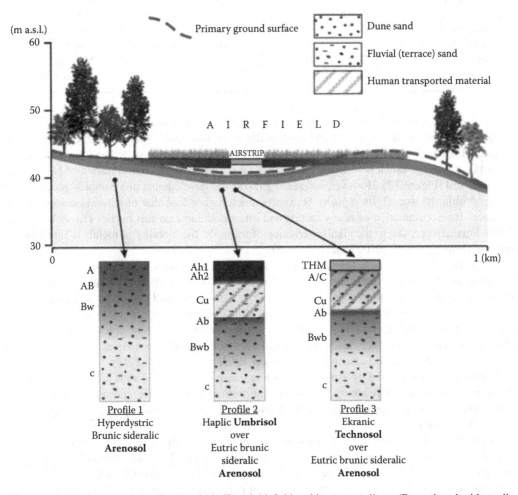

FIGURE 7.2 Technosequence of soils within Toruń Airfield and its surroundings. (Reproduced with modifications from Charzyński P, Świtoniak M, *Soil Sequences Atlas,* Wydawnictwo Naukowe Uniwersytetu Mikołaja Kopernika, Toruń, 2014. With permission from Nicolaus Copernicus University Press.)

TABLE 7.1

Ranges of pH (in H₂O) and CaCO₃ Content in Sealed Soils (to 100 cm depth)

Location	pH (in H₂O)	CaCO₃ [%]	Source
Cluj-Napoca, Romania	8.0–9.5	1.6–10.4	Charzyński et al. (2011)
Debrecen, Hungary	7.9–8.3	0.1–11.8	Sándor et al. (2013)
Moscow, Russia	7.5–9.0	0.8–4.0	Stroganova et al. (1998)
Toruń, Poland	7.7–9.5	0.3–7.3	Charzyński et al. (2011, 2013)
Zielona Góra, Poland	7.3–7.8	0.6–2.1	Greinert (2013)

TABLE 7.2

The Average Total Concentrations of Heavy Metals in Sealed Urban Soils

Location	Cr	Zn	Cd	Pb	Cu	Hg	Ni	Sources
	[mg·kg⁻¹]							
Toruń, Poland	2.73	23.8	0.15	25.2	11.8	0.12	3.05	Charzyński et al. (2017)
Debrecen, Hungary	8.00	67.7	<1	10.3	7.10	–	4.58	Sándor et al. (2013)
Hacava, Slovakia	140	20.0	–	159	60.0	2.00	79.0	Sobocká (2013)
Szczecin, Poland	–	28.8	0.29	18.0	8.07	–	8.17	Meller et al. (2013)
Zielona Góra, Poland	–	293	0.58	85.2	33.0	–	9.70	Greinert (2013)

Compared to nonurbanized areas, soils in urban areas are considerably more dust polluted and contain high levels of heavy metals and other pollutants (Oliva and Espinosa 2007; Lu et al. 2009; Chen et al. 2010). The impact of industrial and traffic emissions on the sealed soils is small due to the surface coating, which is reflected in the vertical variation of the concentration of heavy metals in the soil (Figure 7.2). However, soil sealing favors the development of a complex geochemical barrier within the topsoil. Its capacity is usually much higher than that of natural analogs, which enhances the accumulation of heavy metals and other pollutants on this barrier (Kosheleva et al. 2015). Furthermore, due to the alkaline reaction of the soils, the mobility of metals is significantly reduced (Imperato et al. 2003; Plak et al. 2015).

The data in Table 7.2 show that the concentration of heavy metals in sealed soils from Poland, Hungary, and Slovakia is rather low. In the urban area of Szczecin (Poland), Sammel et al. (2013) revealed that the average total content of metals (e.g., Cd, Zn, Ni, and Co) in the soil covered with the bituminous pavement and the parent material (100–140 cm) is similar. However, a high accumulation of heavy metals in these soils may occur incidentally and can be observed in the subsoil (Figure 7.3). This trend is specifically observed in soils from old urban centers exposed to long-term human impact (Puskás and Farsang 2009; Greinert 2003; Sammel et al. 2013).

Contrary to soils covered by asphalt or concrete, the partially sealed soils are more exposed to the transfer of pollutants resulting from water infiltration. For example, the seam material from paved soils may show stronger sorption of Pb and Cd compared to the original construction sand (Nehls et al. 2008).

Greinert (2015) reported a significant contribution of soluble forms of heavy metals in the total pool despite relatively low concentrations of total forms (Figure 7.4). The results obtained by this author demonstrated that over 40% of the potentially mobile form of Pb in the subtotal content was recorded in sealed, industrial, and forest areas, not particularly polluted with this metal. Further, the potential solubility of Cu was lower than that of the two other heavy metals. A relatively high ratio of potentially available to subtotal Zn solubility (i.e., >20%) was observed in soils of roadsides, sealed areas, and parks. This trend may suggest an anthropogenic origin of these elements

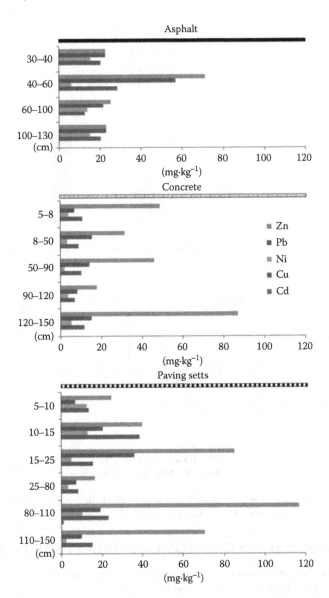

FIGURE 7.3 The content of heavy metals in soils covered with asphalt, concrete slabs, and small-sized paving setts in Zielona Góra based on results by Greinert A, *Studies of Soils in the Zielona Góra Urban Area,* Oficyna Wydawnicza Uniwersytetu Zielonogórskiego, Zielona Góra, 2003.

(Chłopecka et al. 1996; Chojnacka et al. 2005; Ljung et al. 2007; Islam et al. 2015; Szolnoki and Farsang 2013).

The degree of soil sealing is an important factor affecting soil contamination with polycyclic aromatic hydrocarbons (PAHs). In Toruń (Poland), Mendyk and Charzyński (2016) revealed that the concentration of these harmful compounds decreases with increase in the degree of soil sealing, because the completely sealed soils were the least contaminated, and the nonsealed reference soils were the most contaminated (Figure 7.5). Therefore, despite the fact that soil sealing is an extremely negative phenomenon, there are also some positive impacts. The contamination with PAHs was mainly attributed to the combustion of grass, wood, and coal, most likely coming from individual heating systems of single-family residences.

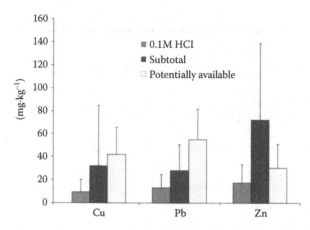

FIGURE 7.4 Copper, lead, and zinc content in sealed soils and their potential solubility. (Modified from Greinert A, *J Soils Sediments,* 15, 1725–1737, 2015.)

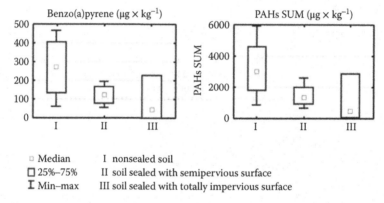

FIGURE 7.5 The differences in Benzo[a]pyrene and sum of PAHs depending on the soil sealing degree. (Modified from Mendyk L, Charzyński P, *Soil Sci Annu* 67(1), 17–23, 2016.)

7.3 IMPACT OF SOIL SEALING ON SOME BIOLOGICAL SOIL PROPERTIES

Soil microorganisms are fundamental to the development and maintenance of the soil ecosystem. They play an important role in the incorporation of organic matter, decomposition, mineralization, and nutrient cycling, as well as in the development and maintenance of soil structure. So far, the research on soil biology has focused mainly on natural and agricultural systems, with insufficient attention paid to those in the urban environment, especially sealed areas. In recent years, however, there has been much stronger interest in urban soils because of the continuously growing urban population (Lorenz and Kandeler 2005).

7.3.1 Effect of Soil Sealing on Microbial Biomass, Activity, and Functional Diversity

Most of the previous studies have focused on effects of sealed areas on water movement (Bhaduri et al. 2001; Peffy and Nawaz 2008), gaseous diffusion (Wiegand and Schott 1999; Kaye et al. 2006), and biodiversity (Savard et al. 2000). However, effects of soil sealing on nutrient cycling and biological components (e.g., diversity and enzymatic activity of soil microorganisms) are rarely investigated (Zhao et al. 2012a; Wei et al. 2013; Piotrowska-Długosz and Charzyński 2015). The sealing of soils prevents the exchange of gases, water, and nutrients between the soil and the atmosphere, which results in a negative effect not only on their physicochemical properties but also on the microbial and enzymatic activity because microorganisms are the main source of enzymes in the soil. Some studies have indicated that the artificial soil sealing in urban areas can significantly affect both the activity and the diversity of soil microorganisms. Generally, the microbial biomass and activities in the open urban soils are comparable to those of the agricultural or forest surface soils. However, the lowest microbial biomass and activities are often observed in highly disturbed and sealed soils (Wei et al. 2014). Another study was conducted on 10 plots that varied in land use (paved roads, residential paved square, residential paved alley [RPA], and grasses area) selected to investigate the effects of impervious surfaces on microbiological properties in urban soils from Nanjing City, China (Wei et al. 2013).

The contents of soil organic carbon (SOC), soil microbial carbon (MBC), and soil microbial nitrogen (MBN) were significantly lower in sealed soils than in open soils. Mean concentrations of SOC, MBC, and MBN in fine earth from the impervious areas at a depth of 0–20 cm were respectively 6.5 $g \cdot kg^{-1}$ and 55.8 $mg \cdot kg^{-1}$, which were significantly lower than those in grassed areas (18.5 $g \cdot kg^{-1}$, 317.9 $mg \cdot kg^{-1}$). Soil MBN data ranged from 5.6 to 79.3 $mg \cdot kg^{-1}$, and varied greatly among land-use types in the sealed soils ($p < .05$) and between the open and sealed sites ($p < .01$). The data of Wei et al. (2013) concerning the carbon content in impervious soils were consistent with those from some previous studies of carbon storage in urban areas (Pouyat et al. 2006; Churkina et al. 2010) which also showed that the C concentration in the soil covered with impervious surfaces is lower than that of open soils (e.g., urban grass or forests). Similar results were obtained by Zhao et al. (2012a), where soil MBC and MBN concentrations significantly varied between the three land covers studied (e.g., the impervious land, forest vegetation, and bare land with no sealing and no vegetation cover). The highest MBC and MBN were observed at a forest site and the lowest at sealed sites. In a study conducted on Ekranic Technosols of Toruń City (NW Poland) (Piotrowska-Długosz and Charzyński 2015), where the age of sealing at impervious sites ranged between 30 and 40 years, the content of soil MBC and MBN were significantly higher in reference soils compared to those in the sealed soils, while the degree of soil sealing (semipervious concrete paving slabs, which allow partial penetration of water and air, and soils sealed with impervious surfaces, such as asphalt and concrete—Figure 7.6) had no significant effect on these properties (Table 7.3). The disturbance of sealed soils was additionally confirmed by significantly lower MBC:SOC and MBN:N total ratios and a higher metabolic quotient (qCO_2) compared to those of the reference soils (Table 7.4). In fact, the increasing qCO_2 values confirmed the impact of site disturbance and organic matter losses (Insam and Haselwandter 1989).

The content of CO_2 released during incubation is an important indicator of microbial activity in both naturally and anthropogenically affected soils (Ryan and Law 2005). Wei et al. (2013, 2014) showed that the accumulation of CO_2 from the open soils was much higher than that from the sealed soils, suggesting that the impervious surface covering the urban soils can significantly reduce the soil microbial activity. Wei et al. (2014) observed that the basal respiration rate in the reference soil was three times higher than at the sealed sites. Likewise, the amount of accumulated CO_2 emissions of the impervious-covered soils during the incubation period, which ranged from 56.4 to 200.3 $mg \cdot kg^{-1}$, was significantly lower compared to 169.1–464.8 $mg \cdot kg^{-1}$ for urban open soils. Thomsen et al. (2008) also observed that the proportion of carbon loss was higher in the covered soil than that from the open sites. Piotrowska-Długosz and Charzyński (2015) reported that soil respiration within 11 days of incubation was significantly higher in the reference soil compared to that of the sealed soil, while the degree of soil sealing had no significant effect. After that period, the significant differences between the studied soils disappeared. Murphy et al. (2011) reported that the amount of CO_2 evolved increased with the increase in SOC and MBC concentrations indicating that sealing of soils reduced the SOC content and resulted in the reduction of soil microbial activity. These trends may cause a decline in the microbial diversity of soil microbial communities (Degens et al. 2000).

Soil biodiversity is defined by the variation in the soil life, from genes to communities, and the variation in soil habitats, from microaggregates to entire landscapes (European Environment Agency 2010). Soil sealing in urban areas strongly reduces microbial diversity caused by the lack of carbon input of plant litter and animal feces (Wei et al. 2013). Further, the effects of sealing on soil microbial diversities varies among types of land use, whereby the road pavement has the

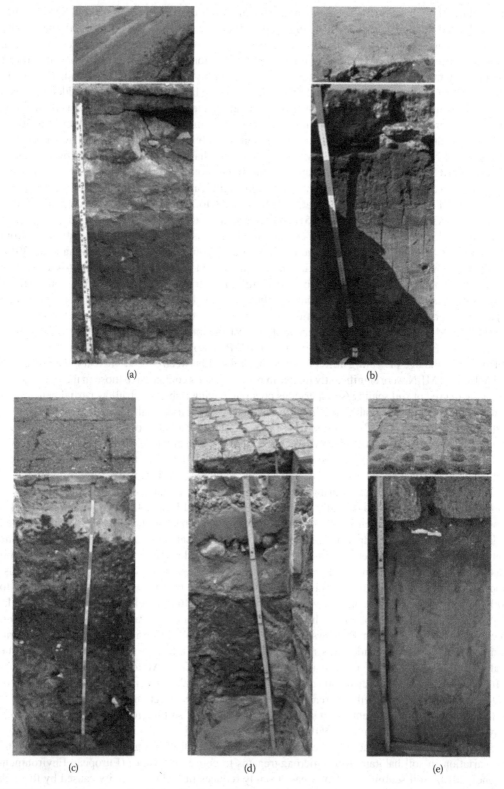

FIGURE 7.6 Examples of sealed soils from Toruń, Poland. (a and b) Impervious surface and (c–e) semipervious surface.

TABLE 7.3
Microbial Biomass Content and Enzymatic Activity as Dependent on the Degree of Soil Sealing

Variable	Degree of Soil Sealing	Mean (±SD)	Range	CV (%)
DHA	C	7.15 (8.11) a	0.90–25.4	113.4
	A	5.59 (7.98) a	0.66–24.7	142.8
	B	4.08 (5.29) a	0.44–18.5	129.6
CAT	C	1.17 (1.15) a	0.065–3.13	97.8
	A	0.84 (0.77) a	0.23–2.44	91.1
	B	0.77 (0.52) a	0.23–1.98	67.3
GLU	C	0.13 (0.10) a	0.02–0.30	81.5
	A	0.10 (0.08) a	0.02–0.21	81.1
	B	0.08 (0.07) a	0.015–0.21	92.1
CEL	C	4.08 (3.44) a	0.24–10.0	84.4
	A	5.18 (7.02) a	0.07–18.9	135.6
	B	2.87 (3.50) a	0.23–11.4	121.0
NR	C	2.79 (0.53) a	1.98–3.99	22.6
	A	2.52 (0.61) a	1.85–3.85	27.8
	B	1.81 (0.30) b	1.29–2.15	16.5
UR	C	4.25 (1.15) a	2.01–6.54	33.8
	A	2.73 (0.66) b	1.85–3.98	29.3
	B	1.97 (0.42) b	1.12–2.85	26.2
FDA	C	21.6 (8.6) a	12.9–35.8	43.0
	A	22.9 (9.1) a	15.3–59.5	50.0
	B	13.7 (3.7) b	8.6–21.0	38.0
MBC	C	63.8 (27.4) a	34.7–111.2	49.0
	A	36.9 (14.6) b	23.5–66.7	38.5
	B	32.0 (9.72) b	26.0–48.4	22.9
MBN	C	15.3 (8.0) a	6.15–29.5	50.2
	A	9.10 (4.0) b	3.20–17.2	44.5
	B	6.75 (2.2) b	4.12–11.7	29.9

Source: Piotrowska-Długosz A, Charzyński P, *J Soils Sediments,* 15, 47–59, 2015.

A, semipervious sites; B, impervious sites; C, control (nonsealed soils); DHA, dehydrogenase activity (mg TPF kg^{-1} 24 h^{-1}); CAT, catalase activity (mM H$_2$O$_2$ kg^{-1} h^{-1}); GLU, β-glucosidase activity (mg pNP kg^{-1}h^{-1}); CEL, cellulase activity; NR, nitrate reductase activity (mg N-NO$_2^-$ kg^{-1} 24 h^{-1}); UR, urease activity (mg N-NH$_4^+$ kg^{-1} h^{-1}); FDA, fluorescein sodium salt hydrolysis (mg of fluorescein kg^{-1} h^{-1}); MBC, microbial biomass carbon (mg kg^{-1}); MBN, microbial biomass nitrogen (mg kg^{-1}); SD, standard deviation; CV, coefficients of variation. Standard deviation is given in parentheses. Different letters in the same column for each property indicate significant effects (at p < .05 level) of soil sealing degree compared with the nonsealed sites.

Note: The value of the property marked with "a" is significantly higher than the value marked with "b"

most negative effect on the soil microbial functional diversity. Shannon diversity (H'), Simpson diversity (D), and substrate richness (S) of paved roads' areas (PR) are significantly lower than those of RPAs and open soils ($p < .05$), indicating that urban sealing, especially sealing due to traffic pavement (PR), can significantly reduce the soil microbial functional diversity. These impacts are probably because of residential activities, which may cause exogenous organic carbon to enter sealed soil through subsurface flow from open soil or by infiltration from the sealing surface. The sealed sites in residential areas can therefore obtain organic carbon more easily than those under the road pavement, resulting in higher soil microbial diversity. Unlike total substrates, Shannon diversity (H') for substrate categories (amine/amides, polymers, carbohydrates, carboxylic acids,

TABLE 7.4

Soil C:N Ratio and Soil Microbial Parameters as Dependent on the Degree of Soil Sealing

Ratio	Degree of Soil Sealing	Mean (±SD)	Ranges
$C_{ORG}:N_{TOT}$	C	14.7 (4.54) a	9.3–22.1
	A	14.1 (3.84) a	9.1–21.1
	B	18.5 (5.58) a	11.1–28.9
MBC:MBN	C	4.28 (1.21) a	3.25–5.06
	A	4.16 (1.03) a	3.44–4.94
	B	4.61 (1.32) a	3.78–5.30
$MBC:C_{ORG}$	C	3.32((1.34) a	0.34–5.95
	A	2.01 (1.13) b	0.25–5.34
	B	1.79 (1.01) b	0.39–3.91
$MBN:N_{TOT}$	C	5.88 (2.56) a	1.46–7.00
	A	3.44 (1.99) b	1.25–6.45
	B	3.68 (1.56) b	1.26–4.29
qCO_2	C	2.29 (0.88) b	0.83–3.35
	A	2.18 (1.01) b	0.81–3.27
	B	3.08 (1.45) a	1.00–5.01

Source: Piotrowska-Długosz A, Charzyński P, *J Soils Sediments,* 15, 47–59, 2015.

A, semipervious sites; B, impervious sites; C, control (nonsealed soils); qCO_2, metabolic quotient; qCO2, mgC-CO_2 $(gCmic)^{-1} h^{-1}$; MBC, microbial biomass carbon (mg kg^{-1}); MBN, microbial biomass nitrogen (mg kg^{-1}); SD, standard deviation; CV, coefficients of variation; Different letters in the same column for each property indicate significant effects (at $p < .05$ level) of soil sealing degree compared with the nonsealed sites.

Note: The value of the property marked with "a" is significantly higher than the value marked with "b"

and amino acids) has significant differences between the sealed and open soils (Figure 7.7). Values of H' for all substrate categories of soils under impervious surfaces are lower than those in grass areas, except for polymers. Significant differences occur between the sealed and open soils in terms of substrate-use rates for amines/amides, polymers, and carbohydrates. Generally, values of H' for amines/amides and carbohydrates are significantly higher in open soils, whereas polymers are used to a great extent in sealed soils ($p < .05$).

The total number of bacteria and fungi, as well as the abundance of selected physiological groups of microorganisms, was determined in the sealed soils and in a reference profile in the study by Charzyński et al. (2013). The largest number of bacteria was found in the upper horizon of the reference soil—$6.61 \cdot 10^{-6}$ cfu·g^{-1}. Fewer bacteria were found in layers immediately beneath the concrete (Figure 7.8). The concrete cover contributed to the factors leading to equal number of bacteria directly under the concrete slabs as that at a depth of 50 cm, both in Ekranic Technosols and unsealed soil. A similar situation was observed in case of fungi. The largest number of fungi was found in the humus horizon of the reference soil—9.0×10^{-4} cfu·g^{-1}, and a much smaller number in the layers immediately beneath the concrete slabs in sealed soils. The number of denitrifiers was reduced in the profiles under the concrete. Also, the number of glucose acidifying bacteria significantly decreased in upper layers at a depth of 20 cm in sealed soils compared to the counterpart horizon of the reference soil (Charzyński et al. 2013).

The process of nitrogen transformation in sealed soil is another noteworthy process (Wei et al. 2014; Zhao et al. 2012a). Zhao et al. (2012a) reported that the soil net potential mineralization and nitrification rates at impervious sites are significantly reduced, indicating that nitrogen transformation is affected because soil sealing cuts off the nutrient supply and causes the reduction of SOM and total N. Changes in physicochemical soil properties (e.g., pH, soil moisture content, and particle size) by soil sealing affect the environment for the microbial growth. A low level of nutrients

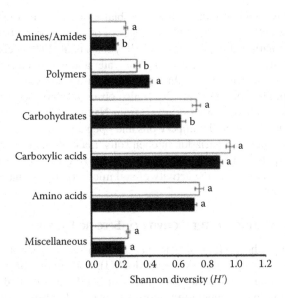

FIGURE 7.7 Shannon diversity index (*H'*) for substrate categories of urban sealed soils (■) and open soils (□). Bars represent ±1 standard error (*n* = 24 for urban sealed soils, *n* = 6 for urban open soils). Differences between the two soils were determined by l.s.d. test, and bars with the same letter are not significantly different at *p* = .05. (Reproduced from Wei Z et al., *Soil Res.*, 51, 59–67, 2013. With permission from CSIRO Publishing.)

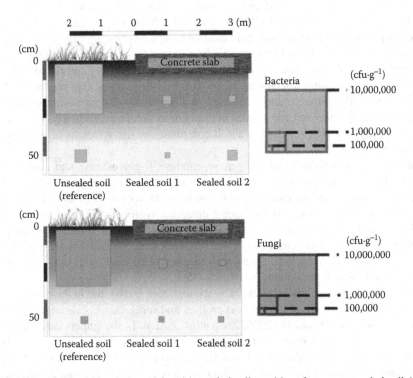

FIGURE 7.8 The number of bacteria and fungi in sealed soils and in reference, unsealed soil (cfu, colony forming units). (Reproduced with modifications from Charzyński P, Świtoniak M, *Soil Sequences Atlas*, Wydawnictwo Naukowe Uniwersytetu Mikołaja Kopernika, Toruń, 2014. With permission from Nicolaus Copernicus University Press.)

and alteration of the environment reduces the microbial biomass, which reduces mineralization. Changes in soil enzyme activities, induced by changes in other soil properties, also cause a reduction in nitrogen transformation at impervious sites (Reynolds et al. 1985; Geisseler et al. 2010).

Additionally, soil sealing prevents the exchange of water and gas and results in the formation of dry and oxygen-poor soil, which finally inhibits the growth of nitrifying bacteria and affects the nitrification (Zhao et al. 2012a). Wei et al. (2014) also observed that soil net nitrogen mineralization differed after 35 days of incubation of a soil from Nanjing, China. However, the differences in nitrogen mineralization between the impervious and open soils were not statistically significant ($p = .07$) probably due to the high variation in open soils. Lack of plant litter input and imbalance between inputs and outputs of nutrients in the soils underlying the impervious surfaces may be the most important drivers reducing the activity of soil microorganisms and microbially mediated processes (Zeng et al. 2009).

7.3.2 Effect of Soil Sealing on the Activity of Specific Enzymes

Soil enzymes involved in the cycling of nutrients are appropriate indicators of the disturbance and development of the soil ecosystem (Wang et al. 2011). Soil sealing in urban areas affects soil enzymes mainly by reducing the quantity and quality of SOM content, by deteriorating the enzymatic reaction conditions (e.g., temperature, moisture, pH, and substrate concentration) thereby affecting the activity of soil enzymes (Kaye et al. 2006).

Zhao et al. (2012a) observed that impervious sites had the lowest soil urease activity compared to that of the bare land and forest sites due to an increasing level of NH_4^+ in the sealed soil layers. The urease synthesis is regulated by soil nitrogen and is repressed when the preferred nitrogen source (e.g., $N-NH_4^+$) is available (Mobley et al. 1995). Furthermore, Zhao et al. (2012a) observed that the urease activity in the sealed soils is affected by SOM, N total, MBC and MBN, pH, and C/N ratio, as was also reported by Reynolds et al. (1985). On the other hand, impervious sites have the highest protease activity compared to that under bare land and forest, suggesting that protease synthesis is not inhibited by sealing of soils, and protein substrates are not limited in the soil. Zhao et al. (2012a) observed that in the impervious sites the protease activity was negatively correlated with SOM and MBN but was not significantly affected by NH_4^+-N in the soil, as was also reported by Geisseler and Horwath (2008). The nitrate reductase activity at a depth of 0–20 cm was lower in sealed soils compared to that under bare land and similar in impervious and forest soils. The activity of nitrate reductase is regulated by NO_3^--N content and it contributes to the synthesis of NH_4^+-N in sealed soils (Zhao et al. 2012a).

Some of the enzyme activities (dehydrogenase, catalase, β-glucosidase, and cellulases) studied by Piotrowska-Długosz and Charzyński (2015) in Toruń (NW Poland) were not statistically different between the reference and sealed soils, and between the degrees of soil sealing (Table 7.3). In the same study, the nitrate reductase and fluorescein diacetate hydrolysis (FDAH) activities were significantly higher in the semi-impervious and reference soils than at the impervious sites, while the highest soil urease activity was determined in the reference soils, followed by the semi-impervious and impervious sites. Soil urease and nitrate reductase activities were moderately varied and characterized by a coefficient of variation (CV) of 16%–35%.

In open urban soils (e.g., gardens, parks, forests, and bare lands), the highest enzymatic activity is usually observed in the top layer of soil profiles and it regressively decreases in the deeper horizons (e.g., Lorenz and Kandeler 2005; Zhao et al. 2012a). In the impervious sealed soil profile, however, the differences in the soil enzyme activity between different soil depths become increasingly blurred due to decline in soil structure (Zhao et al. 2012b). Thus, the urease activity decreased with the increase in depth in bare land and forest, while it increased with the increase in soil depth at impervious sites (Zhao et al. 2012a). With regard to the soil protease and nitrate reductase activities in the same soils there was no significant difference between different depths at impervious sites, which was different from that under other land-cover types (e.g., bare land and forest soils).

7.3.3 FACTORS AFFECTING BIOLOGICAL PROPERTIES OF SEALED SOILS

The biological properties of urban soils are mainly affected by the vegetation, especially the accumulation of plant litter at the topsoil layer of lawns and forest, which are selected mostly as reference soils. The reduced availability of carbon and substrates may be the main cause of the reduced MBC content and activity (Nannipieri and Eldor 2009). On the other hand, vegetation-covered soils receiving frequent inputs of carbon substrates, nutrients, and water would maintain a healthy soil microbial community (Wang et al. 2011). Plant litter provides the habitat, food, and energy for microorganisms in the soil. The topsoil is removed in the course of infrastructure development, which leads to reduction in the SOM content (Harris et al. 1999). Additionally, the impermeable materials reduce the supply of organic debris (e.g., tree leaves and other plant debris), which further reduces the input to the SOM pool. Moreover, moisture stress, alkaline environment, and poor ventilation caused by soil sealing may all contribute to the reduction in the microbial biomass content and activity (Tripathi and Singh 2009). On the other hand, destruction of vegetation and alteration of soil nutrient status after soil sealing are the main factors affecting the synthesis and activities of soil enzymes (Chakrabarti et al. 2000).

The level of microbial activities in sealed soils is usually associated with soil physicochemical properties (Wei et al. 2014; Piotrowska-Długosz and Charzyński 2015). In the study of impervious, covered soil in Nanjing City in China (Wei et al. 2014), results of a canonical correlation analysis showed that the first canonical correlation among the physicochemical soil properties, SOC transformation, and microbial activities were highly significant ($r = 0.95$ and 0.94, respectively) with a good fit of $p < .01$, indicating a strong association between them. The highest correlation was determined between water-soluble organic carbon, total organic carbon, and total N and the first canonical variable for microbial activities, suggesting that these properties had the most significant effects on soil microbial activities (Wei et al. 2014). Piotrowska-Długosz and Charzyński (2015) observed a relatively low correlation between C_{ORG} and N_{TOT} and the first canonical variable for ME (microbial biomass, enzyme activities) did not support the commonly known importance of these properties in the formation of microbial biomass and activity.

Piotrowska-Długosz and Charzyński (2015) reported that soil moisture reserve is significantly reduced at semipervious and impervious sites compared to that under open soils. Canonical correlation analysis showed a strong effect of soil moisture on microbial biomass and enzymatic activity, which is also confirmed by the regression analysis presented in Figure 7.9. Positive correlations are found between water content and catalase and β-glucosidase activities as well as between MBC content in the reference soils and at the semipervious sites (except for β-glucosidase in the semipervious soils). However, no significant relationship was observed between the studied properties at the impervious sites. In fact, the sufficient water content in soil is one of the most important factors affecting the soil microbiological and enzymatic processes (Gianfreda and Ruggiero 2006). The sealing of the soil surface makes it impermeable to the water flow; consequently, the water regime of the underlying soils is severely altered. This not only reduces the soil moisture content but also lowers the water table in some urban areas, which in turn reduces the rate of chemical reaction and microbial activity and diversity (Scalenghe and Ajmone Marsan 2009).

Compaction of sealed soils may be an important factor in the reduction of soil microbial functional diversity since it increases the bulk density, decreases porosity and macropore continuity, and creates unfavorable conditions for soil microorganisms (Jensen et al. 1996). Soil compaction may also alter the spatial distribution and availability of organic substances and nutrients and reduce soil water availability and aeration (Breland and Hansen 1996). Many studies have reported high bulk densities in urban soils (Scharenbroch et al. 2005; Edmondson et al. 2011). Jim (1998) observed that soil bulk density values exceeded 1.6 g cm^{-3}, which is considered the upper threshold for unimpaired root growth (Mullins 1991). Piotrowska-Długosz and Charzyński (2015) reported that half of the soil samples had bulk density higher than 1.6 g cm^{-3} (up to 1.75 g cm^{-3}), and there were no significant differences between reference, semipervious, and impervious sites. These results indicate that soils selected as reference soils (mainly lawns, but also forest and roads without vegetation cover) had a similar degree of compaction as that of sealed soils due to a wide range of technogenic activities.

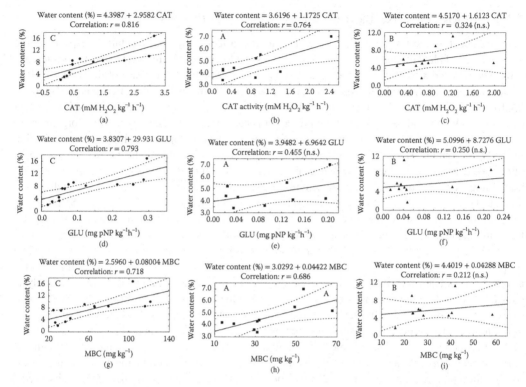

FIGURE 7.9 Linear regression between water content (%) and (a–c) catalase (CAT) activity in nonsealed soils (C), semipervious sites (A), and impervious sites (B); (d–f) β-glucosidase (GLU) activity in nonsealed soils (C), semipervious sites (A), and impervious sites (B); (g–i) MBC content in nonsealed soils (C), semipervious sites (A), and impervious sites (B); $p < .05$. (From Piotrowska-Długosz A, Charzyński P, *J Soils Sediments*, 15, 47–59, 2015.)

Compaction of urban soils may reduce the soil microbial functional diversity (Wei et al. 2013). The sealed soils under road pavement have the lowest microbial functional diversity and more severe compaction as compared to other sealed soils. Moderate soil compaction can increase the soil microbial diversity and the total organic carbon, while more severe compaction can reduce these properties (Pengthamkeerati et al. 2011), while some sealed soils in the neighborhood may have different microbial functional diversity. This trend may be due to the fact that these soils were not native and were brought from other places, and the microbial biodiversity in these soils could be affected via a different mechanism. Compaction and the associated reduction in soil porosity create unfavorable conditions for soil organisms (earthworms) living in the soil surface. Soil compaction damages the structure of earthworm tunnels and kills many of them. Alteration of soil aeration and humidity status due to soil compaction can also seriously affect the activity of soil organisms. Oxygen reduction can modify microbial activity, favoring microbes that can survive under anaerobic conditions (Turbé et al. 2010). Anaerobiosis alters the types and distribution of all organisms found in the remaining part of the soil food web.

7.4 VEGETATION COLONIZATION AND SUCCESSION IN SEALED AREAS

Covering the soil surface with cement or asphalt concrete significantly reduces the possibility of vegetation establishment. Construction of sidewalks often consists of several layers of foundation under the proper surface (Randrup et al. 2001). Therefore, seedlings have limited access to the soil surface and often are completely isolated from it. However, systematic floristic studies conducted in urban centers for more than 40 years indicate the existence of a large group of plants that can colonize pavements' cracks and curbs (Brandes 1995). Plants that colonize sidewalks face extremely difficult conditions. Pedestrian traffic and cars compress the soil, which impairs the properties of water retention and reduces the access of air (Lundholm 2013). Plants growing in the cracks of sidewalks are exposed to long periods of drought and short periods of water stagnation (Láníková and Lososová 2009). They must also endure extreme temperatures (Yang and Zhao 2016), industrial and oil pollution, salt from street deicing, and above all trampling, which is the main factor limiting their growth. Soils in the crevices of hard surfaces in urban centers are often characterized by the increased content of nutrients and calcium carbonate (Kent et al. 1999; Wheater 2002; Lundholm 2013). Plants become established in pavements from seeds or other propagules contained in the base material underneath the hard surface, or brought by wind or on shoes and car tires. They can also grow from runners and rhizomes of plants growing outside the space covered with a hard surface. Difficult habitat conditions created by treading allow the growth of species with specific adaptation only. Most often those are plants with leaf rosettes or short stems, often creeping and having extensive root systems (Lisci et al. 2003; Láníková and Lososová 2009). The first stage of succession in the cracks of sidewalks and roads are communities of lichens and bryophytes. They settle in the surface gaps, which is followed by the process of organic and inorganic matter accumulation (i.e., atmospheric dust, animal manure, and waste), and consequently the formation of a soil layer around the plants (Lisci et al. 2003; Charzyński et al. 2015). This enables the establishment of vascular plants and the development of more complex plant communities. The type of vegetation growing in the cracks is determined by three factors: (1) the intensity of land use (e.g., trampling and traffic), (2) water availability, and (3) the time elapsed since the crack formation (Figure 7.10).

Dry and intensively used habitats are colonized by communities of lichens, liverworts, and mosses, dominated by *Bryum argenteum* (Figure 7.11). Vascular plants encroach on the intensively trampled habitats with moderate humidity and the *Sagino procumbentis-Bryetum argentei* community develops, common in all European cities (Figure 7.11). *Herniarietum glabrae* may occur in similar conditions. In moderately used areas, typical species may be accompanied by ruderal plants (e.g., *Stellaria*, *Cerastium*, and *Taraxacum* spp). Communities with dominance of annual and perennial vascular plants develop in conditions with a moderate human impact (e.g., common *Poëtum*

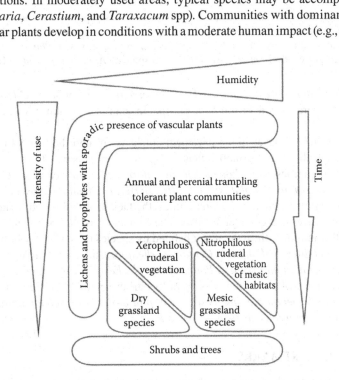

FIGURE 7.10 Block diagram presenting the relationship between the vegetation types of urban pavements and the habitat conditions.

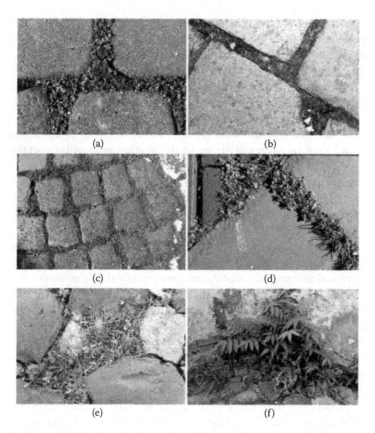

FIGURE 7.11 Examples of vegetation in pavement cracks. (a) Bryophytes community (Spała, Poland), (b) *Sagino procumbentis-Bryetum argentei* (Venice, Italy), (c) *Poëtum annuae* (Firenze, Italy), (d), *Poëtum annuae* with *Stellaria media* (Bergamo, Italy), (e) species rich community with *Plantago coronopus* (Roma, Italy), and (f) ruderal species (*Cirsium arvense*, *Taraxacum* sp.) with *Ailanthus altissima* (Toruń, Poland).

annuae, Figure 7.11) and *Eragrostio minoris-Polygonetum arenastri* or *Sclerochloo-Coronopodion squamati* (the latter characteristic of southern Europe) instead of communities with a large proportion of bryophytes.

The formation of the soil layer in the pavement cracks or around the plants in conditions with minor human impact allows the establishment of species less resistant to trampling, characteristic of ruderal communities (e.g., *Polygonum aviculare* agg., *Matricaria discoidea*, *Plantago major*, *Digitaria ischaemum*, *Diplotaxis muralis*, *Berteroa incana*, *Capsella bursa-pastoris*, *Sonchus asper*, and *Conyza canadensis*). Sometimes, the lack of use of paved surfaces leads to the development of a soil layer that completely covers the cement or asphalt pavement.

Depending on the humidity, the habitat is inhabited by species typical of dry or mesic grasslands (e.g., *Saxifraga tridactylites*, *Sedum acre*, *Centaurea rhenana*, and *Potentilla argentea*) at dry sites and species such as *Achillea millefolium*, *Poa pratensis* s.l., *Festuca rubra*, and *Trifolium repens* at moderately wet locations. Slots on unused sidewalks harbor seedlings of trees, among which the most common are *Betula pendula*, *Populus* spp., *Acer platanoides*, *A. pseudoplatanus*, and, spreading in European cities, *Ailanthus altissima* (Figure 7.11).

7.5   CONCLUDING REMARKS

Sealed soils are an essential component of urban landscapes and play a crucial role in urban ecosystems. Compared to unsealed soils, Ekranosols exert much lesser impact on the functioning of ecosystems in urban areas and provide fewer ecological functions. Sealed soils are not involved in a number of soil-forming processes; however, they should be treated the same way as the nearby unsealed soils since they are also a component of the urban pedosphere. They should be researched and charted, because their characteristics affect the ecological situation in urban areas. Sealed soils are also important because they store information about the history of the urban ecosystem and act as the archive of the human settlement history in a given area.

The data presented in the literature demonstrate that artificial sealing by impervious surfaces such as asphalt and concrete, which are very common in urban areas, can result in the alteration of physical and chemical soil properties, as well as the deterioration of microbial activity and its functional diversity. Profiles of the sealed soils have both natural horizons and technogenic layers. Moreover, the degree of soil sealing significantly affects any further soil development. The differences recorded in raw data between impervious and semi-impervious sites are remarkable.

Thus, sealing of soils can be considered as some kind of hibernation and they are ready to be exposed and restored to their original functions. On the one hand, the nearly complete isolation of the soil has mainly a negative impact on the functioning of the soil environment, but on the other hand, it may also contribute to protection of the soil from various contaminants such as heavy metals and PAHs. On the other hand, a permeable or semipermeable pavement system would be helpful to limit the consequences of soil sealing as it allows the exchange of materials and energy between the soil and the atmosphere.

REFERENCES

Ad-Hoc-Arbeitsgruppe Boden (2005) *Bodenkundliche Kartieranleitung.* 5. Aufl., Stuttgart: E. Schweizerbart'sche Verlagsbuchhandlung.

Bhaduri M, Minner M, Tatalovich S, Harbor J (2001) Long-term hydrologic impact of urbanization: A tale of two models. *J Water Res Plan* ASCE 127, 13–19.

Brandes D (1995) The flora of old town centres in Europe. In: Sukopp H, Numata M, Huber A (Eds) *Urban Ecology as the Basis of Urban Planning.* Amsterdam, the Netherlands: SPB Academic Publishing, pp. 49–58.

Breland TA, Hansen S (1996) Nitrogen mineralization and microbial biomass as affected by soil compaction. *Soil Biol Biochem* 28, 655–663.

Burghardt W (2001) Soils of low age as specific features of urban ecosystem. In: Sobocka J (Ed) *Soil Anthropization VI.* Bratislava, Slovakia: Soil Science and Conservation Research Institute, pp. 11–18.

Chakrabarti K, Sarkar B, Chakraborty A, Banik P, Bagchi DK (2000) Organic recycling for soil quality conservation in a sub-tropical plateau region. *J Agron Crop Sci* 184, 137–142.

Charzyński P, Bednarek R, Błaszkiewicz J (2011) Morphology and properties of ekranic technosls in Toruń and Cluj-Napoca. *Soil Sci Annu* 62(2), 48–53 (in Polish with English abstract).

Charzyński P, Bednarek R, Mendyk Ł, Świtoniak M, Pokojska-Burdziej A, Nowak A (2013) Ekranosols of Torun Airfield. In: Charzyński P, Hulisz P, Bednarek R (Eds) *Technogenic Soils of Poland.* Toruń: PSSS, pp. 173–190.

Charzyński P, Hulisz P, Bednarek R, Piernik A, Winkler M, Chmurzyński M (2015) Edifisols—A new soil unit of technogenic soils. *J Soils Sediments* 15(8), 1675–1686, DOI:10.1007/s11368-014-0983-4.

Charzyński P, Plak A, Hanaka A (2017) Influence of the soil sealing on the geoaccumulation index of heavy metals and various pollution factors. *Environ Sci Pollut Res* 24(5), 4801–4811.

Charzyński P, Świtoniak M (2014) Pleistocene terraces of the Toruń Basin on the border of the urban area. In: Świtoniak M, Charzyński P (Eds.) *Soil Sequences Atlas.* Toruń: Wydawnictwo Naukowe Uniwersytetu Mikołaja Kopernika, pp. 125–139.

Chen X, Xia XH, Zhao Y, Zhang P (2010) Heavy metal concentrations in roadside soils and correlation with urban traffic in Beijing, China. *J Hazard Mat* 181, 640–646.

Chłopecka A, Bacon JR, Wilson MJ, Kay J (1996) Forms of cadmium, lead and zinc in contaminated soils from southwest Poland. *J Environ Qual* 25, 69–79.

Chojnacka K, Chojnacki A, Górecka H, Górecki H (2005) Bioavailbility of heavy metals polluted soils to plants. *Sci Total Environ* 337, 175–186.

Churkina G, Brown DG, Keoleian G (2010) Carbon stored in human settlements: The conterminous United States. *Glob Chang Biol* 16, 143–153.

Couch C, Petschel-Held G, Leontidou L (2007) *Urban Sprawl in Europe: Landscape, Land-Use Change and Policy.* London: Blackwell, p. 275.

Degens BP, Schipper LA, Sparling GP, Vukovic MV (2000) Decreases in soil C reserves in soils can reduce the catabolic diversity of soil microbial communities. *Soil Biol Biochem* 32, 189–196.

Edmondson JL, Davies ZG, McCormack SA, Gaston KJ, Leake JR (2011) Are soils in urban ecosystems compacted? A citywide analysis. *Biol Lett* 7, 771–774.

EU Technical Report-2011-050, 2011. *Final Report: Overview of Best Practices for Limiting Soil Sealing and Mitigating Its Effects in EU-27*. European Commission, EU, p. 227.

European Commission (2006) *Thematic Strategy for Soil Protection*, COM (2006) 231 final, 22.9.2006. EC, Brussels, EU, p 12.

European Environment Agency (2010) *Europe: Prices, Taxes and Use Patterns*. EEA Report 4, Copenhagen, EU, p 64.

European Environment Agency (2011) *Urban Soil Sealing in Europe*. http://www.eea.europa.eu/articles/urban-soil-sealing-in-europe.

Florea N, Munteanu I (2012) *Romanian System for Soil Classification* (Sistemul Român de Taxonomie a Solurilor - SRTS), Ed Sitech, Craiova. p. 206 (in Romanian).

Geisseler D, Horwath WR (2008) Regulation of extracellular protease activity in soil in response to different sources and concentrations of nitrogen and carbon. *Soil Biol Biochem* 40, 3040–3048.

Geisseler D, Horwath WR, Joergensen RG, Ludwig B (2010) Pathways of nitrogen utilization by soil microorganisms—A review. *Soil Biol Biochem* 42, 2058–2067.

Gianfreda L, Ruggiero P (2006) Enzyme activities in soil. In: Nannipieri P, Smalla K (Eds) *Nucleic Acids and Proteins in Soil*. Germany: Springer, Berlin Heidelberg, pp. 20–25.

Greinert A (2003) *Studies of Soils in the Zielona Góra Urban Area*. Zielona Góra: Oficyna Wydawnicza Uniwersytetu Zielonogórskiego (in Polish).

Greinert A (2013) Technogenic soils in Zielona Góra In: Charzyński P, Markiewicz M, Świtoniak M (Eds) *Technogenic Soils Atlas*. Toruń, Poland: Polish Society of Soil Science, pp. 141–164.

Greinert A (2015) The heterogeneity of urban soils in the light of their properties. *J Soils Sediments* 15, 1725–1737, DOI:10.1007/s11368-014-1054-6.

Harris RW, Clark JR, Matheny NP (1999) *Arboriculture: Integrated Management of Landscape Trees, Shrubs, and Vines*. Englewood Cliffs, NJ: Prentice-Hall.

Imperato M, Adamo P, Naimo D, Arienzo M, Stanzione D, Violante P (2003) Spatial distribution of heavy metals in urban soils of Naples city (Italy). *Environ Pollut* 124, 247–256.

Insam H, Haselwandter K (1989) Metabolic quotient of soil microflora to plant succession. *Oecologia* 79, 174–178.

Islam MS, Ahmed MK, Al-Mamun MH (2015) Metal speciation in soil and health risk due to vegetables consumption in Bangladesh. *Environ Monit Assess* 187, 288–303.

IUSS Working Group WRB (2007) *World References Base for Soil Resources 2006*. Update 2007, World Soil Resources Reports, 103, FAO, Rome.

IUSS Working Group WRB (2015) *World Reference Base for Soil Resources 2014*, Update 2015. International soil classification system for naming soils and creating legends for soil maps. World Soil Resources Reports No. 106. FAO, Rome.

Jensen LS, McQueen DJ, Shepherd TG (1996) Effects of soil compaction on N mineralization and microbial-C and N. I. Field measurement. *Soil Till Res* 38, 175–188.

Jim CY (1998) Impacts of intensive urbanization on trees in Hong Kong. *Environ Conserv* 25, 146–159.

Kaye JP, Groffman PM, Grimm NB, Baker LA, Pouyat RV (2006) A distinct urban biogeochemistry? *Trends in Ecol Evol* 21, 4, 192–199.

Kent M, Stevens RA, Zhang L (1999) Urban plant ecology patterns and processes: A case study of the flora of the City of Plymouth, Devon, U.K. *J Biogeogr* 26, 1281–1298.

Kosheleva NE, Kasimov NS, Vlasov DV (2015) Factors of the accumulation of heavy metals and metalloids at geochemical barriers in urban soils. *Eurasian Soil Sci* 48, 476–492.

Láníková D, Lososová Z (2009) Vegetace sešlapávaných stanovišť (Polygono arenastri-Poëtea annuae) Vegetation of trampled habitats. In: Chytrý M (Ed). *Vegetace České Republiky. 2, Ruderální, Plevelová, Skalní a Suťová Vegetace (Vegetation of the Czech Republic. 2, Ruderal, Weed, Rock and Scree Vegetation*. Praha: Academia, pp. 43–72.

Lisci M, Monte M, Pacini E (2003) Lichens and higher plants on stone: A review. *Int Biodeterior Biodegrad* 51, 1–17.

Ljung K, Oomen A, Duits M, Selinus O, Berglund M (2007) Bioaccessibility of metals in urban playground soils. *J Environ Sci Health A Tox Hazard Subst Environ Eng* 42, 1241–1250.

Lorenz K, Kandeler E (2005) Biogeochemical characterization of urban soil profiles from Stuttgart, Germany. *Soil Biol Biochem* 37, 1373–1385.

Lundholm J (2013) Vegetation of urban hard surfaces. In: Niemelä J (Ed) *Urban Ecology: Patterns, Processes and Applications*. Oxford-New York: Oxford University Press, pp. 93–102.

Meller E, Malinowski R, Niedźwiecki E, Malinowska K, Kubus M (2013) Technogenic soils in Szczecin. In: Charzyński P, Markiewicz M, Świtoniak M (Eds) *Technogenic Soils Atlas*. Toruń, Poland: Polish Society of Soil Science, pp. 93–111.

Mendyk Ł, Charzyński P (2016) Soil sealing degree as factor influencing urban soil contamination with polycyclic aromatic hydrocarbons (PAHs). *Soil Sci Annu* 67(1), 17–23.

Mobley HL, Island MD, Hausinger RP (1995) Molecular-biology of microbial ureases. *Microbiol Rev* 59, 451–480.

Mullins CE (1991) Physical properties of soils in urban areas. In: Bullock P, Gregory PJ (Eds) *Soils in the Urban Environment*. Cambridge: Blackwell Scientific Publications, pp. 87–118.

Murphy DV, Cookson WR, Braimbridge M, Marschner P, Jones DL, Stockdale EA, Abbott LK (2011) Relationships between soil organic matter and the soil microbial biomass (size, functional diversity, and community structure) in crop and pasture systems in a semi-arid environment. *Soil Res* 49, 582–594.

Nannipieri P, Eldor P (2009) The chemical and functional characterization of soil N and its abiotic components. *Soil Biol Biochem* 41, 2357–2369.

Nehls T, Jozefaciuk G, Sokolowska Z, Hajnos M, Wessolek G (2008) Filter properties of seam material from paved urban soils. *Hydrol Earth Syst Sci* 12, 691–702.

Nestroy O (2006) Soil sealing in Austria and its consequences. *Ecohydrol Hydrobiol* 6(1–4), 171–173.

Oliva SR, Espinosa AJF (2007) Monitoring of heavy metals in topsoils, atmospheric particles and plant leaves to identify possible contamination sources. *Microchem J* 86, 131–139.

Peffy T, Nawaz R (2008) An investigation into the extent and impacts of hard surfacing of domestic gardens in an area of Leeds, United Kingdom. *Landsc Urban Plan* 86, 1–13.

Pengthamkeerati P, Motavalli PP, Kremer RJ (2011) Soil microbial activity and functional diversity changed by compaction, poultry litter and cropping in a claypan soil. *Appl Soil Ecol* 48, 71–80.

Piotrowska-Długosz A, Charzyński P 2015. The impact of the soil sealing degree on microbial biomass, enzymatic activity, and physicochemical properties in the Ekranic Technosols of Toruń (Poland). *J Soils Sediments* 15, 47–59.

Plak A, Chodorowski J, Melke J, Bis M (2015) Influence of land use on the content of select forms of Cd, Cr, Cu, Ni, Pb, and Zn in urban soils pol. *J Environ Stud* 24(6), 2577–2586.

Polish Soil Classification (2011) *Soil Sci Annu* 62(3), 193 (in Polish with English summary).

Pouyat RV, Yesilonis ID, Nowak DJ (2006) Carbon storage by urban soils in the United States. *J Environ Qual* 53, 1566–1575.

Prokofyeva TV, Martynenko IA, Ivannikov FA (2011) Classification of Moscow soils and parent materials and its possible inclusion in the classification system of Russian soils. *Eurasian Soil Sci* 44, 561–571.

Puskás I, Farsang A (2009) Diagnostic indicators for characterizing urban soils of Szeged, Hungary. *Geoderma* 148, 267–281.

Randrup TB, McPherson EG, Costello LR (2001) A review of tree root conflicts with sidewalks, curbs, and roads. *Urban Ecosyst* 5(3), 209–225.

Reynolds CM, Wolf DC, Armbruster JA (1985) Factors related to urban hydrolysis in soils. *Soil Sci Soc Am J* 49, 104–108.

Ryan MG, Law BE (2005) Interpreting, measuring, and modeling soil respiration. *Biogeochemistry* 73, 3–27.

Sammel A, Chorągwicki Ł, Niedźwiecki E, Meller E, Malinowski R. (2013) Morphological features and chemical properties of soils sealed with bituminous surface (ekranosols) within Obrońców Stalingradu street in Szczecin. *Folia Pomer Univ Technol Stetin Agric Aliment Pisc Zootech* 307(28), 75–90 (in Polish with English abstract).

Sándor G, Szabo G, Charzyński P, Szynkowska E, Novak TJ, Świtoniak M (2013) Technogenic soils in Debrecen. In: Charzyński P, Markiewicz M, Świtoniak M (Eds) *Technogenic Soils Atlas*. Toruń, Poland: Polish Society of Soil Science, pp. 35–75.

Savard JP, Clergeau P, Mennechez G (2000) Biodiversity concepts and urban ecosystems. *Landsc Urban Plan* 48, 131–142.

Scalenghe R, Ajmone Marsan F (2009) The anthropogenic sealing of soil in urban areas. *Landsc Urban Plan* 90, 1–10.

Scharenbroch BC, Lloyd JE, Johnson ML (2005) Distinguishing urban soils with physical, chemical and biological properties. *Pedobiologia* 49, 283–296.

Shishov LL, Tonkonogov VD, Lebedeva II, Gerasimova MI (Eds) (2004) *Classification and Diagnostics of Soils of Russia*. Smolensk: Oecumena, p. 343 (in Russian).

Sobocká J (2013) Technogenic soils in Slovakia. In: Charzyński P, Markiewicz M, Świtoniak M (Eds) *Technogenic Soils Atlas*. Toruń, Poland: Polish Society of Soil Science, pp. 75–93.

Societas Pedologica Slovaca (2014) *Morfogenetický klasifikačný systém pôd Slovenska*. Bazálna referenčná taxonómia. Druhé upravené vydanie. Bratislava: NPPC–VÚPOP Bratislava, pp. 1–96. ISBN: 978-80-8163-005-7.

Soil Survey Staff (2014) *Keys to Soil Taxonomy*, 12th ed. Washington, DC: USDA-Natural Resources Conservation Service.

Stroganova M, Myagkova A, Prokofieva T, Skvortsova I (1998) *Soils of Moscow and Urban Environment*. Moscow: PAIMS.

Stroganova MN, Prokofieva TV (1995) Influence of road surface on urban soil (Влияние дорожного покрытия на городские почвы). Bulletin of Moscow University, series 17, Soil Science, N2, 3–11 (in Russian).

Szolnoki Z, Farsang A (2013) Evaluation of metal mobility and bioaccessibility in soils of urban vegetable gardens using sequential extraction. *Water Air Soil Pollut* 224, 1737–1752.

Thomsen IK, Kruse T, Bruun S, Kristiansen SM, Petersen SO, Jensen LS, Holst MK, Christensen BT (2008) Characteristics of soil carbon buried for 3300 years in a Bronze Age burial mound. *Soil Sci Soc Am J* 72(5), 1292–1298.

Tripathi N, Singh RS (2009) Influence of different land uses on soil nitrogen transformation after conversion from an Indian dry tropical forest. *Catena* 77, 216–223.

Turbé A, De Toni A, Benito P, Lavelle P, Ruiz N, Van der Putten WH, Labouze E, Mudgal S (2010) *Soil Biodiversity: Functions, Threats and Tools for Policy Makers*. European Commission, DG Environment, p. 254.

Wang M, Marked B, Shen W, Chen W, Peng C, Ouyang Z (2011) Microbial biomass carbon and enzyme activities of urban soils in Beijing. *Environ Sci Pollut Res* 18, 958–967.

Wei Z, Wu S, Zhou S, Li J, Zhao Q (2014) Soil organic carbon transformation and related properties in urban soil under impervious surfaces. *Pedosphere* 24, 1, 56–64.

Wei Z, Wu S, Zhou S, Lin C (2013) Installation of impervious surface in urban areas affects microbial biomass, activity (potential C mineralisation) and function diversity of the fine earth. *Soil Res* 51, 59–67.

Wesolek G (2008) Sealing of soils In: Marzluff JM, Shulenberger E, Endlicher W, Alberti M, Bradley G, Ryan C, Simon U, Zumbunnen C (Eds) *Urban Ecology*. Springer Science, Business Media LLC, pp. 161–179, ISBN: 978-0-387-73411-8.

Wheater CP (2002) *Urban Habitats*. London-New York: Routledge.

Wiegand J, Schott B (1999) The sealing of soils and its effect on soil-gas migration. *Nuovo Cimeno Soc It Fisica C* 22, 449–455.

Xiao R, Su S, Zhang Zh, Qi J, Jiang D, Wu J (2013) Dynamics of soil sealing and soil landscape patterns under rapid urbanization. *Catena* 109, 1–12.

Yang X, Zhao L (2016) Diurnal thermal behavior of pavements, vegetation and water pond in a hot-humid city. *Buildings* 6(1), 2 DOI:10.3390/buildings6010002.

Zeng DH, Hu YL, Chang SX, Fan ZP (2009) Land cover change effects on soil chemical and biological properties after planting Mongolian pine (*Pinus sylvestris* var. *mongolica*) in sandy lands in Keerqin, northeastern China. *Plant Soil* 317, 121–133.

Zhao D, Feng L, Wang R, Qingrui R (2012a) Effect of soil sealing on the microbial biomass, N transformation and related enzymes activities at various depths of soils in urban area of Beijing, China. *J Soils Sediments* 12, 519–539.

Zhao D, Li F, Wang R (2012b) The effects of different urban land use patterns on soil microbial biomass nitrogen and enzyme activities in urban area of Beijing, China. *Acta Ecol Sin* 32, 144–149.

8 Contaminants in Urban Soils: Bioavailability and Transfer

Dorothy S. Menefee and Ganga M. Hettiarachchi

CONTENTS

8.1 INTRODUCTION

Soil contamination with potentially toxic trace elements and anthropogenic organic pollutants is a widespread problem in urban areas worldwide. Major concerns are potentially toxic heavy (atomic weights greater than that of iron or density greater than 5 g cm^{-3}) elements (lead [Pb], arsenic [As], cadmium [Cd]), and organic contaminants such as polycyclic aromatic hydrocarbons (PAHs), antibiotics (from waste application), and petroleum-based compounds (Lourenco et al. 2010; Brown et al. 2016; Chaney and Ryan 1994; Kachenko and Singh 2006; Yang 2010; Taylor and Lovell 2014; Chen et al. 2007, 2014; Wortman and Lovell 2013). These types of contamination are common in urban areas which may conflict with the growing trend of urban agriculture. Urban agriculture increases food security and provides greenspace, jobs, and fresh produce to urban communities (see Figure 8.1); however, without knowledge of urban soil contamination, there can be some potential risks (Wortman and Lovell 2013; Smith 2009; Lee-Smith 2010; Halloran and Magid 2013).

FIGURE 8.1   An urban garden in Indianapolis, IN.

8.2   OVERVIEW

In order to understand soil contamination issues, it is important to first know the scope of the problem and the risks associated with soil contamination.

8.2.1   LEAD

Soil contamination with Pb is a widespread problem and is one of the most common soil issues in urban areas that is associated with risks to human health. Almost half of the US Environmental Protection Agency's (EPA) Superfund sites have Pb contamination as one of the causes for concern (Hettiarachchi and Pierzynski 2004; Brown et al. 2016; McBride et al. 2014). Much of this contamination is due to historical use of leaded paint and leaded gasoline that released Pb-containing dust. That dust was then deposited on the nearby ground and incorporated into the soil (McClintock 2015; Yang 2010; Chaney and Ryan 1994). Another common source of Pb contamination is from mining activities and spills, as well as from industrial processing of mined material, such as smelting (Perez-de-Mora et al. 2006; Kachenko and Singh 2006). As an example of how pervasive Pb can be, a study in Boston, MA, found that 88% of gardens in Boston had Pb contamination, with soil Pb concentration above the EPA limit for residential yard soil of 400 mg kg^{-1} (Clark et al. 2006). This high level of soil lead in the city is not unique to Boston. Lead and other potentially toxic metal(loid)s tend to be much more common in the soil of urban areas than of their surrounding countryside, particularly in older and more industrial urban regions, which are often poorer parts of cities (McClintock 2015; Henry et al. 2015). A survey of cities across the United States found that 52% of lots with buildings constructed before 1978 had elevated Pb in the nearby soil (McClintock 2015). Urban soil contamination is not just a problem in the United States, but in cities around the world. A survey of soils in various cities across China found an average Pb content of about 65% (Yang 2010). A survey of Pb in garden soils in Australia found 100% of gardens in Boolaroo to exceed national safety limits. While this was an extreme case, about 30% of garden sites in Sydney and Port Kembla also had elevated Pb (Kachenko and Singh 2006). Soil Pb contamination in urban areas has been identified as a problem in Brazil and Europe (Lourenco et al. 2010; Perez-de-Mora et al. 2006; Levonmäki et al. 2006; Cruz et al. 2016).

Lead exposure poses human health risks which are particularly significant risk for children. Young children (particularly kids under 3 years of age) absorb four to five times as much Pb from their environment as adults (Chaney and Ryan 1994; Henry et al. 2015). Lead has serious impacts on brain development in young children, which can cause lifelong cognitive problems even at low exposure levels (Hettiarachchi and Pierzynski 2004; McClintock 2015). Blood concentrations as low as 5.0 μg/dL have been linked to learning difficulties and decreased IQ (Henry et al. 2015; World Health Organization 2016; Hettiarachchi and Pierzynski 2004). Children exposed to Pb are also likely to develop behavioral problems such as shortened attention span and antisocial behaviors. Due to the decreased IQ and behavioral problems, Pb-exposed children are less likely to graduate high school or college (World Health Organization 2016; McClintock 2015). Lead exposure is responsible for approximately 600,000 cases of intellectual disability annually (World Health Organization 2016). Much higher levels of Pb exposure are associated with anemia, kidney damage, and high blood pressure (Hettiarachchi and Pierzynski 2004; World Health Organization 2016). Regulations on use of Pb in products like gasoline and paint have significantly decreased the incidence of Pb exposure high enough to cause systemic problems; however, cognitive Pb-associated problems are still a public concern. The current Centers for Disease Control and Prevention (CDC) reference value for blood Pb in children is 5.0 μg/dL, above which there is an increased risk for decreased IQ (Henry et al. 2015; McClintock 2015). A significant amount of child Pb poisoning cases (as reported to the CDC), 40%–45%, is thought to come from contaminated soil or dust in industrial and urban areas (Mielke et al. 1983). More recent studies suggest that remediating soil Pb can significantly help to lower the blood Pb levels of children (Lanphear et al. 2003).

Lead contamination poses environmental risks as well. Elevated soil Pb is correlated with a decrease in microbial biomass and microbial activity (Wang et al. 2007, 2010; Gai et al. 2011; Liao et al. 2007). Lead above background concentration levels has also been shown to inhibit or decrease the activity of soil microbial enzymes (Hinojosa 2004; Makoi and Ndakidemi 2008). Given a particular Pb concentration and bioavailability, some microbial groups are more affected than others. Multiple studies have reported that bacteria, particularly actinomycetes and gram-positive bacteria, are harmed by available soil Pb more than other microbial groups (Wang et al. 2007, 2010; Abaye et al. 2005; Liao et al. 2007; Yang 2010). Meanwhile, there is usually a minimal effect of Pb on soil fungi and gram-negative bacteria, sometimes even an increase in population following Pb exposure (Yang 2010; Liao et al. 2007; Abaye et al. 2005; Wang et al. 2007, 2010). This difference in response to Pb from different microbial groups is likely due to differences in cell walls. Gram-positive bacteria (including actinomycetes) have very simple cell walls that bind easily to metals, like Pb, while gram-negative bacteria and fungi have more complex cell walls that likely have better mechanisms for protecting the cell from harmful metals (Wase 1997; Abaye et al. 2005). This change of the soil microbial community can affect nutrient cycling pathways and the ability of the soil to support plant and animal life. Ecosystem damage from lead contamination is another important thing to consider when working in Pb contaminated soils.

8.2.2 Arsenic

Arsenic (As), like Pb, is a naturally occurring element that is elevated in some regions due to geologic processes; S-containing minerals and some Fe oxide minerals have a greater tendency to have As substitution and inclusion. Regions where ground and surface waters are derived from rock formations rich in As minerals will often have water that is high in As. Arsenic-containing rock is found in regions throughout Asia, Europe, and the Americas. In regions where As-containing water is the primary source of drinking or irrigation water, human health risks emerge (Smedley and Kinniburgh 2002). Arsenic is also elevated in some regions due to a number of human activities. Arsenic has been used in a number of products, including pesticides, herbicides, wood treatment, wood preservatives, and animal feed additives, all of which can contaminate soil (Wang and Mulligan 2006; Mandal and Suzuki 2002). Arsenic can also be introduced into soils via mining and industrial processing of mined material, like smelting and foundries (Perez-de-Mora et al. 2006; Mandal and Suzuki 2002; Han et al. 2003). Arsenic in urban soils is a potential hazard to human health. Arsenic is a known human carcinogen; chronic exposure to As (even at low levels) increases risks of developing liver, bladder, skin, and lung cancers (caused by the inhalation of As-containing dust) and is classified by the US EPA as a "Group A" carcinogen (Mandal and Suzuki 2002; U.S. Environmental Protection Agency 2004). Other health problems caused by As exposure are skin irritation and damage, gastrointestinal problems, kidney and liver disease, nerve damage, and increased risk of miscarriage and birth defects (Mandal and Suzuki 2002; National Institute of Neurological Disorders and Stroke 2016).

Arsenic contamination can damage ecological functioning as well. Multiple studies have found that soil contaminated with As, as with many other heavy elements, often has a decrease in the biodiversity and activity of soil microorganisms (Hinojosa 2004; Wang et al. 2010). Arsenic does not affect all microbial groups equally. Fungi have greater resistance to As than bacteria (Wang et al. 2010; Oliveira and Pampulha 2006; Kaloyanova 2007). Bacteria involved with N cycling, nitrifying bacteria and nitrogen fixing bacteria, are particularly sensitive to As when compared to other bacterial groups (Kaloyanova 2007; Oliveira and Pampulha 2006; Yeates et al. 1994). The decline in bacteria that are important for N-cycling can have detrimental effects on plants that depend on these bacteria. Plants are also harmfully affected by As contamination, often having slower growth rates and lower biomass (Yeates et al. 1994; Claassens 2006; Szakova et al. 2007). Arsenic soil contamination was found to have detrimental effects on earthworm populations through increased mortality and lower average body weight in contaminated soils (Sizmur et al. 2011). A similar study also found As to be detrimental to earthworms as well as nematodes and roundworms (Yeates et al. 1994). A diverse and stable microbial population is vital for nutrient cycling, which is something that growing plants are dependent on—when soil pollution is high enough to harm the soil microbial community, the ability of that soil to support plant growth is seriously impaired.

8.2.3 Cadmium

Cadmium is a widespread urban soil pollutant. In urban areas, much of the Cd contamination is from a variety of industrial activities: plastic manufacturing and incineration, battery manufacture and disposal, dye/pigment production, and processing (mining and smelting) of zinc and copper metal ores (Lu et al. 2007; Franz et al. 2008; Kaji 2012; Chaney and Ryan 1994). Cadmium is a particular concern because it is often more soluble and thus more bioavailable than other potentially toxic metals (Sauve and Hendershot 1997a; Weber and Hrynczuk 2000). Currently, Cd soil contamination is a common urban soil problem in regions where metal ore smelting was a major economic activity. Surveys of lakeside and riverbed sediments across the United States have found a trend of increasing Cd concentrations in urban areas (Rice 1999; Mahler et al. 2006). Changes in regulations have resulted in significant reductions on the deposition of new Cd into urban soils (Mahler et al. 2006). Cadmium contamination from mining and smelting activities has become very common problem in urban China (Qiu et al. 2011; Zhao et al. 2016; Chen et al. 2014; Yang 2010). A study of urban areas in Australia found Cd contamination to be quite common, especially near smelting activities. One city with heavy mining and smelting industry, Boolaroo, in New South Wales, had 50% of sampled sites with Cd contents above national limits for soil (Kachenko and Singh 2006). Long-term application of contaminated organic (i.e., sewage sludge) and inorganic (i.e., high Cd rock phosphate) can also lead to an accumulation of Cd in soil (Lombi et al. 2001).

Cadmium causes harm to human health. Ingested Cd replaces Zn (an essential micronutrient) in biological pathways and when consumed leads to chronic kidney disease, osteoporosis, and gastrointestinal issues (Kaji 2012; Lu et al. 2007; World Health Organization 2010). The first well-documented case of Cd toxicity in humans was the outbreak of a disease called Itai-Itai in zinc mining and smelting regions of Japan, which was later found to be chronic Cd toxicity (Kaji 2012). Due to the widespread nature of Cd contamination and the high bioavailability of Cd, consumption of food grown on Cd-rich soils is an important pathway for human Cd exposure. According to the World Health Organization, food is the most common route of Cd exposure in nonsmokers with rice (due to plant uptake of Cd under reduced conditions), organ meats (i.e., liver, kidney) (due to the tendency of consumed metals to accumulate in specific organs in the animal), and seafood (due to contaminated waters) being responsible for most of the food-related Cd exposure (World Health Organization 2010).

In addition to damage to human health, Cd can impact ecosystems and wildlife. Erosion of Cd-contaminated soil has caused harm to fish and aquatic crustaceans (Bandara et al. 2011; Satarug et al. 2003). A study on tubifid sludge worms found that Cd contamination decreased their population and reproductive success (Delmotte et al. 2007). A study on invertebrate populations in Cd contaminated soils found that annelid worms (i.e., earthworms) and terrestrial crustaceans (i.e., pill bugs and wood lice) experienced significant declines with soil Cd contamination (Hunter et al. 1987). Cadmium can have harmful effects on soil microbes as well (Khan et al. 2010; Yang 2010; Wang et al. 2010). Like Pb, the groups of soil microbes most sensitive to Cd are gram-positive bacteria and actinomycetes while gram-negative bacteria and fungi are more resistant (Wang et al. 2010; Yang 2010). It seems likely that once again the differences in cell walls are driving the different responses (Abaye et al. 2005; Wase 1997).

8.2.4 Polycyclic Aromatic Hydrocarbons

PAHs are a class of persistent organic pollutants that are comprised of two or more fused aromatic carbon rings. Some of these compounds are known to be carcinogens and/or mutagens (Li et al. 2014; Rengarajan et al. 2015). While there are hundreds of different PAH compounds, 16 of them have been classified as priority pollutants by the US EPA for their carcinogenetic nature (Lorenzi et al. 2011). These compounds are typically formed during the combustion of fossil fuels and during incineration processes and are released as airborne particulate matter (PM). Naturally occurring PAHs are much less prevalent and are mainly due to fires and volcanic activity (Li et al. 2014; Lorenzi et al. 2011; Rengarajan et al. 2015). The airborne PAHs are then deposited on the ground where they are incorporated into the soil (Wilcke 2000; Evans et al. 2014; McBride et al. 2014). PAH concentration is typically the highest in urban areas and along roadsides, where combustion of fossil fuels is the highest (Li et al. 2014; Khan and Cao 2012; Woodhead et al. 1999; Wilcke 2000; Chen et al. 2016). Since PAHs are a class of many different compounds, they vary greatly in molecular weight and other properties; larger PAH compounds fall out of the air quicker and thus tend to be more concentrated closer to the source while smaller compounds are more widely distributed over a broader area (Li et al. 2014; Plachá et al. 2009; Yunker et al. 2002). Larger compounds also tend to be more persistent in the environment and have greater potential to be carcinogenic (Cerniglia 1992; Plachá et al. 2009; Canadian Council of Ministers of the Environment 1999). Smaller molecular weight compounds, on the other hand, tend to be more mobile and acutely toxic, especially to aquatic organisms (Boonchan et al. 2000; Canadian Council of Ministers of the Environment 1999). In addition to causing cancer, PAH exposure has been linked to skin/eye irritation, endocrine disruption, male infertility, and increased risk of birth defects (Rengarajan et al. 2015; Evans et al. 2014; Han et al. 2011).

PAH contamination is widespread in many developed countries with economies that are largely based on fossil fuel consumption (Khan and Cao 2012; Li et al. 2014; Wilcke 2000; Woodhead et al. 1999; Lorenzi et al. 2011; Plachá et al. 2009; Canadian Council of Ministers of the Environment 1999; Rengarajan et al. 2015; Han et al. 2011). A study of soils in Delaware found that nearly all sites had PAH levels elevated above background levels (Zhang et al. 2013). Multiple studies of PAH contaminated soils in urban China have also found that the PAH concentration exceeds commonly held safety guidelines (Wang et al. 2012; Khan and Cao 2012; Chen et al. 2016). A survey of PAH exposure in the United Kingdom found that up to 90% of human PAH exposure came from contaminated soils (Wilcke 2000).

8.2.5 OTHER CONTAMINANTS

There are many possible soil contaminants and it would be nearly impossible to cover every single one; however, it is possible to cover some of the larger groups of contaminant types. Other organic compounds, such as chlorinated hydrocarbons, industrial solvents, and pharmaceuticals, can also become soil contaminants (U.S. Environmental Protection Agency 2011; Pierzynski et al. 2005). These compounds are almost everywhere and can sometimes be hard to characterize and predict distribution. Every day over 70,000 different synthetic organic compounds are used in the United States for everything from industrial solvents to nail polish. While many of these compounds are indeed useful, they can also have undesirable effects on ecosystem function and human health, especially when disposed of improperly (Pierzynski et al. 2005).

In addition to organic contaminants, there are many other potentially toxic inorganic compounds and elements that can end up in soils, such as Zn, Cu (while Zn and Cu are micronutrients, they can be phytotoxic in sufficiently large concentrations), Cr, and Hg (U.S. Environmental Protection Agency 2011). Elevated Zn, Cu, and Hg are usually the results of human activities like mining and smelting; these elements often co-occur with Pb (Lombi et al. 2001; Huang et al. 2013; Facchinelli et al. 2001). Chromium contamination is usually due to local geology, similar to that of As (Facchinelli et al. 2001; Huang et al. 2013). These types of contaminants are likely to become more of a concern as new types of waste (i.e., e-waste) are being disposed of. For example, antimony is a potentially hazardous metalloid that can leach into soils from the heating of plastics during manufacturing, recycling, and solar degradation (Cheng et al. 2010). Antimony also makes its way into soils during mining, smelting, coal burning, and incineration activities. Antimony is genotoxic, meaning it damages DNA, which can cause birth defects and cancers (Hockmann et al. 2014; Wu et al. 2011).

8.3 FATE AND TRANSFER PATHWAYS

Fate and transport of soil contaminants depends largely on the nature of the contaminant but also on the nature of the soil itself (Tu et al. 2013). This section aims to explore the prominent fate and transfer pathways for important soil contaminants. Some of the major processes of transformation and transportation of mineral soil contaminants are leaching, sorption, formation of secondary minerals, erosion, uptake, and dissolution.

8.3.1 LEAD

Once introduced to a soil environment, Pb is particularly prone to sorption processes and the formation of secondary minerals (Tu et al. 2013; Sauve and Hendershot 1997a; Scheckel et al. 2013; Rooney et al. 2007; Hettiarachchi and Pierzynski 2004). Lead sorbs (i.e., binds) readily to binding sites on clay minerals and organic matter (Chaney and Ryan 1994; Cruz et al. 2016; Levonmäki et al. 2006; Smith 2009; Brown et al. 2012). Some work suggests the majority of the lead sorption is through specific or chemisorption to the O atoms in the clay mineral, hinner sphere adsorption, which is relatively permanent (Strawn and Sparks 1999). As such the mineralogy, texture, and organic matter content of the soil greatly determines the fate and transport of Pb (Tu et al. 2013; Cruz et al. 2016; Rooney et al. 2007; Hettiarachchi and Pierzynski 2004). The formation of Pb phosphates and Pb carbonate minerals is fairly common in Pb_containing soils and can reduce Pb mobility (Hettiarachchi and Pierzynski 2004; Rooney et al. 2007; Sauve and Hendershot 1997a). Lead phosphate minerals typically form when solution P is high, and Pb carbonate minerals tend to form under higher pH conditions. Ultimately, since Pb is highly sorbed to clays, Fe–Mn minerals, and SOC, the mobility of soil Pb is usually very low. However, in very acidic conditions (soil pH 4 or less), many of the Pb minerals are more prone to dissolution and soil Pb becomes more mobile (Chaney and Ryan 1994; Levonmäki et al. 2006; Rooney et al. 2007; Sauve and Hendershot 1997). A study on lead leaching on an acidic (pH 4.1) peat soil found leaching to be a transport pathway via both free Pb (Pb^{2+}) and Pb bound to soluble soil organic matter (SOM) (Levonmäki et al. 2006). Another study in Brazil found that Pb leaching was significant only in areas with very sandy soils (Cruz et al. 2016). Since soil properties are important in determining the fate and transport of Pb, management practices that alter these properties can reduce Pb mobility (Hettiarachchi and Pierzynski 2004).

Lead has limited capacity to transfer to plant tissues; plants typically only uptake Pb when it is in the soil solution and even then in typically small quantities (Chaney and Ryan 1994; Hettiarachchi and Pierzynski 2004; U.S. Environmental Protection Agency 2014). A study on the uptake of metals by wheat crops found that Pb had a much lower plant uptake than Zn or Cd (Weber and Hrynczuk 2000). Even with the limited capacity for Pb uptake, plant Pb uptake level does vary by type of crop grown. Crops harvested for fruits and seeds (i.e., tomatoes) will tend to have the lowest Pb tissue concentration while those harvested for the roots (i.e., carrot) will have the most (McBride et al. 2014; Attanayake et al. 2015; McBride 2013; Brown et al. 2016; Wortman and Lovell 2013). As a result, human risk from lead-contaminated soil is mostly linked to direct contact with and consumption of soil (Brown et al. 2016; Attanayake et al. 2015; McBride et al. 2014; Henry et al. 2015). In adults, most direct contact and consumption of soil are through dust on unwashed produce and in the home. With children hand-to-mouth play, toy-to-mouth play, and directly consuming soil (pica) are common exposure routes for Pb_contaminated soil (Mielke et al. 1983; Attanayake et al. 2015; McBride et al. 2014; Brown et al. 2016; Chaney and Ryan 1994; U.S. Environmental Protection Agency 2014).

8.3.2 ARSENIC

Arsenic exists in multiple oxidation states, which have different fate and transport pathways: As(-III), As(+0), As(+III), and As(+V). The first two are uncommon in nature and are not important for this discussion (Zhao et al. 2010). Arsenate, As (+V), is the most oxidized form of As and readily sorbs to Fe–Mn minerals and forms As sulfur minerals. Arsenate is more typically strongly sorbed to soil or in insoluble minerals and is thus less mobile. Arsenite (As +III) is the more reduced of the As forms common in nature and is less readily sorbed and thus more readily mobile (Mandal and Suzuki 2002; Zhao et al. 2010a; Wang and Mulligan 2006; Garcia-Sanchez et al. 2010). Due to the affinity of As (+V) for Fe-oxide minerals and the effect of oxidation state on mobility, the redox potential of the soil is very important for determining the fate and transport of As (Wang and Mulligan 2006; Zhao et al. 2010; Mandal and Suzuki 2002). Under oxidized conditions, As is generally in the oxidized form, As (+V), and is sorbed to clays, Fe–Mn minerals, and SOC (Sizmur et al. 2011; Han et al. 2004; Attanayake et al. 2015; Zhao et al. 2010a). Arsenate is also commonly found in As–Fe secondary minerals such as arsenopyrite and scorodite (Brown et al. 2012; Zhao et al. 2010; Garcia-Sanchez et al. 2010; Han et al. 2004). Arsenic bound to Fe minerals and clays is typically less soluble and thus less mobile (Zhao et al. 2010; Garcia-Sanchez et al. 2010). Arsenic bound to organic compounds is typically more mobile and can be leached and is available for plant uptake (Wang and Mulligan 2006; Han et al. 2004; Zhao et al. 2010). However, when conditions in the soil become reduced and oxygen is limited, many anaerobic soil microbes will attack the Fe minerals (i.e., biologically mediated reductive dissolution of Fe minerals) that As is sorbed to or bound in, thereby releasing reduced As into the soil solution (Garcia-Sanchez et al. 2010; Zhao et al. 2010). The combination of organic-bound As and saturated conditions makes peat soils very prone to As leaching (Miller et al. 2010; Wang and Mulligan 2006).

A study in 2007 found that leaching was the primary mechanism for arsenic translocation in soils (Chen et al. 2007). Similarly, a study of land application of wastewater found that the saturated conditions increased As mobility and leaching (Aryal and Reinhold 2015). It is this increase in mobility of As under reduced conditions that make crops that require saturated conditions, like rice, much more susceptible to As than dryland crops (Meharg and Rahman 2003; Zhao et al. 2010).

8.3.3 CADMIUM

Cadmium exists in the soil solution as a divalent cation, Cd^{2+}, and in the soil solid phase as bound to or part of carbonate, phosphate, Fe–Mn oxides, organo-mineral complexes, and soil clay minerals (Barrow 2000; Appel and Ma 2002; Loganathan et al. 2012). Cadmium in the soil solution is readily mobile and thus prone to leaching and capable of plant uptake (Satarug et al. 2003; Barrow 2000). The soil factors that most influence Cd mobility are thus those that influence the ability of Cd to either adsorb or precipitate out of solution (Barrow 2000; Loganathan et al. 2012; Appel and Ma 2002).

Predicting Cd fate and transport can be more difficult than other soil metals. A modeling study in 2008 found that attempts to model Cd transport in soils had very little success, especially when compared to the modeling attempt for Pb, which was largely accurate (Hutzell and Luecken 2008a). The mobility of Cd largely depends on the availability of potential adsorption sites, which largely depends on soil pH and concentrations of minerals and organic compounds that Cd adsorbs to (U.S. Department of Health and Human Services 2012; De Livera et al. 2011; Simmler et al. 2013; Barrow 2000). Cadmium is more adsorbed and less mobile in soils that are less acidic; Cd availability increases considerably at pH 5.0 (Barrow 2000; U.S. Department of Health and Human Services 2012). While Cd has a leaching potential, it still easily accumulates in soils due to its affinity for cation exchange sites. Cadmium does adsorb to cation binding sites, although it is less tightly bound to them than lead and other metals (Zhao et al. 2009; Chen et al. 2013). Cadmium forms complexes with recalcitrant organic matter either as part of the soil humic substances fraction or from added OM sources such as biosolids and composts. Cadmium complexed with SOM is relatively stable over time, suggesting that Cd complexed and sorbed into soil is associated with the more recalcitrant fractions of SOM or with inorganic minerals (Hettiarachchi et al. 2003b; Li et al. 2001).

8.3.4 POLYCYCLIC AROMATIC HYDROCARBONS

Once in the soil, PAHs have many of the same transformation and translocation pathways of other contaminants; however, as they are organic contaminants they can also be decomposed by microbes (Cerniglia 1992; Kawasaki et al. 2012; Chen et al. 2016). The major transport and transformation pathways for PAHs are sorption, leaching, erosion, plant uptake, volatilization, and microbial decomposition (Wilcke 2000). What processes dominate varies greatly both by soil type and by the molecular weight of the PAH compound (Khan and Cao 2012; Wilcke 2000). Sorption processes involve the sorption of PAHs to clays and SOM, which increases the capacity of a soil to retain PAHs while decreasing bioavailability (Zhang et al. 2010; Wilcke 2000). PAH compounds are generally more readily adsorbed to SOC which have much more limited ability to absorb to clays (Khan and Cao 2012a; Zhang et al. 2010; Wilcke 2000). Leaching potential of PAH compounds depends greatly on their molecular weight. Small compounds that are readily soluble are prone to leaching, while leaching is rare for heavier compounds (Reilley et al. 1996; Plachá et al. 2009; Li et al. 2014; Yunker et al. 2002). Plant uptake of PAHs varies considerably with both PAH compound and plant type. Typically, plant uptake of PAHs is very low; lower molecular weight PAHs are often more plant-available than heavier PAHs (Attanayake et al. 2015). Brassicas and leafy vegetables like spinach tend to have higher uptake of PAHs, especially of lower molecular weight compounds. The uptake level of these vegetables occasionally crosses safety thresholds for small children (Wang et al. 2012; Khan and Cao 2012). When compared to leafy vegetables, root vegetables uptake less PAHs and much of what they do uptake is typically stored in the peel, which is easily removed and discarded (Wang et al. 2012; Zohair et al. 2006; Attanayake et al. 2015). A significant portion of human–plant PAH exposure comes not from the plant tissues, but from PAH laden dust or sediment on the surfaces of the plant (Wilcke 2000; Attanayake et al. 2015).

The most important pathway for PAH removal from soils is decomposition by microorganisms (De Nicola et al. 2015; Boonchan et al. 2000; Cerniglia 1992; Chen et al. 2015). The degradation processes of PAHs are thus correlated with soil properties that affect microbial growth, such as SOC content and climate (Cerniglia 1992). PAH accumulation in soils tend to be more of a problem in regions and soils where microbial activity is low, typically due to cold temperatures or anaerobic conditions (Li et al. 2014; Cerniglia 1992). The chemistry of each individual PAH compound affects its decomposition rate, with larger PAHs being more resistant to microbial attack (Cerniglia 1992; Plachá et al. 2009; Boonchan et al. 2000). An incubation study on microbial degradation of high-molecular-weight PAHs found that differing compounds of the same molecular weight may degrade at different rates (e.g., pyrene degrades much faster than chrysene). Much of this is due to differences in molecular structure that makes some compounds easier to break down (Boonchan et al. 2000).

8.3.5 OTHER CONTAMINANTS

Other organic contaminants follow similar trends to PAH compounds in their main fate and transport pathways. Many organic contaminants are preferentially adsorbed by SOC rather than clay (Fabietti et al. 2010; Schlebaum 1998; Liu et al. 2012). The main pathway through which most organic contaminants leave the soil is via microbial decomposition. Many compounds that are potentially toxic to humans, plants, and animals are potential energy sources for soil microorganisms (Liu et al. 2012; Patterson et al. 2010; Semple et al. 2007). The exact nature of how the fate and transport pathways work for organic contaminants depends greatly on the contaminant in question. Some organic contaminants tend to remain in the soil for long periods of time because microbes are unable to degrade them. Contaminants with more complex structures and contaminants that are tightly adsorbed to soil humic substances are often difficult for microorganisms to break down (Semple et al. 2007; Liu et al. 2012; Patterson et al. 2010; Schlebaum 1998).

Many inorganic contaminants are metals or metalloids and follow similar trends to Pb, Cd, and As. For many metals and metalloids, sorption/desorption reactions are important in determining the dominant fate and transport pathways. In addition to pH, redox status can directly (due to changing less bioavailable to more bioavailable forms such as Cr(III) to Cr(VI)) or indirectly (due to reductive dissolution of Fe and Mn minerals) affect the fate and transport of these metals and metalloids (Schaider et al. 2014; McLaren et al. 2005; Sauve and Hendershot 1997). It is important to remember that each element has its own unique reaction pathways that will determine how it will behave in a soil environment.

8.4 BIOAVAILABILITY

8.4.1 LEAD

Bioavailability of soil Pb to humans is unusually low and generally tied to Pb speciation (Hettiarachchi and Pierzynski 2004; Attanayake et al. 2015). Even without the use of amendments, most lead in soils is highly sorbed (all mechanisms) to soil minerals; it is not easily absorbed by the human digestive tract (Chaney and Ryan 1994; Freeman et al. 1996; Henry et al. 2015). Rat feeding tests (which is an acceptable model for human digestive bioavailability) often put the bioavailability of soil lead to the human gut as less than 10% (Zia et al. 2011; Freeman et al. 1994). Lead phosphate minerals, which are readily formed, tend to have low human bioavailability (Henry et al. 2015; Sauve and Hendershot 1997; Scheckel et al. 2013). However, certain soil conditions can increase Pb bioavailability to humans, particularly low pH and the formation of certain easily dissolvable solids (such as cerussite, $PbCO_3$) under acidic conditions (i.e., human stomach). Low pH increases the solubility of lead which increases its bioavailability (Oliver et al. 1999; Yang et al. 2016; Chaney and Ryan 1994; Sauve and Hendershot 1997). Some Pb minerals are more easily dissolved, particularly Pb oxides and Cerussite (Hettiarachchi and Pierzynski 2004; Zhang and Ryan 1999; Smith et al. 2011).

Lead bioavailability also greatly depends on the organism in question. While human bioavailability is generally fairly similar to the bioavailability to other animals like rats or young swine, it is quite different from plant or microbe bioavailability (Chaney and Ryan 1994; Hettiarachchi and Pierzynski 2004; Attanayake et al. 2015). Even within the realm of human bioavailability, there is considerable variation from person to person for a variety of reasons. Children tend to uptake Pb more than adults (Chaney and Ryan 1994; Henry et al. 2015; McClintock 2015; Mielke et al. 1983). Fasting prior to consumption of Pb leads to higher Pb uptake. During fasting, the digestive tract pH decreases, which increases the solubility of any Pb in the digestive track; animal feeding studies have shown that Pb dissolution can take place in the stomach, allowing for absorption in the intestines (Hettiarachchi and Pierzynski 2004). Even a seemingly small change in pH can significantly alter human Pb uptake (Sauve and Hendershot 1997a; Chaney and Ryan 1994; Brown et al. 2003a; Oliver et al. 1999). Additionally, people who are not fasting will have, presumably, food in their digestive system, which often contains substances that can react with Pb. Lead can form complexes with P, Fe-oxides, and organic matter even inside the digestive tract of an animal (Chaney and Ryan 1994; Brown et al. 2003a; Hettiarachchi et al. 2003a; Zhang and Ryan 1999; Freeman et al. 1996; Davis et al. 1994; Juhasz et al. 2014).

Since there are multiple ways of defining bioavailability, there are multiple methods of accessing it. Bioavailability assessment can be broken down into two large groups: *in vitro* and *in vivo*. *In vitro* is the use of proxy extraction solutions and chemical fractionation to make assumptions about bioavailability based on chemistry. *In vivo* is the use of live organisms to determine actual bioavailability (Brown et al. 2003a; Zia et al. 2011; Scheckel et al. 2009; Smith et al. 2011). *In vivo* is generally accepted as more accurate but is generally more difficult and expensive to perform. For human bioavailability, the *in vivo* model often uses rodents or pigs as proxies (Scheckel et al. 2009; Smith et al. 2011; Hettiarachchi et al. 2003a; Freeman et al. 1994). Human *in vitro* bioavailability is often accessed with artificial gastric methods such as the physiologically based extraction test (PBET), *in vitro* bioaccessability (IVBA), and methods set by the US EPA (Smith et al. 2011; Lestan and Finzgar 2007; Brown et al. 2003a; Obrycki et al. 2016). Plant and microbial bioavailability does not have a set model like with human bioavailability. Instead, bioavailability is often assumed based on chemical fractionation, Pb mineralogy, or exchangeable Pb (Mao et al. 2016; Rooney et al. 2007; McBride 2013). Plant bioavailability is also assessed using *in vivo* methods much more frequently than with animal and human bioavailability as *in vivo* methods for plants are much less expensive and less complicated to carry out (Attanayake et al. 2015; Mao et al. 2016; Hettiarachchi and Pierzynski 2004; McBride et al. 2014).

8.4.2 ARSENIC

Arsenic bioavailability both to humans and to other organisms greatly depends on its oxidation state. Arsenic in the form of As +III has greater solubility and greater bioavailability than As in the As +V state (Zhao et al. 2010; Mandal and Suzuki 2002). Since As can be taken up readily by plants under certain growing conditions, it is important to consider human bioavailability from both direct soil consumption and from high-As crops (i.e., crops grown under saturated conditions, such as rice production or irrigated with As-containing water) (Bastías and Beldarrain 2016). Most of the soils that are in urban garden areas are not regularly waterlogged or flooded. In this situation, the human As bioavailability from direct soil exposure is typically fairly low (Zhao et al. 2010; Garcia-Sanchez et al. 2010; Kumpiene et al. 2009). Soil to human *in vitro* bioavailability of As has been reported between 0.6% and 20%, the wide range of values being due to differences in soil and As mineralogy, soil texture, and other soil properties (Liu et al. 2016; Yoon et al. 2016; Attanayake et al. 2015; Defoe et al. 2014). Much of the variability in human to soil bioavailability comes from differences in soil properties like pH, texture, and mineralogy (Garcia-Sanchez et al. 2010; Miller et al. 2010; Yang et al. 2007; Brown et al. 2012). While As bioavailability in terrestrial soils is generally low, As in plant tissues often has much higher human bioavailability. As a result, most human As exposure comes from food grown in a high As environment, flooded contaminated soils, or areas irrigated with As-contaminated water (Clemente et al. 2016; Bastías and Beldarrain 2016; Gillispie et al. 2015; Zhao et al. 2010; Meharg and Rahman 2003; Wu et al. 2011). As in plant, tissues tend to be in the more bioavailable As +III form and in organo-As compounds (Clemente et al. 2016; Zhao et al. 2010). Processing and cooking methods can make a difference in the bioavailability of As in food (Yager et al. 2015; Signes-Pastor et al. 2012). A study on cooking methods found that the most common rice cooking method (parboiling) can enhance As bioavailability by as much as 30% when compared to a nonparboil (Signes-Pastor et al. 2012). Clemente et al. (2016) reported As bioavailability in white rice samples to be around 95% and in brown rice samples 75%. A similar range for rice bioavailability has been reported by others (Yager et al. 2015; Signes-Pastor et al. 2012). The differences in bioavailability reported here are likely due to the higher presence of micronutrients like Fe and Mn in the brown rice, which can complex with As and reduce human bioavailability, even inside the human digestive tract (Bastías and Beldarrain 2016; Clemente et al. 2016).

As bioavailability and uptake mechanisms to plants also depends on the oxidation state, plants uptake both As +III and As +V, albeit with different mechanisms (Zhao et al. 2010; Wang and Mulligan 2006; Aryal and Reinhold 2015; Gillispie et al. 2015). In terrestrial systems, typically more relevant to urban gardeners, As +V enters plants via P uptake pathways; in anaerobic systems, As +III can enter the plant through Si and NH_4^+ channels. The active Si channels of rice combined with its cultivation in flooded fields greatly increases rice As uptake (Zhao et al. 2010; Sanglard et al. 2016).

Bioavailability measurement for As is similar to Pb, with *in vitro* models being based on artificial gastric solutions (Bastías and Beldarrain 2016; Clemente et al. 2016). Arsenic bioavailability can also be accessed via use of tissue cultures (Yoon et al. 2016). Finally, human As bioavailability can be studied via observational studies. While unable to determine exact causality, these studies measure As uptake in actual humans. One such study in Hunan Province, China, looked at As contents of soils, rice, and voluntary human hair samples. They found that 40% of people had elevated As in their hair samples (Wu et al. 2011).

8.4.3 CADMIUM

Cadmium in soil may be taken up by plants in fairly significant quantities. As such, it is important to understand the bioavailability of Cd in soils and plants (Satarug et al. 2003; Chaney 2015; Qiu et al. 2011; Franz et al. 2008; Chen et al. 2007). Direct soil consumption Cd bioavailability can be highly variable and is dependent on soil characteristics (Zhao et al. 2016; Xia et al. 2016; Balal Yousaf et al. 2016a; Aziz et al. 2015). Soils that are high in compounds that can sorb Cd tend to have lower bioavailability. Iron oxides, soil clays, soil carbonates, biochar, and Zn can all complex with Cd and reduce its bioavailability (Oliver et al. 1998; De Livera et al. 2011; Chaney 2015; Barrow 2000; Fan MeiRong et al. 2012; Balal Yousaf et al. 2016). Bioavailability of Cd in food is typically low but can also be fairly variable and can depend on the nutritional status of the person and on food preparation methods (Fu and Cui 2013; Reeves and Chaney 2008; Chaney 2015). Iron, Zn, and Ca can all reduce Cd uptake inside the human body (Reeves and Chaney 2008; Chaney 2015; Chaney and Ryan 1994). The method used to determine Cd bioavailability can also have a large impact on the results (Reeves and Chaney 2008; Aziz et al. 2015). Most *in vitro* models for Cd intake use a simple gastric solution that does not take into account complexities that occur *in vivo* (Reeves and Chaney 2008; Aziz et al. 2015; Chaney 2015). However, there are a few cases where human Cd bioavailability and human Cd exposure are higher than usual. First is with rice. Rice is particularly prone to accumulating Cd due to the way it is grown (i.e., flooded soils); additionally, the Cd in rice is typically more bioavailable due to the low Fe and Zn content of rice (World Health Organization 2010; Chaney 2015; Reeves and Chaney 2008; Chaney and Ryan 1994). So, Cd is capable of bioaccumulation and can move up the food chain. When Cd_contaminated soil is used to grow forages to feed livestock, the Cd will move from the soil to the plant to the animal where it will be accumulated in the liver and kidneys (Franz et al. 2008; LoganatHan et al. 2003; World Health Organization 2010). As such organ meats can have high Cd contents; fortunately these are not a staple for most people and there is very little Cd accumulation in the muscle meats (Franz et al. 2008). This may be a slight concern for those wanting to raise urban livestock (usually chickens or goats), which is common in developing countries and like urban gardening, is a growing trend (McClintock 2015; Halloran and Magid 2013; Rogus and Dimitri 2015). While not a food crop, tobacco is commonly consumed by humans and is capable of accumulating high amounts of Cd in its tissues. Additionally, tobacco is generally not eaten, but smoked, which greatly increases bioavailability (Satarug et al. 2003; World Health Organization 2010).

Cadmium bioavailability is typically determined with *in vitro* gastric models (Intawongse and Dean 2006; Fu and Cui 2013; Reeves and Chaney 2008). However, these models often fail to completely capture Cd absorption processes in human bodies (Reeves and Chaney 2008; Chaney 2015). Investigations into better ways to model Cd uptake is needed and *in vitro* models should be verified with *in vitro* feeding trials using rats or young swine (Chaney 2015; Hutzell and Luecken 2008b).

8.4.4 POLYCYCLIC AROMATIC HYDROCARBONS

Bioavailability of PAHs varies with the exact PAH compound (Chen et al. 2015; Zohair et al. 2006; Plachá et al. 2009). PAHs in soil tend to degrade over time and are often highly sorbed to SOC ` and as a result direct soil contact is not a major pathway for human PAH exposure (Attanayake et al. 2015; Peters et al. 2015; Ounnas et al. 2009). Inhalation of PAH-laden dust and consumption of PAH-contaminated food are more important human exposure pathways for PAHs (Ramesh et al. 2004; Khan and Cao 2012; Lorenzi et al. 2011). Soil properties can also influence PAH bioavailability. Soils with higher organic carbon tend to have more PAHs sorbed and thus not bioavailable (Khan and Cao 2012). PAH bioavailability can be accessed using both *in vitro* and *in vitro* models successfully (Peters et al. 2015; Ounnas et al. 2009; Costera et al. 2009; Ramesh et al. 2004; Attanayake et al. 2015).

8.5 REDUCING TRANSFER IN MILDLY CONTAMINATED URBAN SOILS

Remediation or proper management of sites contaminated with potentially toxic metal(loid)s and organic contaminants, such as PAHs, can reduce risks to human and ecosystem health. There are two major categories of remediation methods: *in situ* and *ex situ*. *In situ* remediation is the remediation of the impacted soil in place, while *ex situ* is the removal of the soil to be remediated off site. This section primarily focuses on *in situ* methods. Table 8.1 summarizes various soil amendments used for *in situ* stabilization of soil contaminants, their mode of stabilization, and selected references. Figure 8.2 shows growth of garden crops in nonamended contaminated soil (Figure 8.2a and b) in comparison with contaminated soil receiving soil amendments (Figure 8.2c and d).

There are a number of methods that have been considered for *in situ* remediation. One of the main benefits of *in situ* remediation is that it is much more cost-effective than *ex situ* remediation, especially when the effected site is large in area and the contamination load is moderate (Hettiarachchi and Pierzynski 2004). Remediation treatment can reduce human risks from exposure to contaminated soil and can improve soil ecological functioning. However, even with remediation, it is possible that affected sites will not be as productive as sites that were never contaminated (Hinojosa 2004; De Mora et al. 2005).

Most *in situ* remediation methods involve using soil amendments and establishing plant growth on affect sites in order to stabilize them (chemical, physical, and phytostabilization). The idea behind stabilization efforts is to reduce transport of contaminated soils and reduce contaminant bioavailability/mobility. The reduction of bioavailability and mobility makes the contaminants much less of a threat to humans and ecosystems and prevents the contamination from spreading to other areas via erosion, leaching, or food chain (Vangronsveld et al. 2009). Encouraging perennial vegetation on contaminated sites is phytostabilization. The roots help to keep the soil in place and

TABLE 8.1

Summary of Various Soil Amendments Used for *In Situ* Stabilization of Soil Contaminants

Summary of Amendments Used in Remediation

Amendment Type	Examples	Contaminant Effected	Mode of Action	References
Composts	Cotton compost, municipal waste compost	As, Pb, Cd, Zn, Cu, Cr, PAHs	Formation of stable, insoluble organo-metal complexes	Tejada et al. (2006), de Mora et al. (2005), Juang et al. (2012), Zeng et al. (2015), Taiwo et al. (2016), and Brown et al. (2012)
Biosolids	Composted and noncomposted biosolids	Pb, Zn, Cd, As (results vary), PAHs	Formation of stable, insoluble organo-metal complexes. Formation of inorganic complexes with Fe and P in biosolids	Attanayake et al. (2015), Brown et al. (2003b), Defoe et al. (2014), Basta et al. (2001), Basta (1999), and Farfel et al. (2005)
Lime	Agricultural lime, oyster shell	Cd, Pb	Increase in pH decreases contaminant solubility. Formation of CO_3 complexes.	U.S. Department of Health and Human Services (2012), Rooney et al. (2007), Sauve and Hendershot (1997b), Pierzynski et al. (2004), Chaney and Ryan (1994), Levonmäki et al. (2006a), El-Azeem et al. (2013), and Hong et al. (2010)
Phosphorus	TSP, apatite	Pb, Cd	Formation of insoluble P–Pb complexes	Sauve and Hendershot (1997b), Gu et al. (2016), Hettiarachchi et al. (2003a), and Scheckel et al. (2013)
Iron_rich products	Ferrihydrite, Fe-containing biosolids	Pb, As, Cd	Formation of Fe–metal complexes	Vangronsveld et al. (2009), Yang et al. (2007), Chen et al. (2011), Sneath et al. (2013), and Brown et al. (2003a)
Other organic amendments	Biochar, peat	Cd, As, Pb	Complexation with element of concern	Simmler et al. (2013), Chen et al. (2011), Lu et al. (2014), Gu et al. (2016), Balal Yousaf et al. (2016b), Zeng et al. (2015), and Yang et al. (2016)

reduce erosion. Ideal plants for stabilization are ones that are well rooted, do not accumulate metals in their tissues, and are well suited to the area. Often on more highly contaminated sites, such as mine tailing disposal sites, the soil will have to be modified prior to planting as initial conditions are too harsh to allow for plant growth (Mendez and Maier 2008). Establishing plants on heavily contaminated sites can be challenging as many of these sites have properties that inhibit plant growth. Low organic carbon, low nutrients, presence of contaminants, and acidity often inhibit plants grown in mine waste sites (generally more contaminated than typical urban sites) and urban brownfields (often selected for urban garden sites). In addition, toxicity from whatever contaminant is impacting the site can also hamper plant growth. In order to establish plant growth on contaminated sites, it is often necessary to amend the soil with organic matter (i.e., composts, manures), liming agents, and fertilizers (Mendez and Maier 2008).

Another potential method is phytoextraction, which uses plants that uptake the metals, which are then harvested and then disposed of appropriately. Phytoextraction has a few difficulties associated with it. First is the tendency of most plants to accumulate low amounts of metals in their tissues, and second is that of the plants that do accumulate metals in their tissues, most are slow growing or produce low biomasses (McGrath and Zhao 2003; Hettiarachchi and Pierzynski 2004). The feasibility

FIGURE 8.2 Growth of garden crops in contaminated soil (a) in comparison with uncontaminated and remediated soil receiving soil amendments (b)–(d).

of phytoextraction depends on the contaminant in question and the degree of contamination. More soluble metals like Cd and Zn can possibly be remediated via phytoextraction in mild to moderately contaminated sites. However, other metals like Pb and Cr have limited feasibility for phytoextraction with current technologies (McGrath and Zhao 2003; Hettiarachchi and Pierzynski 2004). There is work on genetically modifying plants to create more ideal hyperaccumulators but this technology is not yet developed enough for use in widespread *in situ* remediation (McGrath and Zhao 2003; Chaney and Ryan 1994).

A final possibility for reducing human exposure from food grown in contaminated soils (especially for contaminants that are more plant-available like Cd and As) is to select or breed crop cultivars that do not uptake contaminants of concern. Encouraging gardeners to plant low-accumulating plants and cultivars can reduce human transfer risk. From a future development perspective, there is the possibility of specifically breeding plants that have very low metal uptake (Qiu et al. 2011).

8.5.1 Lead

The most practical methods for managing (or remediation) Pb in urban soils is to use appropriate amendments to stabilize the soil and prevent or minimize direct exposure (mainly ingestion) of Pb-contaminated soil to humans. Methods of extraction and soil removal are generally impractical due to the high cost and the sheer number of sites impacted by Pb (Hettiarachchi and Pierzynski 2004). While Pb cannot be truly removed from the soil with this method, it can be made immobile and nonbioavailable through changes in soil chemistry that encourage the sorption of lead to SOC and soil minerals (Smith 2009; Hettiarachchi and Pierzynski 2004; Henry et al. 2015).

There are a number of soil amendments that can be added to reduce Pb availability and improve the ability of the soil to support stable plant growth. Organic amendments such as composts, manures, and biochar can help Pb-impacted soils in a number of ways. Organic carbon compounds in native soils are capable of complexing Pb and reducing its availability. This process can be enhanced by adding organic matter to soils as part of the remediation process (Yang et al. 2016; Attanayake et al. 2015; Perez-de-Mora et al. 2006; Simmler et al. 2013). Organic amendments are often added in large enough volumes to also provide a dilution effect (Attanayake et al. 2015). Finally, organic amendments add nutrients and improve the ability of a soil to support plant growth, which can help to establish permanent vegetation to prevent movement of Pb-contaminated soil via erosion processes (Mendez and Maier 2008). In addition to organic matter, agricultural lime can be very helpful in reducing risks from soil Pb. As Pb bioavailability, both to plants and to humans, is correlated with pH, raising the pH of soils can help to reduce Pb-related risks (Mendez and Maier 2008; Levonmäki et al. 2006; Chaney and Ryan 1994). Lead readily forms unavailable complexes with P and thus the addition of P through rock phosphate or manures can decrease Pb availability (Mendez and Maier 2008; Scheckel et al. 2013; Hettiarachchi et al. 2003a). Finally, Fe or Mn rich amendments have also been shown to reduce the bioavailability of Pb in soils to plants and humans (Brown et al. 2003a, 2012; Hettiarachchi et al. 2003a; Hettiarachchi and Pierzynski 2004).

While organic amendments like composts are often great for remediating Pb-contaminated sites, it is important to keep in mind that these materials can also be a source of lead. While Pb in composts is generally low, testing organic amendments before application can alleviate such concerns (Chaney and Ryan 1994; Hettiarachchi and Pierzynski 2004). More careful separation of compostable yard and food scraps from other wastes (i.e., metals and plastics) will help by creating safer composts (Smith 2009).

8.5.2 ARSENIC

Much of the remediation of As-contaminated sites uses similar principles to that of Pb; however, since As chemistry is not the same as Pb, there are some differences. Of notable concern are sites that are co-contaminated with both Pb and As. Some of the Pb remediation strategies can actually increase As bioavailability, so care needs to be taken in sites that are co-contaminated (Henry et al. 2015). Phosphorus amendments can reduce Pb bioavailability in soils but they have the opposite effect on As (Henry et al. 2015; Brown et al. 2012). Organic amendments (composts, etc.) can reduce As bioavailability through forming organo-As complexes and by diluting As in the soil (De Mora et al. 2005; Brown et al. 2012; Chen et al. 2011; Huq et al. 2008; Attanayake et al. 2015; Brown et al. 2012; Chen et al. 2011; Huq et al. 2008; Defoe et al. 2014). However, organic amendments can also increase the bioavailability of As; some forms of organo-As are bioavailable (Zhao et al. 2009; Sizmur et al. 2011). Poorly chosen organic amendments can also introduce significant amounts of P which out-competes As for exchange sites, thereby increasing As availability (Brown et al. 2012; Henry et al. 2015). Iron mineral amendments generally reduce bioavailability of As by encouraging the formation of As–Fe complexes (Kumpiene et al. 2009; Yang et al. 2007; Brown et al. 2012; Sneath et al. 2013; Chen et al. 2011).

Since As is more bioavailable in its reduced form, management practices that reduce the incidence of waterlogging will reduce As bioavailability (Kumpiene et al. 2009; Zhao et al. 2010a). Ponded or waterlogged conditions can be prevented in garden soils by installing drainage, using mulches, building raised beds, and avoiding gardening in poorly drained locations (Royal Horticultural Society 2016). Plant uptake of As can also be managed by selecting plants and cultivars with low As uptake (Zhao et al. 2010; Xiao et al. 2012; Pillai et al. 2010).

8.5.3 CADMIUM

Cadmium can be remediated using many of the same techniques as Pb and As. The addition of various soil amendments can reduce Cd bioavailability and thus reduce risks to human gardeners. Biochars are an organic amendment made from organic materials, such as straw, that are then charred in anoxic conditions. They have been shown to be effective at absorbing soils contaminants like Cd (Balal Yousaf et al. 2016b; Sneath et al. 2013; Lu et al. 2014; Simmler et al. 2013). Other organic amendments like composts can also be effective at absorbing Cd effectively, thus reducing bioavailability (Balal Yousaf et al. 2016; Zeng et al. 2015; Juang et al. 2012). Controlling soil pH via liming is another cost-effective way for urban gardeners to reduce the potential risks of soil Cd (Chaney and Ryan 1994; El-Azeem et al. 2013; Hong et al. 2010). Mineral and inorganic amendments, like silicate minerals, have been shown to reduce Cd bioavailability due to mineral complexation and cation exchange (Zhao et al. 2016; Gu et al. 2016; Hou et al. 2014).

Cadmium uptake by plants can vary considerably between different species and between different cultivars of the same species. This is often due to differences in root structure and nutrient uptake. Gardeners can reduce their Cd-related risks by selecting plant species and cultivars that have low Cd uptake (Qiu et al. 2011; Li et al. 2005; Daud et al. 2009; Xiao et al. 2015).

8.5.4 POLYCYCLIC AROMATIC HYDROCARBONS

PAHs are organic contaminants and they can be mineralized by soil bacteria, and thus completely removed from the soil ecosystem. Much of PAH remediation is about encouraging the microbial activities that will enhance the breakdown of the PAHs (Cerniglia 1992; Chen et al. 2016; Boonchan et al. 2000). The rate that PAHs are degraded depends on the PAH in question and the soil conditions. As a general rule, lower molecular weight PAHs tend to degrade faster than larger PAHs (Boonchan et al. 2000; Plachá et al. 2009; Cerniglia 1992; Attanayake et al. 2015). Soil conditions that increase microbial activities will also increase PAH degradation. Actively growing plants exude compounds into the soil near the roots, creating a surge of microbial activity near plant roots. Not surprisingly, PAH degradation is greatly enhanced in the rhizosphere when compared to the bulk soil (Boonchan et al. 2000; Kawasaki et al. 2012; Reilley et al. 1996; Chen et al. 2016; De Nicola et al. 2015). PAHs can also be sequestered and rendered nonavailable by absorbing to SOM (Ounnas et al. 2009; Khan and Cao 2012; Wilcke 2000; Zhang et al. 2010). Under optimal microbial growth conditions (planted and ideal moisture/temperature), as much as 85%–100% of PAHs can be mineralized in a growing season (Chen et al. 2016).

8.6 SUMMARY

The potential exposure pathway of concern is direct exposure of contaminated urban soils to humans. If managed properly, the pathway from contaminated soil to plant to human can be kept at a minimum. Contaminant concentrations can be diluted by the addition of compost and contaminant bioavailability can be reduced via addition of appropriate soil amendments. Research has also shown that, in general, concentrations of contaminants in produce harvested from contaminated urban soils are low.

REFERENCES

Abaye, D.A., K. Lawlor, P.R. Hirsch, and P.C. Brookes. 2005. Changes in the microbial community of an arable soil caused by long-term metal contamination. *European Journal of Soil Science* 56(1): 93–102.

Appel, C. and L. Ma. 2002. Concentration, pH, and surface charge effects on cadmium and lead sorption in three tropical soils. *Journal of Environmental Quality* 31(2): 581–589.

Aryal, N. and D.M. Reinhold. 2015. Reduction of metal leaching by poplars during soil treatment of wastewaters: Small-scale proof of concept studies. *Ecological Engineering* 78: 53–61.

Attanayake, C.P., G.M. Hettiarachchi, S. Martin, and G.M. Pierzynski. 2015. Potential bioavailability of lead, arsenic, and polycyclic aromatic hydrocarbons in compost-amended urban soils. *Journal of Environmental Quality* 44(3): 930–944.

Aziz, R., M.T. Rafiq, T. Li, D. Liu, Z. He, P.J. Stoffella, K. Sun et al. 2015. Uptake of cadmium by rice grown on contaminated soils and its bioavailability/toxicity in human cell lines (Caco-2/HL-7702). *Journal of Agricultural and Food Chemistry* 63(13): 3599–3608.

Balal, Y., L. GuiJian, W. RuWei, M.Z. Rehman, M.S. Rizwan, M. Imtiaz, G. Murtaza, and A. Shakoor. 2016. Investigating the potential influence of biochar and traditional organic amendments on the bioavailability and transfer of Cd in the soil-plant system. *Environmental Earth Sciences* 75(5): 374.

Bandara, J.M., H.V. Wijewardena, Y.M. Bandara, R.G. Jayasooriya, and H. Rajapaksha. 2011. Pollution of River Mahaweli and farmlands under irrigation by cadmium from agricultural inputs leading to a chronic renal failure epidemic among farmers in NCP, Sri Lanka. *Environmental Geochemistry and Health* 33: 439–453.

Barrow, N.J. 2000. *Cadmium in Soils and Plants*, edited by M.J. Mclaughlin and B.R. Singh, Vol. 96. Dordrecht, the Netherland: Kluwer Academic Publishers, 1999. Hardcover, 271, ISBN 0 7923 5843 0.

Basta, N.T. 1999. Bioavailability of heavy metals in strongly acidic soils treated with exceptional quality biosolids. *Journal of Environmental Quality* 28(2): 633–638.

Basta, N.T., R. Gradwohl, K.L. Snethen, and J.L. Schroder. 2001. Chemical immobilization of lead, zinc, and cadmium in smelter-contaminated soils using biosolids and rock phosphate. *Journal of Environmental Quality* 30(4): 1222.

Bastías, J. M. and T. Beldarrain. 2016. Arsenic translocation in rice cultivation and its implication for human health. *Chilean Journal of Agricultural Research* 76(1): 114–122.

Boonchan, S., M.L. Britz, and G.A. Stanley. 2000. Degradation and mineralization of high-molecular-weight polycyclic aromatic hydrocarbons by defined fungal-bacterial cocultures. *Applied and Environmental Microbiology*. 66: 1007–1019.

Brown, S.L., I. Clausen, M.A. Chappell, K.G. Scheckel, M. Newville, and G.M. Hettiarachchi. 2012. High-iron biosolids compost-induced changes in lead and arsenic speciation and bioaccessibility in co-contaminated soils. *Journal of Environmental Quality* 41(5): 1612–1622.

Brown, S., R.L Chaney, J.G Hallfrisch, and Q. Xue. 2003. Effect of biosolids processing on lead bioavailability in an urban soil. *Journal of Environmental Quality* 32(1): 100–108.

Brown, S.L., R.L. Chaney, and G.M. Hettiarachchi. 2016. Lead in urban soils: A real or perceived concern for urban agriculture? *Journal of Environmental Quality* 45(1): 26–36.

Canadian Council of Ministers of the Environment. 1999. *Canadian Sediment Quality Guidelines for the Protection of Aquatic Life: Polycyclic Aromatic Hydrocarbons (PAHs).*Canadian environmental quality guidelines, Canadian Council of Ministers of the Environment, Winnipeg.

Cerniglia, C.E. 1992. Biodegradation of polycyclic aromatic hydrocarbons. *Biodegradation* 4(3): 331–338.

Chaney, R. 2015. How does contamination of rice soils with Cd and Zn cause high incidence of human Cd disease in subsistence rice farmers? *Current Pollution Reports* 1(1): 13–22.

Chaney, R.L. and J.A. Ryan. 1994. *Risk Based Standards for Arsenic, Lead and Cadmium in Urban Soils.* Frankfurt: Dechema.

Chen, W., A.C. Chang, and L. Wu. 2007. Assessing long-term environmental risks of trace elements in phosphate fertilizers. *Ecotoxicology and Environmental Safety* 67(1): 48–58.

Chen, R., A. de Sherbinin, C. Ye, and G. Shi. 2014. China's soil pollution: Farms on the frontline. *Science (New York, N.Y.)* 344(6185): 691.

Chen, W., S. Lu, C. Peng, W. Jiao, and M. Wang. 2013. Accumulation of Cd in agricultural soil under long-term reclaimed water irrigation. *Environmental Pollution*. 178: 294–299.

Chen, Z-X., H-G. Ni, X. Jing, W-J. Chang, J.-L. Sun, and H. Zeng. 2015. Plant uptake, translocation, and return of polycyclic aromatic hydrocarbons via fine root branch orders in a subtropical forest ecosystem. *Chemosphere* 131: 192–200.

Chen, F., M. Tan, J. Ma, S. Zhang, G. Li, and J. Qu. 2016. Efficient remediation of PAH-metal co-contaminated soil using microbial-plant combination: A greenhouse study. *Journal of Hazardous Materials* 302: 250–261.

Chen, D., X. XianJu, L. DeQin, Z. YongChun, S. XiaoHou, and H. Juan. 2011. Remediation of paddy soil contaminated by arsenic, cadmium and lead with amendments. *Jiangsu Journal of Agricultural Sciences* 27(6): 1284–1288.

Chen, H., N. Wu, Y. Wang, Y. Gao, and C. Peng. 2011. Methane fluxes from Alpine wetlands of Zoige Plateau in relation to water regime and vegetation under two scales. *Water, Air and Soil Pollution* 217 (1–4):173–183.

Cheng, X., H. Shi, C.D. Adams, and Y. Ma. 2010. Assessment of metal contaminations leaching out from recycling plastic bottles upon treatments. *Environmental Science and Pollution Research International* 17(7): 1323–1330.

Claassens, S. 2006. Soil microbial properties in coal mine tailings under rehabilitation. *Applied Ecology and Environmental Research* 4(1): 75–83.

Clark, H.F., D.J. Brabander, and R.M. Erdil. 2006. Sources, sinks, and exposure pathways of lead in urban garden soil. *Journal of Environmental Quality* 35(6): 2066–2074.

Clemente, M.J., V. Devesa, and D. Vélez. 2016. Dietary strategies to reduce the bioaccessibility of arsenic from food matrices. *Journal of Agricultural and Food Chemistry.* 64: 923–931.

Costera, A., C. Feidt, M-A. Dziurla, F. Monteau, B. Le Bizec, and G. Rychen. 2009. Bioavailability of polycyclic aromatic hydrocarbons (PAHs) from soil and hay matrices in lactating goats. *Journal of Agricultural and Food Chemistry* 57(12): 5352.

Cruz, A.C.F., L.M. Buruaem, A.P.C. Rodrigues, and W.T.V. Machado. 2016. Using a tiered approach based on ecotoxicological techniques to assess the ecological risks of contamination in a subtropical estuarine protected area. *Science of the Total Environment* 544: 564-573.

Daud, M.K., Y. Sun, M. Dawood, Y. Hayat, M.T. Variath, Y-X. Wu, and Raziuddin. 2009. Cadmium-induced functional and ultrastructural alterations in roots of two transgenic cotton cultivars. *Journal of Hazardous Materials* 161(1): 463–473.

Davis, A., M.V. Ruby, and P.D. Bergstrom. 1994. Factors controlling lead bioavailability in the Butte mining district, Montana, USA. *Environmental Geochemistry and Health* 16 (3–4):147–157.

De Livera, J., M.J. McLaughlin, G.M. Hettiarachchi, J.K. Kirby, and D.G. Beak. 2011. Cadmium solubility in paddy soils: Effects of soil oxidation, metal sulfides and competitive Ions. *Science of the Total Environment* 409(8): 1489–1497.

De Mora, A.P., J.J. Ortega-Calvo, F. Cabrera, and E.M. 2005. Changes in enzyme activities and microbial biomass after "in situ" remediation of a heavy metal-contaminated soil. *Applied Soil Ecology* 28(2): 125–137.

De Nicola, F., D. Baldantoni, L. Sessa, F. Monaci, R. Bargagli, and A. Alfani. 2015. Distribution of heavy metals and polycyclic aromatic hydrocarbons in holm oak plant-soil system evaluated along urbanization gradients. *Chemosphere* 134: 91.

Defoe, P.P., G.M. Hettiarachchi, C. Benedict, and S. Martin. 2014. Safety of gardening on lead- and arsenic-contaminated urban brownfields. *Journal of Environmental Quality* 43(6): 2064.

Delmotte, S., F.J.R. Meysman, A. Ciutat, A. Boudou, S. Sauvage, and M. Gerino. 2007. Cadmium transport in sediments by tubificid bioturbation: An assessment of model complexity. *Geochimica et Cosmochimica Acta* 71(4): 844–862.

El-Azeem, S.A.M.A., M. Ahmad, A.R.A. Usman, K. KwonRae, O. SangEun, L. SangSoo, and O. YongSik. 2013. Changes of biochemical properties and heavy metal bioavailability in soil treated with natural liming materials. *Environmental Earth Sciences* 70(7): 3411–3420.

Evans, N.P., M. Bellingham, R.M. Sharpe, C. Cotinot, S.M. Rhind, C. Kyle, H. Erhard et al. 2014. Reproduction Symposium: Does grazing on biosolids-treated pasture pose a pathophysiological risk associated with increased exposure to endocrine disrupting compounds? *Journal of Animal Science* 92(8): 3185–3198.

Fabietti, G., M. Biasioli, R. Barberis, and F. Ajmone-Marsan. 2010. Soil contamination by organic and inorganic pollutants at the regional scale: The case of Piedmont, Italy. *Journal of Soils and Sediments* 10(2): 290–300.

Facchinelli, A., E. Sacchi, and L. Mallen. 2001. Multivariate statistical and GIS-based approach to identify heavy metal sources in soils. *Environmental Pollution* 114(3): 313–324.

Fan, M., L. Lin, L. YuLin, W. JianHong, T. Jie, and H. Bo. 2012. Effects of red mud application on rice growth and transformation of cadmium forms in Cd-contaminated paddy soils. *Plant Nutrition and Fertilizer Science* 18(2): 390–396.

Farfel, M.R., A.O. Orlova, R.L. Chaney, P.S.J. Lees, C. Rohde, and P.J. Ashley. 2005. Biosolids compost amendment for reducing soil lead hazards: A pilot study of Orgro® amendment and grass seeding in urban yards. *Science of the Total Environment* 340(1): 81–95.

Franz, E., L. Van Raamsdonk, D.F. Van, and P. Römkens. 2008. A chain modeling approach to estimate the impact of soil cadmium pollution on human dietary exposure. *Journal of Food Protection* 71(12): 2504–2513.

Freeman, G.B, J.A. Dill, J.D. Johnson, P.J. Kurtz, F. Parham, and H.B. Mathews. 1996. Comparative absorption of lead from contaminated soil and lead salts by weanling Fischer 344 rats. *Fundamental and Applied Toxicology* 33(1): 109–119.

Freeman, G.B, J.D. Johnson, S.C. Liao, P.I. Feder, A.O. Davis, M.V. Ruby, R.A. Schoof et al. 1994. Absolute bioavailability of lead acetate and mining waste lead in rats. *Toxicology* 91(2): 151–163.

Fu, J. and Y. Cui. 2013. In vitro digestion/Caco-2 cell model to estimate cadmium and lead bioaccessibility/bioavailability in two vegetables: The influence of cooking and additives. *Food and Chemical Toxicology* 59: 215–221.

Gai, N., Y. Yang, T. Li, J. Yao, F. Wang, and H. Chen. 2011. Effect of lead contamination on soil microbial activity measured by microcalorimetry. *Chinese Journal of Chemistry* 29(7): 1541–1547.

Garcia-Sanchez, A., P. Alonso-Rojo, and F. Santos-Frances. 2010. Distribution and mobility of arsenic in soils of a mining area (Western Spain). *Science of the Total Environment* 408(19): 4194–4201.

Gillispie, E., T. Sowers, O. Duckworth, and M. Polizzotto. 2015. Soil pollution due to irrigation with arsenic-contaminated groundwater: Current state of science. *Current Pollution Reports* 1(1): 1–12.

Gu, J., H. Zhou, Y. Wu, W. Zhu, Z. Zou, P. Peng, M. Zeng et al. 2016. Synergistic control of combined amendment on bioavailability and accumulation of Cd and as in rice paddy soil. *China Environmental Science* 36(1): 206–214.

Halloran, A. and J. Magid. 2013. Planning the unplanned: Incorporating agriculture as an urban land use into the Dar es Salaam master plan and beyond. *Environment & Urbanization* 25(2): 541–558.

Han, F.X., W.L. Kingery, H.M. Selim, P.D. Gerard, M.S. Cox, and J.L. Oldham. 2004. Arsenic solubility and distribution in poultry waste and long-term amended soil. *Science of the Total Environment* 320(1): 51–61.

Han, F., Y. Su, D. Monts, M. Plodinec, A. Banin, and G. Triplett. 2003. Assessment of global industrial-age anthropogenic arsenic contamination. *Naturwissenschaften* 90(9): 395–401.

Han, X., N. Zhou, Z. Cui, M. Ma, L. Li, M. Cai, Y. Li et al. 2011. Association between urinary polycyclic aromatic hydrocarbon metabolites and sperm DNA damage: A population study in Chongqing, China. *Environmental Health Perspectives* 119(5): 652.

Henry, H., M.F. Naujokas, C. Attanayake, N.T. Basta, Z. Cheng, G.M. Hettiarachchi, M. Maddaloni et al. 2015. Bioavailability-based in situ remediation to meet future lead (Pb) standards in urban soils and gardens. *Environmental Science & Technology* 49(15): 8948.

Hettiarachchi, G.M. and G.M. Pierzynski. 2004. Soil lead bioavailability and in situ remediation of lead-contaminated soils: A review. *Environmental Progress* 23(1): 78–93.

Hettiarachchi, G., G. Pierzynski, F. Oehme, O. Sonmez, and J. Ryan. 2003a. Treatment of contaminated soil with phosphorus and manganese oxide reduces lead absorption by Sprague-Dawley rats. *Journal of Environmental Quality* 32(4): 1335–1345.

Hettiarachchi, G.M., J.A. Ryan, R.L. Chaney, and C.M. La Fleur. 2003b. Sorption and desorption of cadmium by different fractions of biosolids-amended soils. *Journal of Environmental Quality* 32(5): 1684–1693.

Hinojosa, A.B. 2004. Microbiological rates and enzyme activities as indicators of functionality in soils affected by the Aznalcollar toxic spill. *Soil Biology & Biochemistry* 36(10): 1637–1644.

Hockmann, K., M. Lenz, S. Tandy, M. Nachtegaal, M. Janousch, and R. Schulin. 2014. Release of antimony from contaminated soil induced by redox changes. *Journal of Hazardous Materials* 275: 215–221.

Hong, C.O., S.Y. Kim, J. Gutierrez, V.N. Owens, and P.J. Kim. 2010. Comparison of oyster shell and calcium hydroxide as liming materials for immobilizing cadmium in upland soil. *Biology and Fertility of Soils* 46(5): 491–498.

Hou, Q-H., A-Z. Ma, Y. Li, X-L. Zhuang, Z-H. Bai, X-K. Zhang, and G-Q. Zhuang. 2014. Assessing the effect of phosphate and silicate on Cd bioavailability in soil using an *Escherichia coli* cadAp::luc-based whole-cell sensor. *Environmental Science-Processes & Impacts* 16(4): 890–896.

Huang, L-M., C-B. Deng, N. Huang, and X-J. Huang. 2013. Multivariate statistical approach to identify heavy metal sources in agricultural soil around an abandoned Pb–Zn mine in Guangxi Zhuang Autonomous Region, China. *Environmental Earth Sciences* 68(5): 1331–1348.

Hunter, B.A., M.S. Johnson, and D.J. Thompson. 1987. Ecotoxicology of copper and cadmium in a contaminated grassland ecosystem.2. Invertebrates. *Journal of Applied Ecology* 24(2): 587–599.

Huq, S.M.I., S. Al-Mamun, J.C. Joardar, and S.A. Hossain. 2008. Remediation of soil arsenic toxicity in ipomoea aquatica, using various sources of organic matter. *Land Contamination & Reclamation* 16(4): 333–341.

Hutzell, W.T. and D.J. Luecken. 2008. Fate and transport of emissions for several trace metals over the United States. *Science of the Total Environment* 396(2): 164–179.

Intawongse, M. and J.R. Dean. 2006. Uptake of heavy metals by vegetable plants grown on contaminated soil and their bioavailability in the human gastrointestinal tract. *Food Additives and Contaminants* 23(1): 36–48.

Juang, K-W., P-C. Ho, and C-H. Yu. 2012. Short-term effects of compost amendment on the fractionation of cadmium in soil and cadmium accumulation in rice plants. *Environmental Science and Pollution Research* 19(5): 1696–1708.

Juhasz, A.L., D. Gancarz, C. Herde, S. McClure, K.G. Scheckel, and E. Smith. 2014. In situ formation of pyromorphite is not required for the reduction of in vivo Pb relative bioavailability in contaminated soils. *Environmental Science and Technology* 48(12): 7002–7009.

Kachenko, A. and B. Singh. 2006. Heavy metals contamination in vegetables grown in urban and metal smelter contaminated sites in Australia. *Water, Air, and Soil Pollution; An International Journal of Environmental Pollution* 169(1): 101–123.

Kaji, M. 2012. Role of experts and public participation in pollution control: The case of Itai-Itai disease in Japan. *Ethics in Science and Environmental Politics (ESEP)* 12(2): 99–111.

Kaloyanova, N. 2007. Effect of arsenic contamination on the microbial changes of two paddy soils. *Pochvoznanie, Agrokhimiya i Ekologiya* 41(1): 37–43.

Kawasaki, A., E. Watson, and M. Kertesz. 2012. Indirect effects of polycyclic aromatic hydrocarbon contamination on microbial communities in legume and grass rhizospheres. *Plant and Soil: An International Journal on Plant-Soil Relationships* 358(1): 169–182.

Khan, S. and Q. Cao. 2012. Human health risk due to consumption of vegetables contaminated with carcinogenic polycyclic aromatic hydrocarbons. *Journal of Soils and Sediments* 12(2): 178–184.

Khan, S., A.E-L. Hesham, M. Qiao, S. Rehman, and J-Z. He. 2010. Effects of Cd and Pb on soil microbial community structure and activities. *Environmental Science and Pollution Research* 17(2): 288–296.

Kumpiene, J., D. Ragnvaldsson, L. Lovgren, S. Tesfalidet, B. Gustavsson, A. Lattstrom, P. Leffler et al. 2009. Impact of water saturation level on arsenic and metal mobility in the Fe-amended soil. *Chemosphere* 74(2): 206–215.

Lanphear, B.P., P. Succop, S. Roda, and G. Henningsen. 2003. The effect of soil abatement on blood lead levels in children living near a former smelting and milling operation. *Public Health Reports* 118(2): 83–91.

Lee-Smith, D. 2010. Cities feeding people: An update on urban agriculture in equatorial Africa. *Environment and Urbanization* 22(2): 483–499.

Lestan, D. and N. Finzgar. 2007. Relationship of soil properties to fractionation, bioavailability and mobility of Pb and Zn in soil. *Plant Soil and Environment* 53 (5): 225–238.

Levonmäki, M., H. Hartikainen, and T. Kairesalo. 2006a. Effect of organic amendment and plant roots on the solubility and mobilization of lead in soils at a shooting range. *Journal of Environmental Quality* 35(4): 1026–1031.

Li, K., T. Chen, X. Bi, and Q. Wang. 2014. Soil contamination by polycyclic aromatic hydrocarbons at natural recreational areas in Delaware, USA. *Environmental Earth Sciences* 72 (2): 387–398.

Li, Z.W., L.Q. Li, G.X. Pan, and J. Chen. 2005. Bioavailability of Cd in a soil-rice system in China: Soil type versus genotype effects. *Plant and Soil* 271 (1–2):165–173.

Li, Z., J.A. Ryan, J.L. Chen, and S. Al-Abed. 2001. Adsorption of cadmium on biosolids-amended soils. *Journal of Environmental Quality* 30(3): 903–911.

Liao, M., C-L. Chen, L-S. Zeng, and C-Y. Huang. 2007. Influence of lead acetate on soil microbial biomass and community structure in two different soils with the growth of Chinese cabbage (*Brassica chinensis*). *Chemosphere* 66(7): 1197–1205.

Liu, Y., J. Ma, H. Yan, Y. Ren, B. Wang, C. Lin, and X. Liu. 2016. Bioaccessibility and health risk assessment of arsenic in soil and indoor dust in rural and urban areas of Hubei province, China. *Ecotoxicology and Environmental Safety* 126: 14–22.

Liu, F., J. Wu, G-G. Ying, Z. Luo, and H. Feng. 2012. Changes in functional diversity of soil microbial community with addition of antibiotics sulfamethoxazole and chlortetracycline. *Applied Microbiology and Biotechnology* 95(6): 1615–1623.

Loganathan, P., M.J. Hedley, N.D. Grace, J. Lee, S.J. Cronin, N.S. Bolan, and J.M. Zanders. 2003. Fertiliser contaminants in New Zealand grazed pasture with special reference to cadmium and fluorine: A review. *Australian Journal of Soil Research* 41(3): 501.

Loganathan, P., P. Vigneswaran, J. Kandasamy, and R. Naidu. 2012. Cadmium sorption and desorption in soils: A review. *Critical Reviews in Environmental Science and Technology* 42(5): 489–533.

Lombi, E., F.J. Zhao, S.J. Dunham, and S.P. McGrath. 2001. Phytoremediation of heavy metal-contaminated so natural hyperaccumulation versus chemically enhanced phytoextraction. *Journal of Environmental Quality* 30: 1919–1926.

Lorenzi, D., J.A. Entwistle, M. Cave, and J.R. Dean. 2011. Determination of polycyclic aromatic hydrocarbons in urban street dust: Implications for human health. *Chemosphere* 83(7): 970–977.

Lourenco, R.W., P.M.B. Landim, A.H. Rosa, J.A.F. Roveda, A.C.G. Martins, and L.F. Fraceto. 2010. Mapping soil pollution by spatial analysis and fuzzy classification (Report). *Environmental Earth Sciences* 60(3): 495.

Lu, L-T., I-C. Chang, T-Y. Hsiao, Y-H. Yu, and H-W. Ma. 2007. Identification of pollution source of cadmium in soil: Application of material flow analysis and a case study in Taiwan. *Environmental Science and Pollution Research* 14 (1): 49–59.

Lu, K., X. Yang, J. Shen, B. Robinson, H. Huang, D. Liu, N. Bolan et al. 2014. Effect of bamboo and rice straw biochars on the bioavailability of Cd, Cu, Pb and Zn to *Sedum plumbizincicola*. *Agriculture Ecosystems & Environment* 191: 124–132.

Mahler, B.J., P.C. Van Metre, and E. Callender. 2006. Trends in metals in urban and reference lake sediments across the United States, 1970 to 2001. *Environmental Toxicology and Chemistry* 25(7): 1698–1709.

Makoi, J.H.J.R. and P.A. Ndakidemi. 2008. Selected soil enzymes: Examples of their potential roles in the ecosystem. *African Journal of Biotechnology* 7(3): 181–191.

Mandal, B.K. and K.T. Suzuki. 2002. Arsenic round the world: A review. *Talanta* 58(1): 201–235.

Mao, X., F.X. Han, X. Shao, K. Guo, J. McComb, Z. Arslan, and Z. Zhang. 2016. Electro-kinetic remediation coupled with phytoremediation to remove lead, arsenic and cesium from contaminated paddy soil. *Ecotoxicology and Environmental Safety* 125: 16–24.

McBride, M.B. 2013. Arsenic and lead uptake by vegetable crops grown on historically contaminated orchard soils. *Applied and Environmental Soil Science* 2013: Article ID 283472.

McBride, M.B., H.A. Shayler, H.M. Spliethoff, R.G. Mitchell, L. Marquez-Bravo, G.S. Ferenz, J. Russell-Anelli et al. 2014. Concentrations of lead, cadmium and barium in urban garden-grown vegetables: The impact of soil variables. *Environmental Pollution* 194: 254–261.

McClintock, N. 2015. A critical physical geography of urban soil contamination. *Geoforum* 65: 69–85.

McGrath, S.P. and F-J. Zhao. 2003. Phytoextraction of metals and metalloids from contaminated soils. *Current Opinion in Biotechnology* 14(3): 277–282.

McLaren, R.G., L.M. Clucas, and M.D. Taylor. 2005. Leaching of macronutrients and metals from undisturbed soils treated with metal-spiked sewage sludge. 3. Distribution of residual metals. *Australian Journal of Soil Research* 43(2): 159–170.

Meharg, A.A. and M.M. Rahman. 2003. Arsenic contamination of Bangladesh paddy field soils: implications for rice contribution to arsenic consumption. *Environmental Science and Technology* 37 (2): 229–234.

Mendez, M.O. and R.M. Maier. 2008. Phytostabilization of mine tailings in arid and semiarid environments—An emerging remediation technology. *Environmental Health Perspectives* 116(3): 278–283.6p.

Mielke, H.W., J.C. Anderson, K.J. Berry, P.W. Mielke, R.L. Chaney, and M. Leech. 1983. Lead concentrations in inner-city soils as a factor in the child lead problem. *American Journal of Public Health* 73(12): 1366.

Miller, F.S., K.L. Kilminster, B. Degens, and G.W. Firns. 2010. Relationship between metals leached and soil type from potential acid sulphate soils under acidic and neutral conditions in Western Australia. *Water Air and Soil Pollution* 205 (1–4):133–147.

National Institute of Neurological Disorders and Stroke. 2016. Peripheral Neuropathy Fact Sheet. National Institute of Health. http://www.ninds.nih.gov/disorders/peripheralneuropathy/detail_peripheralneuropathy.htm (accessed 22 March 2016).

Obrycki, J.F., N.T. Basta, K. Scheckel, B.N. Stevens, and K.K. Minca. 2016. Phosphorus amendment efficacy for in situ remediation of soil lead depends on the bioaccessible method. *Journal of Environmental Quality* 45(1): 37–44.

Oliveira, A. and M.E. Pampulha. 2006. Effects of long-term heavy metal contamination on soil microbial characteristics. *Journal of Bioscience and Bioengineering* 102(3): 157–161.

Oliver, D.P., M.J. McLaughlin, R. Naidu, L.H. Smith, E.J. Maynard, and I.C. Calder. 1999. Measuring Pb bioavailability from household dusts using an in vitro model. *Environmental Science & Technology* 33(24): 4434–4439.

Oliver, D.P., K.G. Tiller, A.M. Alston, G.D. Cozens, and R.H. Merry. 1998. Effects of soil pH and applied cadmium on cadmium concentration in wheat grain. *Australian Journal of Soil Research* 36(4): 571.

Ounnas, F., S. Jurjanz, M.A. Dziurla, Y. Guiavarc'h, C. Feidt, and G. Rychen. 2009. Relative bioavailability of soil-bound polycyclic aromatic hydrocarbons in goats. *Chemosphere* 77(1): 115–122.

Patterson, B.M., M. Shackleton, A.J. Furness, J. Pearce, C. Descourvieres, K.L. Linge, F. Busetti et al. 2010. Fate of nine recycled water trace organic contaminants and metal (loid)s during managed aquifer recharge into a anaerobic aquifer: Column studies. *Water Research* 44(5): 1471–1481.

Perez-de-Mora, A., P. Burgos, E. Madejón, F. Cabrera, P. Jaeckel, and M. Schloter. 2006. Microbial community structure and function in a soil contaminated by heavy metals: Effects of plant growth and different amendments. *Soil Biology & Biochemistry* 38(2): 327–341.

Peters, R.E., M. Wickstrom, and S.D. Siciliano. 2015. The bioavailability of polycyclic aromatic hydrocarbons from different dose media after single and sub-chronic exposure in juvenile swine. *Science of the Total Environment* 506: 308–314.

Pierzynski, G.M., J.L. Heitman, P.A. Kulakow, G.J. Kluitenverg, and J. Carlson. 2004. Revegetation of waste fly ash landfills in a semiarid environment. *Rangeland Ecology & Management* 57(3): 312–319.

Pierzynski, G.M., T.J. Sims, and G.F. Vance. 2005. *Soils and Environmental Quality.* 3rd ed. Boca Raton, FL: CRC Press.

Pillai, T.R., W. Yan, H.A. Agrama, W.D. James, A.M.H. Ibrahim, A.M. McClung, T.J.Gentry et al. 2010. Total grain-arsenic and arsenic-species concentrations in diverse rice cultivars under flooded conditions. *Crop Science* 50(5): 2065–2075.

Plachá, D., H. Raclavská, D. Matýsek, and M. Rümmeli. 2009. The polycyclic aromatic hydrocarbon concentrations in soils in the Region of Valasske Mezirici, the Czech Republic. *Geochemical Transactions* 10(1): 1–21.

Qiu, Q., Y. Wang, Z. Yang, J. Xin, J. Yuan, J. Wang, and G. Xin. 2011. Responses of different Chinese flowering cabbage (*Brassica parachinensis* L.) cultivars to cadmium and lead exposure: Screening for Cd+Pb pollution-safe cultivars. *CLEAN Soil, Air, Water* 39(11): 925–932.

Ramesh, A., S.A. Walker, D.B. Hood, M.D. Guillén, K. Schneider, and E.H. Weyand. 2004. Bioavailability and risk assessment of orally ingested polycyclic aromatic hydrocarbons. *International Journal of Toxicology* 23 (5): 301–333.

Reeves, P.G. and R.L. Chaney. 2008. Bioavailability as an issue in risk assessment and management of food cadmium: A review. *Science of the Total Environment* 398 (1–3):13–19.

Reilley, K.A., M.K. Banks, and A.P. Schwab. 1996. Dissipation of polycyclic aromatic hydrocarbons in the rhizosphere. *Journal of Environmental Quality* 25(2): 212–219.

Rengarajan, T., P. Rajendran, N. Nandakumar, B. Lokeshkumar, P. Rajendran, and I. Nishigaki. 2015. Exposure to polycyclic aromatic hydrocarbons with special focus on cancer. *Asian Pacific Journal of Tropical Biomedicine* 5(3): 182–189.

Rice, K.C. 1999. Trace-element concentrations in streambed sediment across the conterminous United States. *Environmental Science & Technology* 33(15): 2499–2504.

Rogus, S. and C. Dimitri. 2015. Agriculture in urban and peri-urban areas in the United States: Highlights from the Census of Agriculture. *Renewable Agriculture and Food Systems* 30(1): 64–78.

Rooney, C.P., R.G. McLaren, and L.M. Condron. 2007. Control of lead solubility in soil contaminated with lead shot: Effect of soil pH. *Environmental Pollution* 149(2): 149–157.

Royal Horticultural Society. 2016. Waterlogging and Flooding. Web Page URL:https://www.rhs.org.uk/advice/profile?PID=235 (Accessed: Feb 2016)

Sanglard, L.M.V.P., K.C. Detmann, S.C.V. Martins, R.A. Teixeira, L.F. Pereira, M.L. Sanglard, A.R. Fernie et al. 2016. The role of silicon in metabolic acclimation of rice plants challenged with arsenic. *Environmental and Experimental Botany* 123: 22–36.

Satarug, S., J.R. Baker, S. Urbenjapol, M. Haswell-Elkins, P.E.B. Reilly, D.J. Williams, and M.R. Moore. 2003. A global perspective on cadmium pollution and toxicity in non-occupationally exposed population. *Toxicology Letters* 137 (1–2):65–83.

Sauve, S. and W. Hendershot. 1997. Solubility control of Cu, Zn, Cd and Pb in contaminated soils. *European Journal of Soil Science* 48(2): 337–346.

Schaider, L.A., D.B. Senn, E.R. Estes, D.J. Brabander, and J.P. Shine. 2014. Sources and fates of heavy metals in a mining-impacted stream: Temporal variability and the role of iron oxides. *Science of the Total Environment* 490: 456–466.

Scheckel, K.G., R.L. Chaney, N.T. Basta, and J.A. Ryan. 2009. Advances in assessing bioavailability of metal (loid)s in contaminated soils. *Advances in Agronomy* 104 104: 1–52.

Scheckel, K.G., G.L. Diamond, M.F. Burgess, J.M. Klotzbach, M. Maddaloni, B.W. Miller, C.R. Patridge et al. 2013. Amending soils with phosphate as means to mitigate soil lead hazard: A critical review of the state of the science. *Journal of Toxicology and Environmental Health. Part B Critical Reviews* 16(6): 337–380.

Schlebaum, W. 1998. *Organic Contaminants in Soil: Desorption Kinetics and Microbial Degradation.* Wageningen, Netherlands: Landbouwuniversiteit Wageningen (Wageningen Agricultural University).

Semple, K.T., K.J. Doick, L.Y. Wick, and H. Harms. 2007. Microbial interactions with organic contaminants in soil: Definitions, processes and measurement. *Environmental Pollution* 150(1): 166–176.

Signes-Pastor, A.J., S.W. Al-Rmalli, R.O. Jenkins, A.A. Carbonell-Barrachina, and P.I. Haris. 2012. Arsenic bioaccessibility in cooked rice as affected by arsenic in cooking water. *Journal of Food Science* 77(11): T201–T206.

Simmler, M., L. Ciadamidaro, R. Schulin, P. Madejón, R. Reiser, L. Clucas, P. Weber et al. 2013. Lignite reduces the solubility and plant uptake of cadmium in pasturelands. *Environmental Science & Technology* 47(9): 4497.

Sizmur, T., B. Palumbo-Roe, and M.E. Hodson. 2011. Impact of earthworms on trace element solubility in contaminated mine soils amended with green waste compost. *Environmental Pollution* 159(7): 1852–1860.

Smedley, P.L. and D.G. Kinniburgh. 2002. A review of the source, behaviour and distribution of arsenic in natural waters. *Applied Geochemistry* 17(5): 517–568.

Smith, S.R. 2009. A critical review of the bioavailability and impacts of heavy metals in municipal solid waste composts compared to sewage sludge. *Environment International* 35(1): 142–156.

Smith, E., I.M. Kempson, A.L. Juhasz, J. Weber, A. Rofe, D. Gancarz, R. Naidu et al. 2011. In vivo–in vitro and XANES spectroscopy assessments of lead bioavailability in contaminated periurban soils. *Environmental Science and Technology* 45(14): 6145–6152.

Sneath, H.E., T.R. Hutchings, and F.A.A.M. de Leij. 2013. Assessment of biochar and iron filing amendments for the remediation of a metal, arsenic and phenanthrene co-contaminated spoil. *Environmental Pollution* 178: 361–366.

Strawn, D.G. and D.L. Sparks. 1999. The use of XAFS to distinguish between inner- and outer-sphere lead adsorption complexes on montmorillonite. *Journal of Colloid and Interface Science* 216(2): 257–269.

Szakova, J., P. Tlustos, W. Goessler, D. Pavlikova, and E. Schmeisser. 2007. Response of pepper plants (*Capsicum annum* L.) on soil amendment by inorganic and organic compounds of arsenic. *Archives of Environmental Contamination and Toxicology* 52(1): 38–46.

Taiwo, A.M., A.M. Gbadebo, J.A. Oyedepo, Z.O. Ojekunle, O.M. Alo, A.A. Oyeniran, O.J. Onalaja et al. 2016. Bioremediation of industrially contaminated soil using compost and plant technology. *Journal of Hazardous Materials* 304: 166–172.

Taylor, J.R. and S.T. Lovell. 2014. Urban home food gardens in the global north: Research traditions and future directions. *Agriculture and Human Values* 31(2): 285–305.

Tejada, M., C. Garcia, J.L. Gonzalez, and M.T. Hernandez. 2006. Use of organic amendment as a strategy for saline soil remediation: Influence on the physical, chemical and biological properties of soil. *Soil Biology & Biochemistry* 38(6): 1413–1421.

Tu, C-L, T-B. He, C-Q. Liu, and X-H. Lu. 2013. Effects of land use and parent materials on trace elements accumulation in topsoil. *Journal of Environmental Quality* 42(1): 103–110.

U.S. Department of Health and Human Services. 2012. Toxicological Profile for Cadmium. PDF Document.

U.S. Environmental Protection Agency. 2004. Chemicals Evaluated for Carcinogenic Potential. PDF Document.

U.S. Environmental Protection Agency. 2011. Reusing Potentially Contaminated Landscapes: Growing Gardens in Urban Soils. PDF Document. https://www.atsdr.cdc.gov/toxprofiles/tp5.pdf (Accessed: Jan 2016)

U.S. Environmental Protection Agency. 2014. Technical Review Workgroup Recommendations Regarding Gardening and Reducing Exposure to Lead-Contaminated Soils. PDF Documenthttps://clu-in.org-download/misc/urban_gardening_fact_sheet.pdf Accessed: Jan 2016.

Vangronsveld, J., R. Herzig, N. Weyens, J. Boulet, K. Adriaensen, A. Ruttens, T. Thewys et al. 2009. Phytoremediation of contaminated soils and groundwater: Lessons from the field. *Environmental Science and Pollution Research International* 16(7): 765–794.

Wang, S. and C.N. Mulligan. 2006. Occurrence of arsenic contamination in Canada: Sources, behavior and distribution. *Science of the Total Environment* 366 (2–3):701–721.

Wang, Y., J. Shi, H. Wang, Q. Lin, X. Chen, and Y. Chen. 2007. The influence of soil heavy metals pollution on soil microbial biomass, enzyme activity, and community composition near a copper smelter. *Ecotoxicology and Environmental Safety* 67(1): 75–81.

Wang, Y., Z. Tian, H. Zhu, Z. Cheng, M. Kang, C. Luo, J. Li et al. 2012. Polycyclic aromatic hydrocarbons (PAHs) in soils and vegetation near an e-waste recycling site in South China: Concentration, distribution, source, and risk assessment. *Science of the Total Environment* 439: 187–193.

Wang, F., J. Yao, Y. Si, H. Chen, M. Russel, K. Chen, Y. Qian et al. 2010. Short-time effect of heavy metals upon microbial community activity. *Journal of Hazardous Materials* 173 (1–3):510–516.

Wase, D.A.J. 1997. *Biosorbents for Metal Ions*, In: General bacterial sorption processes, edited by M.M.Urrutia, pp. 67–86, and

Fungi as biosorbents, edited by A.Kapoor and T.Viraraghavan, pp. 87 - 114. London: CRC Press.

Weber, R. and B. Hrynczuk. 2000. Effect of leaf and soil contaminations on heavy metals content in spring wheat crops. *Nukleonika* 45(2): 137–140.

Wilcke, W. 2000. Polycyclic aromatic hydrocarbons (PAHs) in soil—A review. *Journal of Plant Nutrition and Soil Science* 163:229–248.

Woodhead, R.J., R.J. Law, and P. Matthiessen. 1999. Polycyclic aromatic hydrocarbons in surface sediments around England and Wales, and their possible biological significance. *Marine Pollution Bulletin* 38(9): 773–790.

World Health Organization. 2010. Exposure to Cadmium: A major public health concern. PDF Document.http://www.who.int/ipcs/features/cadmium.pdf?ua=1 (Accessed Jan. 2016)

World Health Organization. 2016. Lead Poisoning and Health. http://www.who.int/mediacentre/factsheets/fs379/en/ (accessed 6 January 2016).

Wortman, S.E. and S.T. Lovell. 2013. Environmental challenges threatening the growth of urban agriculture in the United States. *Journal of Environmental Quality* 42(5): 1283–1294.

Wu, F., Z. Fu, B. Liu, C. Mo, B. Chen, W. Corns, and H. Liao. 2011. Health risk associated with dietary co-exposure to high levels of antimony and arsenic in the world's largest antimony mine area. *Science of the Total Environment* 409(18): 3344–3351.

Xia, Q., C. Peng, D. Lamb, M. Mallavarapu, R. Naidu, and J.C. Ng. 2016. Bioaccessibility of arsenic and cadmium assessed for in vitro bioaccessibility in spiked soils and their interaction during the Unified BARGE Method (UBM) extraction. *Chemosphere* 147: 444–450.

Xiao H., H. JaiFang, and G. JiuLan. 2012. Analysis of arsenic and cadmium concentrations in vegetables and soils in Guiyang suburbs. *Guizhou Agricultural Sciences* 2012 (9):229–231.

Xiao, Q., M.H. Wong, L. Huang, and Z. Ye. 2015. Effects of cultivars and water management on cadmium accumulation in water spinach (Ipomoea aquatica Forsk.). *Plant and Soil* 391 (1–2):33–49.

Yager, J.W., T. Greene, and R.A. Schoof. 2015. Arsenic relative bioavailability from diet and airborne exposures: Implications for risk assessment. *Science of the Total Environment* 536: 368–381.

Yang, L. 2010. A review of heavy metal contaminations in urban soils, urban road dusts and agricultural soils from China. *Microchemical Journal* 94 (2): 99–107.

Yang, L., R.J. Donahoe, and J.C. Redwine. 2007. In situ chemical fixation of arsenic-contaminated soils: An experimental study. *Science of the Total Environment* 387 (1–3):28–41.

Yang, X., J. Liu, K. McGrouther, H. Huang, K. Lu, X. Guo, L. He et al. 2016. Effect of biochar on the extractability of heavy metals (Cd, Cu, Pb, and Zn) and enzyme activity in soil. *Environmental Science and Pollution Research* 23(2): 974–984.

Yeates, G.W., V.A. Orchard, T.W. Speir, J.L. Hunt, and M.C.C. Hermans. 1994. Impact of pasture contamination by copper, chromium, arsenic timber preservative on soil biological-activity. *Biology and Fertility of Soils* 18(3): 200–208.

Yoon, Y., S. Kim, Y. Chae, S-W. Jeong, and Y-J An. 2016. Evaluation of bioavailable arsenic and remediation performance using a whole-cell bioreporter. *Science of the Total Environment* 547: 125–131.

Yunker, M.B., R.W. Macdonald, R. Vingarzan, R.H. Mitchell, D. Goyette, and S. Sylvestre. 2002. PAHs in the Fraser River basin: A critical appraisal of PAH ratios as indicators of PAH source and composition. *Organic Geochemistry* 33(4): 489–515.

Zeng, G., H. Wu, J. Liang, S. Guo, L. Huang, P. Xu, Y. Liu et al. 2015. Efficiency of biochar and compost (or composting) combined amendments for reducing Cd, Cu, Zn and Pb bioavailability, mobility and ecological risk in wetland soil. *RSC Advances* 5(44): 34541–34548.

Zhang, Y., P. Jiang, Y. Li, J. Wu, K. Xu, S. Hill, and H. Wang. 2013. Chemistry of decomposing mulching materials and the effect on soil carbon dynamics under a *Phyllostachys praecox* bamboo stand. *Journal of Soils and Sediments* 13(1): 24–33.

Zhang, P.C. and J.A. Ryan. 1999. Transformation of Pb(II) from cerussite to chloropyromorphite in the presence of hydroxyapatite under varying conditions of pH. *Environmental Science & Technology* 33(4): 625–630.

Zhang, J., J-M. Séquaris, H-D. Narres, H. Vereecken, and E. Klumpp. 2010. Effect of organic carbon and mineral surface on the pyrene sorption and distribution in Yangtze River sediments. *Chemosphere* 80(11): 1321–1327.

Zhao, X., D. Dong, X. Hua, and S. Dong. 2009. Investigation of the transport and fate of Pb, Cd, Cr(VI) and as(V) in soil zones derived from moderately contaminated farmland in Northeast China. *Journal of Hazardous Materials* 170 (2–3):570–577.

Zhao, F., S. Mcgrath, and A.A. Meharg. 2010. Arsenic as a food chain contaminant: Mechanisms of plant uptake and metabolism and mitigation strategies. *Annual Review of Plant Biology*. 61: 535–559.

Zhao, D., Y. Xu, L. Wang, and X. Liang. 2016. Effects of sepiolite on stabilization remediation of heavy metal-contaminated soil and its ecological evaluation. *Frontiers of Environmental Science & Engineering* 10(1): 85–92.

Zia, M.H., E.E. Codling, K.G. Scheckel, and R.L. Chaney. 2011. In vitro and in vivo approaches for the measurement of oral bioavailability of lead (Pb) in contaminated soils: A review. *Environmental Pollution* 159(10): 2320–2327.

Zohair, A., A-B. Salim, A.A. Soyibo, and A.J. Beck. 2006. Residues of polycyclic aromatic hydrocarbons (PAHs), polychlorinated biphenyls (PCBs) and organochlorine pesticides in organically-farmed vegetables. *Chemosphere* 63(4): 541–553.

9 Optimizing the Hydrologic Properties of Urban Soils

Mary G. Lusk and Gurpal S. Toor

CONTENTS

9.1 INTRODUCTION

Soil–water relationships are important in all landscapes because the movement of water into and through a soil affects water and nutrient availability for plant roots, the incidence of soil erosion, and the degree to which surface runoff occurs (Fried 2012; Morgan 2009; Huffman et al. 2012). In natural landscapes such as grasslands and forests, biological activity builds the soil organic matter (SOM) content, which, in turn, promotes aggregate and macropore formation leading to high surface infiltration rates. The plant detritus on the soil surface also dissipates raindrop energy, thus moderating the destructive potential of soil erosion. In these vegetated landscapes, there are sufficient soil macropores and soil porosity to support water storage and movement through the soil. On the other hand, urban landscapes are a much different story. For example, the hydrologic properties of urban soils are often characterized by loss of water retention, restricted water drainage, and increased susceptibility to erosion and surface runoff (Craul 1994; Pavao-Zuckerman 2008) (Figures 9.1 and 9.2).

Development of land from vegetated to urban landscapes causes loss of upper soil layers and stripping of natural vegetation, and requires use of fill material from construction sites—all these activities destroy soil structure and reduce pore space. The poor soil structure and low porosity that often characterize urban soils lead to compaction, and thus increased bulk density, with reduced water infiltration and increased surface runoff (De Kimpe and Morel 2000;

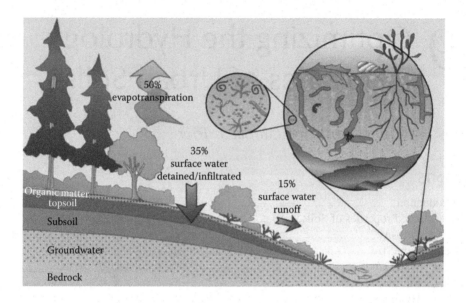

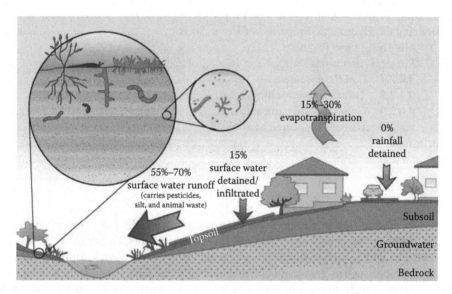

FIGURE 9.1 Natural soil (top) and urban soil (bottom). Urbanization reduces soil biological activity and compacts topsoil, leading to reduced soil aggregation and increased soil bulk density. These in turn lead to increased runoff and reduced infiltration in the urban soil. (Reprinted from King County, Washington Solid Waste Division. With permission.)

Craul 1994; Huffman et al. 2012). The results of this are seen in poorly drained areas of the urban landscape where water pools at the soil surface and in the compromised vigor of urban trees and shrubs experiencing water stress from poor water infiltration and low water-holding capacity in the root zone (Figure 9.3). Furthermore, once water movement into the urban sub-surface is restricted, then the local hydrologic regime shifts from one in which subsurface flow dominates to one in which overland flow dominates. This means less groundwater recharge and more surface runoff, which leads to water quality impairment issues in urban landscapes (see for example, Lusk and Toor, 2016 a,b; Yang and Toor, 2016,2017).

FIGURE 9.2 Runoff from a compacted urban soil. (Courtesy of nih.gov.)

Due to the potential problematic hydrologic properties of urban soils, they must be evaluated and often require some form of reconditioning or amending before urban agriculture (UA), residential lawns, or green infrastructure can be implemented. The objectives of this chapter are to (1) provide a brief review of the hydrologic properties of soils and how those properties are affected by urbanization and (2) discuss actions that land managers can use to optimize the hydrologic properties of problematic urban soils.

FIGURE 9.3 These tree roots could not penetrate the compacted urban soil and were forced to spread out laterally instead of downward. (Reprinted from Bassuk et al., *Using CU-Structural Soil in the Urban Environment,* Urban Horticultural Institute, Cornell University, NewYork, 2005. With permission.)

9.2 HYDROLOGIC PROPERTIES OF SOILS

The basic hydrologic properties of soils are explained below and summarized in Table 9.1.

9.2.1 POROSITY

Porosity is the volume of soil pores that can be filled by water and/or air. Total soil porosity is made up of pores of all sizes and shapes. A typical mineral soil has a porosity value of 40%–60%. Porosity is inversely proportional to bulk density. Compacted soils have higher bulk densities and thus lower porosity than loose, porous soils.

9.2.2 PORE SIZE

Soil pores vary in size. Small pores are called micropores and are 0.08 mm or less in diameter, whereas larger pores greater than 0.08 mm are referred to as macropores (Brady and Weil 1996). The macropores drain freely by gravity and are important for easy movement of water through soil. In contrast, micropores require suction to drain. The presence of SOM encourages aggregation and macropore formation. Soil compaction destroys macropores. For optimum plant vigor, macropores should occupy 20%–25% of soil volume (Bakker and Hidding 1970).

9.2.3 WATER CONTENT

The term "water content" refers to all water held in a soil. It can be expressed gravimetrically as g water/g soil or volumetrically as cm^3 water/cm^3 soil, or as percent of total porosity. When all the pores in a soil are filled with water, the soil water content is said to be at *saturation*. After a soil has been saturated and allowed to drain freely for 24–48 hours, it reaches a condition called *field*

TABLE 9.1

The Basic Hydrologic Properties of Soils

Hydrologic Property	Reported Range[a]	References
Porosity	40%–60% typical; as low as 25%–30% for very compacted soils	Brady and Weil (1996)
Pore size	<2–200 μm	Hillel (2013), Richard et al. (2001), and Luxmoore (1981)
Volumetric water content[a]	12% sandy loam 30% silt loam 35% clay	Topp (1971) and Alexandrou and Earl (1998)
Infiltration rate	20–30 mm/hr sandy loam 5–10 mm/hr clay loam 1–5 mm/hr clay	Bayabil et al. (2013) and Jabro et al. (2016)
Hydraulic conductivity	0.4–1.4 μm/sec	Rienzner and Gandolfi (2014)
Permeability	10^{-1} cm/sec sand 10^{-5} silt Less than 10^{-7} clay	Saxton et al. (1986) and Iversen et al. (2001)

[a]Reported values are at field capacity.

capacity, which is the water content held in a soil after excess water has drained away (Brady and Weil 1996). The matric potential at field capacity ranges from about −0.1 to −0.3 bar.

9.2.4 INFILTRATION RATE

The rate at which water enters the soil surface is called infiltration rate. Infiltration is largely a function of soil texture, with water moving more quickly through the large pores of sandy soils than through the smaller pores of clay soils. Organic-rich soils with stable soil structure promote infiltration, whereas compaction reduces the potential for infiltration. Water unable to infiltrate the soil surface has the potential to become runoff water.

9.2.5 HYDRAULIC CONDUCTIVITY

The hydraulic conductivity of a soil is a measure in length per unit time of how easily water moves through a soil when subjected to a hydraulic gradient. It is directly proportional to the permeability of the soil and the soil texture, with sand and coarse sand having rapid to very rapid (42–141 μm/sec) hydraulic conductivity rates and clay soils having much slower rates (0.4–1.4 μm/sec) (Brady and Weil 1996).

9.2.6 PERMEABILITY

Permeability is a measure of how well a porous media transmits a fluid. Whereas hydraulic conductivity is a property of both the soil and the fluid moving through it, permeability is an intrinsic property of the soil only. Permeability is measured typically in units of cm/sec. Coarse-textured soils have higher permeability than fine-textured soils (Brady and Weil 1996).

9.3 URBANIZATION EFFECTS ON THE HYDROLOGIC PROPERTIES OF SOILS

The urbanization process involves a number of activities that affect the hydrologic properties of urban soils (Craul 1994; Jinling et al. 2005). During development, surface soil may be scraped away and mixed with other soil, stockpiled, and then reapplied or transported to another location.

This modifies and weakens the soil structure and decreases permeability. Loss of natural vegetation in favor of the built environment decreases SOM, which leads to less soil aggregation and smaller soil pores (Craul 1985). In general, there exists a positive relationship between SOM and the maintenance of stable aggregates that promote soil water infiltration (Bronick and Lal 2005; Jastrow et al. 1998). Loss of vegetation may also leave portions of the urban soil bare and susceptible to crust formation, which reduces water infiltration rate (Pagliai, Vignozzi, and Pellegrini 2004). Traffic from urban vehicles and heavy machinery both during and after construction also leads to soil compaction and consequent reductions in porosity, permeability, and hydraulic conductivity. From all of these disturbances, it is not surprising that urban soils behave much differently hydrologically than their undisturbed counterparts. Table 9.2 summarizes the major effects of urbanization on the basic soil hydrologic properties, as well as on the soil physical properties (e.g., bulk density) that have significant influence on soil–water relationships.

Jim and Ng (2000) observed that the mean porosity of an urban soil in Hong Kong was 32.8%, compared to 49.0% for its undeveloped counterpart. The distribution of pore sizes in a soil—and especially the presence of macropores—was shown to be as important as bulk density in influencing infiltration rates in urban soils in two separate studies in Minneapolis, MN and Melbourne, Australia (Ossola, Hahs, and Livesley 2015; Olson et al. 2013). Several other studies have well documented that increase in compaction in urban soils decreases water infiltration rate. Yang and Zhang (2011) observed that 17% of soils around Nanjing, China, had an infiltration rate of less than 5 mm/hr, and that soils with the lowest infiltration rates were also the most compacted. Likewise, Gregory et al. (2006) observed that infiltration rates in compacted urban construction site soils in Florida were 8–188 mm/hr, whereas infiltration rate was 255–262 mm/hr in an adjacent undisturbed pasture and forest soils—a 77%–99% reduction in infiltration due to urbanization. Urban parkland soils in Toronto, Canada, were observed

TABLE 9.2
Physical and Hydrologic Properties That Affect Soil–Water Relationships and Their General Responses to Urbanization

Soil Property	General Response to Urbanization	Comments	References
Bulk density	Increased	Results from incorporation of construction debris, fill material, and compaction	Millward et al. (2011)
Compaction	Increased	Results from heavy machinery, vehicle traffic, and foot traffic	Gregory et al. (2006)
Aggregation	Decreased	Loss of soil structure due to compaction and organic matter reductions	Bronick and Lal (2005) and Jastrow et al. (1998)
Porosity	Decreased	Reduced by loss of organic matter and compaction	Jinling et al. (2005)
Number of macropores	Decreased	Limited by weakening of soil structure due to loss of organic matter and compaction	Craul (1985), Ossola et al. (2015), and Olson et al. (2013)
Saturated hydraulic conductivity	Slower	Limited by weakening of soil structure and compaction of subsurface soils	Craul (1994)
Surface infiltration	Slower	Limited by surface crusting and/or decreases in soil permeability	Pagliai (2004), Yang and Zhang (2011), and Gregory et al. (2006)
Available water content	Decreased	Limited by increased stormwater runoff from low-infiltration, compacted soils	Rawls et al. (2003)

to have bulk densities as high as 1.8 g/cm³ as a result of compaction from foot traffic, with water infiltration consequently retarded to the point of surface pooling (Millward, Paudel, and Briggs 2011). Because SOM has a positive relationship with soil aggregation, porosity, and surface area, its loss due to urbanization has been shown to reduce the plant-available water content of soils (Rawls et al. 2003).

9.4 OPTIMIZING THE HYDROLOGIC PROPERTIES OF URBAN SOILS

Careful planning and development can prevent some of the less than favorable hydrologic properties of urban soils before they become a problem. For example, low-impact development (LID) practices are employed to minimize negative impacts on soil properties during development. In many urban areas, though, the damage has already been done by decades of conventional development, with the consequent impacts that soil compaction, loss of soil structure, and loss of SOM have on site hydrology. Where water infiltration and retention problems already occur in urban soils, actions such as soil restoration or rehabilitation are necessary to improve soil hydrologic properties.

Table 9.3 summarizes typical values of porosity and hydraulic conductivity for a variety of soil textures. These values represent what is usually considered optimal and/or most feasible and are the values that should be worked toward in any soil restoration or rehabilitation process. Following Table 9.3 is a discussion of site activities that can optimize soil hydrologic properties in urban areas. The activities are grouped into two categories: (1) site design and planning and (2) physical alterations.

9.4.1 Site Design and Planning

Optimizing the hydrologic properties of soils in a planned urban development begins with improved site design. One of the best management practices (BMPs) of LID is to design the site for cluster development, which clusters properties on smaller portions of total available land. Cluster development leaves protected open space for vegetation or UA. In these protected spaces, soil remains largely undisturbed, such that changes to porosity, permeability, conductivity, and infiltration are minimized (Gregory et al. 2006). The absence of soil disturbance and the maintenance of SOM on undeveloped portions of the site favor well-structured soils that can intercept, infiltrate, and store stormwater. Hood et al. (2007) observed large differences in runoff volumes from traditional and clustered urban sites. For small storms (rainfall depth <25.4 mm), the traditional sites lost on average

TABLE 9.3
Typical Values of Porosity and Hydraulic Conductivity in Different Soil Textures

Soil Texture	Porosity	Hydraulic Conductivity (cm/sec)
Sand	0.437	3.25×10^{-3}
Loamy sand	0.437	8.3×10^{-4}
Sandy loam	0.453	3×10^{-4}
Loam	0.463	9.4×10^{-5}
Silt loam	0.501	1.8×10^{-4}
Sandy clay loam	0.398	4.2×10^{-5}
Clay loam	0.464	2.8×10^{-5}
Silty clay loam	0.471	2.8×10^{-5}
Sandy clay	0.430	1.7×10^{-5}
Silty clay	0.479	1.4×10^{-5}
Clay	0.475	8.3×10^{-6}

Source: Adapted from Rawls et al., *Journal of Hydraulic Engineering,* 109, 62–70, 1983.
Note: These values represent reasonable targets for optimization goals in urban landscapes.

1.06 mm of runoff, while the LID cluster development lost only 0.10 mm on average as runoff. In the same study, large storms (rainfall depth >25.4 mm) produced on average 10.01 mm runoff compared to an average 1.68 mm of runoff in the LID site. Difference in runoff volume between the two sites was attributed to increased infiltration in the LID site.

Another feature of LID design is incorporation of bioswales and urban forest patches within the urban landscape matrix (Figure 9.4). A vegetated and mulched bioswale in the urban area of Charlotte, NC, reduced peak stormwater flow by at least 96% for storms less than 40 mm depth (Hunt et al. 2008). Trees in the urban landscape intercept rainfall and encourage infiltration through their contribution to SOM (Xiao and McPherson 2016; Stovin, Jorgensen, and Clayden 2008).

FIGURE 9.4   Bioswales, like this in Denver, CO, and other vegetated patches in the urban landscape help infiltrate rainwater and encourage optimal soil hydrologic properties by improving soil organic matter, porosity, and structure. (Courtesy of epa.gov.)

Thus, urban trees play a key role in mitigating stormwater flows by intercepting rainfall in the canopy and by helping set up the conditions that promote stormwater infiltration. Scharenbroch et al. (2016) showed that trees in urban bioswales could be responsible for 46%–72% of water storage in the bioswales, thus reducing runoff volumes. Black oak (*Quercus velutina*) and red maple (*Acer rubrum*) trees planted in a compacted urban clay loam soil increased water infiltration rates by an average of 153%, compared to an unplanted control (Bartens et al. 2008). The increased water infiltration rate in the tree-planted soil was due to root activity of the trees.

Maintaining zones of native vegetation on a site can also impact overall site hydrology. Johnston et al. (2016) tested how soil hydrologic properties responded when urban turfgrass was replaced with strands of native prairie grass in the Midwestern United States. Compared to that of turfgrass urban lawns, the prairie grass lawns at 0–15 cm depth had 25% greater SOM, 15% less resistance to water infiltration, 10% lower bulk density, and 33% greater saturated hydraulic conductivity. Encouraging developers to leave portions of the native vegetation at urban sites and educating urban residents about benefits of turfgrass alternatives are excellent tools for enhancing hydrologic functions in urban landscapes.

Another aspect of site design and planning involves careful construction practices. Weeks and months of passage of heavy equipment over a site have profound influence on soil compaction. This is often followed by repeated daily passage of vehicles and foot traffic, which maintains soil compaction. Thus, future water infiltration and retention on the site can be improved by employing construction practices that reduce soil compaction. Before construction, "no disturbance" and "minimal disturbance" zones should be planned in construction drawings and delineated with flags or fences on the site. These will be areas where equipment passage is prevented or limited. While the site is under construction, a 20-cm layer of wood chip mulch over the paths of light vehicles can help to reduce compaction of the underlying soil; for heavy equipment paths, plywood can be placed over the mulch layer for further protection against compaction (Johnson 1999; Tyner et al. 2011). Lichter et al. (1994) noted that mulch covers on equipment paths reduced compaction but did not eliminate it completely. In their study, the preconstruction soil bulk density increased from 1.42 to 1.59 g/cm^3 when the vehicle paths were unprotected but to only 1.51 g/cm^3 when the paths were covered with wood chip mulch.

9.4.2 Physical Alterations

Increased soil compaction through construction activities and loss of organic matter at urban sites is often a given and likely occurs to some extent no matter how much care is taken during the development phase. Urban land managers and residents often must employ practices to ameliorate soil compaction and improve water storage in soils. Various physical alteration practices to achieve more optimal soil–water dynamics include (1) soil replacement, (2) tillage and subsoiling, (3) compost amendments, and (4) use of engineered media, which are discussed in the following sections.

9.4.2.1 Soil Replacement

Soil replacement is an intense and expensive option for rehabilitating the hydrology of urban landscapes. It involves excavating the compacted, problematic soil at the site and replacing it with more optimal material (Craul 1992). It is most appropriate for small areas, such as sidewalk cutouts, residential lots, and parking lot islands. It is especially beneficial when the problematic soil is also contaminated with urban pollutants (e.g., lead, organic contaminants) and the intended use for the site is an urban garden. Of course, the potential replacement material should be investigated to ensure it is free from contaminants and that it has the desired hydrologic properties (Table 9.2). If a suitable natural material cannot be obtained, engineered soils (structural soils, discussed below) can be used as the replacement choice.

9.4.2.2 Tilling and Subsoiling

Compacted and poorly drained urban soils can be remediated with deep tillage or subsoiling, which uses large machinery to till up to 60–90 cm depth of soil profile (Loper et al. 2010). It is not only expensive but also impractical to till in many existing urban areas because of the buried utility lines. However, this practice can be applied at selected construction sites before installation of utility lines and landscape plants. Olson et al. (2013) studied tilled and untilled urban residential soils in Minnesota, MN, and observed an up to twofold increase in soil infiltration rates after the soils were tilled. They added that results were most positive in newly developed areas as opposed to older developments. They attributed this to disruption of natural water pathways, such as earthworm and tree root channels, in the established developments. Also, more areas in the new sites are accessible to large tillage machinery as utility lines and trees have not yet been established in new developments.

9.4.2.3   Compost Amendments

Soil amendments with compost are some of the most applied and most studied solutions to urban soil problems (Cogger 2005; Pitt et al. 1999; Pavao-Zuckerman 2008). Compared to soil replacement or tillage, adding compost or mulch to soil is a less expensive and less intensive management choice. It is also suitable for areas with established trees or buried utility lines. At new construction sites, compost additions can accompany tillage for positive hydromodification of the site (Figure 9.5). For example, in their study of the effects of tilling compacted soils, Olson and Gulliver (2013) observed that tillage alone improved soil infiltration up to twofold but that tillage plus compost application was even more effective, increasing infiltration up to more than fivefold as compared to the control soil. The beneficial effects of compost are attributed to simply diluting highly compacted soils with less dense organic materials, which increases porosity and decreases bulk density (Krull et al. 2004). Cogger (2005) gave a review of important works that have studied compost as a treatment for compacted urban soils, and a summary of those studies is presented in Table 9.4. All the studies showed positive effects on at least one of the following urban soil properties: bulk density, porosity,

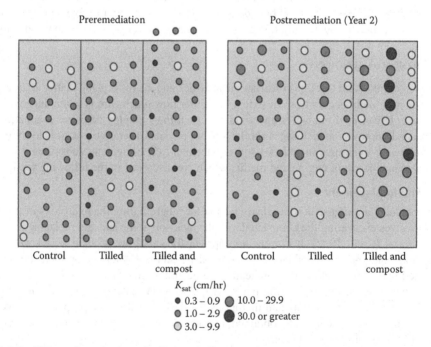

FIGURE 9.5 Tillage alone (preremediation) and tillage with compost (postremediation, year 2) can help increase the hydraulic conductivity (K_{sat}) of a compacted urban soil, which would allow more water to infiltrate the soil and reduce surface runoff. (With permission from John Gulliver, also see Olson et al., *Journal of Environmental Management*, 117:85–95, 2013.)

TABLE 9.4
Summary of Studies That Report on the Effects of Compost on Soil–Water Dynamics

Study	Compost Material	Application Notes	Organic Matter	Bulk Density	Infiltration	Porosity	Aggregation
Cox et al. (2001)	Manure coal ash	One application at 110 Mg/ha; silt loam soil	Positive	Positive	Positive	–[a]	Positive
Tester (1990)	Biosolids	Three treatments at 60, 120, and 240 Mg/ha	Positive	Positive	–	–	–
Mamo et al. (2000)	Municipal solid waste	One application at 270 Mg/ha	–	Positive	–	–	–
Pagliai et al. (2004)	Municipal solid waste	5-yearly treatments of 24 Mg/ha	Positive	–	–	–	Positive
Giusquiani et al. (1995)	Biosolids-municipal solid waste	2-yearly treatments of 75 and 225 Mg/ha	–	–	–	Positive	Positive
Oldfield et al. (2014)	Biosolids plus wood	Shallow tillage followed by 2.5 m³/100 m²	Positive	Positive	Positive	Positive	–

Source: Adapted from Cogger, *Compost Science & Utilization*, 13, 243–251, 2005.
[a] Not reported.

infiltration, aggregate stability, and available water content. Materials that can be used for compost amendments include municipal biosolids, yard wastes, and manures. These materials should be well composted and have C:N ratios at 20:1 or less, otherwise microbes will incorporate available N into their biomass, rendering it unavailable for plant growth (Cogger 2005). Most studies have found the optimum amount of compost addition to be 2–5 cm, incorporated into the top 15–20 cm of soil (Craul 1992; Cogger 2005). Craul (1999) added that compost addition rates above 50% of the soil volume actually had detrimental effects due to the settling and waterlogging of urban soils.

9.4.2.4 Use of Engineered Media

One unfortunate irony of the urban landscape is that while soil compaction is detrimental to the hydrology of a site, it may be useful in terms of the structural and load-bearing needs of urban development (Grabosky and Bassuk 1995). Hence, sustainable urban development from a water perspective can be at odds with the structural requirements for foundational soils. Urban trees are especially sensitive to soil compaction stress (Rahman et al. 2011; Smiley et al. 2006; Buhler, Kristoffersen, and Larsen 2007). One solution to this in recent decades has been the development of engineered materials, or structural soils that can be compacted to meet the load-bearing require-ments of urban environments while also being penetrable to urban tree roots (Figure 9.6). The advantage of this from a hydrological perspective is that urban tree roots promote stormwater infil-tration, and healthy urban trees play a significant role in intercepting rainfall before it becomes runoff (Bartens et al. 2008). Structural soils work because they are composed of a matrix of angular stones that distribute the load through stone-to-stone contact and have gaps between them. The gaps are filled with high-quality mineral soil, usually clay loam, with high water-holding capacity (Bassuk, Grabosky, and Trowbridge 2005).

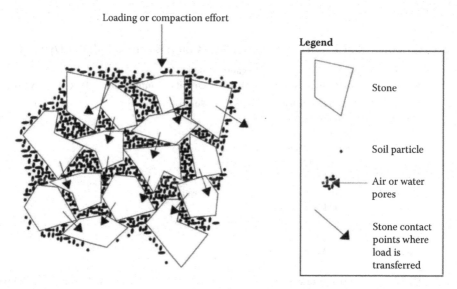

FIGURE 9.6 The composition of structural soil, which contains soil particles, soil, and air. (Reprinted from Bassuk et al., *Using CU-Structural Soil in the Urban Environment,* Urban Horticultural Institute, Cornell University, New York, 2005. With permission.)

Comparison of macropore porosity among native silt loam soil and two different structural soils (Carolina stalite and CU-Soil™) showed that macropore porosity was only 2.2% in the native soil as compared to 31%–39% in both structural soils (Haffner 2008). This resulted in nearly 60 times greater infiltration rate in the structural soils than the native silt loam. However, the enhanced infiltration rates of structural soils may be problematic if the soils must also function as filters for pollutants in the urban stormwater. A number of studies have shown that structural soils may transmit water too rapidly for adequate attenuation of various pollutants (Hinman 2009; Hunt and Lord 2006; Xiao 2008).

9.5 CONCLUSIONS

One of the challenges of sustainable urban development is to reduce the water quantity and quality impacts of stormwater runoff. Because the urbanization process often removes vegetation and leads to compacted soils, urban soils have numerous problematic hydrologic properties. Their typically increased bulk densities lead to reduced porosity, while reductions in organic matter reduce porosity and weaken soil structure—all of which lead to retarded infiltration into urban soils and the problems associated with urban runoff. LID practices, such as cluster development, maintenance of urban forests, and "no disturbance" zones in the landscape can be a first line of protection against urban activities that negatively impact soil–water dynamics. For existing sites that already have problems with water infiltration and/or storage, soil rehabilitation practices can be put into place. These practices include soil replacement, tillage, compost amendments, and the use of engineered media. All of these practices positively influence soil hydrology but vary in their suitability for specific sites. While soil replacement is very intensive and most practical for small areas only, tillage and compost can be carried out at larger scales, and are especially useful when implemented just after construction but before installation of landscape plants.

REFERENCES

Alexandrou, A, and R Earl. 1998. The relationship among the pre-compaction stress, volumetric water content and initial dry bulk density of soil. *Journal of Agricultural Engineering Research* 71 (1): 75–80.

Bakker, JW, and AP Hidding. 1970. The influence of soil structure and air content on gas diffusion in soils. *Netherlands Journal of Agricultural Science* 18:37–48.

Bartens, J, SD Day, JR Harris, JE Dove, and TM Wynn. 2008. Can urban tree roots improve infiltration through compacted subsoils for stormwater management? *Journal of Environmental Quality* 37 (6): 2048–2057.

Bassuk, N, J Grabosky, and P Trowbridge. 2005. *Using CU-Structural Soil in the Urban Environment.* New York, NY: Urban Horticultural Institute, Cornell University.

Bayabil, HK, JC Lehmann, B Yitaferu, C Stoof, and TS Steenhuis. 2013. Hydraulic properties of clay soils as affected by biochar and charcoal amendments. In Wolde Mekuria (ed), Rainwater management for resilient livelihoods in Ethiopia: Proceedings of the Nile Basin Development Challenge science meeting, Addis Ababa, 9–10 July 2013. NBDC Technical Report 5. Nairobi, Kenya: International Livestock Research Institute.

Brady, NC, and RR Weil. 1996. *The Nature and Properties of Soils.* Upper Saddle River, NJ: Prentice-Hall.

Bronick, CJ, and R Lal. 2005. Soil structure and management: A review. *Geoderma* 124 (1): 3–22.

Buhler, O, P Kristoffersen, and SU Larsen. 2007. Growth of street trees in Copenhagen with emphasis on the effect of different establishment concepts. *Arboriculture and Urban Forestry* 33 (5):330.

Cogger, CG 2005. Potential compost benefits for restoration of soils disturbed by urban development. *Compost Science & Utilization* 13 (4): 243–251.

Cox, D, D Bezdicek, and M Fauci. 2001. Effects of compost, coal ash, and straw amendments on restoring the quality of eroded Palouse soil. *Biology and Fertility of Soils* 33 (5): 365–372.

Craul, PJ. 1985. A description of urban soils and their desired characteristics. *Journal of Arboriculture* 11 (11): 330–339.

Craul, PJ. 1992. *Urban Soil in Landscape Design.* Hoboken, NJ: John Wiley & Sons.

Craul, PJ. 1994. The nature of urban soils: Their problems and future. *Arboricultural Journal* 18 (3): 275–287.

Craul, PJ 1999. *Urban Soils: Applications and Practices.* Hoboken, NJ: John Wiley & Sons.

De Kimpe, CR, and J-L Morel. 2000. Urban soil management: A growing concern. *Soil Science* 165 (1): 31–40.

Fried, M. 2012. *The Soil-Plant System: In Relation to Inorganic Nutrition.* Amsterdam: Burlington Elsevier Science.

Giusquiani, PL, M Pagliai, G Gigliotti, D Businelli, and A Benetti. 1995. Urban waste compost: Effects on physical, chemical, and biochemical soil properties. *Journal of Environmental Quality* 24 (1): 175–182.

Grabosky, J, and N Bassuk. 1995. A new urban tree soil to safely increase rooting volumes under sidewalks. *Journal of Arboriculture* 21:187–187.

Gregory, JH, MD Dukes, PH Jones, and GL Miller. 2006. Effect of urban soil compaction on infiltration rate. *Journal of Soil and Water Conservation* 61 (3): 117–124.

Haffner, EC. 2008. *Porous Asphalt and Turf: Exploring New Applications through Hydrological Characterization of CU Structural Soil® and Carolina Stalite Structural Soil.* Ithaca, NY: Cornell University.

Hillel, D. 2013. *Introduction to Soil Physics.* Cambridge, MA: Academic Press.

Hinman, C. 2009. *Bioretention Soil Mix Review and Recommendations for Western Washington.* Washington State University. Available at: http://www.ecy.wa.gov/programs/wq/stormwater/bsmresultsguidelines.pdf

Hood, MJ, JC Clausen, and GS Warner. 2007. *Comparison of Stormwater Lag Times for Low Impact and Traditional Residential Development.* Hoboken, NJ: Wiley Online Library.

Huffman, RL, DD Fangmeier, WJ Elliot, and SR Workman. 2012. Infiltration and runoff. In *Soil and Water Conservation Engineering, Seventh Edition*, 81–113. St. Joseph, MI: American Society of Agricultural and Biological Engineers.

Hunt, WF, JT Smith, SJ Jadlocki, JM Hathaway, and PR Eubanks. 2008. Pollutant removal and peak flow mitigation by a bioretention cell in urban Charlotte, NC. *Journal of Environmental Engineering* 134 (5): 403–408.

Hunt, WF, and WG Lord. 2006. *Bioretention Performance, Design, Construction, and Maintenance.* Raleigh, NC: NC Cooperative Extension Service.

Iversen, BV, P Moldrup, P Schjønning, and P Loll. 2001. Air and water permeability in differently textured soils at two measurement scales. *Soil Science* 166 (10): 643–659.

Jabro, JD, WM Iversen, WB Stevens, RG Evans, MM Mikha, and BL Allen. 2016. Physical and hydraulic properties of a sandy loam soil under zero, shallow and deep tillage practices. *Soil and Tillage Research* 159:67–72.

Jastrow, JD, RM Miller, and J Lussenhop. 1998. Contributions of interacting biological mechanisms to soil aggregate stabilization in restored prairie. *Soil Biology and Biochemistry* 30 (7): 905–916.

Jinling, Y, Z Ganlin, Z Yuguo, R Xinling, and H Yue. 2005. Application and comparison of soil compaction indexes in the evaluation of urban soils. *Transactions of the Chinese Society of Agricultural Engineering* 5:012.

Jim, CY and JYY Ng. 2000. Soil porosity and associated properties at roadsite treepits in urban Hong Kong. *Soils of Urban, Industrial, Traffic, and Mining Areas III: The Soil Quality and Problems: What Shall We Do.* 12:629–34.

Johnson, GR. 1999. *Protecting Trees from Construction Damage: A Homeowner's Guide.* University of Minnesota, Minneapolis, MN: Extension Service.

Johnston, MR, NJ Balster, and J Zhu. 2016. Impact of residential prairie gardens on the physical properties of urban soil in Madison, Wisconsin. *Journal of Environmental Quality* 45 (1): 45–52.

Krull, ES, JO Skjemstad, JA Baldock, and Commonwealth Scientific. 2004. *Functions of Soil Organic Matter and the Effect on Soil Properties.* Cooperative Research Centre for Greenhouse Accounting. GRDC report. Project CSO 00029.

Lichter, JM, and PA Lindsey. 1994. The use of surface treatments for the prevention of soil compaction during site construction. *Journal of Arboriculture* 20:205–209.

Loper, S, AL Shober, C Wiese, GC Denny, CD Stanley, and EF Gilman. 2010. Organic soil amendment and tillage affect soil quality and plant performance in simulated residential landscapes. *HortScience* 45 (10): 1522–1528.

Lusk, MG, and GS Toor. 2016a. Biodegradability and molecular composition of dissolved organic nitrogen in urban stormwater runoff and outflow water from a stormwater retention pond. *Environmental Science & Technology* 50:3391–3398. DOI: 10.1021/acs.est.5b05714

Lusk, MG, and GS Toor. 2016b. Dissolved organic nitrogen in urban streams: biodegradability and molecular composition studies. *Water Research* 96:225–235. DOI: 10.1016/j.watres.2016.03.060

Luxmoore, RJ 1981. Micro-, meso-, and macroporosity of soil. *Soil Science Society of America Journal* 45 (3): 671–672.

Mamo, M, JF Moncrief, CJ Rosen, and TR Halbach. 2000. The effect of municipal solid waste compost application on soil water and water stress in irrigated corn. *Compost Science & Utilization* 8 (3): 236–246.

Millward, AA, K Paudel, and SE Briggs. 2011. Naturalization as a strategy for improving soil physical characteristics in a forested urban park. *Urban Ecosystems* 14 (2): 261–278.

Morgan, RPC. 2009. *Soil Erosion and Conservation.* Hoboken, NJ: John Wiley & Sons.

Oldfield, EE, AJ Felson, SA Wood, RA Hallett, MS Strickland, and MA Bradford. 2014. Positive effects of afforestation efforts on the health of urban soils. *Forest Ecology and Management* 313:266–273.

Olson, NC, JS Gulliver, JL Nieber, and M Kayhanian. 2013. Remediation to improve infiltration into compact soils. *Journal of Environmental Management* 117:85–95.

Ossola, A, AK Hahs, and SJ Livesley. 2015. Habitat complexity influences fine scale hydrological processes and the incidence of stormwater runoff in managed urban ecosystems. *Journal of Environmental Management* 159:1–10.

Pagliai, M, N Vignozzi, and S Pellegrini. 2004. Soil structure and the effect of management practices. *Soil and Tillage Research* 79 (2): 131–143.

Pavao-Zuckerman, MA 2008. The nature of urban soils and their role in ecological restoration in cities. *Restoration Ecology* 16 (4): 642–649.

Pitt, R, J Lantrip, R Harrison, CL Henry, and D Xue. 1999. *Infiltration through Disturbed Urban Soils and Compost-Amended Soil Effects on Runoff Quality and Quantity.* National Risk Management Research Laboratory.

Rahman, MA, JG Smith, P Stringer, and AR Ennos. 2011. Effect of rooting conditions on the growth and cooling ability of *Pyrus calleryana.* *Urban Forestry & Urban Greening* 10 (3): 185–192.

Rawls, WJ, DL Brakensiek, and N Miller. 1983. Green-Ampt infiltration parameters from soils data. *Journal of Hydraulic Engineering* 109 (1): 62–70.

Rawls, WJ, YA Pachepsky, JC Ritchie, TM Sobecki, and H Bloodworth. 2003. Effect of soil organic carbon on soil water retention. *Geoderma* 116 (1): 61–76.

Richard, G, I Cousin, JF Sillon, A Bruand, and J Guérif. 2001. Effect of compaction on the porosity of a silty soil: Influence on unsaturated hydraulic properties. *European Journal of Soil Science* 52 (1): 49–58.

Rienzner, M, and C Gandolfi. 2014. Investigation of spatial and temporal variability of saturated soil hydraulic conductivity at the field-scale. *Soil and Tillage Research* 135:28–40.

Saxton, KE, WJ Rawls, JS Romberger, and RI Papendick. 1986. Estimating generalized soil-water characteristics from texture. *Soil Science Society of America Journal* 50 (4): 1031–1036.

Scharenbroch, BC, J Morgenroth, and B Maule. 2016. Tree species suitability to bioswales and impact on the urban water budget. *Journal of Environmental Quality* 45 (1): 199–206.

Smiley, ET, L Calfee, BR Fraedrich, and EJ Smiley. 2006. Comparison of structural and noncompacted soils for trees surrounded by pavement. *Arboriculture and Urban Forestry* 32 (4):164.

Stovin, VR, A Jorgensen, and A Clayden. 2008. Street trees and stormwater management. *Arboricultural Journal* 30 (4): 297–310.

Tester, CF. 1990. Organic amendment effects on physical and chemical properties of a sandy soil. *Soil Science Society of America Journal* 54 (3): 827–831.

Topp, GC. 1971. Soil water hysteresis in silt loam and clay loam soils. *Water Resources Research* 7 (4): 914–920.

Tyner, JS, DC Yoder, BJ Chomicki, and A Tyagi. 2011. A review of construction site best management practices for erosion control. *Transactions of the ASABE* 54 (2): 441–450.

Xiao, Q. 2008. *Urban Runoff Pollutants Removal of Three Engineered Soils*. Davis, CA: Center for Urban Forest Research. Pacific Southwest Research Station, Land, Air & Water Resources, UC Davis.

Xiao, Q, and EG McPherson. 2016. Surface water storage capacity of twenty tree species in Davis, California. *Journal of Environmental Quality* 45 (1): 188–198.

Yang, J-L, and G-L Zhang. 2011. Water infiltration in urban soils and its effects on the quantity and quality of runoff. *Journal of Soils and Sediments* 11 (5): 751–761.

Yang, Y, and GS Toor. 2016. δ15N and δ18O reveal the sources of nitrate-nitrogen in urban residential stormwater runoff. *Environmental Science & Technology* 50:2881–2889. DOI: 10.1021/acs.est.5b05353

Yang, Y, and GS Toor. 2017. Sources and mechanisms of nitrate and orthophosphate transport in urban stormwater from residential catchments. *Water Research* 112:176–184. DOI: 10.1016/j.watres.2017.01.039

10 Making Soils from Urban Wastes

Sally Brown

CONTENTS

10.1 INTRODUCTION

Every person makes waste. In urban areas, these wastes are classified as either solid or liquid. Solid waste is the general term for what goes out in the trash. Liquid wastes consist of any and all materials that go down the drain or toilet. Household drains as well as drains from industries and business use water to carry wastes to centralized wastewater treatment plants. Cities have collected and treated these "wastes" with the primary goal of protecting public health. Centralized collection also maintains order in urban areas. Memories of garbage strikes are a reminder of the disorder that results when these collection systems fail (http://1981.nyc/christmas-trash-strike-1981/).

Over time, we have gradually started to recognize the value of certain of these "wastes." Historically, solid waste was a single category. Materials were set into a single bin, collected, and then taken to landfills. Currently, recycling of glass, select plastics, metals, and clean paper, all previously considered as part of solid waste, is common practice in most of the US, Europe, Australia and other nations. Much of the remaining wastes, including components of both the solid and liquid streams, are increasingly recognized for their value either as soil amendments or as the foundation for building soil (Brown 2016; Brown et al. 2012; Penninsi 2012). An overview of urban residuals with their potential value for soils is shown in Table 10.1. This chapter focuses on the soil-building potential of urban residuals. Descriptions and characterization of the organic components of both waste streams are discussed. Examples of cities, communities, and individuals who are making soil from waste are given. Potential end uses for these soils within urban areas are provided.

TABLE 10.1

Different Types of Urban Wastes That Have Potential as Soil Amendments

Material	Quantity per Capita (kg dry weight)	Potential Uses	Concerns	Legal Restrictions	Stabilization Process
Yard waste	74	Soil conditioner	Characteristics can vary widely seasonally, may contain pathogens and herbicide residues	Generally none	Can be chipped and used as mulch if feedstocks are woody. Can be feedstock in compost mixtures or composted directly
Soiled paper	37	Soil conditioner	High carbon, no nutrients	Generally none	Can be feedstock for compost mixtures
Woody waste	86	Soil conditioner/ mulch	Some urban wood can contain cadmium, chromium, and arsenic	Generally none	Can be chipped and used as mulch, can be feedstock in compost mixtures
Food scraps	34	High in required plant nutrients, high in carbon, source of all required plant micronutrients	Can contain pathogens, can be a vector attraction (i.e., rodents), highly unstable without treatment	Generally part of the municipal solid waste stream. Some areas encourage decentralized or home composting. Some municipalities cocollect with yard waste and produce compost. Decentralized composting may be prohibited	Composting or anerobic digestion followed by drying or composting
Municipal biosolids	30	Highly consistent, high in nitrogen and phosphorus, high in organic matter, source of all required plant micronutrients	Public acceptance concerns, will contain typically parts per billion concentration of a wide range of compounds	All are municipally managed. Are not allowed to be used for urban agriculture until they have been treated to destroy all pathogens and met metal limits	Range of processes can result in a pathogen-free product. These include composting, pellitization, pasteurization, and air drying over time

Source: Brown, S. and N. Goldstein, *Sowing Seeds in the City: Ecological and Municipal Considerations*, Dordrecht: Springer, 2016.

10.2 SOIL FROM WASTEWATER

Centralized wastewater treatment plants use a range of processes to separate the water used to convey the solids from the solids themselves. Gravity settling, followed by aerobic and finally anaerobic digestion are the most common techniques used by larger municipalities (Metcalf and Eddy et al. 2013). In this process, the solids are transformed. Human waste is a major component of these solids. They also include food scraps, food processing wastes, personal care products, and cleaning products. After settling, the remaining treatment processes are biological. Microbes use the carbon in the waste as an energy source with a significant portion of the carbon released as CO_2 or CH_4. The nutrients in the water are also used by the microorganisms and can become incorporated into the microbial biomass. The solids remaining at the end of treatment consist of spent microbial biomass. In order for these solids to be suitable for use in soil, they have to meet requirements for pathogen reduction as well as limits for metal concentrations (US EPA 1993). Once treated to acceptable levels, the solids are referred to as biosolids. Characteristics of biosolids from select municipalities are shown in Table 10.2. Each person produces about 30 kg of biosolids per year (Figure 10.1). Currently, about half of the biosolids generated in the United States or

FIGURE 10.1 Each person produces wastes on a daily basis that have potential value for soils.

TABLE 10.2
Examples of the Nutrient and Carbon Contents of Different Soil Amendments Produced from Urban Residuals

	Total Carbon (g kg⁻¹)	Total Nitrogen (g kg⁻¹)	Total Phosphorus (g kg⁻¹)	Total Potassium (g kg⁻¹)
Dewatered biosolids (Cogger et al. 2013)	270	44	33.8	2.8
Pelletized biosolids	–	44.8	16	1.4
Composted biosolids	455	19	8.9	–
	173	12.5	6.9	–
Biosolids based potting soil	310	11.1	6.5	1.9
Food/yard compost (Sullivan et al. 2002)	11.7	2.6	10.5	54
Yard compost (Barker 2001)	7	1.6	2.6	310

Sources: Cogger, C.G. et al., *J. Environ. Qual.*, 42, 516–522, 2013; Sullivan, D.M. et al., *Soil Sci. Soc. Am. J.*, 66, 154–161, 2002; Barker, A.V., *Commun. Soil Sci. Plant Anal.*, 32, 1841–1860, 2001.

5 million tons are beneficially used with the majority of use as a substitute for synthetic fertilizers on agronomic crops (NEBRA 2007). The remainder is landfilled or incinerated.

10.3   SOIL FROM SOLID WASTE

Yard waste and food scraps are the two main components of municipal solid waste (MSW) that have value for soils. These materials can be stabilized through composting to produce valuable soil amendments (Table 10.2). Together, these comprise 28% of the solid waste stream. Yard waste diversion began in the 1990s. Currently, a number of states have bans on landfilling yard waste or offer separate collection and composting. Per capita quantities of yard waste will vary by region and season. However, a gross estimate on yard waste per person is 100 kg with a current recycling rate of 58% (https://www.epa.gov/sites/production/files/2015-09/documents/2012_msw_fs.pdf). Characteristics of yard waste will also vary based on the time of year and part of the country. The material typically consists of dryer material that has high carbon and low nutrient content. Woody biomass and leaves fall into this category. Wetter yard wastes such as grass clippings are typically higher in nutrients. Over 95% of the food scraps generated are currently landfilled. Each person generates approximately 124 wet kg of food waste annually (Figure 10.1) (Buzby and Hyman 2012). The EPA recently revised its waste reduction model (WARM) to reflect that landfilling food scraps generates significant quantities of CH_4 (Brown 2016; US EPA 2014). The model also quantified benefits associated with diverting these materials from landfills to composting facilities. A credit of 0.24 Mg CO_2/wet Mg of food waste was provided for soil carbon storage for composted food scraps. The model also acknowledges that compost use results in other benefits including increased soil water-holding capacity and nutrient recycling but does not quantify these benefits. Food waste diversion from landfills has recently started to increase. Select states have enacted or are planning to enact bans on landfilling commercial and/or residential food scraps (Yepsen 2014). A number of communities have recently enacted collection programs for food scraps (Yepsen 2016). Select cities provide for cocollection of food scraps and yard waste. Other cities have changed the solid waste rules to allow for community composting of these materials (Brown and Goldstein 2016). As of 2015, 2.74 million households across 19 states have access to municipal food waste collection (Figure 10.2).

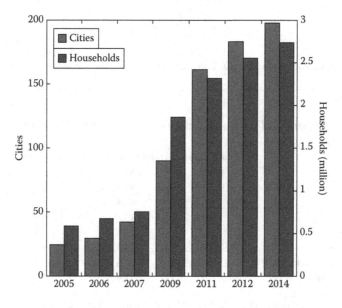

FIGURE 10.2   Number of households and cities that offer centralized collection of food scraps.

10.4 BENEFITS ASSOCIATED WITH USE OF THESE MATERIALS

Greenhouse gas emission avoidance is considered to be the most significant benefit realized when municipal biosolids and the organic fraction of MSW are diverted from landfills or combustion facilities (Brown et al. 2008, 2010; Brown 2016; US EPA 2014). As both biosolids and food scraps are high in moisture, carbon, and nutrients, landfilling these materials will generate significant quantities of CH_4. Burning these materials is typically not a source of energy as the energy required to dry them is similar to or greater than their inherent fuel value (Brown et al. 2010). In addition, burning biosolids has been shown to emit significant quantities of N_2O. However, each of these materials has significant benefits when added to soils.

When diverted from disposal or combustion, the organic fractions of MSW and biosolids can be made into products suitable for use on urban soils (Figure 10.3). These products typically include composts, soil conditioners, or soil blends. The benefits associated with use of products produced using biosolids or MSW feedstocks are well documented. Biosolids are commonly used as a replacement for synthetic fertilizers for agronomic crops, providing both N and P as well as all of the required micronutrients (Rigby et al. 2016). Composts derived from food scraps will also be a significant source of plant nutrients (Recycled Organics Unit 2006). When added to soil, these amendments typically increase soil carbon concentrations and associated carbon storage (Brown et al. 2011; Brown and Cotton 2011; Cogger et al. 2013a,b; Li and Evanylo 2013; Tian et al. 2009). They also increase the nutrient status of the soil (Brown et al. 2011; Brown and Cotton 2011; Cogger et al. 2013a,b). Improved soil physical properties including bulk density, aggregation, and soil water-holding capacity are also commonly associated with use of residuals-based amendments (Albaladejo et al. 2008; Brown et al. 2011; Brown and Cotton 2011; Khaleel et al. 1981; McIvor et al. 2012; Reynolds et al. 2015). There is limited research on the use of these materials in urban soils; however, it would be expected that benefits observed in these soils would be similar to those reported for agricultural and mined soils. McIvor et al. (2012) added two biosolids composts to soils in an urban area and saw increased organic matter,

FIGURE 10.3 Transforming municipal biosolids and food/yard wastes into soil amendments and composts.

total N and available P, and infiltration rates as well as reduced bulk density. Researchers have also studied the impact of biosolids and yard/food composts on metal availability in urban soils. Brown et al. (2003, 2012) found that use of a high Fe biosolids compost reduced Pb availability in both *in vitro* and *in vivo* bioavailability tests. Changes in Pb speciation were seen when similar composts were added to a high-Pb soil (Brown et al. 2012). A growing body of work has looked at the impact of these amendments on contaminants in urban soils used for growing food. Dilution of total soil metal, yield increases, and reductions in plant uptake of Pb and As were generally seen with use of compost amendments (Attanayake et al. 2014, 2015; Brown et al. 2016; Defoe et al. 2014; Fleming et al. 2013). A recent review recommended use of composts and biosolids-based soil products as a means to both enrich soils and insure protection of public health for urban agriculture (Brown et al. 2016a).

The benefits associated with the use of recovered organics will vary based on the characteristics of the final product and the targeted end use. Some examples of responses to composted products for turf and landscape are shown in Table 10.3. Products that have value as soil conditioners and balanced fertilizers will have different benefits and potential types of use than those that have low fertility value (Recycled Organics Unit 2006). Recent research has indicated that soil carbon sequestration potential is greatest when amendments with balanced nutrients are added to soil (Kirkby et al. 2014). This suggests that adding biosolids or food scrap-based materials to soil will result in higher carbon storage than wood or carbon-based mulches. However, high-carbon amendments have value for end uses such as surface mulches. Use of mulches can reduce transpiration, soil erosion, and decrease irrigation demands. Benefits associated with diversion of organics from disposal will also vary based on the methane generation potential of the material (US EPA 2014).

TABLE 10.3

Examples of Responses of Turf Grass and Landscape Plants to Amendments Derived from Municipal Residuals

Study	Amendment	Crop	Control	Amended
Lawns				
Cogger et al. (2013)	Biosolids	Turf	–	Increased yield compared with fertilizer during 10 years of amendment addition and for 9 years postamendment addition
Loschinkohl and Boehm (2001)	Biosolids compost	Turf	2.1 g m^{-2}	7.6 g m^{-2}
		3 grass cultivars	3	3.4
			1.8	3.8
Landscape Plants				
Cogger et al. (2008)	Compost incorporated, bark mulch	Landscape plants		Growth response for first 4 years of trial
De Lucia and Cristiano (2015)	Composted biosolids in planting hole	Woody hedges	Urban soils	
			15%, 30%, 45% compost	Growth response, most pronounced at 30% rate

10.5 OBSTACLES TO WASTE DIVERSION AND RECYCLING

As stated, the organic components of MSW have traditionally been collected in combination with the remainder in a single bin. Over time, different components of MSW have been recognized for their value and recycling programs have been initiated. Recycling programs, including those for organics, can vary by collection systems, associated fees, end use of materials, and types of materials accepted. In the simplest cases, drop-off points for organics or local neighborhood composting sites require the least changes in municipal infrastructure and can offer limited landfill diversion and compost production. For example, Thompkins County, New York is now offering drop-off points for food scraps (Goldstein 2016). Individuals or organizations can also accept food scraps for composting, provided that this practice is allowed within municipal codes.

For broader participation rates, centralized collection and composting is likely required. Many cities offer some type of collection of yard waste with some also expanding to include food scrap collection. The City of Seattle offers weekly cocollection of yard waste and food scraps (http://www.seattle.gov/util/MyServices/FoodYard/HouseResidents/WhatsAccepted/index.htm). Food scrap collection is offered for both residential and commercial properties. Atlanta offers weekly collection of yard waste but not food scraps (http://www.atlantaga.gov/index.aspx?page=491). Other communities offer only seasonal collection of yard waste (http://www.philadelphiastreets.com/recycling/home-base-residential) or collection of commercial and not residential organics (http://www.mass.gov/eea/agencies/massdep/recycle/reduce/food-waste-ban.html). The decision to offer centralized collection of food scraps and yard waste year round is complex. In order to have a successful program, the generators—people who make waste—have to be educated on how to separate appropriate organics from the MSW stream. Separate containers for household collection and then for curbside have to be provided. It is also desirable to require compostable food packaging to reduce the amount of contamination. Once the material is at the curb, specialized trucks have to be used to collect it. Food scraps can be wet and leaky and odorous trucks will not be welcome. This need for new collection vehicles and potentially for multiple collections is a large capital expense and operating expense. In some cases, trucks have been designed that can simultaneously collect food and yard wastes and recyclables. It is also possible to transition to every other week collection of traditional MSW as volumes are significantly reduced. Once the organics are collected, they have to be treated (Morris et al. 2014). Treatment options include anaerobic digestion and/or composting. These processes require some level of capital expenditure and also available land. Finally, the finished material has to be sold or distributed. The costs associated with these processes may be significantly higher than landfill tip fees. If that is the case, the relatively low cost of disposal can provide a significant incentive to maintain the status quo. Raising the tip fee at a landfill is one way to create an immediate incentive for diversion of organics. The Landfill Directive imposed a significant tax on waste at landfills in the European Union in 1999. This effectively increased composting and anaerobic digestion of organics (European Environmental Agency 2009).

10.6 USES OF URBAN RESIDUAL-BASED SOILS AND PRODUCTS

There are many potential end uses for composts, fertilizers, and soil amendments produced from urban residuals. These include home gardens, turf fertilization, urban agriculture, green stormwater infrastructure, tree plantings, and manufactured topsoils. While there is currently no unified national guidance on incorporating use of residuals into urban design, there are some noteworthy examples where the value of these materials has been recognized. The Sites initiative provides guidance for maintaining and building sites (http://www.sustainablesites.org/). It was designed to provide guidance and certification for exterior spaces in the same way that the Leadership in Energy and Environmental Design (LEED) program has provided a structure for sustainable building (http://www.usgbc.org/leed). The Sites program recommends compost use as a means to restore and maintain soil quality. In Washington State, the Soils for Salmon program (http://www.soilsforsalmon.org/)

sets standards for new home construction that require compost addition to new landscapes. This is done for erosion control, reduced stormwater runoff with resulting improvements in water quality, and reduced landscape requirements for fertilizers and water. In an innovative program, the city of Petaluma, CA, has a program that provides compost, cardboard, and mulch to homeowners to replace lawns with native plants (http://cityofpetaluma.net/wrcd/mulch-madness-program.html). "Mulch Madness" has converted 21,500 m^2 of lawns to native landscapes using this approach.

Urban agriculture is expanding across the United States. As with conventional agriculture, healthy soil is a critical foundation for growing food. Because of concerns about both soil quality and soil safety, use of composts or other residual-based soil amendments is common in urban agriculture (Brown and Goldstein 2016). There are three potential models for access to sufficient compost to support urban agriculture: compost can be provided to gardens by the traditional waste management infrastructure within a municipality, the municipality can provide training and some funds to start local collection and composting operations, and finally collection and composting can be done without any municipal support. An example of each case is given below.

10.6.1 Soils and Mulch Provided by City—Tacoma, WA

Tacoma is a moderate-sized city in Washington State. It has a history of soil contamination as a result of a metal smelter that operated within the city for several decades. Aerial deposition from the smelter has elevated soil Pb and As concentrations across the city. The mayor of the city of Tacoma has set increasing the number of community gardens as one of her primary goals. In order to accomplish this goal, she has made use of a number of the residuals that the city manages. Gardens have been established on city-owned property (Figure 10.4). Waste cardboard has been diverted from the landfill and used as a barrier to potentially contaminated soil in the gardens. Woody waste, also diverted from the landfill, is mulched and placed on top of the cardboard. Raised beds are used to grow plants in most of the gardens. Topsoil, produced from the city's biosolids, aged wood, and sawdust from lumber mills, is provided to all gardeners for use as a growing media (McIvor and Brown 2016).

The program is housed in the conservation district. Gardens have been established both within the city and the county. The program has seen the number of community gardens grow from 5 to

FIGURE 10.4 A compost site in New York City near the Gowanus Canal.

almost 80 since it began. A critical factor to the success has been the availability of biosolids-based topsoil. The soil is exceptionally fertile and makes it easier for first-time growers to be successful. As the biosolids used to produce it are regulated, the topsoil is routinely tested for metals and pathogens. Gardeners with concerns about the safety of the native soil are able to use the tested material to grow their food. The city also composts its yard waste and is considering expanding the program to include residential and commercial food scraps as well. The wastewater treatment plant manages both the biosolids-based soil products and the yard waste compost. Material is available free or at minimal cost to local residents who come to the plant. The program also delivers the products to homeowners.

10.6.2 Compost Training Provided by City—New York, NY

The New York City Department of Sanitation does not collect and compost food scraps. In the absence of a centralized collection program, they provide training to individuals and groups on how to compost (Goldstein 2013). The goal of the training is to set up neighborhood composting operations. This program was started in 1993 as a means to build support and public awareness for waste diversion and composting. There are currently more than 200 community sites and 8–10 midsize composting operations across the five boroughs. The program is run through the Department of Sanitation's Bureau of Waste Prevention, Reuse and Recycling (BWPRR). A survey conducted in the early 1990s found that the constituents most likely to compost were those who were already interested in gardening. This continues to be the case with the majority of the community compost sites located alongside community gardens. About 60 people are trained each year through the department to become master composters. The staff also assist local composters with managing their composting operations. The program has expanded to include a Local Organics Recovery Program (LORP). Started in 2012, the goal of the program is to provide additional drop-off sites for food scraps and to make sure that these materials are composted within the city. As the city moves toward centralized collection of food scraps, community composting operations are viewed as an important component of the larger diversion infrastructure. It is expected that these local programs and trainings will continue.

10.6.3 Collection and Composting outside Municipal Infrastructure—Growing Power

In the absence of centralized collection and composting, community groups and organizations have started to collect and compost food scraps locally. Growing Power was founded by Will Allen as a way to provide fresh food and employment to people in Milwaukee, WI (http://www.growingpower.org/). The organization set up a farm and has worked to teach members of the community how to become farmers. A critical factor in the success of the farm has been the composting program used to make soil for use on the farm. Allen views clean soil as the missing link for producing food in cities. He is suspicious of urban soil as likely full of contaminants. He also notes that much of the conventionally farmed soil is lacking critical nutrients. His solution is to compost residential and commercial food scraps and yard waste. The finished composts are used as the sole growing media for the farm. He received a MacAurthur Genius grant for his work. Growing Power now has farms in three cities, all using compost that they have produced from urban residuals.

10.7 COMPOST IN GREEN STORMWATER INFRASTRUCTURE

Due to the prevalence of impermeable surfaces in urban areas, stormwater management is a significant concern. With the lack of sufficient surface soil to allow for infiltration, stormwater flows over surfaces to combined sewer systems or directly to urban surface waters. Overland flow over pavement, asphalt, and vacant lots results in contamination of the stormwater by both dissolved and suspended materials (McIntyre et al. 2015). Green infrastructure is increasingly recognized as an alternative to gray or engineered methods for stormwater treatment. The goal of these systems is to divert stormwater from overland flow to specially designed bioretention cells where the water is filtered through specially designed soils. Here natural processes are used to filter contaminants from rainwater, and in certain cases, increase the quantity of stormwater that infiltrates into the ground. The focus of these systems will likely vary by region. For example, in the Pacific Northwest, the primary concerns about untreated stormwater are its impact on spawning salmon and other aquatic life (McIntyre et al. 2015, 2016). In other regions, eutrophication of lakes and streams is the critical focus (Ohio Lake Erie Phosphorus Task Force 2010). For regions with combined sewer and stormwater systems, the potential for combined sewer overflows is a driving factor (Wang et al. 2013).

There are a wide range of green stormwater systems ranging from rain gardens in private homes to systems in parking lots. In the simplest of these systems, the infiltration rate of the soil is enhanced and water infiltration into the soil and subsoil is the goal. In the most engineered of these "green" systems, a bioretenion media is used to filter rainwater and then the rainwater is piped directly into a stream or river. Benefits of these systems include reduced costs in comparison to gray systems, decreased "flashiness" of urban streams, reduced flooding potential, and additional green spaces in urban areas (Center for Neighborhood Technology 2010). While recommendations for the soil in these systems vary, a majority of these include a requirement for use of compost derived from municipal residuals. Research has also shown that compost is effective for supporting plant growth on these systems and filtering metals and polycyclic aromatic hydrocarbons from stormwater (Brown et al. 2016b; Diblasi et al. 2009).

Seattle Public Utilities installed rain gardens on the parking strip in a neighborhood as a trial in 2001 (Figure 10.5). The "SeaStreets" program included reconfiguring the strips to encourage flow of stormwater into the bioretention systems. The systems were constructed using a 60/40 by volume blend of sand and yard waste compost. The strips were planted with native species and have been allowed to grow over time. Overland flow from the treated areas has been eliminated as a result of

FIGURE 10.5   A bioretention system for stormwater in Seattle, WA.

these systems. The program is now expanding and this approach for stormwater treatment is now recommended by the Washington State Department of Ecology with compost use included in the guidelines (City of Seattle 2011).

10.8 SUMMARY

Each day, in every city, residents produce wastes that can be transformed into valuable products for soils. To date, a large fraction of these materials have been disposed of in landfills or incinerated. This value of transforming wastes comes in part from the avoidance of greenhouse gas emissions associated with conventional disposal practices. It also stems from the benefits associated with use of these materials on soils. However, changing from disposal to beneficial use is difficult. In order for this change to be successful, public education campaigns are required to gain acceptance, increase participation, and reduce contamination. Changes are also required in municipal collection and treatment with capital expenses for vehicles and treatment sites. In some cases, decentralized composting is an alternative to municipal disposal.

There are a wide range of uses for these products in urban areas including urban agriculture, landscaping, turf fertilization, reducing soil contaminant availability, and green stormwater infrastructure. Expanding the use of these materials will result in increased soil literacy for city dwellers as well as improved soils, soil safety, and soil function in cities.

REFERENCES

Albaladejo, J., J. Lopez, C. Boix-Fayos, G.G. Barbera, and M. Martinez-Mena Akala. 2008. Long-term effect of a single application of organic refuse on carbon sequestration and soil physical properties. *J. Environ. Qual.* 37:2093–2099.

Attanayake, C.P., G.M. Hettiarachchi, A. Harms, D. Presley, S. Martin, and G.M. Pierzynski. 2014. Field evaluations on soil plant transfer of lead from an urban garden soil. *J. Environ. Qual.* 43:475–487.

Attanayake, C.P., G.M. Hettiarachchi, S. Martin, and G.M. Pierzynski. 2015. Potential bioavailability of lead, arsenic, and polycyclic aromatic hydrocarbons in compost-amended urban soils. *J. Environ. Qual.* 44:930–944.

Barker, A.V. 2001. Evaluation of composts for growth of grass sod. *Commun. Soil Sci. Plant Anal.* 32:1841–1860.

Brown, S. 2016. Greenhouse gas accounting for landfill diversion of food scraps and yard waste. *Compost Sci.* 24(1): 11–19.

Brown, S., A. Carpenter, and N. Beecher. 2010. Calculator tool for determining greenhouse gas emissions for biosolids processing and end use. *Environ. Sci. Technol.* 44:9505–9515.

Brown, S., R. Chaney, J. Hallfrisch, and Q. Xue. 2003. Effect of biosolids processing on lead bioavailability in an urban soil. *J. Environ. Qual.* 32:100–108.

Brown, S.L., R.L. Chaney, and G.M. Hettiarachchi. 2016a. Lead in urban soils: A real or perceived concern for urban agriculture? *J. Environ. Qual.* 45:26–36.

Brown, S.L., I. Clausen, M.A. Chappell, A. Williams, K.G. Scheckel, M. Neuville, and G.M. Hettiarachchi. 2012. Lead and As speciation and bioaccessibility in compost amended contaminated soils. *J. Environ. Qual.* 41:1612–1622.

Brown, S.L., A. Corfman, K. Mendrey, K. Kurtz, and F. Grothkopp. 2016b. Stormwater bioretention systems—Testing the phosphorus saturation index and compost feedstocks as predictive tools for system performance. *J. Environ. Qual.* 45(1): 98–106.

Brown, S. and M. Cotton. 2011. Changes in soil properties and carbon content following compost application: Results of on-farm sampling. *Compost Sci. Util.* 19:88–97.

Brown, S. and N. Goldstein. 2016. The role of organic residuals in urban agriculture. In S.L. Brown, K. McIvor, and E. Snyder (Eds.), *Sowing Seeds in the City: Ecological and Municipal Considerations*. Dordrecht: Springer.

Brown, S., C. Krueger, and S. Subler. 2008. Greenhouse gas balance for composting operations. *J. Environ. Qual.* 37:1396–1410.

Brown, S., K. Kurtz, A. Bary, and C. Cogger. 2011. Quantifying benefits associated with land application of organic residuals in Washington State. *Environ. Sci. Technol.* 45:7451–7458.

Buzby, J.C. and J. Hyman. 2012. Total and per capita value of food loss in the United States. *Food Policy.* 37:561–570.

Center for Neighborhood Technology. 2010. The value of green infrastructure: A guide to recognizing its economic, environmental and social benefits. http://www.cnt.org/publications/the-value-of-green-infrastructure-a-guide-to-recognizing-its-economic-environmental-and (Accessed 16 June 2015).

City of Seattle. 2011. City of Seattle standard specification. Div-7 storm drain, sanitary and combined sewers, water mains and related structures. http://www.seattle.gov/util/Engineering/StandardSpecsPlans/index. htm (Accessed 30 December 2014).

Cogger, C.G., A.I. Bary, A.C. Kennedy, and A.M. Fortuna. 2013a. Long-term crop and soil response to biosolids applications in dryland wheat. *J. Environ. Qual.* 42(6): 1872–1880.

Cogger, C.G., A.I. Bary, E.A. Myhre, and A. Fortuna. 2013b. Biosolids applications to tall fescue have long-term influence on soil nitrogen, carbon and phosphorus. *J. Environ. Qual.* 42:516–522.

Cogger, C., R. Hummel, J. Hart, and A. Barr. 2008. Soil and redosier dogwood response to incorporated and surface-applied compost. *Hort. Sci.* 43:2143–2150.

Defoe, P.P., G.M. Hettiarachchi, C. Benedict, and S. Martin. 2014. Safety of gardening on lead- and arsenic-contaminated urban brownfields. *J. Environ. Qual.* 43:2064–2078.

De Lucia, B. and G. Cristiano. 2015. Composted amendment affects soil quality and hedges performance in the Mediterranean urban landscape. *Compost Sci. Util.* 23:48–57.

Diblasi, C.J., H. Li, A.P. Davis, and U. Ghosh. 2009. Removal and fate of polycyclic aromatic hydrocarbon pollutants in an urban stormwater bioretention facility. *Environ. Sci. Technol.* 43:494–502.

European Environmental Agency. 2009. Diverting waste from landfill. EEA Report No 7/2009. ISSN 1725–9177. http://www.eea.europa.eu/publications/diverting-waste-from-landfill-effectiveness-of-waste-management-policies-in-the-european-union (Accessed 22 May 2017).

Fleming, M.Y.T., P. Zhuang, and M.B. McBride. 2013. Extractability and bioavailability of Pb and As in historically contaminated orchard soil: Effects of compost amendments. *Environ. Pollut.* 177:90–97.

Goldstein, N. 2013. Community composting in New York City. *Biocycle.* 54(11):22.

Goldstein, N. 2016. Residential food scraps drop off. *Biocycle.* 57:29.

Khaleel, R., K.R. Reddy, and M.R. Overcash. 1981. Changes in soil physical properties due to organic waste applications: A review. *J. Environ. Qual.* 10:133–141.

Kirkby, C.A., A.E. Richardson, L.J. Wade, J.B. Passioura, G.D. Batten, C. Blanchard, and J.A. Kirkegaard. 2014. Nutrient availability limits carbon sequestration in arable soils. *Soil Biol. Biochem.* 68:402–409.

Li, J. and G.K. Evanylo. 2013. The effects of long-term application of organic amendments on soil organic carbon accumulation. *Soil Sci. Soc. Am. J.* 77:964–973.

Loschinkohl, C. and M.J. Boehm. 2001. Composted biosolids incorporation improves turfgrass establishment on disturbed urban soil and reduces leaf rust severity. *Hort Sci.* 36:790–794.

McIntyre, J.K., J.W. Davis, C. Hinman, K.H. Macneale, B.F. Anulacion, N.L. Scholz, J.D. Stark. 2015. Soil bioretention protects juvenile salmon and their prey from the toxic impacts of urban stormwater runoff. *Chemosphere.* 132:213–219.

McIntyre, J.K., R.C. Edmunds, M.G. Redig, E.M. Mudrock, J.W. Davis, J.P. Incardona, J.D. Stark, and N.L. Scholz. 2016. Confirmation of stormwater bioretention treatment effectiveness using molecular indicators of cardiovascular toxicity in developing fish. *Environ. Sci. Technol.* 50:1561–1569.

McIvor, K. and S. Brown. 2016. A case study: Integrating urban agriculture into the municipal infrastructure in Tacoma, WA. In S.L Brown, K. McIvor, and E. Snyder (Eds.), *Sowing Seeds in the City: Ecological and Municipal Considerations.* Springer.

McIvor, K., C. Cogger, and S. Brown. 2012. Effects of biosolids-based soil products on soil physical and chemical properties in urban gardens. *Compost Sci. Util.* 20:199–206.

Metcalf and Eddy, G. Tchobanoglous, H.D. Stensel, R. Tsuchihasi, and F. Burton. 2013. *Wastewater Engineering: Treatment and Resource Recovery*, 5th edition. New York, NY: McGraw-Hill. p. 2048.

Morris, J., S. Brown, H.S. Mathews, and M. Cotton. 2014. Evaluation of climate, energy, and soils impacts of selected food discards management systems. Oregon Department of Environmental Quality. http://www.oregon.gov/deq/LQ/Documents/SWdocs/FoodReport.pdf (Accessed August 2016).

North East Biosolids and Residuals Association (NEBRA). 2007. *A National Biosolids Regulation, Quality, End Use & Disposal Survey.* Tamworth, NH: NEBRA.

Ohio Lake Erie Phosphorus Task Force. 2010. Ohio Lake Erie Phosphorus Task Force Final Report. Ohio Environmental Protection Agency. http://epa.ohio.gov/portals/35/lakeerie/ptaskforce/Task_Force_Final_Report_April_2010.pdf (Accessed August 2016).

Penninsi, E. 2012. Water reclamation going green. *Science.* 337:674–676.

Recycled Organics Unit. 2006. Life Cycle Inventory and Life Cycle Assessment for Windrow Composting Systems. Sydney, Australia: The University of New South Wales. http://www.epa.nsw.gov.au/resources/warrlocal/060400-windrow-assess.pdf (Accessed 9 January 2016).

Reynolds, W.D., C.F. Drury, C.S. Tan, and X.M. Yang. 2015. Temporal effects of food waste compost on soil physical quality and productivity. *Canadian J. Soil Sci.* 95(3):251–268.

Rigby, H., B.O. Clarke, D.L. Pritchard, B. Meehan, F. Beshah, S.R. Smith, and N.A. Porter. 2016. A critical review of nitrogen mineralization in biosolids-amended soil, the associated fertilizer value for crop production and potential for emissions to the environment. *Sci. Total Environ.* 541:1310–1338.

Sullivan, D.M., A.I. Bary, D.R. Thomas, S. Fransen, and C.G. Cogger. 2002. Food waste compost effects on fertilizer nitrogen efficiency, available nitrogen and tall fescue yield. *Soil Sci. Soc. Am. J.* 66:154–161.

Tian, G., T.C. Granato, A.E. Cox, R.I. Pietz, C.R.Jr. Carlson, and Z. Abedin. 2009. Soil carbon sequestration resulting from long-term application of biosolids for land reclamation. *J. Environ. Qual.* 38:61–74.

US EPA. 1993. 40 CFR Part 503. Standards for the use or disposal of sewage sludge; final rules. *Fed. Reg.* 58(32): 9248–9415.

US EPA. 2014. Solid waste management and greenhouse gasses. Documentation for greenhouse gas emission and energy factors used in the waste reduction model (WARM). http://epa.gov/epawaste/conserve/tools/warm/SWMGHGreport.html

Wang, R., M.J. Eckelman, and J.B. Zimmerman. 2013. Consequential environmental and economic life cycle assessment of green and gray stormwater infrastructures for combined sewer systems. *Environ. Sci. Technol.* 47:11189–11198.

Yepsen, R. 2014. Taking states' pulse on residential food waste collection. *Biocycle.* 55(9):39.

Yepsen, R. 2016. Biocycle nationwide survey: Residential food waste collection in the U.S. *Biocycle.* 56(1):53.

11 Properties of Soils Affected by Highways

Michele Bigger

CONTENTS

11.1 INTRODUCTION

In North America, roadways connect citizens, providing a framework for community planning, development, and function (Coffin 2007; Sipes and Sipes 2013). Roadsides, which are one of the most viewed landscapes in North America (Morrison 1999), support roadway infrastructure and provide increasingly apparent environmental benefits (Steinfeld et al. 2007; Armstrong et al. 2011; Harrison 2014; WSDOT 2016). The United States (US) and Canada have over 7.6 million km of public roadways (Transport Canada 2016; USDT-FHWA 2016a), with more than 402,350 km dedicated to highway infrastructure (Transport Canada 2012; USDT-FHWA 2016b). In 2014, Americans traveled over 4.8 trillion vehicle kilometers, approximately 70% of which occurred on urban roadways (USDT-FHWA 2016c).

Titled land acquired to build a public road is known as right-of-way (ROW) land (USDT-FHWA 2012; Sipes and Sipes 2013). Typical linear ROW land defines the travel way, including the road, roadside, and associated infrastructure (Sipes and Sipes 2013). US ROW land is estimated to occupy more than 24 million ha (Sipes and Sipes 2013), of which 8.3% is estimated to belong to the American National Highway System (ANHS) (USDT-FHWA and Volpe Ctr 2010). An estimated 68% of the ANHS ROW land is unpaved, with widths of up to 327 m (and a mean of 53 m) (USDT-FHWA and Volpe Ctr 2010).

ROW landscapes are challenging and complex systems (Forman and Alexander 1998; Trombulak and Frissell 2000; Coffin 2007; Steinfeld et al. 2007). To achieve ROW operational and functional objectives, constructed roadway systems and natural roadside systems must work together (Steinfeld et al. 2007; WSDOT 2016). The indigenous and imported soils of ROW land support these systems directly and indirectly. Existing soil conditions and soil manipulations can impact the entire ROW; thus, it is important to understand the soil properties that support the roadway and the roadside. This chapter discusses soil properties measured under and along roadways, with an emphasis on the roadside landscape. It is not intended to provide a technical design or review of engineering principles or practices within ROW environments. For more technical guidelines, readers are referred to Sabatini et al. (2002), USDT-FHWA (2002), Christopher et al. (2006), Samtani and Nowatzki (2006), AASHTO (2010), and AASHTO (2011). Similarly, numerous reference manuals, standards, and specifications are available from the websites of both federal and state department(s) of transportation.

11.2 ROAD AND ROADSIDE OVERVIEW

ROW soils are disturbed during construction, rehabilitation, reconstruction, and roadway use. Roadway construction and design techniques often focus on reducing water near the roadway and on proper soil compaction of pavement foundation materials (WTIC 1995; Christopher et al. 2006). Both of these design strategies are important for the performance and longevity of roadways, regardless of whether the travel surface is paved (i.e., concrete or asphalt) or unpaved (i.e., gravel) (AASHTO 1993; Christopher et al. 2006). Roadways are designed as compacted horizontal layers also known as courses (WTIC 1995; Christopher et al. 2006). Supporting the travel way, subsurface courses span the entire travel surface and may extend beyond the adjoining shoulder (WTIC 1995; Christopher et al. 2006). The cross-sections of these courses form a uniform trapezoidal pavement support structure, including (from bottom to top) a prepared subgrade, subbase, and base (WTIC 1995; Christopher et al. 2006; ODOT 2013). Roadway systems with concrete surfaces often only have one supporting course between the surface and the prepared subgrade (Christopher et al. 2006). Subbase and base courses are typically constructed of naturally occurring aggregate materials; however, soil stabilizers, thickened asphalt, and concrete may also be used (WTIC 1995; Christopher et al. 2006). The subgrade course supports the base and subbase courses and provides a foundation for roadway construction. This subgrade course is generally composed of compacted natural soil or fill material placed on top of existing native soil or on an engineered embankment (Christopher et al. 2006). Hereafter, if the subgrade, subbase, and base courses are not directly mentioned, they will be jointly referred to as "support courses."·

Roadway profiles may vary in both depth and composition based on climate and site conditions, such as existing subgrade, type of road, road life expectancy, road design method, material selection, and design vehicle capacity (Christopher et al. 2006; Sullivan 2006). The profile depth of supportive courses have ranged from 15 to 81 cm thick, while pavement surface courses have had ranges of 8–28 cm thick (Lee et al. 2001; Christopher et al. 2006; Sullivan 2006; Lee et al. 2010). The accepted theoretical lifespan of highway infrastructure is approximately 20 years (AASHTO 1993).

To accommodate roadway construction design needs, the topography and soil profiles of roadsides are often altered. Soil profiles can develop compacted subsoils and rehabilitated soil surfaces (Steinfeld et al. 2007; Nelson Brown et al. 2010; WSDOT 2016). Roadside land is often used for construction staging, storing materials, transporting materials on site, and facilitating construction activities. Independent of roadway construction, roadside soils may be directly disturbed due to landscape restoration, site grading, construction of drainage features and/or slope support structures, installation of utilities, and revegetation. Revegetation of disturbed roadside land serves both functional and ecological purposes (Steinfeld et al. 2007). To support revegetation, rehabilitation of roadside soil may include decompaction, the redistribution or import of topsoil, the addition of soil amendments, and the implementation of soil stabilization measures (Craul 1992; Steinfeld et al. 2007; WSDOT 2016).

After construction, ongoing maintenance and general road use continue to impact these environments. In 2003, an estimated 82% of land in the continental United States was less than 1000 m away from a road (Ritters and Wickham 2003). Disturbances impact the ROW environment at local, landscape, and ecological scales (Forman and Alexander 1998; Coffin 2007). Roadside landscapes are often highly exposed, which results in drought conditions; altered soil physical, chemical, biological, and temperature properties; high levels of compaction; poor hydraulic properties; and the presence of anthropogenic artifacts in the soil (Craul 1992; Forman and Alexander 1998; Jim 1998; Gale and Biesboer 2004; Coffin 2007; Haynes et al. 2013). These alterations are not static, but vary in time and space (i.e., length and distance from the road as well as depth) (Craul 1992; Forman and Alexander 1998; Coffin 2007), resulting in highly heterogeneous soil environments similar to those found in urban soils (Craul 1992; Trowbridge and Bassuk 2004; Beniston and Lal 2012). Although variable, minimum disturbance from roads has been estimated to affect 15%–22% of the land in the United States (Forman and Alexander 1998; Forman 2000; Ritters and Wickham 2003).

11.3 CASE STUDY SOILS

Figures 11.1 through 11.4 show the studied sites and tree growth from 2011 to 2014; the case study sites and procedures are based on those given by Bigger (2015). The study soils and sites mentioned throughout this chapter were taken from two afforestation sites (sites 1 and 2). Each

FIGURE 11.1   Site 1 in 2011.

FIGURE 11.2   Site 1 in 2014.

FIGURE 11.3   Site 2 in 2011.

FIGURE 11.4   Site 2 in 2014.

site consisted of one roughly trapezoidal plot, with a squared-off end to the south, measuring 0.06 ha. The sites were located along Hamilton Road (OH 317), where it intersects US Interstate Highway (I)-270 in Gahanna, Ohio. OH 317 is a four-lane, north–south roadway with a posted speed limit of 72.4 km hr^{-1}. Sites 1 and 2 are located on the east and west sides of OH 317, respectively. A bridged portion of OH 317 over I-270 separates the sites by approximately 350 m; the nonbridge portions of OH 317 have gravel shoulders. Each site is located 9.1 m from the edge of the pavement and both sites slope, having high gradients near OH 317 and lower gradients farther from OH 317. The average slopes are 8.2% and 6.9% for sites 1 and 2, respectively. Site 1 was not protected from prevailing winds, and the water within and immediately surrounding site 1 drained through the site and was collected beyond the lower portion of the plot. Conversely, site 2 was protected from prevailing winds by mature coniferous trees and no water detention areas were observed near the site. The eastern portions of I-270 were completed in 1975 (Tebben, 2010), and the I-270 ramps to and from OH 317 were modified in 2010; these modifications had no known effects on either site. In 2012, the daily vehicle volume on OH 317 was 27,200 cars, and the ramps for I-270 East and I-270 West accommodated 15,500 and 7,000 cars, respectively (Personal Communication, Rob Wendling, City of Gahanna, 2012). Before planting the plots, both sites were maintained as mowed turf.

In June 2011, 199 and 194 trees were planted at sites 1 and 2, respectively. Four tree species were included: red maple (*Acer rubrum*), white-barked Himalayan birch (*Betula jacquemontii*), hackberry (*Celtis occidentalis*), and Japanese tree lilac (*Syringa reticulata*). The trees were planted directly as bare roots or in containers. Most of the containers had a volume of 11.8 L, with a height of 22.5 cm, and a diameter of 25.5 cm. At the time of planting, the trees were 46–320 cm tall (average 172.5 cm) with calipers of 6.7–19.2 mm (average 12.1 mm) measured 15 cm from the planting surface. Following plot location at each site all of the existing plant materials within the plot boundaries were killed, plots were rototilled, and organic matter (OM) was incorporated to a depth of 25–30 cm. The OM included the existing plant material at each site and additional aged wood chips. Next, 46-cm diameter holes were pre-augured 1.8 m apart for planting trees. The holes were backfilled with existing soil following tree placement. After planting, both sites received supplemental water as needed through July 2011. All of the trees were fertilized once using Osmocote Pro 22-3-8 Plus Minors with Poly S® nitrogen at a rate of 48 g per tree, which was hand raked into the soil to a depth of 3.8 cm. The plots were mulched with shredded hardwood to a depth of 7.6 cm in late August 2011 and were maintained by the City of Gahanna as deemed necessary between June 2011 and fall 2014. Plot maintenance mainly included mechanical and chemical weed control.

Core soil samples were collected after plot preparation and tree installation once per year from 2011 to 2014. Samples were obtained among the trees and along the edges of the plots. One sample set for each site contained 24 individual soil sampling locations (ISSLs) and represented an area of 8.6–38.4 m² within each site. Fewer samples were taken in 2011; thus, data from these samples were combined with data from samples taken in 2012 to form a complete sample set. Full sample sets were taken in 2013 and 2014. Missing data for any sample location for a given sampling date were replaced by taking the average of at least two of the nearest sampled locations for the corresponding year. Samples within an ISSL were taken no less than 30 cm from the previous sampling collection point. All samples were analyzed for bulk density, gravimetric water, and gravel content (USDA et al. 2001; Grossman and Reinsch 2002; Topp and Ferre 2002). Samples taken in 2014 were also analyzed to determine the 1:1 soil/water electrical conductivity (Rhoades 1996; NRCS-NSSC 2009a), 1:1 soil/water pH (Thomas 1996; NRCS-NSSC 2009b), particle size (Kilmer and Alexander 1949; Glendon and Or 2002), carbon (C) content in the sand fraction following particle size analysis (Nelson and Sommers 1996), and soil organic matter (SOM) content (Nelson and Sommers 1996; Combs and Nathan 1998). Total porosity, volumetric moisture content, air porosity, gravel corrections, and the total carbon stock (TCS) were calculated from the analyzed results using the equations of Lal and Shukla (2004) and Lorenz and Lal (2012). A particle density of 2.65 Mg m⁻³ was assumed when necessary for calculations. All samples were stratified at three depths: 0–15 cm, 15–30 cm, and 30–45 cm. The analysis was divided based on the distance from the edge of the pavement as either near or far, and 12 ISSLs comprised each distance range. The closest samples were 9.1 m from the edge of the pavement to approximately 16.0 m, while samples farther from the road were taken between 16.5 m and 25.6 m from the edge of the pavement. The results for bulk density, gravel content, total porosity, volumetric moisture content, and air porosity with and without gravel were averaged across the yearly sample sets. R version 3.2.4 (R Core Team 2016) with packages Car (Fox and Weisberg 2011), Pastecs (Grosjean and Ibanez 2014) Pgirmess (Giraudoux 2016), Psych (Revelle 2015), and Userfriendlyscience (Peters 2016) was used for statistical analysis. The presented correlations are examples of the systematic effects of roadside tree survival related to selected soil properties. These correlations were based on tree survival across all species taken approximately 1 year (2012), 2 years (2013), and 3 years (2014) after planting. Each year after planting, tree survival was calculated as the number of trees surviving within an ISSL divided by the total number of trees planted within the ISSL. Due to the unbalanced experimental design, survival was proportionally weighted by the number of trees installed in a single ISSL divided by the total number of trees installed in all 12 ISSLs for the corresponding distance range. It is important to acknowledge that correlations are generalized as a deciduous tree population and that different relationships may be expected on an individual species basis.

11.4 PHYSICAL SOIL PROPERTIES

Physical soil properties stratified by depth are shown in Table 11.1. The use of the highway, the existing soil composition and construction activities and materials all contribute to the heterogeneity of roadside soils. Soil disturbance within the ROW begins when the site is prepared for construction by moving soil and materials and by performing excavation (Craul 1992). Soil disturbance affects the distribution of soil particles, the soil type, the sizes of the soil particles, and the soil aggregate composition, structure, and stability (Craul 1992; Jim 1998; Lal and Shulka 2004). Soil disturbance occurs within the immediate roadway area and, to a lesser extent, within the roadside. Subsequently, other physical properties, such as soil bulk density, strength, and porosity, are impacted. A theoretical healthy soil is typically described as consisting of 50% solids and 50% pore space (Lal and Shukla 2004; Trowbridge and Bassuk 2004). However, soils that are disturbed by construction may consist of as much as 90% solids (Lal and Shukla 2004).

11.4.1 Soil Particle Components

Construction performance is attributed to the selection of coarse, well-graded, angular, aggregate mixtures for the base and subbase courses over finer soil materials (WTIC 1995; Christopher et al. 2006; USDA-NRCS 2016). Construction materials may be naturally occurring, recycled, and/or man-made (WTIC 1996; Skinner 2008). Coarse soil materials result in inherent and maintainable strength, sustainable drainage, resilience to climate changes, and resistance to erosive forces (WTIC 1995; Christopher et al. 2006; USDA-NRCS 2016). Due to the substantial use of soils with coarse fragments, roadway construction soils are classified using either the American Association of State Highway and Transportation Officials (AASHTO) classification system or the Unified Soil Classification System (USCS). Both systems include larger coarse fragments (≤7.6 cm) and integrate Atterberg's limits to include a soil plasticity measure (Christopher et al. 2006; USDA-NRCS 2016). The definitions and requirements for soil particle components used for roadway construction are typically found in the specifications of state and/or federal departments of transportation.

In construction areas, topsoil is typically removed due to its lack of stability for constructed features (Craul 1992). Roadside topsoil may be native to the site and stockpiled, manufactured, or imported from elsewhere (Steinfeld et al. 2007). ROW topsoil is valuable and often sold; thus, the depth of topsoil replaced often depends on availability and cost (Craul 1992, Venner 2004; Steinfeld et al. 2007). Manufactured topsoil is typically composed of loamy naturally occurring soil material, OM, and/or soil amendments, which are used to adjust the soil nutrient content, pH, and biological function (Steinfeld et al. 2007; WSDOT 2016). The minimum recommended and specified topsoil depths for roadside vegetation have ranged from 10 to 20 cm (Claassen and Zasoski 1994; IOWADOT 2016); however, depending on the roadside environment and on revegetation needs, greater topsoil depths have been used (Armstrong et al. 2011). Conversely, topsoil may not be recommended for extreme gradients, depending on the surface and subgrade soil texture and soil strength (Venner 2004; Steinfeld et al. 2007).

Highway roadside soils can have sand contents greater than 50%, with some roadside soils containing up to 80% sand (Gale and Biesboer 2004; Persyn et al. 2004; Nelson Brown et al. 2011; Haan et al. 2012; Bouchard et al. 2013; Eimers et al. 2015). Highway roadsides have also been found to contain mainly fine particles (<25% sand) (Parsons et al. 2001; Trammell et al. 2011; Gordon 2014; McGrath and Henry 2015). At the study sites, the average soil texture across different soil depths was loam (NRCS-NSSC 2016). However, site 1 had contained more coarse-textured particles, while site 2 contained more fine-textured particles and was similar to clay or silt loams (NRCS-NSSC 2016). Significant differences occurred between sites that were within a similar distance from the road at every depth for all three soil particle measures. Within the site, only one significant difference was observed, which was a greater clay content located at a soil depth of 15–30 cm farther from the road at site 2. The sand content at site 1 ranged from 30% to 67%, with a median range from

TABLE 11.1

Physical Soil Property Means or Medians[a] and Ranges for Three Soil Depths (0–15, 15–30, and 30–45 cm) and Two Distances from the Road (Near, Far) in Two Sites (1, 2) at the Intersection of Interstate 270 and Hamilton Road (OH 317) in Gahanna, Ohio

Physical Soil Properties

Dry Bulk Density (ρ_b) (Mg m⁻³) and *Dry Bulk Density Corrected for Gravel (ρ_{bc}) (Mg m⁻³)*

Distance Site	0–15 cm Mean/Median	Min–Max Range	15–30 cm Mean/Median	Min–Max Range	30–45 cm Mean/Median	Min–Max Range	0–15 cm Mean/Median	Min–Max Range	15–30 cm Mean/Median	Min–Max Range	30–45 cm Mean/Median	Min–Max Range
Near												
1	1.2 A[b],a[c]	0.9–1.6	1.6 A,a	1.5–1.7	1.6[k] A,a	1.4–1.7	1.1 A,a	0.8–1.5	1.4 A,a	1.2–1.6	1.4 A,a	1.1–1.6
2	1.3 A,a	1.1–1.5	1.5 A,a	1.4–1.7	1.6[k] A,a	1.5–1.9	1.2 A,a	1.1–1.5	1.4 A,a	1.2–1.7	1.5 A,a	1.4–1.7
Far												
1	1.2 A,a	1.0–1.4	1.5 A,a	1.3–1.7	1.6[k] A,a	1.4–1.7	1.1 A,a	0.9–1.3	1.3 A,a	1.1–1.5	1.3 A,a	1.0–1.6
2	1.1 A,b	1.0–1.3	1.4 A,a	1.1–1.6	1.5[k] A,a	1.2–1.6	1.1 A,a	1.0–1.3	1.4 A,a	1.1–1.6	1.5 A,a	1.2–1.7

Total Porosity (f_t) (m³ m⁻³, %) and *Total Porosity Corrected for Gravel (f_{tc}) (m³ m⁻³, %)*

Distance Site	0–15 cm Mean/Median	Min–Max Range	15–30 cm Mean/Median	Min–Max Range	30–45 cm Mean/Median	Min–Max Range	0–15 cm Mean/Median	Min–Max Range	15–30 cm Mean/Median	Min–Max Range	30–45 cm Mean/Median	Min–Max Range
Near												
1	54.0 A,a	40.0–67.0	40.0 A,a	35.0–45.0	39.0[k] A,a	34.0–47.0	58.0 A,a	45.0–69.0	48.0 A,a	39.0–54.0	47.0 A,a	40.0–58.0
2	51.0 A,b	43.0–57.0	43.0 A,a	35.0–49.0	41.0[k] A,a	30.0–44.0	54.0 A,a	44.0–60.0	47.0 A,a	36.0–55.0	43.0 A,a	37.0–48.0
Far												
1	56.0 A,a	47.0–64.0	42.0 A,a	35.0–51.0	40.0[k] A,a	36.0–49.0	60.0 A,a	52.0–66.0	50.0 A,a	42.0–60.0	49.0 A,a	42.0–62.0
2	57.0 A,a	50.0–62.0	46.0 A,a	40.0–57.0	42.0[k] A,a	38.0–54.0	58.0 A,a	53.0–63.0	47.0 A,a	40.0–58.0	45.0 A,a	38.0–56.0

(Continued)

TABLE 11.1 (Continued)

Physical Soil Property Means or Medians[a] and Ranges for Three Soil Depths (0–15, 15–30, and 30–45 cm) and Two Distances from the Road (Near, Far) in Two Sites (1, 2) at the Intersection of Interstate 270 and Hamilton Road (OH 317) in Gahanna, Ohio

Distance / Site	Physical Soil Properties					
	Mean/Median	Min–Max Range	Mean/Median	Min–Max Range	Mean/Median	Min–Max Range
Gravel Content on a Volume Basis (cm³ cm⁻³, %)						
Near						
1	14.0[l] A,a	7.0–21.0	30.0[w] A,a	22.0–38.0	34.0[lw] A,a	24.0–47.0
2	9.0[l] A,a	5.0–17.0	16.0[w] B,a	5.0–26.0	14.0[lw] B,a	5.0–34.0
Far						
1	12.0[l] A,a	6.0–21.0	31.0[w] A,a	24.0–39.0	34.0[lw] A,a	21.0–47.0
2	3.0[l] B,b	1.0–11.0	6.0[w] B,b	3.0–14.0	4.0[w] B,b	2.0–21.0
Silt Content (%)						
Near						
1	40.0[k] B,a	31.0–47.0	36.0[k] B,a	33.0–43.0	37.0[k] B,a	22.0–39.0
2	46.0[k] A,a	43.0–51.0	44.0[k] A,a	33.0–48.0	44.0[k] A,a	41.0–53.0
Far						
1	44.0[k] B,a	39.0–45.0	37.0[k] B,a	29.0–44.0	36.0[k] B,a	30.0–41.0
2	50.0[k] A,a	47.0–58.0	46.0[k] A,a	40.0–55.0	45.0[k] A,a	42.0–62.0
Sand Content (%)						
Near						
1	46.0[k] A,a	34.0–49.0	45.0[k] A,a	35.0–55.0	45.0[k] A,a	38.0–67.0
2	32.0[k] B,a	22.0–56.0	32.0[k] B,a	24.0–41.0	32.0[k] B,a	23.0–39.0
Far						
1	44.0[k] A,a	30.0–57.0	40.0[k] A,a	33.0–43.0	46.0[k] A,a	37.0–51.0
2	24.0[k] B,a	19.0–30.0	27.0[k] B,a	24.0–33.0	22.0[k] B,a	16.0–39.0
Clay Content (%)						
Near						
1	18.0[l] A,a	16.0–24.0	16.0[k] B,a	7.0–21.0	18.0[w] B,a	11.0–23.0
2	22.0[l] A,b	11.0–34.0	20.0[k] A,a	16.0–26.0	25.0[w] A,a	20.0–32.0
Far						
1	19.0[l] A,a	14.0–26.0	17.0[k] A,a	16.0–22.0	19.0[w] B,a	16.0–22.0
2	30.0[l] B,a	20.0–40.0	21.0[k] A,a	17.0–27.0	29.0[w] A,a	17.0–38.0

[a] Mean reported for parametric ANOVA unless: (w) Welch's corrected ANOVA, mean reported, (l) natural log transformation performed on data prior to analysis, back-transformed means reported, (k) Kruskal–Wallis test performed, median reported. $n = 12$/site/distance/depth; Tukey–Kramer adjustment with all pairwise comparisons, $p \leq .05$.

[b] Means or medians with unlike capital letters are statistically different within distance from the road between sites.

[c] Means or medians with unlike lowercase letters are statistically different between distances from the road within site.

40% to 46%; the sand content range for site 2 was 16%–56%, with a median range of 22%–32%. The silt content ranged from 22% to 47% and from 33% to 62%; and the clay content ranged from 7% to 26% and from 11% to 40% for sites 1 and 2, respectively. At site 1, the mean and median silt contents ranged from 36% to 44% and the clay content ranged from 16% to 19%. At site 2, the mean and median silt and clay contents ranged from 44% to 50% and from 20% to 30%, respectively. Maximum sand contents of more than 50% were observed more frequently at site 1. Sand content values of less than 25% only occurred at site 2. The median sand contents at site 1 were 13%–24% greater than those at site 2 at a similar soil depth and at a similar distance from the road. Moreover, the median or mean clay content of ≥20% was only observed at site 2, with maximum clay content measures exceeding 30% beginning at a depth of 15 cm, regardless of the distance from the road.

Less than 50% median sand content values at these study sites were likely due to the distance (9.1 m) from the road that the samples were obtained and natural occurring soil particle composition. Lower sand contents (<10%) were found 3 km away at another roadside site (Gordon 2014). Excessive sand or gravel contents often occur within 3 m of the roadway edge (Gale and Biesboer 2004; Persyn et al. 2004; Nelson Brown et al. 2011; Bouchard et al. 2013; Eimers et al. 2015), although exceptions do exist (Haan et al. 2012). Because the soil particle size distribution supports functional soil dynamics (Lal and Shulka 2004), it is important to note that small (nonsignificant) increases in the percentage of fine material may result in a substantial influx of finer particles compared with similar changes in the percentage of coarse particles (Spomer 1983; Trowbridge and Bassuk 2004).

11.4.2 Coarse Fragments

Coarse fragment (gravel, >2 mm, US No. 10 sieve) locations, concentrations, types, shapes, and positions are known to vary within soils and play significant roles in soil dynamics (Poesen and Lavee 1994; Lal and Shukla 2004; Christopher et al. 2006). Gravels also have physical properties (e.g., density, strength, and porosity) that affect their function within the road and roadside environment (Poesen and Lavee 1994; Lal and Shukla 2004; Christopher et al. 2006). Thus, the volumetric gravel content is important for defining spatial relationships within the soil matrix and significant gravel contents are identified for soils with >15% by volume (Lal and Shukla 2004). Gravel may affect soil porosity, hydrological processes at the soil surface or within the soil, soil erosion, soil thermal properties, plant growth, and the overall soil substrate density and strength (Poesen and Lavee 1994; Lal and Shukla 2004). For vegetation, gravel within a soil matrix is noted to both hinder and create opportunities for root exploration and the access of roots to essential plant resources (i.e., air, nutrients, and water) (Perry 1994; Jim 1998; Day et al. 2010).

Gravel may occur in roadside soil due to construction activities (inadvertently or as fill material; WTIC 1995; Christopher et al. 2006; ODOT 2013). On constructed, sloped roadside soil surfaces, gravel or cobble may be placed as mulch or supportive riprap (Armstrong et al. 2011). Gravel may also be indigenous to the roadside environment (Skousen and Venable 2007) or may result from the transfer of indigenous rocks disturbed during the construction of roadway infrastructure. Negligible amounts of coarse fragments were reported along highway roadsides in Ontario, Canada (McGrath and Henry 2015), and Gordon (2014) reported average coarse fragment compositions of 9%–13% across three profile depths at an Ohio highway roadside site. Conversely, Bouchard et al. (2013) observed a decrease of at least 10% in average bulk density along the roadsides of two highways after correcting for coarse fragments. This trend suggests that these soils potentially had greater gravel contents. Similarly, urban roadside tree pits in Hong Kong were reported to contain an average of more than 40% gravel (Jim 1998).

The volumetric gravel content was consistently different between the study sites that were located at similar distances from the road. Individual samples contained 6%–47% gravel at site 1 and 1%–34% gravel at site 2. At site 1, average gravel contents of ≥30% occurred for both distance ranges at soil depths of 15–30 cm, and the gravel content increased with depth. However, no significant differences were found between the distances considered at site 1. The highest average gravel content observed at site 2 was 16% at a soil depth of 15–30 cm near the road. All of the volumetric gravel contents farther from the road at site 2, regardless of depth, were ≤6%. At site 2, the low gravel volume contents farther from the road significantly contrasted with the gravel contents near the road at each depth.

The gravel content observed at both study sites varied in shape, color, and type (i.e., natural and man-made). To a greater extent at site 1, platy shale was encountered. Shale, which is commonly found in Ohio, is a sedimentary rock composed of silt and clay that has a laminate structure (ODGS 2006; ODOT 2013). The shale located at these sites was generally black in color. Although the shale was not tested for durability in this study, in Ohio, shale is often classified as a "nondurable" material because it deteriorates when pressure is applied or when weathering occurs (ODOT 2013). At the study sites, a greater quantity of coarse soil materials (i.e., sand and gravel) was frequently observed closer to the road, and fine soil materials were observed farther from the road, although these differences were generally not significant. These trends are consistent with road construction practices, the sloping terrain, and water or wind erosion, which are all common in roadside environments (WTIC 1995; Forman and Alexander 1998; Lal and Shukla 2004).

11.4.3 BULK DENSITY

Naturally occurring soils do not typically meet the required engineering strength and stability standards for supporting road infrastructure (Parsons et al. 2001); these standards are commonly attained by compaction (Parsons et al. 2001; Christopher et al. 2006). Coarse soil materials facilitate compaction by fitting tightly together, increasing soil strength and stability, while reducing pore space and permeability within granular substrate mixes (Christopher et al. 2006). Engineers specify soil relative compaction requirements for supportive courses as percentages, which are based on the road design needs as well as laboratory testing of the maximum dry (bulk) soil density and the corresponding optimal soil moisture content of the supportive course substrates (Lal and Shukla 2004; Christopher et al. 2006). Soil compaction tests are conducted based on the Proctor methods, which are standardized by AASHTO and ASTM (Parsons et al. 2001; Christopher et al. 2006). Then, the supportive courses are compacted using static or dynamic methods, depending on the soil particle characteristics and consistency (Parsons et al. 2001; Lal and Shukla 2004). The minimum compaction requirements for roadway supportive course materials are typically 95%–100% of the maximum dry density (Parsons et al. 2001; Christopher et al. 2006; ODOT 2013; USDT-FHWA 2014; WSDOT 2016). Compaction requirements vary due to location, material used, design load, and type of construction (Parsons et al. 2001; Christopher et al. 2006; ODOT 2013). The bulk densities specified for road base and subbase construction materials in Ohio are between 1.4 Mg m^{-3} (finer soils) and 2.2 Mg m^{-3} (coarse soils), with minimum compaction requirements ranging from 98% to 102% for coarse and fine soils, respectively (ODOT 2013). The embankment foundation densities at seven highway sites in eastern Kansas ranged from 1.6 Mg m^{-3} to 1.8 Mg m^{-3}, with an average relative compaction of 97% (range: 84%–104%) for soil depths of 40–420 cm, approximately 1–7 years following construction (mean: 2.9 years) (Parsons et al. 2001).

Direct forms of roadside compaction occur during earthwork activities (Steinfeld et al. 2007; WSDOT 2016). Roadside compaction may also be an indirect result of the movement of commonly used large equipment during or after construction (Randrup and Dralle 1997). Similarly, compaction also occurs from the lengthy storage of heavy construction materials (Lichter and Lindsey 1994a; WSDOT 2016). Lichter and Lindsey (1994a) observed that eight passes with a front end loader increased the bulk density of a silty loam soil at a depth of 0–10 cm by more than 10% (1.4 Mg m^{-3} to 1.6 Mg m^{-3}); however, at greater depths, the bulk density increased by less than 3%. These compaction effects, resulting from large equipment, are known to have long residence times and bulk density increases with increasing soil depth (Brady 1990; Vora 1998; Trombulak and Frissell 2000; Lal and Shukla 2004).

The soil dry bulk density (ρ_b) and the soil bulk density corrected for gravel (ρ_{bc}) were generally similar between the two study sites, depending on soil depth and the distance from the road. Between the study sites, the ρ_b differed by <6.5% and the ρ_{bc} differed by <10%. The soil density increased with increasing soil depth, regardless of the gravel content. The dry ρ_b for site 1 ranged from 0.9 to 1.7 Mg m^{-3}, with mean and median ρ_b values ranging from 1.2 to 1.6 Mg m^{-3}. The ρ_b at site 2 ranged from 1.0 to 1.9 Mg m^{-3}, with mean and median values of 1.1–1.6 Mg m^{-3}. Although not significant, a higher ρ_b was observed closer to the road at both sites. The dry ρ_b was generally higher at site 1 than at site 2, and the ρ_{bc} was generally higher at site 2 than at site 1. This change in pattern may suggest the important roles of gravel and soil texture in this landscape.

Although similar high ρ_b values found in soils that directly support roadway infrastructure were not expected in the roadside soil samples, the mean ρ_b for both study sites at a soil depth of 30–45 cm was generally less than the corresponding mean bulk densities of surface soil samples obtained from the Washington, DC Mall or along streets in New York (Craul and Klein 1980; Short et al. 1986). As observed at the study sites, optimal root growth is thought to occur at a soil ρ_b of ≤1.4 Mg m^{-3} in loam soils (USDA et al. 2001). Regardless of the distance from the road, average ρ_b and ρ_{bc} values of <1.4 Mg m^{-3} were observed at a soil depth of 0–15 cm at both sites. Average ρ_b values of ≤1.4 Mg m^{-3} extended to a soil depth of 15–30 cm far from the road at site 2. Average ρ_{bc} values of ≤1.4 Mg m^{-3} were observed at both sites and both distances to a soil depth of 15–30 cm, and extended to a soil depth of 30–45 cm for both distances from the road at site 1. In fact, across both sites at a soil depth of 0–15 cm, 31% and 67% of the ISSL measures were ≤1.1 Mg m^{-3} for ρ_b and ρ_{bc}, respectively.

Low soil ρ_b values have also been found on highway roadsides in Kentucky and Michigan to a soil depth of ≤20 cm (Trammell et al. 2011; Haan et al. 2012). Furthermore, a ρ_b of <1.0 Mg m^{-3} was observed along the Trans-Canada Highway in Newfoundland (Karim and Mallik 2008). The lower soil bulk densities observed in Canada and Kentucky may be an artifact of the roadside locations within park or urban forested environments, which receive higher OM inputs. In California, lower soil ρ_b values were also observed on "selectively" graded residential construction sites (Lichter and Lindsey 1994b). This California study suggested that the response was related to the soil characteristics at the site, which reduced compaction levels (Lichter and Lindsey 1994b). With different soil characteristics but similar low ρ_b values between the sites, it was hypothesized that lower ρ_b and ρ_{bc} values potentially resulted from cultivation and the addition of OM at the study sites before planting the trees. Similar site preparation decreased the soil ρ_b to <1.0 Mg m^{-3} in an afforestation study in a New York City park (Oldfield et al. 2014). Significant decreases in ρ_b have also been observed at tilled sites (20–30 cm) with known vehicle disturbance in North Carolina; however, these differences were not as prominent as those observed in New York City (Haynes et al. 2013).

11.4.4 VEGETATION AND SOIL STRENGTH

Roadsides are often revegetated after construction to provide stability and prevent soil erosion (Steinfeld et al. 2007; WSDOT 2016). Roadside vegetation often consists of low-growing, fine-rooted herbaceous materials (i.e., grasses) (Gale and Biesboer 2004; Nelson Brown et al. 2011; Harrison 2014). Herbaceous groundcover provides expedient erosion control, maintains a safe off-road space for errant cars, and can be feasibly implemented and maintained (Gale and Biesboer 2004; USDT-FHWA and Volpe Ctr 2010; Nelson Brown et al. 2011; Harrison 2014). ANHS pervious ROW land is estimated to contain 64% turf grass, 11% shrubs, 14% trees, and 11% tree/grass mix (USDT-FHWA and Volpe Ctr 2010; Harrison 2014).

Soil strength, in relation to plant rooting, is a product of soil ρ_b, texture, and moisture content (Lal and Shukla 2004; Day et al. 2010). For plants to grow and survive, roots must elongate, regenerate, and acquire essential resources within the soil (Beeson and Gilman 1992; Struve 2009). Therefore, the growth of plant roots can cause deformations within the soil matrix (Day et al. 2010).

Root exploration within soil is constrained by low soil moisture content (Perry 1994; Watson 1994; Struve 2009), anaerobic soil conditions (Perry 1994), or natural or constructed impenetrable soil (Perry 1994; Day et al. 2010). Thus, roots orientate and grow where resources exist and where soil penetration is easier (Watson and Himelick 1982; Gilman et al. 1987; Perry 1994).

Soil compaction hinders plant growth and productivity (Lal and Shukla 2004; Trowbridge and Bassuk 2004; Day et al. 2010) by augmenting soil structure, decreasing soil porosity, and creating discontinuous pore spaces (Craul 1992; Lal and Shukla 2004; Trowbridge and Bassuk 2004). Discontinuous pore spaces prohibit soil water infiltration, limit gaseous exchange, constrain soil water movement, limit rooting volume, and inhibit soil biological functions (Craul 1992; Lal and Shukla 2004; Trowbridge and Bassuk, 2004; Day et al., 2010). Thus, a maximum soil density of <80% has been suggested for the implementation of roadside vegetation (WSDOT 2016).

Daddow and Warrington (1983) integrated ρ_b and soil texture to express the relationship of root penetration occurring at increased ρ_b within coarser soil textures and vice versa. USDA et al. (2001) classifies root penetration limits as either affected or restricted, where rooting is considered affected in coarse-textured soils at a ρ_b of >1.6 Mg m^{-3} and restricted at >1.8 Mg m^{-3}. Silt and clay loam soils are considered affected at a ρ_b of 1.6 Mg m^{-3} and restricted starting between 1.7 and 1.8 Mg m^{-3} (USDA et al. 2001). In fine-textured soils, roots may be affected with a ρ_b between 1.4 and 1.5 Mg m^{-3} and restricted starting between 1.5 and 1.6 Mg m^{-3} (USDA et al. 2001). Lichter and Lindsey (1994b) observed construction sites requiring 90% relative compaction based on ρ_b surpassed the acceptable ranges for root penetration for many soil texture classes.

Soil strength decreases as the soil moisture content increases (Lal and Shukla 2004; Day et al. 2010). Fine-textured dry soils can have high soil strength; however, after adequate moisture is added, lower soil strength may allow root penetration to occur even at excessive ρ_b levels (Lal and Shukla 2004; Day et al. 2010). Soil strength may begin to affect plant root penetration around 1.5 MPa and restricts rooting at 2.5 MPa (Kees 2005); however, these limits vary by plant species (Day and Bassuk 1994; Day et al. 2010). Lessening soil strength may require excessive amounts of water (Perry 1994; Lal and Shukla 2004; Day et al. 2010). Many roadside landscapes rely on natural sources of water (Steinfeld et al. 2007), where the actual intake and consistent distribution of soil water depend on both the surrounding climate and soils. Under dry conditions, roots may penetrate soils with excessive ρ_b values by passing through macropores or cracks in the soil matrix (Day et al. 2010; Short et al. 1986). This was thought to be the case in Washington, DC, where plants were able to penetrate soils with densities of >1.8 Mg m^{-3} (Short et al. 1986). However, transitions into planting environments with increased soil strength generally have a negative impact on root growth, at least initially (Day et al. 2010). Thus, roadside plant installation practices in Minnesota specify soil strength of <1.4 MPa to a soil depth of 41 cm (MNDOT 2016).

Only bulk densities were measures at the study sites, where the upper range of ρ_b measures exceeded the affected limit at both sites beginning at a soil depth of 15 cm, regardless of the distance range, except at site 2 at a depth of 15–30 cm near the road, where the upper range of ρ_b exceeded the restricted limit. Furthermore, at site 2, even the average ρ_b near the road at this depth was within the affected rooting limit. At site 1, the soil ρ_b potentially affected root penetration in approximately 25% and 33% of ISSLs (n = 24/site) at soil depths of 15–30 cm and 30–45 cm, respectively. At site 2, ρ_b potentially affected root penetration of 25% and 50% of the ISSLs at depths of 15–30 cm and 30–45 cm, respectively (using 1.6 Mg m^{-3} as the limit for site 2). When correcting for gravel, ρ_{bc} may not have impacted root penetration at site 1 at any depth or distance; however, site 2 showed potential effects on root growth at the same distances and depths. McGrath and Henry (2015) reported average ρ_{bc} values in fine highway roadside soils that potentially constrained root growth at the surface and at deeper depths. Furthermore, a significant number of individual sampling sites were characterized by ρ_b values greater than the theoretical root penetration limits (McGrath and Henry 2015). Subsoiling is a common practice for alleviating compaction and reducing soil strength; however, effectiveness varies and it may offer a relatively short interval of relief (Day and Bassuk 1994). Roadside landscape practices have specified subsoiling to soil depths

>40 cm (MNDOT 2016; WSDOT 2016) and Day and Bassuk (1994), Haynes et al. (2013), and Oldfield et al. (2014) all note the important impact of soil improvement during initial plant establishment.

11.5 SOIL MOISTURE AND POROSITY

For comparison, calculations of the total soil porosity (f_t), volumetric moisture content (θ), and air porosity (f_a) were conducted based on data provided in the literature by Parsons et al. (2001) and Scharenbroch and Catania (2012). The calculated values are noted as "calculated" in this chapter. Scharenbroch and Catania (2012) only included data from four suburban residential street-side plantings. The equations used for these calculations follow those presented in Lal and Shukla (2004), and the particle and water density were assumed to equal 2.65 Mg m^{-3} and 1.0 Mg m^{-3}, respectively, when they were not provided. Other data was converted from their original units for comparison.

Failure of roadway infrastructure is well connected with soil–water relationships (AASHTO 1993; WTIC 1995; Hassan and White 1996; Christopher et al. 2006). Consequences occur when a soil remains saturated for a long period of time, particularly when the soil is subjected to ongoing vehicle loads (Cyr and Chiasson 1999). Free water at the surface or within supportive pavement courses will decrease the roadway service life and result in deformities and cracking of the pavement surface (AASHTO 1993; Hassan and White 1996; Christopher et al. 2006). Within sloped features, excessive water may cause instability and possible failure (Parsons et al. 2001; Steinfeld et al. 2007). Water dynamics within roadway systems can be particularly important for roads constructed with flexible pavements, in cold climates, and/or on expansive or collapsible soils (Cedergen et al. 1972; Hassan and White 1996; Christopher et al. 2006). Soil hydraulic characteristics may vary with water inputs, the duration of moisture exposure, regional locations, existing soil moisture contents, soil material selection, and the design of the roadway (Cedergen et al. 1972; AASHTO 1993; Hassan and White 1996). Water can erode roadway surfaces, change the volume of the soil material, augment topography, break down and reorganize soil particles (changing the soil gradation, type, and concentration), increase soil weight, alter pressures within the soil matrix, change state based on the thermal surroundings, and decrease soil strength (Hassan and White 1996; Cyr and Chisasson 1999; Christopher et al. 2006; Steinfeld et al. 2007). Thus, roadway design focuses on both preventing the infiltration of water and rapidly draining water within a resilient constructed system (AASHTO 1993; Christopher et al. 2006).

Water inhabits roadway systems as a result of construction methodologies, the use of porous materials, including pavements, environmental exposure, and roadway use (Cedergen et al. 1972; Hassan and White 1996; Rainwater et al. 1999). The *in situ* soil moisture contents measured in seven highway embankments at soil depths of 40–420 cm in Kansas ranged from 9.7% at 280 cm to 26.1% at 240 cm, with an average soil moisture content of 19.6% (Parsons et al. 2001). Water typically enters a roadway support course vertically through openings in the sealed surface (pavement) or horizontally as the roadside soil becomes saturated (Cedergen et al. 1972; Cyr and Chiasson 1999). Significant water can enter roadways through pavement cracks, which leads to more rapid subsequent saturation than lateral water movement (Cyr and Chiasson 1999; Christopher et al. 2006). The upward movement of water through capillary action from groundwater sources or changes in the state of the water already present in the support courses can also alter the water content (Hassan and White 1996; Christopher et al. 2006). Furthermore, the flow within the soil matrix may be hindered when restrictive layers exist within the soil due to physical barriers, such as laminate rock, developed soil pans, or a change in the overall soil material texture within the constructed soil layers that affect hydraulic pressures (Lal and Shukla 2004; Steinfeld et al. 2007).

Thus, most bases and subbases must balance two tasks: to provide uniform stability for carrying loads and to provide adequate drainage (Hall and Shiraz 2009; IOWA 2013). Greater permeability is achieved by using large open-graded aggregate materials for optimum pore architecture (Christopher et al. 2006; ACPA 2008; Hall and Shiraz 2009). Drainage may also be improved by adding supplemental drainage infrastructure between or within the layers supporting the pavement

(Christopher et al. 2006; ACPA 2008). In the case of concrete road systems, subbases during the 1990s commonly had permeabilities in excess of 2400 m day^{-1} (Hall and Shiraz 2009). However, the loss of stability within these pavement systems had become a concern (ACPA 2008). More recently, the recommended range of permeability for permeable supportive courses was set as 152–244 m day^{-1} (Hall and Shiraz 2009) or lower (15–107 m day^{-1}), depending on the design of the roadway system (ACPA 2008). For additional stability, granular base and subbase materials may be stabilized with lightweight concrete, asphalt, or by using chemical binding treatments (Christopher et al. 2006). Furthermore, finer particles (<US No. 200 sieve) are typically part of engineered granular support layers because they increase stability by decreasing the void ratio (Cyr and Chiasson 1999; Christopher et al. 2006). However, the specified percentages of fine materials generally remains ≤10% because fine materials increase water retention, decrease permeability, and increase the susceptibility of the material to climate and environmental factors (Cyr and Chiasson 1999; Christopher et al. 2006; IOWA 2013).

Substrate gradation, particle size distribution, and type of material significantly impact the permeability of bases and subbases (Bennert and Maher 2005; IOWA 2013). Commonly used mixtures of soils for subbase and base courses that were collected from multiple regions of New Jersey had permeability values ranging from 8.7 to 17 m day^{-1} and 37.5 to 52 m day^{-1}, respectively (Bennert and Maher 2005). Within the particle gradations, the permeability of the New Jersey subbase material ranged from >110 m day^{-1} for the coarsest material to <2 m day^{-1} for the finest material (Bennert and Maher 2005). In pure sand, Ahmed et al. (1997) found that a 3% increase in fine particles (<US No. 200 sieve) (from 3% to 6%) decreased the hydraulic conductivity by 57%. Similarly, in pure stone, a 2.5% increase in fine material (from 7.5% to 10%) decreased the hydraulic conductivity by 44% (Ahmed et al. 1997).

Due to the water avoidance strategies implemented in roadway construction, water quantity and quality affect the roadside landscape by impacting the slope or surface stability, runoff volume and content, and plant–soil–water relationships. Slopes and surfaces along roadsides may become unstable due to poor soil strength and/or particle distributions differences between soil layers, in conjunction with terrain geometry and soil water content (Steinfeld et al. 2007). On slopes, cohesive soils result in deeper interior impacts and noncohesive soils impact shallower depths (Steinfeld et al. 2007). Greater differences between the compositions of soil layers and hydraulic properties can result in a division or boundary between soil layers (Steinfeld et al. 2007). Separation may occur when unbalanced soil weight and/or pore pressure occur in one layer, typically topsoil, that overcomes the strength of the underlying layer (Steinfeld et al. 2007).

Planting soil surfaces generally improves their stability (Gray and Sotir 1996; Steinfeld et al. 2007; WSDOT 2016); however, planting can also have negative impacts (Gray and Sotir 1996; Steinfeld et al. 2007; Nelson Brown et al. 2010). These impacts generally depend on a plant's rooting habit (Gray and Sotir 1996; Steinfeld et al. 2007). Plants can decrease the slope stability if rooting only occurs in the topsoil and not in the compacted subgrade, which would further differentiate the boundary layer (Gray and Sotir 1996; Nelson Brown et al. 2010). On roadsides in Rhode Island, 25%–88% of native and ornamental grass roots (by weight) were located in the upper 7.6 cm of soil (Nelson Brown et al. 2010). In addition, most tree roots are typically located in the upper 38 cm of the soil (Watson and Himelick 1982; Gilman et al. 1987). The physical soil environment and the need for resources may supersede genetically driven root growth habits in roadside environments (Watson and Himelick 1982; Gilman et al. 1987), particularly for species known to demonstrate plasticity within their roots (Fischer and Brinkley 2000). By the end of the study period, root architecture was observed at the soil surface of both study sites.

Urban areas, with large impervious surface area, have greater rates of runoff than underdeveloped land parcels (Line and White 2007; Kramer 2013). The effects of stormwater runoff on soil depend on the speed of impact and movement of the water across the land, terrain gradient, travel distance, compaction, surface strength, soil texture, and infiltration rate (Lal and Shukla 2004; Steinfeld et al. 2007). Significant decreases in infiltration rate can occur in response to compaction during construction activities (Gregory et al. 2006; Pitt et al. 2008; Haynes et al. 2013). Gordon (2014) measured *in situ* infiltration at a soil depth of 50 cm along an Ohio highway roadside with finer textured soils and observed a low average infiltration rate of approximately 0.01 cm hr^{-1}. Cyr and Chiasson (1999) observed roadside hydraulic conductivities of between 0.1 and 2.8 cm hr^{-1} under saturated and coarse soil conditions. Excessive water can result in the development of rills and gullies on the soil surface (Steinfeld et al. 2007). Rills were observed at both study sites, but were more prominent at site 1.

Roadside soil physical, biological, and chemical properties may be influenced by the quality of stormwater from adjacent impervious road surfaces (Kramer 2013). Pollution may originate from two sources: toxic or excessive concentrations of substances or the transformation of the existing water (e.g., increasing the water temperature) (Strassler et al. 1999; Kayhanian et al. 2007). Excessive substances may include heavy metals, hydrocarbons, salts, dissolved and suspended solids (i.e., eroded soil sediments), nutrients, direct products from vehicles, litter, and bacteria (Strassler et al. 1999; Kramer 2013). Pollutants can attach to and remain on pervious or impervious surfaces until a water event. The concentrations of pollutants in runoff can vary with vehicle volume, road type, season, moisture event (type, duration, and volume), distance from roadway, type of exposed particles near the roadway, monitoring method, location, and surrounding environmental characteristics (Coffin 2007; Kayhanian et al. 2007).

The relationships between roadside vegetation and water are highly dependent on the soil porosity and the ratio of water filled to air-filled pores over time (Lal and Shukla 2004; Trowbridge and Bassuk 2004). Soil aeration and water content are inversely related (Lal and Shukla 2004) and are a function of soil texture and compaction, which control the amount and size of soil pores (Lal and Shukla 2004; Trowbridge and Bassuk 2004). An air-filled soil porosity of ≤10% impairs root growth and survival (Glinski and Stepniewski 1985; Craul 1992; Kelsey 1994; Lal and Shukla 2004). Adequate soil aeration may be compromised because of compaction alone (Lal and Shukla 2004; Day et al. 2010). However, water may be retained in soil profiles depending on the site design, excessive fine particle soil content, and high SOM content and if a barrier exists in the soil that results in a higher water table (Craul 1992; Lal and Shukla 2004; Steinfeld et al. 2007). It has been suggested that root stress is lessened in a compacted, unsaturated soil compared to a compacted, saturated soil (Day et al. 2010).

In contrast, coarse-textured soils with high porosity and low ρ_b may drain quickly, resulting in drought conditions. The types of high-level environmental exposure that commonly occur along roadways include increased temperatures; solar radiation; and wind exposure, altered wind patterns, and lower atmospheric and soil moisture contents (Trombulak and Frissell 2000; Coffin 2007; Gale and Biesboer 2004). These conditions may be exacerbated by ROW design, site aspect, soil and pavement color, buried debris or existing gravel pans, and low SOM concentrations (Craul 1992; Jim 1998; Lal and Shukla 2004; Haynes et al. 2013). Root growth in response to adverse environmental conditions can be species specific (Trowbridge and Bassuk 2004; Day et al. 2010). For example, some species have predispositions or adaptations to rocky terrain (Perry 1994) or wet bottomlands (Day et al. 2010), enabling their growth in either less than ideal soil and/or soil moisture conditions, while other species have very particular soil moisture needs (Trowbridge and Bassuk 2004).

11.5.1 Total Porosity

The data in Tables 11.1 and 11.2 show soil porosity results for the study sites. The total soil porosity (f_t) for site 1 ranged from 34% to 67%, with mean and median porosities ranging from 39% to 56%. The f_t at site 2 ranged from 30% to 62%, with mean and median concentrations of 41%–57%. At both sites, f_t decreased with depth and increased slightly by correcting for gravel (f_{tc}). These higher porosities were unexpected in this landscape and are more akin to agricultural soils (Lal and Shukla 2004). These values are thought to be the result of site preparation. At a soil depth of 0–15 cm, the average f_t and f_{tc} were comparable to the calculated f_t from four Illinois suburban street-side sites

TABLE 11.2

Biological and Chemical Soil Property Means or Medians[a] and Ranges for Three Soil Depths (0–15, 15–30, and 30–45 cm) and Two Distances from the Road (Near, Far) in Two Sites (1, 2) at the Intersection of Interstate 270 and Hamilton Road (OH 317) in Gahanna, Ohio

Biological and Chemical Soil Properties

Distance Site	0–15 cm Mean/Median	0–15 cm Min–Max Range	15–30 cm Mean/Median	15–30 cm Min–Max Range	30–45 cm Mean/Median	30–45 cm Min–Max Range
Moisture Content (θ) (m³ m⁻³, %)						
Near						
1	28.0 A,[b],[bc]	18.0–37.0	23.0 B,a	19.0–31.0	22.0[k] A,a	18.0–32.0
2	29.0 A,a	19.0–38.0	30.0 A,a	19.0–39.0	27.0[k] A,a	22.0–33.0
Far						
1	35.0 A,a	27.0–50.0	25.0 A,a	18.0–32.0	26.0[k] A,a	22.0–30.0
2	27.0 B,a	21.0–31.0	28.0 A,a	22.0–36.0	30.0[k] A,a	22.0–34.0
Air Porosity (fₐ) (m³ m⁻³, %)						
Near						
1	26.0 A,a	11.0–36.0	17.0 A,a	10.0–26.0	14.0[k] A,a	5.0–29.0
2	22.0 A,a	6.0–36.0	14.0 A,a	6.0–28.0	12.0[k] A,a	6.0–22.0
Far						
1	21.0 B,a	10.0–34.0	17.0 A,a	8.0–29.0	14.0[k] A,a	7.0–24.0
2	30.0 A,a	19.0–39.0	18.0 A,a	5.0–36.0	14.0[k] A,a	7.0–30.0
Moisture Content Corrected for Gravel (θ_c) (m³ m⁻³, %)						
Near						
1	26.0 A,b	15.0–36.0	20.0 B,a	16.0–24.0	21.0 B,a	15.0–29.0
2	28.0 A,a	17.0–37.0	28.0 A,a	15.0–38.0	26.0 A,a	21.0–32.0
Far						
1	33.0 A,a	24.0–48.0	22.0 B,a	15.0–29.0	22.0 B,a	18.0–26.0
2	26.0 B,a	20.0–30.0	27.0 A,a	21.0–35.0	28.0 A,a	20.0–34.0
Air Porosity Corrected for Gravel (f_ac) (m³ m⁻³, %)						
Near						
1	32.0 A,a	13.0–47.0	28.0 A,a	16.0–39.0	24.0[k] A,a	14.0–44.0
2	26.0 A,a	8.0–41.0	19.0 A,a	7.0–39.0	19.0[k] A,a	9.0–26.0
Far						
1	27.0 A,a	12.0–39.0	28.0 A,a	18.0–43.0	27.0[k] A,a	19.0–43.0
2	32.0 A,a	23.0–40.0	20.0 A,a	6.0–37.0	12.0[k] A,a	7.0–32.0

(Continued)

TABLE 11.2 (Continued)
Biological and Chemical Soil Property Means or Medians[a] and Ranges for Three Soil Depths (0–15, 15–30, and 30–45 cm) and Two Distances from the Road (Near, Far) in Two Sites (1, 2) at the Intersection of Interstate 270 and Hamilton Road (OH 317) in Gahanna, Ohio

Biological and Chemical Soil Properties

Distance Site	pH[a] Mean/Median	Min-Max Range	Mean/Median	Min-Max Range	Mean/Median	Min-Max Range	Organic Matter (%) Mean/Median	Min-Max Range	Mean/Median	Min-Max Range	Mean/Median	Min-Max Range	Electroconductivity (EC) (dS m^{-1}) Mean/Median	Min-Max Range	Mean/Median	Min-Max Range	Mean/Median	Min-Max Range	Total Carbon Stock (Mg C ha^{-1}) Mean/Median	Min-Max Range	Mean/Median	Min-Max Range	Mean/Median	Min-Max Range
Near																								
1	7.8^{k} A,a	7.4–8.3	8.1^{k} A,a	7.6–8.8	7.9^{k} A,a	7.4–8.8	7.0^{k} A,a	4.1–15.2	3.2^{lw} A,a	1.9–4.9	3.3^{lw} A,a	2.2–4.6	0.6^{k} A,a	0.4–2.3	0.6^{k} A,a	0.4–1.8	0.6^{k} A,a	0.3–2.2	175.5 A,a	120–301	89.0^{k} A,a	52–146	116.4^{k} A,a	51–170
2	7.7^{k} A,a	7.4–8.3	7.8^{k} A,a	7.5–8.3	7.6^{k} A,a	7.2–8.5	6.6^{k} A,a	2.7–9.2	4.4^{lw} A,a	1.9–8.0	2.9^{w} A,a	2.3–3.5	0.6^{k} A,a	0.4–0.9	0.6^{k} A,a	0.3–1.2	0.6^{k} A,a	0.3–1.5	205.2 A,a	112–281	109.4^{k} A,a	50–251	80.2^{k} A,a	62–109
Far																								
1	7.7^{k} A,a	7.5–8.3	7.8^{k} A,a	7.6–8.7	7.9^{k} A,a	7.4–8.2	8.1^{k} A,a	2.5–11.1	3.8^{lw} A,a	2.5–7.1	3.9^{w} A,a	3.3–4.9	0.6^{k} A,a	0.5–1.7	0.6^{k} A,a	0.3–1.6	0.6^{k} A,a	0.5–1.3	180.1 A,a	88–244	103.6^{k} A,a	70–165	128.8^{k} A,a	66–180
2	7.4^{k} B,b	7.2–7.7	7.4^{k} A,a	6.3–8.1	7.6^{k} A,a	7.1–8.1	8.4^{k} A,a	5.5–11.4	3.4^{lw} A,a	2.7–4.5	2.8^{w} B,a	2.2–4.9	0.3^{k} B,b	0.3–0.5	0.3^{k} B,b	0.2–0.4	0.2^{k} B,b	0.1–0.4	207.8 A,a	55–384	90.9^{k} A,a	59–293	68.5^{k} A,a	52–199

[a] Mean reported for parametric ANOVA unless: (w) Welch's corrected ANOVA, mean reported, (l) natural log transformation performed on data prior to analysis, back-transformed means reported, (k) Kruskal–Wallis test performed, median reported, pH values are back-transformed from log transformations. n = 12/site/distance/depth; Tukey–Kramer adjustment with all pairwise comparisons, $p \leq .05$.

[b] Means or medians with unlike capital letters are statistically different within distance from the road between sites.

[c] Means or medians with unlike lowercase letters are statistically different between distances from the road within site.

(Scharenbroch and Catania 2012) for both study sites and distance ranges from the road. Similarly, at a soil depth of 30–45 cm, the median f_t was comparable to the calculated f_t for four Kansas highway embankments at a soil depth of 40–50 cm (Parsons et al. 2001) for both study sites and distance ranges from the road. Lower average f_t ($\leq$35%) values were observed along urban street-side planting spaces in New York and Hong Kong at the soil surface (0–10 cm) and at deeper depths (10–35 cm) (Craul and Klein 1980; Jim 1998).

11.5.2 Volumetric Moisture Content and Air Porosity

Greater differences between sites were observed after dividing f_t between the volumetric moisture content (θ) and air porosity (f_a). Within site 1, θ was consistently higher farther from the road, which was also true for site 2 at soil depths of 15–30 cm and 30–45 cm. This difference corresponds with the construction, slope, and greater fine particle concentrations farther from the road. The soil θ ranged from 18% to 50% at site 1 (site mean: 27%) and from 19% to 39% at site 2 (site mean: 28%). Site 2 had a higher θ than site 1 for all distance ranges and soil depths, except for the 0–15 cm soil depth farther from the road; this location was the only significant difference for θ between sites at similar distance ranges. Volumetric soil moisture content corrected for gravel (θ_c), slightly decreased at both sites at all distance ranges and soil depths, exposing a number of significant differences between the sites. Scharenbroch and Catania (2012) measured the θ to a depth of 20 cm and obtained average values that were 6%–25% higher than the average values found for θ or θ_c at either study site and distance range at a soil depth of 0–15 cm. At both sites and distance ranges the median θ at a soil depth of 30–45 cm was lower than the calculated θ along four Kansas highway embankments that were sampled at a soil depth of 40–50 cm (Parsons et al. 2001). The average θ and θ_c at a soil depth of 0–15 cm at both sites and all distance ranges were similar to the θ at field capacity along three Minnesota roadsides (Friell et al. 2014). However, the study samples were not all thought to be at field capacity due to the long sampling time (multiple weeks each year), as water holding capacity would be affected by differing roadside exposure and soil properties.

Neither air porosity (f_a) nor corrected (f_{ac}) resulted in any significant differences between the distance ranges from the road within each of the two study sites. Both f_a and f_{ac} generally decreased with depth, which would correspond with the increase in ρ_b or ρ_{bc}. The soil f_a ranged from 5% to 36% and from 5% to 39% at sites 1 and 2, respectively. The average f_a and median f_{ac} at a soil depth of 30–45 cm at both sites and at all distances exceeded 10%; however, the mean f_a was $\leq$14%, and the median f_{ac} ranged from 12% to 27%. The calculated air porosity values at three of the four suburban street-side sites in Illinois (Scharenbroch and Catania 2012) were similar to both site averaged f_a at the shallowest depths for all distance ranges. However, these calculated values were based on the gravimetric water content rather than on the given θ values (Scharenbroch and Catania 2012).

Site 1 had a minimum range for f_a ($\leq$10%) starting at a soil depth of 15 cm near the road and at all soil depths farther from the road. At site 2, all distance ranges and soil depths, except for the 0–15 cm soil depth farther from the road, resulted in minimum f_a ranges of $\leq$10%. Site 1 had consistently more f_a and f_{ac} compared with site 2 closer to the road. This trend did not continue farther from the road because site 2 had greater f_a than site 1 at shallower soil depths. As expected, f_a increased with the correction for gravel for both sites at all distance ranges and depths because the volume of rock was replaced by air. Excluding gravel did not affect the minimum ranges of f_{ac} at site 2, with five of six depth and distance combinations resulting in <10% f_a. Just under half (11 of 24) of the distance and depth combinations had ratios of θ:f_a or corrected θ_c:f_{ac} ratios that favored water by 10%–20%. In two instances, correcting for gravel would have reversed and favored air-filled porosity (at site 1 for the 15–30 cm soil depth for both distance ranges).

Considering all distances from the road, at a soil depth of 0–15 cm, 4.2% and 8.3% of the ISSLs were characterized by $\leq$10% f_a at sites 1 and 2, respectively. Furthermore, the number of samples with f_a $\leq$10% increased with depth, accounting for 12.5% and 33.3% of the ISSLs at sites 1 and 2, respectively, at a soil depth of 15–30 cm and 20.8% and 41.7% of the ISSLs at sites 1 and 2, respectively, at a soil depth of 30–45 cm. In the correct soil environment, plants can easily root to these depths, which may be particularly true for most of the study trees that were planted in fully rooted containers with a depth of 22.5 cm. Along roadsides in Rhode Island, 15 of 23 grass species maintained between 1% and 15% of their root mass at depths of 68.4–76.0 cm (Nelson Brown et al. 2010).

11.6 SOIL BIOLOGICAL AND CHEMICAL PROPERTIES

The biological and chemical soil property results for the study sites are shown in Table 11.2. Roadside soils typically have altered soil chemistry with high pH values, salt contents, heavy metal contents, low nutrient contents, low OM contents, and decreased biological activity due to exposure to roadway construction materials, disturbance, and use (Craul 1992; Jim 1998; Gale and Biesboer 2004; Trowbridge and Bassuk 2004; Day et al. 2010; Haynes et al. 2013). This discussion is limited to the roadside levels of pH, electroconductivity (EC), SOM, and TCS. Although all four soil properties either directly or indirectly impact nutrient availability (Brady 1990; Lal and Shukla 2004; Trowbridge and Bassuk 2004), further measures of soil nutrient levels along roadsides are warranted as they directly affect vegetation establishment and growth (Brady 1990; Trowbridge and Bassuk 2004; Steinfeld et al. 2007). EC and OM also have important implications for soil physical properties (Lal and Shukla 2004; Fay and Shi 2012). Moreover, soil properties of TCS and total carbon are becoming of greater interest due to the need for understanding carbon dynamics and the potential for increasing ecological benefits in large ROW landscapes (USDT-FHWA and Volpe Ctr 2010; Bouchard et al. 2013; Ament et al. 2014).

11.6.1 pH

Similar to urban soils, alkaline pH levels in roadside landscapes are typically associated with the use and weathering of construction materials (Trowbridge and Bassuk 2004), including the use of more neutral, yet alkaline amendments for revegetation (Trowbridge and Bassuk 2004; Nelson Brown et al. 2011). pH levels can also vary based on geographic location, weathering potential, amount and type of environmental exposure, distance from the road, and soil parent material (Brady 1990; Jim 1998; Bryson and Barker 2002; Day et al. 2010). The accepted optimal pH range for plant growth is 5.5–6.2 (Kelsey 1994). However, pH levels ≥8.0 are common in urban and roadside environments (Jim 1998; Day et al. 2010). The median soil pH at site 1 reached 8.0 at a soil depth of 15–30 cm near the road, while the average soil pH at site 2 did not exceed 7.8. However, the maximum soil pH ranges for both sites and for most of the depth and distance combinations exceeded 8.0. Site 1 had consistently higher soil pH values than site 2. Generally, higher pH values were found closer to the road. At site 2, the 0–15 cm soil depth farther from the road had a pH that was three times less (pH 7.4) than this depth closer to the road. Gordon (2014) found similar and more neutral average soil pH values at a nearby Ohio roadside, and pH values of ≥7.8 were reported along highways in Minnesota (Gale and Biesboer 2004; Friell et al. 2014) and Michigan (Haan et al. 2012). pH values of <7.0 were found along highways in Newfoundland and Ontario in Canada (Karim and Mallik 2008; Eimers et al. 2015), North Carolina (Bouchard et al. 2013), Rhode Island (Nelson Brown et al. 2011), and Kentucky (Trammell et al. 2011).

The soils surrounding the study sites are classified as Cardington, Bennington, and Eldean with the study sites classified as Udorthents (NRCS-WSS 2015). Non-Udorthent soils range in pH from 5.9 to 6.5 at a soil depth of 0–15 cm and from 6.1 to 6.6 at a soil depth of 45 cm (NRCS-WSS 2015). The occurrence of alkaline soil pH levels at the study sites likely resulted from the incorporation of construction materials and subsequent weathering. Tolerance to elevated pH levels is generally species specific due to interference with the availability and solubility of some essential plant nutrients, heavy metals, and biological soil activity (Brady 1990; Trowbridge and Bassuk 2004; Day et al.

2010). Visible chlorosis symptoms, which are typical of high pH, cause manganese deficiencies in red maple (*Acer rubrum*) (Altland 2016) and were evident at both sites.

11.6.2 Salts and EC

Road maintenance during winter months in cold climates includes the use of anti-icing and deicing methods that incorporate abrasives (i.e., sand) and liquid or solid salts (i.e., magnesium, sodium, or calcium chloride), often in combination with plowing (Fay and Shi 2012; Environment Canada 2013). The United States and Canada apply multiple millions of tons of salt to roadways annually (Environment Canada 2013; The Salt Institute 2014). Salts are transported to the roadside through water runoff and/or as aerial spray (Nelson-Brown et al. 2011; Fay and Shi 2012). Salt concentrations found along roadsides are variable and depend on the season, wind, terrain gradient, type of road, vehicle volume, delivery mechanism, speed of delivery, distance from the road, soil properties, and number of dilutions (Health Canada 2001; Bryson and Barker 2002; Nelson Brown et al. 2011; Fay and Shi 2012). Salts typically have greater impacts on the upper 45 cm of soil profiles (Nelson Brown et al. 2011).

Salt affects soil by altering the soil structure, reorganizing organic and inorganic particles, forming soil crusts, and increasing compaction (Health Canada 2001; Nelson Brown et al. 2011; Fay and Shi 2012). Subsequently, salt can decrease soil permeability, aeration, and water holding capacity (Nelson Brown et al. 2011; Fay and Shi 2012), impact soil pH and soil fertility, and can mobilize heavy metals (Health Canada 2001; Nelson Brown et al. 2011; Fay and Shi 2012). The delivery of salt to the roadside affects the above- and belowground plant biomass; however, plant roots are generally more tolerant of salt than plant shoots (Day et al. 2010). Most detrimental plant effects result from desiccation and toxicity at the cellular level (Day et al. 2010), inhibiting physiological processes (Health Canada 2001; Fay and Shi 2012).

Soil EC measures were consistently higher at site 1 than at site 2. Within site 2, significant differences were found between the distance ranges for soil depths of >15 cm. At these locations, measurements closer to the road were two times higher than measurements farther from the road for each soil depth. The EC at site 1 ranged from 0.3 to 2.3 dS m^{-1} (site mean: 0.8 dS m^{-1}). At site 2, the EC ranged from 0.1 to 1.5 dS m^{-1} (site mean: 0.5 dS m^{-1}). Farther from the road, significant differences occurred between the sites at all depths. Site 1 had an EC that was more than two times greater than that of site 2 at soil depths of 0–15 and 15–30 cm and almost three times greater at a soil depth of 30–45 cm. Soils are typically classified as saline if they have an EC of 4 dS m^{-1} (Lal and Shukla 2004); however, differences between nonsaline and slightly saline soils have been noted as >2.0 dS m^{-1} (Brady 1990; Lal and Shukla 2004). Sensitive plants may be affected by EC values of <2.0 dS m^{-1} (Maas and Hoffman 1977; Trowbridge and Bassuk 2004), but the species chosen for these sites had some level of salt tolerance (Gilman and Watson 1994; Trowbridge and Basuk 2004; Wennerberg 2006; USFS et al. 2012). Therefore, most EC levels found at these sites were low and should not have hindered the tree performance. Sampling distance from the road and date may have affected the EC results; samples were taken in June, and spring rainfall events potentially facilitated the leaching of salts from the soil that accumulated during the winter. These results are comparable to findings along roadsides in Massachusetts (Bryson and Barker 2002), Ohio (Gordon 2014), Minnesota (Friell et al. 2014), Kentucky (Trammell et al. 2011), and Ontario, Canada (Eimers et al. 2015). Many of these studies were conducted during the summer or fall. The persistence of plant-harming salt levels was observed to dissipate after early spring and the lack of soil fertility rather than the salt concentration was reasoned for vegetation loss 5 m from the pavement along two Rhode Island highways (Nelson Brown et al. 2011). The higher EC and pH values at site 1 are thought to be a response to greater exposure to climatic conditions because no protective vegetative buffer exists at this site. In addition, the higher measurements may have been impacted by the drainage patterns within and around the site.

11.6.3 SOIL ORGANIC MATTER

SOM is important for improving excessive soil compaction, aeration, nutrient cycling, hydrological, and biological functions (Brady 1990; Lal and Shukla 2004; Trowbridge and Bassuk 2004). Thus, organic amendments are generally recommended for use along roadsides (Steinfeld et al. 2007; WSDOT 2016). The concentration of SOM ranged from 1.9% to 15.2% at site 1 and from 1.9% to 11.4% at site 2. SOM generally decreased with depth and was significantly different within the study sites. The only significant difference between sites occurred at the soil depth of 30–45 cm, farther distance from the road, where the SOM content at site 2 was 1.1% less than that at site 1. Most of the SOM contents were within or exceeded the accepted 2%–5% criteria for healthy soils (Lal and Shukla 2004; Trowbridge and Bassuk 2004). The median concentration of SOM at these sites at a soil depth of 0–15 cm was similar to the median concentration found near forested Kentucky interstates at a similar depth (Trammell et al. 2011) and the average concentrations of suburban streetside landscapes in Illinois at a soil depths of 0–20 cm (Scharenbroch and Catania 2012). However, these sites had higher SOM concentrations than those observed along Minnesota highways at similar depths (Gale and Biesboer 2004; Friell et al. 2014). The benefits of increasing the SOM levels in roadside soils have been observed to persist for multiple years following initial organic amendment application, depending on the type of amendment used (Nelson Brown et al. 2011).

11.6.4 TOTAL CARBON STOCK

For comparison, the TCS was calculated from data provided in the literature by Sharenbroch and Catania (2012). The calculations were performed using equations found in Lorenz and Lal (2012), which are noted as "calculated" in the text and only included suburban residential street-side locations. Other data from Trammell et al. (2011) and Bouchard et al. (2013) were converted from their original units.

No significant differences in TCS were observed within or between the study sites at any soil depth or for any distance range. At site 1, the TCS initially decreased from a soil depth of 0–15 cm to a soil depth of 15–30 cm and then increased at a soil depth 30–45 cm. At site 2, the TCS decreased with depth. These TCS trends were consistent for both sites at each distance range. At site 1, the TCS ranged from 51.0 to 301.0 Mg C ha^{-1}, with mean and median TCS values of 103.6 –175.5 Mg C ha^{-1}. Site 2 had a similar TCS range of 50.0 – 384.0 Mg C ha^{-1} and a wider range of mean and median TCS values (spanning from 68.5 to 207.8 Mg C ha^{-1}). Both sites and all distances from the road and soil depths with minimum TCS ranges and median values of <100 Mg C ha^{-1} were similar to either the calculated TCS values for suburban street-side sites in Illinois (Scharenbroch and Catania 2012) or the median TCS values found along two forested Kentucky highways (Trammell et al. 2011). However, the minimum ranges, means, and medians for all distances and depths were greater than the averages found at corresponding soil depths in urban Columbus, Ohio parks (Lorenz and Lal 2012) or highways in North Carolina (Bouchard et al. 2013). The high TCS levels at the study sites may be similar to those proposed in certain Ohio forested soils; however, Tan et al. (2004) considered TCS to a depth of 1 m, whereas TCS at the study sites was determined to a depth of 45 cm.

The high TCS measurements across both sites and the greater TCS at site 1 for greater depths may result from several factors. A lack of differentiation exists between the forms of organic and inorganic carbon determined by using the loss-on-ignition methodology (Schumacher 2002). The easily fractured shale and asphalt rubble observed at the sites could impact the carbon concentrations. Second, higher ρ_{bc} measures would affect the calculated TCS for these study sites. Alternate forms of carbon and ρ_{bc} values resulted in higher TCS measures in Columbus city parks (Lorenz and Lal 2012). This hypothesis may be supported by the increases in the gravel content and ρ_{bc} with increasing depth at site 1. The trend of TCS decreasing with depth was more expected. The incorporation of organic amendments during site preparation, the decomposition of the applied wood mulch, and location of greater root activity result in greater OM contents and, consequently, greater TCS contents at shallower depths.

11.7 RELATIONSHIPS BETWEEN TREE SURVIVAL AND ROADSIDE SOIL PROPERTIES

The correlation results are summarized in Table 11.3. All significant correlations at both sites occurred at or above a soil depth of 30 cm and may indicate where root activity primarily occurred within the soil profile. The decreased survival rate at site 1 farther from the road was related to the

TABLE 11.3
Significant Correlations between Weighted Yearly Deciduous Tree Survival (%, Pooled Over Species[a]) and Selected Soil Properties by Distance from the Road (Near, Far) and Soil Depth (Depth) (0–15, 15–30, and 30–45 cm) for Sites 1 and 2 at Interstate 270 and Hamilton Road (OH 317) in Gahanna, Ohio

Survival

Site	Distance to Road	Depth (cm)	Soil Property[b]	Rel. +/−[c]	Time(s) When Significant[d]	Time: Correlation and Significance[e]
1	Far	0–15	θ	−	1 Yr	1: $r = -0.58$*
1	Far	15–30	θ	−	2 Yr	2: $r = -0.59$*
1	Far	0–15	Clay	−	2, 3 Yrs	2: $r_s = -0.63$*
						3: $r_s = -0.61$*
1	Near	0–15	ρ_b	−	2, 3 Yrs	2: $r = -0.71$**
						3: $r = -0.62$*
1	Near	0–15	f_t	+	2, 3 Yrs	2: $r = 0.69$**
						3: $r = 0.60$*
1	Near	15–30	Gravel volume	+	2, 3 Yrs	2: $r = 0.75$**
						3: $r = 0.73$**
2	Near	0–15	Sand	−	2, 3 Yrs	2: $r = -0.66$*
						3: $r = -0.65$*
2	Near	0–15	Silt	+	1, 2, 3 Yrs	1: $r = 0.64$*
						2: $r = 0.76$**
						3: $r = 0.78$**
2	Near	0–15	θ	−	1, 2, 3 Yrs	1: $r = -0.73$**
						2: $r = -0.66$*
						3: $r = -0.59$*
2	Near	0–15	f_a	+	1 Yr	1: $r = 0.63$*
2	Near	15–30	f_a	+	1, 2, 3 Yrs	1: $r = 0.79$**
						2: $r = 0.67$**
						3: $r = 0.62$*

a   Tree species include red maple *Acer rubrum*, white-barked Himalayan birch *Betula jacquemontii*, hackberry *Celtis occidentalis*, and Japanese tree lilac *Syringa reticulata*.

b   Soil properties, as calculated means, include dry bulk density (ρ_b, Mg m^{-3}), total porosity (f_t, m^3 m^{-3}, %), volumetric moisture content (θ, m^3 m^{-3}, %), air porosity (f_a, m^3 m^{-3}, %), gravel volume content (cm^3 cm^{-3}, %), and sand, silt, or clay contents (%).

c   Rel: negative or positive relationship found.

d   Survival was conducted for 3 years following installation in June 2011 (1, 2, 3 Yr(s)).

e   (r) Pearson correlation coefficient reported, (r_s) Spearman's correlation coefficient reported, significant differences are at $p \le .05$ (*), $p \le .01$ (**) < $n = 12$ samples/site/distance/depth.

increased θ at a soil depth of 0–15 cm during the first year and at a soil depth of 15–30 cm during the second year. Farther from the road at side 1, the maximum θ at was 50% at a soil depth of 0–15 cm and 32% at a soil depth of 15–30 cm (Table 11.2). Mathematically, given maximum f_t, the range values of f_a would be greater than the critical 10% limit required to sustain plant growth. However, the minimum range of f_t may result in <10% f_a given a higher sustained θ. Furthermore, the minimum ranges of the calculated f_a for site 1 farther from the road are 10% and 8% at soil depths of 0–15 cm and 15–30 cm, respectively (Table 11.2). The negative relationship at a deeper depth observed during the second year may suggest that roots were exploring the soil volume as they established; however, this observation may also indicate that soils at a depth of 15–30 cm may be retaining water. This finding is supported by the below-average precipitation in 2013 (year 2) (USDC-NOAA 2015a, b, data not shown); however, increased θ continued to effect survival. Thus, this relationship may emphasize the importance of the longevity of soil moisture in the roadside landscape. Ponding of water after precipitation events and general extended soil moisture was observed at site 1, particularly in the lower, back and southern portions of the site. However, soil hydrological properties were not monitored on a regular basis. Prolonged saturated conditions potentially resulted from the buried shale at this site and/or the greater fine particle content compared with the sand content when gravel content was excluded. Greater soil–water studies, including infiltration properties, are warranted for sites such as these. Similarly, the negative relationship between survival and clay content at site 1 farther from the road at a soil depth of 0–15 cm may be explained by the abilities of clay particles to attract and retain soil moisture, increasing θ.

The tree survival at site 1 closer to the road at a soil depth of 0–15 cm was positively correlated with f_t, but negatively correlated with ρ_b. At site 1, the maximum ρ_b was slightly lower (1.6 Mg m^{-3}) than the threshold for affecting root penetration in coarse soils (Table 11.1). The total porosity ranged from 40% to 67%, with a mean of 54% (Table 11.1). However, the ratio of the water to air content differed for the minimum and maximum ranges: the ratio was approximately 60:40 (water:air) for the minimum range and 50:50 for the maximum range (Table 11.2). Therefore, a lower ρ_b results in a higher f_t, and this range may be closer to an even water to air porosity ratio, depending on θ.

The increased tree survival at a soil depth of 15–30 cm at site 1 near the road was positively correlated with the gravel content. At site 1, the maximum range of ρ_b (1.7 Mg m^{-3}) potentially affected root penetration, and the minimum f_a was only 10% (Tables 11.1 and 11.2). At greater depths, higher gravel contents may be beneficial. Shale and other coarse fragments with nonuniform shapes that are easily fissured (ODOT 2013) may create pore spaces or change drainage patterns within the soil matrix to benefit the trees (Perry 1994; Day et al. 2010). The species in this study are known to adapt to rocky or gravelly environments, which may have been advantageous for their survival (Walters and Yawney 1990; Trowbridge and Bassuk 2004; MOBOT 2015).

Near the road and at soil depths of 0–15 cm and 15–30 cm at site 2, the minimum range for f_a was 6%, which was well under the critical limit of 10% for plant growth. Thus, the positive correlation between tree survival and increasing f_a is not surprising. This result also corresponds to the negative correlation with θ at a soil depth of 0–15 cm as water fills pore spaces that were previously filled with air.

However, the negative correlation between tree survival and increasing sand content and the positive correlation between tree survival and increased silt content at site 2 may seem to contradict previous findings regarding θ. Closer to the road at site 2, the soil particle contents were uniform across all depths, with mean and median clay contents of more than 20% at each depth (Table 11.1). The increased sand content in fine-textured soils may increase drainage, aeration, and rooting volume by creating larger pore spaces (Spomer 1983; Lal and Shukla 2004; Trowbridge and Bassuk 2004). Greater silt contents can help reduce soil strength and increase f_a because silt has larger particle size than clay; however, greater silt contents would help retain soil moisture (Lal and Shukla 2004). Thus, the correlations with sand and silt contents may be impacted by seasonal precipitation. Generalized precipitation trends (above or below average) were identified by dividing the year into the growing season (May–August), fall (September–December), and spring

(January–April). Precipitation trends were calculated by totaling the differences between the actual monthly precipitation and the 30-year normals of the corresponding months monitored at Port Columbus International Airport for a given time of the year (USDC-NOAA 2015a, b, data not shown). The trees were planted during the growing season of 2011 with below-average precipitation. The low amount of precipitation was offset by water supplemented by the city before the sites were mulched. The above-average precipitation occurred throughout the following fall and spring (2012). During year 1, which received adequate water, significant correlations were not observed between tree survival and sand content. As mentioned, the precipitation was above average in the spring of 2012; however, beginning in May 2012 and continuing through the 2012 growing season (year 2), precipitation was below average. The increased sand content at site 2 near the road at a soil depth of 0–15 cm potentially hindered the ability of the soil to retain moisture during these drought conditions and could be related to the decreased survival rate in year 2. The above-average precipitation occurred in the fall of 2012, but the below-average precipitation occurred during the spring and in May 2013. In the third year, precipitation was above average in June–August (2013) and through the spring of 2014; however, based on the amount of cumulative precipitation, it took 5 months (June–October 2013) to overcome the deficit that occurred during the previous 5 months (January–May 2013). The lower precipitation during the more active part of the growing season may explain the continued negative correlation between tree survival and sand content in the third year. This result also corresponds to greater tree survival with greater silt contents, for years 2 and 3, because silt particles have greater capacity to retain soil moisture in comparison to sand particles.

Interestingly, the tree survival at site 2 was not significantly correlated with soil density, although many mean, median, and maximum ρ_b and ρ_{bc} results exceeded the limits for disrupting root growth. It is plausible that the greater amounts of clay and silt and their water retention properties decreased the soil strength and allowed root penetration. Conversely, the area of site 1 that was closer to the road at a soil depth of 0–15 cm was negatively correlated with ρ_b during years 2 and 3, although the average and maximum bulk densities were less than the limits for constrained root penetration. This area of site 1 also had the largest gravel and sand contents. The resulting correlation may suggest the impacts of coarser roadside soils, particularly if moisture is limited.

11.8 CONCLUSIONS

Highway ROW land is a significant part of the North American landscape. ROW areas are complex systems with varied anthropogenic, atmospheric, climate, and terrestrial inputs. The functions of soils may differ within ROW land; however, road and roadside systems are intertwined and affect one another. ROW soils are heterogeneous, as shown by the variations of the soils at the study site with location and depth. Such variation largely involves the soil particle composition or the calculated volumetric soil moisture content, with different soil properties affecting roadside tree survival in different ways. Important practices for resilient roadside landscapes and vegetation establishment include the following: (1) a systematic understanding of the existing site conditions, including anthropogenic inputs, environmental exposure, plant material, and soil properties. The study sites emphasized the importance of soil composition and its influences on compaction, air, and water relationships. (2) Conducting thorough site preparation activities to alleviate potential soil and site constraints and to promote accessible soil resources for optimal plant growth. Initial roadside preparation may be beneficial for plant establishment and result in longer term ecological benefits. Soil exploration and preparation should be conducted to a depth that supports the rooting capacity of the vegetation. For woody plant material, this depth at a minimum is recommended to be 30–45 cm; however, greater depths may be beneficial depending on the circumstances (MNDOT 2016; WSDOT 2016). (3) Designing sites to protect beneficial soil and water resources. This practice includes understanding ROW stresses and creating healthy soil environments within the soil matrix and at the surface. (4) Selecting plant materials that relate to and benefit the existing roadside soil and site conditions. Roadside landscape design should include planting selections based on the functional qualities (Berketo 2013) that synergistically and systematically benefit the complete ROW environment.

ACKNOWLEDGMENTS

Salaries and research support were provided in part by state and federal funds appropriated to the Ohio Agricultural Research and Development Center (OARDC), Department of Horticulture and Crop Science, The Ohio State University; Vineland Research and Innovation Centre; Landscape Ontario Horticultural Trades Association; the Ontario Farm Innovation Program; and the Ohio Nursery and Landscape Association. Acknowledgement and appreciation is extended to Dr. Hannah Mathers, Dr. Rattan Lal, and staff and colleagues at The Ohio State University: Mr. Rob Wendling, Mr. Michael Andrako, and park and recreation staff with the City of Gahanna; Mr. Scott Lucas of the Ohio Department of Transportation; and Mr. Chad Flowers of Bird-Houk < a Division of OHM who all provided technical assistance with the project. Gratitude is also extended to all reviewers of the manuscript. This paper is produced from the dissertation that was submitted by M. Bigger as part of the requirements for a PhD in horticulture.

REFERENCES

Ahmed, Z., T.D. White, and T. Kuczek. 1997. Comparative field performance of subdrainage systems. *J. Irrig. Drain. Eng.* 123(3):194–201.

Altland, J. 2016. Chlorosis in field grown maples. Article of weed management in nursery crops, North Willamette Research and Extension Center, Oregon State University. http://oregonstate.edu/dept/nurseryweeds/feature_articles/maple_chlorosis/maple_chlorosis.html (Accessed April 27, 2016).

Ament, R., J. Begley, S. Powell, and P. Stoy. 2014. Roadside vegetation and soils on federal lands—Evaluation of the potential for increasing carbon capture and storage and decreasing carbon emissions. Final Report. http://www.westerntransportationinstitute.org/documents/reports/4W3748_Final_Report.pdf (Accessed August 4, 2014).

American Association of State Highway and Transportation Officials (AASHTO). 1993. *AASHTO Guide for Design of Pavement Structures, 1993.* Washington, DC: The Association.

American Association of State Highway and Transportation Officials (AASHTO). 2010. *A Policy on Geometric Design of Highways and Streets (The Green Book).* 6th ed. Washington, DC: The Association.

American Association of State Highway and Transportation Officials (AASHTO). 2011. *Roadside Design Guide.* 4th ed. Washington, DC: The Association.

American Concrete Pavement Association (ACPA). 2008. Free-draining daylighted subbases. TS204.8P. http://www.acpa.org/free-downloads/ (Accessed July 10, 2016).

Armstrong, A., T.C. Roberts, and R. Christians. 2011. Current and innovative solutions to roadside revegetation using native plants a domestic scan report. FHWA-WFL/TD-11-001. http://nativerevegetation.org/pdf/B1422_Roadside_revegetation_Report_complete.pdf (Accessed May 3, 2016).

Beeson, R.C. and E.F. Gilman. 1992. Diurnal water stress during landscape establishment of Slash Pine differs among three production methods. *J. Arboric.* 18(6):218–287.

Beniston, J. and R. Lal. 2012. Improving soil quality for urban agriculture in the north central US. In *Carbon Sequestration in Urban Ecosystems*, ed. R. Lal and B. Augustin, 279–313. Dordrecht: Springer.

Bennert, T. and A. Maher. 2005. The development of a performance specification for granular base and subbase material. FHWA-NJ-2005-003. http://ntl.bts.gov/lib/55000/55700/55791/FHWA-NJ-2005-003.PDF (Accessed July 10, 2016).

Berketo, P. 2013. Form follows function in highway landscape architecture Uniting function and aesthetics in highway landscape design. *Road Talk* Article 2. http://www.mto.gov.on.ca/english/transtek/roadtalk/rt19-2/RT19-2.pdf (Accessed March 5, 2015).

Bigger, M.M. 2015. Greening the highways: Out-plant survival and growth of deciduous trees in stressful environments. PhD dissertation, The Ohio State University.

Bouchard N.R., D.L. Osmond, R.J. Winston, and W.F. Hunt. 2013. The capacity of roadside vegetated filter strips and swales to sequester carbon. *Ecol. Eng.* 54:227–232.

Brady, N.C. 1990. *The Nature and Properties of Soils.* 10th ed. New York, NY: MacMillan Publishing Co.

Bryson G.M. and A.V. Barker. 2002. Sodium accumulation in soils and plants along Massachusetts roadsides. *Commun. Soil Sci. Plant Anal.* 33:67–78.

Cedergen, H.R., K.H. O'Brien, and J.A. Arman. 1972. Guidelines for the design of subsurface drainage systems for highway structural sections. FHWA-RD-72-30. http://ntl.bts.gov/lib/27000/27000/27042/FHWA-RD-72-30.pdf (Accessed May 4, 2016).

Christopher, B.R., C. Schwartz, and R. Boudreau. 2006. Geotechnical aspects of pavements reference manual. NHI-05-037. http://www.fhwa.dot.gov/engineering/geotech/pubs/05037/ (Accessed April 19, 2016).

Claassen, V.P. and R.J. Zasoski. 1994. The effects of topsoil reapplication on vegetation reestablishment. FHWA-CA-TL-94/18. http://www.dot.ca.gov/hq/LandArch/16_la_design/research/docs/topsoil_reapplication_ocr.pdf (Accessed May 5, 2016).

Coffin, A.W. 2007. From roadkill to road ecology: A review of the ecological effects of roads. *J. Transp. Geogr.* 15:396–406.

Combs, S.M. and M.V. Nathan 1998. Soil organic matter. In *Recommended Chemical Soil Test Procedures for the North Central Region.* M. Nathan and R. Gelderman (Eds.), NCR Publication No. 221, 53–58. Columbia, MO: Missouri Agricultural Experiment Station.

Craul, P.J. 1992. *Urban Soil in Landscape Design.* New York, NY: John Wiley & Sons.

Craul, P.J. and C.J. Klein. 1980. Characterization of streetside soils in Syracuse, New York. *Proceedings of Third Conference Metropolitan Tree Improvement Alliance (METRIA).* 3:88–101.

Cyr, R.Y. and P. Chaisson. 1999. Modeling subsoil drainage systems for urban roadways. *Can. J. Civ. Eng.* 26:799–809.

Daddow, R.L. and G.E. Warrington. 1983. *Growth-Limiting Bulk Densities as Influenced by Soil Texture.* USDA Forest Service, and Watershed Systems Development Group report WSDG-TN-00005. http://forest.moscowfsl.wsu.edu/smp/solo/documents/RPs/daddow_warrington_root_limiting_bulk_density.pdf (Accessed July 2, 2016).

Day, S.D.and N.L. Bassuk. 1994. A review of the effects of soil compaction and amelioration treatments on landscape trees. *J. Arboric.* 20(1):9–17.

Day, S.D., P.E. Wiseman, S.B. Dickinson, and J.R. Harris. 2010. Tree root ecology in the urban environment and implications for a sustainable rhizosphere. *Arboric. Urban For.* 36(5):193–205.

Eimers, M.C., N-K. Croucher, S.M. Raney, and M.L. Morris. 2015. Sodium accumulation in calcareous roadside soils. *Urban Ecosys.* 18:1213–1225.

Environment Canada. 2013. Code of practice: The environmental management of road salts. http://www.ec.gc.ca/sels-salts/default.asp?lang=En&n=F37B47CE-1 (Accessed August 20, 2014).

Fay, L. and X. Shi. 2012. Environmental impacts of chemicals for snow and ice control: State of the knowledge. *Water Air Soil Pollut.* 223:2751–2770.

Fischer, R.F. and D. Brinkley, 2000. *Ecology & Management of Forest Soils.* 3rd ed. New York, NY: John Wiley & Sons.

Forman, T.T. 2000. Estimate of the area affected ecologically by the road system in the United States. *Conserv. Biol.* 14(1):31–35.

Forman, T.T. and L.E. Alexander. 1998. Roads and their major ecological effects. *Annu. Rev. Ecol. Syst.* 29:207–231.

Fox, J. and S. Weisberg. 2011. *An R Companion to Applied Regression.* 2nd ed. Thousand Oaks, CA: Sage. http://socserv.socsci.mcmaster.ca/jfox/Books/Companion.

Friell, J., E. Watkins, and B. Horgan. 2014. Developing salt-tolerant mixtures for use as roadside turf in Minnesota. Report MN/RC 2014-46. http://www.dot.state.mn.us/research/TS/2014/201446.pdf (Accessed May 16, 2015).

Gale, S. and D.D. Biesboer. 2004. The effect of novel soil amendments on roadside establishment of cover crop and native prairie plant species. Report 2004-41. http://www.lrrb.org/media/reports/200441.pdf (Accessed March 9, 2015).

Gilman, E.F., I.A. Leone, and F.B. Flower. 1987. Effects of soil compaction on oxygen content on vertical and horizontal root distribution. *J. Environ. Hort.* 5(1):33–36.

Gilman, E.F. and D.G. Watson. 1994. *Syringa reticulata*: Japanese Tree Lilac. Fact sheet ST-610. http://hort.ufl.edu/database/documents/pdf/tree_fact_sheets/syrreta.pdf (Accessed February 10, 2015).

Giraudoux, P. 2016. pgirmess: Data Analysis in Ecology. R package version 1.6.4. https://CRAN.R-project.org/package=pgirmess.

Glendon, W. and D. Or. 2002. Particle-size analysis. In *Methods of Soil Analysis: Part 4 Physical Methods.* *Agron. Monogr. 5.4*, ed. J.H. Dane and C.G. Topp, 225–293. Madison: Soil Science Society of America.

Glinski, J. and W. Stepniewski. 1985. *Soil Aeration and Its Role for Plants.* Boca Raton, FL: CRC Press.

Gordon, P.E. 2014. Greening Ohio highways: Factors and practices that affect tree establishment. PhD dissertation, The Ohio State University.

Gray, D.H. and R.B. Sotir. 1996. *Biotechnical and Soil Bioengineering Slope Stabilization: A Practical Guide for Erosion Control.* New York, NY: John Wiley & Sons.

Gregory, J.H., M.D. Dukes, P.H. Jones, and G.L. Miller. 2006. Effect of urban soil compaction on infiltration rate. *J Soil Water Conserv.* 61(3):117–124.

Grosjean, P. and F. Ibanez. 2014. pastecs: Package for Analysis of Space–Time Ecological Series. Rpackage version 1.3-18. https://CRAN.R-project.org/packge=pastecs

Grossman, R.B. and T.G. Reinsch. 2002. Bulk density and linear extensibility. In *Methods of Soil Analysis: Part 4 Physical Methods. Agron. Monogr. 5.4*, ed. J.H. Dane and C.G. Topp, 201–225. Madison: Soil Science Society of America.

Haan, N.L., M.R. Hunter, and M.D.Hunter. 2012. Investigating predictors of plant establishment during roadside restoration. *Restoration Ecol.* 20(3):315–321.

Hall, K. and T. Shiraz. 2009. Daylighted permeable bases. CPTP TechBrief. FHWA-HIF-09-009. http://www.fhwa.dot.gov/pavement/concrete/pubs/hif09009/hif09009.pdf (Accessed July 10, 2016).

Harrison, G.L. 2014. Economic impact of ecosystem services provided by ecologically sustainable roadside right of way vegetation management practices. Final Report. http://www.dot.state.fl.us/researchcenter/Completed_Proj/Summary_EMO/FDOT-BDK75-977-74-rpt.pdf (Accessed August 4, 2014).

Hassan, H.F. and T.D. White. 1996. Locating the drainage layer for flexible pavements. FHWA/IN/JHRP-96/14. http://docs.lib.purdue.edu/cgi/viewcontent.cgi?article=1612&context=jtrp (Accessed May 3, 2016).

Haynes, M.A., R.A. McLaughlin, and J.L. Heitman. 2013. Comparison of methods to remediate compacted soils for infiltration and vegetative establishment. *Open J. Soil Sci.* 3(September):225–234. http://file.scirp.org/pdf/OJSS_2013090215094343.pdf.

Health Canada. 2001. Priority substance list assessment report for road salts. Cat No: En40-215/63E. http://www.hc-sc.gc.ca/ewh-semt/pubs/contaminants/psl2-lsp2/road_salt_sels_voirie/index-eng.php#a10 (Accessed August 20, 2014).

Iowa SUDAS Corporation (IOWA). 2013. Geotechnical pavement subbase design and construction. In *Statewide Urban Design and Specifications Design manual.* 6F-1-6F-7. Ames, IA: Iowa State University Institute for Transportation. http://www.iowasudas.org/manuals/design/Chapter06/6F-1.pdf (accessed July 10, 2016).

Iowa Department of Transportation (IOWADOT). 2016. Topsoil replacement. In *Design Manual.* http://www.iowadot.gov/design/dmanual/10a-01.pdf (Accessed May 3, 2016).

Jim, C.Y. 1998. Physical and chemical properties of a Hong Kong roadside soil in relation to urban tree growth. *Urban Ecosyst.* 2:171–181.

Karim M.N. and A.U. Mallik. 2008. Roadside revegetation by native plants I. Roadside microhabitats, floristic zonation and species traits. *Ecol. Eng.* 32:222–237.

Kayhanian, M., C. Suverkropp, A. Ruby, and K. Tsay. 2007. Characterization and prediction of highway runoff constituent event mean concentration. *J. Environ. Manag.* 85:279–295.

Kees, G. 2005. Hand-held electronic cone penetrometers for measuring soil strength. USDA Forest Service Report 2E22E60. http://www.fs.fed.us/t-d/pubs/pdfpubs/pdf05242837/pdf05242837dpi300.pdf (Accessed August 19, 2016).

Kelsey, P. 1994. Evaluating tree planter health: Soil chemical perspectives. In *Landscape Below Ground: Proceedings of an International Workshop of Tree Root Development in Urban Soils*, ed. G.W. Watson and D. Neely, 211–218. Savoy: International Society of Arboriculture.

Kilmer, V.J. and L.T. Alexander. 1949. Methods of making mechanical analysis of soils. *Soil Sci.* 68:15–24.

Kramer, M.G. 2013. *Our Built and Natural Environment: A Technical Review of the Interactions Among Land Use, Transportation, and Environmental Quality*, 2nd ed. EPA 2231K12001. 1-148. Washington, DC: United States Environmental Protection Agency Office of Sustainable Communities. http://www.epa.gov/dced/pdf/b-and-n/b-and-n-EPA-231K13001.pdf. (accessed July 8, 2014).

Lal, R. and M.K. Shukla. 2004. *Principles of Soil Physics.* New York, NY: Marcel Dekker.

Lee, K.W., M.T. Huston, J. Davis, and S. Vajjhalla. 2001. Structural analysis of New England subbase materials and structures. Project No. 94–1. http://www.uvm.edu/~transctr/pdf/netc/netcr26_94-1.pdf (Accessed July 18, 2016).

Lee, H., Y. Kim, and J. Purvis. 2010. Pavement thickness design for local roads in Iowa. IHRB Report TR-586. http://publications.iowa.gov/20070/1/IADOT_tr_586_Pavement_Thickness_Design_Local_Rds_Iowa_2010.pdf (Accessed May 10, 2016).

Lichter, J.M. and P.A. Lindsey. 1994a. The use of surface treatments for the prevention of soil compaction during site construction. *J. Arboric.* 20(4):205–209.

Lichter, J.M. and P.A. Lindsey. 1994b. Soil compaction and site construction: Assessment and case studies. In *Landscape Below Ground: Proceedings of an International Workshop of Tree Root Development in Urban Soils,* ed. G.W. Watson and D. Neely, 194–199. Savoy: International Society of Arboriculture.

Line, D.E. and N.M. White. 2007. Effects of development on runoff and pollutant export. *Water Environ. Res.* 79(2):185–190.

Lorenz, K. and R. Lal. 2012. Carbon storage in some urban forest soils of Columbus, Ohio, USA. In *Carbon Sequestration in Urban Ecosystems,* ed. R. Lal and B. Augustin, 139–158. Dordrecht: Springer.

Maas, E.V. and G.J. Hoffman. 1977. Crop salt tolerance-current assessment. *J. Irrig. Drain. Div. Am. Soc. Civ. Eng.* 103(2):115–134.

McGrath, D.M. and J. Henry. 2015. Getting to the root of tree stress along highways. *Acta Hortic.* 1085:109–118.

Minnesota Department of Transportation (MNDOT) 2016. *Inspection and contract administration manual for MnDOT landscape projects.* 2016 ed. http://www.dot.state.mn.us/environment/pdf/landscapeinspect-manual.pdf (Accessed August 19, 2016).

Missouri Botanical Garden (MOBOT). 2015. Betula utilis var. jacquemontii. http://www.missouribotanicalgarden.org/PlantFinder/PlantFinderDetails.aspx?kempercode=e358 (Accessed April 17, 2015).

Morrison D.G. 1999. Designing roadsides with native plant communities—How to work with a disturbed site and reflect local character. In *Roadside Use of Native Plants,* ed. B. Harper-Lore and M. Wilson, 19–20. Washington, DC: Island Press.

Nelson, D.W. and L.E. Sommers. 1996. Total carbon, organic carbon and organic matter. In *Methods of Soil Analysis Part 3: Chemical Methods. Agron. Monogr. 5.3,* ed. D.L. Sparks, A.L. Page, P.A. Helmke et al., 961–1010. Madison: Soil Science Society of America.

Nelson Brown, R., C. Percivalle, S. Narkiewicz, and S. DeCuollo. 2010. Relative rooting depths of native grasses and amenity grasses with potential for use on roadsides in New England. *HortScience.* 45(3):393–400.

Nelson Brown, R., J. Gorres, and C. Sawyer. 2011. Development of salt tolerant grasses for roadside use. FHWA-RIDOT-RTD-07-2A. http://www.dot.ri.gov/documents/about/research/Salt_Tolerant_Grasses.pdf (Accessed May 5, 2016).

Ohio Division of Geological Survey (ODGS). 2006. Bedrock geologic map of Ohio. Ohio Department of Natural Resources, Division of Geological Survey Map BG-1. http://geosurvey.ohiodnr.gov/portals/geosurvey/PDFs/BedrockGeology/BG-1_8.5x11.pdf (Accessed May 5, 2016).

Ohio Department of Transportation (ODOT). 2013. 2013 construction administration manual of procedures. http://www.dot.state.oh.us/Divisions/ConstructionMgt/OnlineDocs/Pages/2013-Manual-of-Procedures.aspx (Accessed April 19, 2016).

Oldfield, E.E., A.J. Felson, S.A.Wood, R.A. Hallett, M.S. Strickland and M.A. Bradford. 2014. Positive effects of afforestation efforts on the health of urban soils. *For. Ecol. Manage.* 313:266–273.

Parsons, D.L., D.H. Foster, and S.A. Cross. 2001. Compaction and settlement of existing embankments. Report No. K-TRAN: KU-00-8. http://ntl.bts.gov/lib/200207/15cb003.pdf (Accessed May 4, 2016).

Perry, T.O. 1994. Size, design, and management of tree planting sites. In *Landscape Below Ground: Proceedings of an International Workshop of Tree Root Development in Urban Soils,* ed. G.W. Watson and D. Neely, 3–15. Savoy: International Society of Arboriculture.

Persyn, R.A., T.D. Glanville, T.L. Richard, J.M. Laflen, and P.M. Dixon. 2004. Environmental effects of applying composted organics to new highway embankments: Part 1. Interrill runoff and erosion. *Soil Water Div. Am. Soc. Ag. Eng.* 47(2):463–469.

Peters G. 2016. Userfriendlyscience: Quantitative analysis made accessible. R package version 0.4-1. http://CRAN.R-project.org/package=userfriendlyscience.

Pitt, R., S-E. Chen, S.E. Clark, J. Swenson, and C.K. Ong. 2008. Compaction's impacts on urban storm-water infiltration. *J. Irrig. Drain. Eng.* 134:652–658.

Poesen, J. and H. Lavee. 1994. Rock fragments in top soils: Significance and processes. *Catena.* 23:1–28.

R Core Team. 2016. R: A language and environment for statistical computing. R Foundation for Statistical Computing, Vienna, Austria. https://www.R-project.org/.

Rainwater, N.R., R.E. Yoder, E.C. Drumm, and G.V. Wilson. 1999. Comprehensive monitoring system for measuring subgrade moisture conditions. *J. Transp. Eng.* 125(5):439–448.

Randrup, T.B. and K. Dralle. 1997. Influence of planning and design on soil compaction in construction sites. *Landsc. Urban Plan.* 38:87–92.

Revelle, W. 2015. psych: Procedures for Personality and Psychological Research. R package version 1.58. http://CRAN.R-project.org/package=psych.

Rhoades, J.D. 1996. Salinity: Electrical conductivity and total dissolved solids. In *Methods of Soil Analysis: Part 3 Chemical Methods. Agron. Monogr. 5.3,* ed. D.L. Sparks, A.L. Page, P.A. Helmke et al., 417–435. Madison: Soil Science Society of America.

Ritters, K.H. and J.D. Wickham. 2003. How far to the nearest road. *Front. Ecol. Environ.* 1(3):125–129.

Sabatini, P.J., R.C. Bachus, P.W. Mayne, J.A. Schneider, and T.E. Zettler. 2002. *Evaluation of Soil and Rock Properties.* Geotechnical Engineering Circular No. 5. FHWA-IF-02-034. http://isddc.dot.gov/OLPFiles/FHWA/010549.pdf (Accessed July 18, 2016).

The Salt Institute. 2014. Salt institute honors snowfighters for protecting motorists and the environment. Press release, Salt Institute. http://www.saltinstitute.org/press_releases/salt-institute-honors-snowfighters-for-protecting-motorists-and-the-environment-2/ (Accessed August 20, 2014).

Samtani, N.C. and E.A. Nowatzki. 2006a. Soils and foundations reference manual–Volume I. FHWA-NHI–06-088. http://www.fhwa.dot.gov/engineering/geotech/pubs/nhi06088.pdf (Accessed July 19, 2016).

Scharenbroch, B.C. and M. Catania. 2012. Soil quality attributes as indicators of urban tree performance. *Arboric. Urban For* 38(5):214–228.

Schumacher, B.A. 2002. Methods for the determination of total organic carbon (TOC) in soils and sediments. Ecological Risk Assessment Support Center NCEA-C-1282 EMASC-001. https://www.researchgate.net/profile/Brian_Schumacher/publication/237792048_METHODS_FOR_THE_DETERMINATION_OF_TOTAL_ORGANIC_CARBON_(TOC)_IN_SOILS_AND_SEDIMENTS/links/5408715d0cf2bba34c28c90b.pdf (Accessed August 11, 2016).

Short, J.R., D.S. Fanning, M.S. McIntosh, J.E. Foss, and J.C. Patterson. 1986. Soils of the Mall in Washington DC: I. Statistical summary of properties. *Soil Sci. Am. J.* 50:700–705.

Sipes, J.L. and M.L. Sipes. 2013. *Creating Green Roadways Integrating Cultural, Natural, and Visual Resources into Transportation.* Washington, DC: Island Press.

Skinner, R.E. Jr. 2008. Highway design and construction: The innovation challenge. *The Bridge.* 38(2):5–12.

Skousen, J.G. and C.L. Venable. 2007. Establishing native plants on newly-constructed and older reclaimed sites along West Virginia highways. *Land Degrad. Develop.* 19:388–396.

Spomer, L.A. 1983. Physical amendment of landscape soils. *J. Environ. Hort.* 1(3):77–80.

Steinfeld, D.E., S.A. Riley, K.M. Wilkinson, T.D. Landis, and L.E. Riley. 2007. Roadside revegetation: An integrated approach to establishing native plants. FHWA-WFL/TD-07-005. http://nativerevegetation.org/pdf/learn/technical_guide.pdf (Accessed May 3, 2016).

Strassler, E., J. Pritts, and K. Strellec. 1999. Environmental Assessment. In *Preliminary Data Summary of Urban Storm Water Best Management Practices.* EPA-821-R-99-012. 4-1-4-49. Washington, DC: United States Environmental Protection Agency Office of Water. http://water.epa.gov/scitech/wastetech/guide/stormwater/upload/2006_10_31_guide_stormwater_usw_b.pdf (accessed July 21, 2014).

Struve, D.K. 2009. Tree establishment: A review of some of the factors affecting transplant survival and establishment. *Arboric. Urban For.* 35(1):10–13.

Sullivan, D.E. 2006. Materials in use in U.S. interstate highways. U.S. Department of Interior, and U.S. Geological Survey Fact Sheet 2006–3127. http://pubs.usgs.gov/fs/2006/3127/2006-3127.pdf (Accessed July 16, 2014).

Tan, Z., R. Lal, N.E. Smeck, F.G. Calhoun, and B.K. Slater. 2004. Taxonomic and geographic distribution of soil organic carbon pools in Ohio. *Soil Sci. Soc. Am. J.* 68(6):1896–1902.

Tebben, G. 2012. Aug. 20, 1975: I-270 gave alternative way to get around city. *The Columbus Dispatch.* August 20, 2012. Columbus Mileposts. http://www.dispatch.com/content/stories/local/2012/08/20/1-270-gave-alternative-way-to-get-around-city.html (Accessed April, 26 2016).

Thomas, G.W. 1996. Salinity: Soil pH and soil acidity. In *Methods of Soil Analysis: Part 3 Chemical Methods. Agron. Monogr. 5.3,* ed. D.L. Sparks, A.L. Page, P.A. Helmke et al., 475–490. Madison: Soil Science Society of America.

Topp, G.C. and P.A. Ferré. 2002. Water content. In *Methods of Soil Analysis: Part 4 Physical Properties. Agron. Monogr. 5.4,* ed. J.H. Dane and G.C. Topp, 417–422. Madison: Soil Science Society of America.

Trammell, T.L.E., B.P. Schneid, and M.M. Carreiro. 2011. Forest soils adjacent to urban interstates: Soil physical and chemical properties, heavy metals, disturbance legacies, and relationships with woody vegetation. *Urban Ecosyst* 14:525–552.

Transport Canada. 2012. Transportation in Canada 2011: Statistical addendum. https://www.tc.gc.ca/media/documents/policy/Stats-Addend-2011-eng.pdf (Accessed July 3, 2016).

Transport Canada. 2016. Road Transportation. https://www.tc.gc.ca/eng/road-menu.htm (Accessed July 18, 2016).

Trombulak, S.C. and C.A. Frissell. 2000. Review of ecological effects of roads on terrestrial and aquatic communities. *Conserv Biol.* 14(1):18–30.

Trowbridge, P.J. and N.L. Bassauk. 2004. *Trees in the Urban Landscape Site Assessment, Design, and Installation.* Hoboken, NJ: John Wiley & Sons.

United States Department of Agriculture Agricultural Research Services, Natural Resource Conservation Service, and Soil Quality Institute (USDA et al.). 2001. *Soil Quality Test Kit Guide.* http://www.nrcs.usda.gov/Internet/FSE_DOCUMENTS/nrcs142p2_051907.pdf (Accessed November 19, 2014).

United States Department of Agriculture Natural Resources Conservation Service (USDA-NRCS). 2016. National soil survey handbook. Title 430-VI. http://www.nrcs.usda.gov/wps/portal/nrcs/detail/soils/ref/?cid=nrcs142p2_054242 (Accessed April 19, 2016).

United States Department of Agriculture Natural Resources Conservation Service National Soil Survey Center (NRCS-NSSC). 2009a. Electrical conductivity and soluble salts. In *Soil Survey Field and Laboratory Methods Manual.* Soil survey investigations Report No. 51. Ver. 1.0, ed. Burt, R. 213. http://www.nrcs.usda.gov/Internet/FSE_DOCUMENTS/nrcs142p2_052249.pdf (Accessed December 2, 2014).

United States Department of Agriculture, Natural Resources Conservation Service National Soil Survey Center (NRCS-NSSC). 2009b. Soil pH. In *Soil Survey Field and Laboratory Methods Manual.* Soil survey investigations Report No. 51. Ver. 1.0, ed. Burt, R. 182. http://www.nrcs.usda.gov/Internet/FSE_DOCUMENTS/nrcs142p2_052249.pdf (Accessed December 2, 2014).

United States Department of Agriculture Natural Resources Conservation Service National Soil Survey Center (NRCS-NSSC). 2016. Soil texture calculator. http://www.nrcs.usda.gov/wps/portal/nrcs/detail/soils/survey/?cid=nrcs142p2_054167 (Accessed April 28, 2016).

United States Department of Agriculture, Natural Resources Conservation Service Web Soil Survey (NRCS-WSS). 2015. Custom soil resource report Franklin County Ohio. Ver. 13. http://websoilsurvey.sc.egov.usda.gov/WssProduct/f14ezjxt3sjbaejmiuef30x1/f14ezjxt3sjbaejmiuef30x1/20160508_09311809175_67_Chemical_Soil_Properties--Franklin_County_Ohio.pdf (Accessed May 8, 2016).

United States Department of Commerce National Oceanic and Atmospheric Administration (USDNOAA). 2015a. Monthly Summaries of the Global Historical Climatology Network-Daily (GHCN-D). Columbus Port Columbus International Airport, OH US GHCND:USW00014821 2011–2014. NOAA National Climatic Data Center (Accessed May 5, 2015).

United States Department of Commerce National Oceanic and Atmospheric Administration (USDC-NOAA). 2015b. Monthly Summaries of the Global Historical Climatology Network - Daily (GHCN-D). Columbus Port Columbus International Airport, OH US GHCND:USW00014821 1981–2010. NOAA National Climatic Data Center (Accessed May 5, 2015).

United States Department of Transportation Federal Highway Administration (USDT-FHWA). 2002. Construction of pavement subsystem drainage systems reference manual. FHWA-IF-01-014. http://www.fhwa.dot.gov/pavement/pubs/013293.pdf (Accessed May 5, 2016).

United States Department of Transportation Federal Highway Administration (USDT-FHWA). 2012. Future uses of highway right of way. Office of Transportation Policy Studies Report Summary. https://www.fhwa.dot.gov/policy/otps/rowstudyproj.htm (Accessed August 5, 2014).

United States Department of Transportation Federal Highway Administration (USDT-FHWA). 2014. Standard specifications for construction of roads and bridges on federal highway projects. FP-14. https://flh.fhwa.dot.gov/resources/specs/fp-14/fp14.pdf (Accessed July 18, 2016).

United States Department of Transportation, Federal Highway Administration (USDT-FHWA). 2016a. Highway Statistics 2014, Public Road Mileage, Lane Miles, and VMT - Lane Miles 1980–2014. Chart VMT-422. http://www.fhwa.dot.gov/policyinformation/statistics/2014/vmt422c.cfm (Accessed July 18, 2016).

United States Department of Transportation, Federal Highway Administration (USDT-FHWA). 2016b. Highway Statistics 2014, National Highway System Length-2014 Miles Open and Not Open to Traffic. Table HM-30. http://www.fhwa.dot.gov/policyinformation/statistics/2014/hm30.cfm (Accessed July 18, 2016).

United States Department of Transportation, Federal Highway Administration (USDT-FHWA). 2016c. Highway Statistics 2014. Annual vehicle-miles of travel 1980–2014 by functional system national summary. Table VM-202. http://www.fhwa.dot.gov/policyinformation/statistics/2014/vm202.cfm (Accessed July 18, 2016).

United States Department of Transportation Federal Highway Administration (USDT-FHWA), and John A Volpe National Transportation Systems Center (Volpe Ctr). 2010. Carbon sequestration pilot program: Estimated land available for carbon sequestration in the national highway system.Final report. http://ntl.bts.gov/lib/33000/33500/33596/Carbon_Sequestration _Pilot_Program.pdf (Accessed August 4, 2014).

United States Forest Service, Northeastern Area State, Private Forestry, and Urban Forestry South, West Virginia Division of Forestry, and Ohio State University (USFS et al.) 2012. Eastern United States performance plant selector whitebarked Himalayan birch. http://www.fire.sref.info/plants/betula-jacquemontii (Accessed April 20, 2015).

Venner, M. 2004. Construction practices for environmental stewardship. In *Compendium of Environmental Stewardship Practices, Procedures, and Policies for Highway Construction and Maintenance*. NCHRP Project 25-25 (04). ed. M. Venner and L. Zeimer, 4-1-4-93. Washington DC: Center for Environmental Excellence by American Association of State Highway and Transportation Officials (AASHTO). http://environment.transportation.org/environmental_issues/construct_maint_prac/compendium/manual/4_1.aspx (accessed May 3, 2016).

Vora, R.S. 1988. Potential soil compaction forty years after logging in northeastern California. *Great Basin Nat.* 48:117–120.

Walters R.S. and H.W. Yawney. 1990. Red Maple *(Acer rubrum L.)*. In *Silvics of North America: Volume 2. Hardwoods* (Technical Coordinators Burns, R.M. and B.H Honkala), Washington, DC: U.S. Forest Service Agriculture Handbook no. 654. http://www.na.fs.fed.us/spfo/pubs/silvics_manual/volume_2/acer/rubrum.htm (Accessed February 11, 2015).

Washington State Department of Transporation (WSDOT). 2016. Roadside manual. Publication M 25-30.03 http://www.wsdot.wa.gov/publications/manuals/fulltext/M25-30/Roadside.pdf (Accessed May 3, 2016).

Watson, G.W. 1994. Root development after transplanting. In *Landscape Below Ground: Proceedings of an International Workshop of Tree Root Development in Urban Soils,* ed. G.W. Watson and D. Neely, 54–67. Savoy: International Society of Arboriculture.

Watson, G.W. and E.B. Himelick. 1982. Root distribution of nursery trees and its relationship to transplanting success. *J. Arboric.* 8(9):225–229.

Wennerberg, S. 2006. Common hackberry *Celtis occidentalis* L. NCRS plant guide. http://plants.usda.gov/plantguide/pdf/pg_ceoc.pdf (Accessed April 17, 2015).

Wisconsin Transportation Information Center (WTIC). 1995. The basics of a good road. Wisconsin Transportation Bulletin No 19. http://epdfiles.engr.wisc.edu/pdf_web_files/tic/bulletins/Bltn_019_GoodRoad.pdf (Accessed June 30, 2014).

Wisconsin Transportation Information Center (WTIC). 1996. Using recovered materials in highway construction. Wisconsin Transportation Bulletin No 20. http://epdfiles.engr.wisc.edu/pdf_web_files/tic/bulletins/Bltn_020_Recycle.pdf (Accessed June 30, 2014).

12 An Applied Hydropedological Perspective on the Rendering of Ecosystem Services from Urban Soils

William D. Shuster and Stephen Dadio

CONTENTS

12.1 INTRODUCTION

As human populations continue to move away from rural areas and toward conurbations (Barthel and Isendahl 2013), there is a corresponding and altogether necessary interest in the possible benefits supplied to these populations from urban soils. It is helpful to look at urban soils from a foundational standpoint. In the still nascent observation of urban soils, these are indeed unique soils that are composed and structured differently than presettlement or agricultural soils. There is little surprise that urban soils should then function differently. Ecosystem services are usually realized as benefits to humans and nature derived from natural and other capital (Millennium Ecosystem Assessment 2005). Just as soils provide the basis for agroecosystem structure and function, there is a largely untapped reservoir of ecosystem services that can be realized from urbanized soils (Dominati et al. 2010; Bouma and McBratney 2013; Adhikari and Hartemink 2016). The soil ecosystem integrates the water cycle and can furthermore play a role in regulating air quality and climate. Herrmann et al. (2016) further developed the concept of urban soils as a cornerstone of terrestrial ecosystem functioning, such that where urban landscapes were not sealed over, there was higher potential to provide a broad palette of services (Pavao-Zuckerman and Byrne 2009). Although a technical perspective is necessary to classify and assess urban soils for their biogeochemical and hydrologic attributes, a purely practical perspective of soils as an urban natural resource is equally important to decision makers (e.g., city planners, wastewater professionals, governance, etc.). The concept of ecosystem services is one way to engage and communicate with a multicultural, pluralistic society. In this chapter, the objectives are to establish basic definitions in the area of ecosystem services; focus these definitions through a hydropedological lens; offer a combined hydrologic-taxonomic perspective that can be used to identify type, availability, and sustainability of benefits rendered from the urban soil ecosystem; and discuss statistical approaches to quantitative interpretation thereof.

12.2   HOW IS AN URBAN SOIL FORMED?

Ecosystem services derive from soil functions, which are themselves reliant upon the presence or absence of different influences on pedogenesis. The weathering of parent materials through time and climate creates soils whose genesis is further modulated by topography and vegetative cover. This conceptual basis of pedogenesis through agroecosystem processes (e.g., tillage) is relatively well characterized. As a function of time, the traditional notion of soils is based entirely on soil-forming processes (Jenny 1941):

$$\text{Natural Soil}(t) = f(\text{Parent Material, Climate, Time, Topography, Biota}) \qquad (12.1)$$

Aside from these five traditional soil-forming factors, since humans have existed on Earth, soils have been altered by anthropogenesis (Leguédois et al. 2016) and utilized as raw material by human ecosystem engineers. The interpretation of the human experience by Jones et al. (1994) shows how humans are indeed ecosystem engineers by learning how to control the flow of resources to other states and species, thereby creating and sustaining suitable habitat for ourselves. These abilities predicted the sequence of soil disturbance by humans. Early humans as hunter-gatherers were influenced by soils, tracking flora and fauna that ultimately patterned to certain landscapes and soil types. Through the process of trial and error, humans worked with soil and developed agronomic practices, including adding inputs (e.g., manure and irrigation) and landscape manipulation (e.g., tillage) to complement their soils (Barthel and Isendahl 2013). As agriculture became more productive and moved beyond subsistence, food and fiber were available for commerce and trade, and human populations concentrated in centers of economic production. At this point, soils became the structural basis for cities, leading to a new type of disturbance, which created new soil morphological characteristics (e.g., horizonation, texture, structure, and weathering; Capra et al. 2013). Pursuant to this chronology of soil disturbance, Leguédois et al. (2016) built on how anthropogenic influence (e.g., urbanization) was a sixth contributing factor to soil formation. An update to this traditional framing of pedogenesis would include the soil in its prior state—the result of the traditional soil-forming factors—which now acts as the baseline state condition, operated upon by new inputs and alterations associated with human ecosystem engineering:

$$\text{Urban soil}(t_d) = f(\text{Natural soil}(t_0), \text{Time since disturbance}(s)(t_d),$$

$$\text{Mass transfer of soil material, Water inputs, Atmospheric deposition,}$$

$$\text{Vegetation, Debris, Land management}) \qquad (12.2)$$

One basic feature of urban soils is that soil material is moved, added, or removed from their point of origin. Urban soils present as highly heterogeneous mixtures and layers of anthropogenic fill (e.g., imported from cut-and-fill projects) with native soils comprising the subsoils below the depth of development activity or other disturbances. This movement of soil can result from excavation, backfilling, removal, and export of soil. Any one or a combination of these activities marks a disturbance, and a pedogenic reset, which can lead to entirely new and unique soils. For example, US housing booms over the last 40 years saw the creation of a new protocol in creating turfgrass lawns. Whereas excavation for houses was previously limited to the extent of the house and driveway, now the entire development was impacted anthropogenically. To facilitate construction activities, all of the land within a development was stripped of its surface (i.e., A horizon); this material was either stockpiled on site for its eventual return after the development was completed or it was trucked away to provide topsoil for commercial sale. If the topsoil was stockpiled, it was then replaced back to its original location within the soil profile; however, the depth, structure, and tilth were severely altered from this cataclysmic disturbance. Where the topsoil was permanently removed from a development and transported to other sites requiring fill soil as excavation backfill, the B horizon was covered with turf sod.

Urban soils are young, and typically have little development nor differentiation among horizons, and therefore key out as Entisols. Yet, the designation of urban soils as Udorthents truncates a tremendous amount of heterogeneity in how soil—and pedogenesis—is affected by urbanization processes. Standard soil taxonomy can be somewhat general when it comes to classifying and describing an urbanized soil. At this time, a carat (^) prefix to a master horizon and a subordinate designation of "u" are used interactively to indicate evidence of anthropogenic activities, or artifacts, respectively. From a Soil Taxonomy perspective, the minimum required depths to recognize these features for classification purposes are 50 cm. Yet, there are numerous instances of anthropogenic impacts that occur at depths both less and more than 50 cm, though most often involving the topsoil. These general taxonomic designations require some degree of specificity so as to properly reflect the features that make these soils *urban* soils.

The interaction of soil pedogenic processes with the water cycle is known as hydropedology (Lin et al. 2006), and this multidisciplinary perspective can be used to further integrate how to describe urban soils (Shuster et al. 2014a, 2014b), and their interaction with water in the ecological context of structure (tilth, redox status, clay weathering, etc.) and function (primary production, transpiration, etc.). There are many pathways through the soil matrix, connecting aboveground hydrologic processes with both groundwater and surface water. At the soil surface, the specific hydraulics regulate the movement of water into the soil matrix via infiltration, and soil moisture is redistributed via near-surface processes. These surficial characteristics control whether soils will infiltrate rapidly to keep pace with rainfall rate, controlling the formation of runoff by Hortonian infiltration-excess mechanism. If a water table or saturated horizon is present, this means that capacity for additional soil moisture is limited, and runoff can be produced via saturation excess. The presence or absence of vegetation further stratifies the local hydrologic cycle into sources and sinks. Bare soils are exposed to compaction and raindrop-induced development of crusts, heightening runoff potential. Vegetated soils are protected from compaction and evaporation, though transpiration rates will rise as a consequence of plant physiology.

Carrying on with the variables in Equation 12.2, the water cycle, airborne deposition, vegetation, debris, and land management are coupled in the urban context. A reference is made in the water cycle to the broader designation of water inputs, as urbanization brings with it a host of new hydrologic routings, distinct from any sort of native or presettlement water cycle. Rainfall pattern is apportioned into runoff (as baseflow, streamflow, or wastewater), infiltration, groundwater, or streamflow, all of which may interact with soils and affect its functional condition since disturbance (t_d), and thus determines the type and delivery of human benefits such as drinking water or irrigation. The urban subsurface can be composed of mixtures of fill, demolition debris, and native soil (Figure 12.1). The degree to which the existing soil components are composed, packed, and ordered thereby predicts water movement. If poorly packed debris is dominant in the subsurface mixture, there are usually large interstitial volumes among the pieces of debris, which allow water to percolate and drain quickly. However, the water can then encounter native subsoils, the nature of which (texture, structure, bulk density, etc.) completely regulates the percolation of soil water in the transition from unsaturated to saturated formations. In terms of airborne deposition, urban soils will often have a much broader range of, for example, carbon (C) molecules, some of which may be pollutants such as polycyclic aromatic hydrocarbons (PAHs), chlorinated solvents, ash, and black carbon (BC), among other compounds that find their way into the atmosphere (Lohmann et al. 2000; Schifman and Boving 2015). For example, a small city with no industrial production capacity would have different quantities and qualities of deposited material than a city which developed from an early industrial base. Depending on the extent and thoroughness of a demolition process, impervious surface (as driveways, sidewalks, etc.) can be left behind, limiting the range of soil ecosystem services. Alternately, trees at the property line are often left in place and thereby can continue to provide some canopy cover and lend a stabilizing influence on soil structure. After the anthropogenic disturbance on these soils, the surrounding human activities can direct certain soil-forming factors. For example, if a large structure is erected near an urban soil, the anthropogenic construct

FIGURE 12.1 This soil core taken from urban park area aquarium landscape illustrates the extent to which anthropogenic materials (stone, concrete debris, brick) are used as fill material and can constitute the surface horizons. In this case, the anthropogenic materials sit on top of native soil, which occupies the base of the core at approximately 4' (120 cm) depth.

can impact the local climate and topography in this area with regards to shading, wind, and precipitation pattern. Depending on the type of vegetation selected in an urbanized area, those plants can impact the specific role of the biotic-vegetative soil-forming factor. Additionally, there are everyday anthropogenic inputs to urbanized soils such as road salt, grit, automotive fluids, and pet wastes.

12.3 URBAN SOILS AS GENERATORS OF ECOSYSTEM SERVICES

The type and potential for ecosystem services can shift as a soil is urbanized (Table 12.1). The potential to generate these services is, in part, constrained by soil resources. As the scale of interest is expanded, ecosystem services can start to effectively provide for larger scale concerns such as stormwater detention, C sequestration (Pouyat et al. 2007), managed treelots, and facilitating other service-providing land uses (Shuster et al. 2014a). This chapter uses the frame of ecosystem services to catalogue and otherwise identify the key roles in which soil ecosystems serve. The benefits accrued and rendered from soil processes to people as ecosystem services include supporting services such as soil formation and rooting media for primary production, provisioning (e.g., a substrate for plant growth and hence food, water purification, and storage), regulating (purify water via filtration through chemically and biologically active pore spaces, provide a reservoir of microbial diversity), and cultural (recreation through gardening or use of parks, aesthetic, values, spiritual connection, and renewal). However, the potential for ecosystem services and their realization in urban systems can be disjoint (Table 12.1; Haase et al. 2014b). For example, the coupled supporting services of soil formation and primary production are perhaps valued more in an agricultural than urban context. With shifts in land use and management, an urban soil can create an import food economy with low infiltration potential of slaked soils and landscapes with low vegetative cover and high runoff potential, without management, the urban soil is unable to generate provisioning ecosystem services, such as food production, freshwater cycling, etc. Ecosystem services often operate below a threshold of perception, and the day-to-day necessity of ecosystem services within a conurbation is likely underappreciated. The identification and cognition of cultural ecosystem services as aesthetic, spiritual, and recreational experiences may be a gateway to recognition of supporting, provisioning, and regulating ecosystem services (Chan et al. 2012). For this reason, cultural ecosystem services may be the most difficult to quantify, but also the most important.

Given below are some examples of how the functions and ecosystem services of actual urban soils follow from their form. Contemporary post-urban landscapes present with perhaps the most representative examples of urban soils. For example, due to legacy blight (Gordon 2003) and the US

TABLE 12.1

The Patterns and Processes of Urbanization Mediate Soil-Based Ecosystem Services through Its Effects on Soil Form and Function

Type of Service	Function	Role of Urbanization	Outcome
Supporting	Soil formation	Interrupt, alter trajectory of pedogenesis	Unique soils, unique functions
	Primary production	Alter soil competence to support plant life, presence/absence of maintenance	Colonization by ruderal, generalist species
Provisioning	Food, fiber	Shift to import economy for food and fiber, provide space for urban agriculture	Food desert, oasis
	Freshwater	Variable recharge potential, runoff potential; collect, convey water resources	Shift from infiltration predominance to runoff production
	Infrastructure foundations	Emphasis on structural, geotechnical attributes	Compaction, sealing, imperviousness
Regulating	Climate regulation	Loss of thermal regulation, evaporative cooling	Heat islands
	Flood control	Loss of detention capacity vs. vacant land as infiltrative landscapes	Flooding vs. detention
	Water purification/soil toxicity	Replacement by WWTP/DWTP, accumulation of toxics, heavy metals	Municipal service source waters, N, P, OM, bacteria, legacy pollution
Cultural	Aesthetic	Reserves, gardens, landscaping	Patchy oases attractive to persons or the community
	Spiritual	Detachment from natural surroundings, seeking	Urban isolation, provide variety of spaces for worship
	Recreational	Specialized landscapes as parks, unformatted green spaces	May stimulate local economy, personal rejuvenation

DWTP, drinking water treatment plant; WWTP, wastewater treatment plant.

foreclosure crisis (Leitsinger 2011; Martinez-Fernandez et al. 2012), vacant land has ascended as a predominant land cover in many major US urban cities (Bowman and Pagano 2004). The underlying soils developed and were urbanized through a sequence of anthropogenic land disturbances, the most recent being demolition (Figure 12.2). The demolition process is an important agent of change in land use, as it permanently removes a structure, both surface and usually subsurface features, and completely revises the footprint of the developed landscape (USEPA 2013). A building lot within a city is transformed from land that has a majority of impervious surface to one where structures (houses and garages) and surfaces (driveways and sidewalks) are removed, and that is now predominantly pervious. Importantly, this process increases the extent and continuity of open acreage, which are both attributes of green infrastructure (Schilling and Logan 2008; Haase 2008). This new patterning of the landscape offers the prospect of decentralizing a range of ecosystem services throughout a conurbation (Haase et al. 2014a; Green et al. 2016).

FIGURE 12.2 Phases of deconstruction that create contemporary urban land features (e.g., vacant lots), and thereby pattern the development of urban soils (Detroit, MI). French settlers farmed in strips of land that extended inland from the Detroit River. As farming gave way to industrialization, soils provided structural support for buildings and other infrastructure (building in background). As economic circumstances lead to demolition (middle section), soils were disturbed by partial deposition of debris and addition of fill soil. The end product is the vacant landscape (foreground), an urban soil ecosystem with turf as vegetative cover.

Vacant lot soils can act as passive, infiltrative green infrastructure commensurate with their capacity to detain stormwater volume through systematic infiltration, detention, and percolation to groundwater (Gallo et al. 2013). Yet, there is a dearth of data on the hydrology of urban soils, which Shuster et al. (2014a, 2014b) have attempted to rectify by offering a protocol for soil hydrologic assessments. This protocol entails actually studying deep soil cores (i.e., a full taxonomic description) and making correspondent measurements of infiltration and drainage, and its application to different urban core areas that represent each of the major soil orders (Schoeneberger and Wysocki 2005; Dominati et al. 2010; Robinson et al. 2014). A summary of soil cores under different landscape-level influences are shown in Figure 12.3. Overall, pedons present with a great deal of variety in layering. The west side Detroit residential vacant landscape has soil profiles that show little difference among disturbance levels, with fill soils comprising approximately the full A1 horizon. The sandy loam surface horizon is more amenable to infiltration than the fine sandy loam soil found in the adjacent, undisturbed pedon. City records indicate that the west side demolitions are more recent than those on the east side, wherein fill soils were finer and arranged in thin layers with no horizon development in the C1, indicating overall lower potential for infiltration. On the other hand, the undisturbed east side pedon was a dark, loamy, well-drained soil, high in organic matter. These attributes indicate suitability for nutrient cycling, favorable soil moisture and drainage conditions, and thereby good prospects for urban agriculture. Parks land on the west side actually had a mollic epipedon where, in fact, a Mollisol was classified (Shuster, unpublished data), and the inherent high fertility—a supporting and provisioning ecosystem service—sustained a turf cover for this park featuring separate sport and recreational fields. By comparison, a park on the east side of Detroit was heterogeneously composed from finely textured soils, suggesting a piecemeal approach to backfilling and leveling this area, limiting the range of ecosystem services to recreation; this park was decommissioned in recent years.

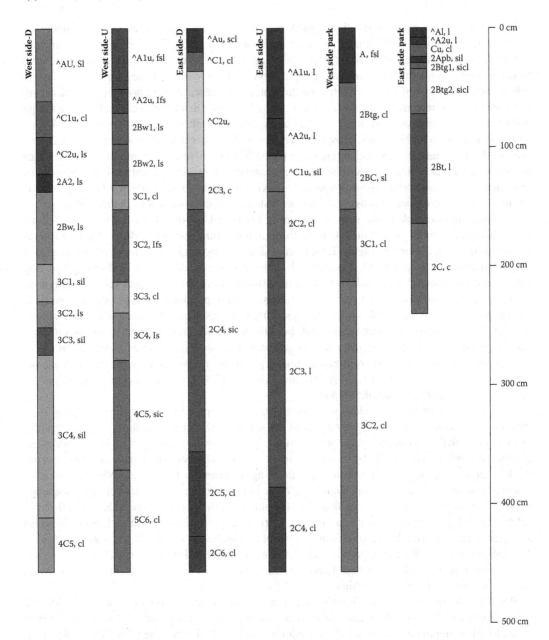

FIGURE 12.3 Deep soil cores from west side, east side, and parks (Detroit, MI) illustrate Munsell color, horizon master, subordinate classification, and texture. The D and U designations are disturbed (where fill soil was used to backfill demolition excavations) and undisturbed (native areas outside of the building envelope), respectively.

Soil measurements made in the Detroit area show that in most cases, the top 90–185 cm (approx. 3'–6') of soil volume has some capacity to move and otherwise store water, before terminating in an impermeable former, ancient lakebed. In practical terms, this data offers planners and designers alike the opportunity to develop specifications and green infrastructure/stormwater management designs that reflect actual rather than assumed conditions. An improved understanding of how soils are layered, and how water flows through these soils can also guide placement of water storage or infiltration in suitable places. From a value-added or city budget perspective, supporting and provisioning services may be helpful with regard to meeting regulatory standards, such as wastewater volume under the Clean Water Act (1972). Continuing with the example of Detroit, many US cities are under orders to reduce discharge of untreated wastewater into receiving waters. These sewer overflows are largely due to storms that overwhelm the capacity of aging combined sewer systems. Detention of stormwater in urban landscapes by plant-soil green infrastructures plays on landscapes with favorable infiltration and drainage rates; and retentive features such as constructed wetlands to leverage soils with poor drainage; and to interactively manage both water quantity and quality. Overall, good field data guides risk management expectation for hydrologic ecosystem services so that water does not flow to places where it is unwanted (e.g., basements and surcharging sewer pipes). The lack of immediate market value for vacant parcels, and coincident imperatives for stormwater management to prevent combined sewer overflows, may furthermore coincide to support longer term public investment in straightforward management of vacant lots for the provision of ecosystem services that are available to a broader demography.

A contrasting example to the soils of the upper Midwest United States is provided by Shuster et al. (2014b), for sandy loam Aridisols in metropolitan Phoenix, AZ, formed out of native soils in the foothills around Phoenix (Figure 12.4, Usery), and moving out into the more level topography where agriculture was practiced. Agricultural land uses have given way to land devoted to residential and commercial areas, which rely on the structural attributes of local soils. As might be appropriate for a semiarid region, xeric urban landscapes are promoted for their water-conserving features (Figure 12.4, Scottsdale and Tempe residential pedons; Kelly 2005), an important ecosystem service in this arid location where high population density and commensurate consumption pattern threatens to outstrip available water resources. The Scottsdale 1 and Tempe (Figure 12.4) pedons had gravelly surface horizons and at times a plastic restrictive layer, giving way to loam and sandy loam under layers, whereas the Scottsdale 2 (Figure 12.4) pedon had a coarse sandy loam in various stages of early horizon development.

For large areas under impervious surface, and traditional piped runoff routing, flooding is an unfortunate outcome. Given the relatively small amount of rainfall that these areas receive annually, one would think that flooding would not be an issue. Yet, stormwater management in the southwestern United States is complicated by the predominant monsoonal precipitation pattern. These storms, while brief, are intense events that can create flooding in urban drainage areas. While evaporation is a predominant component of the hydrologic cycle in this climate, it does not act quickly enough to be of benefit to regulate the stormflow runoff volume. In the residential parcels, the soil physical and hydrologic conditions were suitable for stormwater infiltration and these lots could be net sinks for stormwater runoff. If a sufficient number of residences act in a way to increase stormwater detention recharge on their own property, the burden on centralized stormwater infrastructure such as retention basins should be abated. However, the stormwater collection system instead conveys stormflows entirely to retention basins.

Retention basins are a common stormwater management feature on the Phoenix landscape (e.g., Larson and Grimm 2012), and these basins often serve a dual purpose as recreational park areas, highlighting the degree to which provisioning and aesthetic ecosystem services (Table 12.1) can be rendered in an arid urban setting. Although these retention basins were designed with soils and turf that would promote rapid drawdown, the time for complete infiltration has steadily increased over time. Taxonomic data indicates that anthropogenic impacts on soil formation derived from fine particles deposited during storm events (Figure 12.4, Cactus, Mescal, and Northsight pedons) to create an Ap horizon. From reference pedons taken in the same study, the native soil surface texture would be predicted to be a sandy loam soil. However, soils in retention basins were fine sandy loam soils due to the increased amount of fines (Reheis 2006; Reynolds et al. 2006; Neff et al. 2008), and thus had a lower permeability and were prone to compaction from the recreational services provided in these parks. In this community, the retention basins were utilized as parks and playgrounds. The longer it takes for the retention basin soil to allow a complete drawdown of stormwater runoff volume, there is a correspondent decrease in citizen access to cultural ecosystem services as recreation. This example explicitly illustrates the connection between anthropogenic influences on pedogenesis and consequent changes in hydrology, and its implications for delivery of ecosystem services.

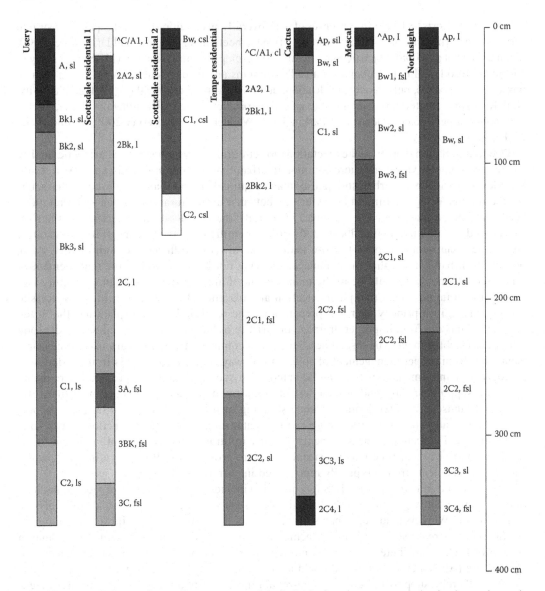

FIGURE 12.4 Pedons reflect soil taxonomy and hydrologic functions across major land-use classes in Phoenix, AZ. Usery is a reference, nonurbanized aridisol; Scottsdale and Tempe illustrate the residential influences on soil layering; and Cactus, Mescal, and Northsight are multipurpose recreational park, retention basin areas.

12.4 QUALIFICATION AND CHARACTERIZATION OF URBAN SOIL ECOSYSTEM SERVICES

The evaluation of urban soils for their functional attributes is necessarily a multivariate endeavor. The specific physical, chemical, biological, and situational aspects of a soil determine its competence to support ecosystem functions. The concept of soil quality developed for agricultural soils has a multivariate foundation that provides an existing body of work that may be extended to describe and integrate attributes of urban soils. Soil quality as originally conceived represents a synthesis of soil biogeochemical and physical characteristics that characterizes the ability of soils to sustain agricultural productivity (Doran and Parkin 1994; Wander and Drinkwater 2000).

Considerable efforts to develop a consensus of what overall soil condition can best support ecosystem functions underpinning agricultural productivity have been made (Wander and Drinkwater 2000; Schindelbeck et al. 2008), in particular, attempts to develop a minimal dataset (MDS) necessary to characterize soil quality that quantifies a soil's strengths and its deficiencies. A MDS can include physical (e.g., texture, bulk density, infiltration, depth of rooting), chemical (e.g., pH, conductivity, N, P, K, organic matter), and biological (e.g., soil respiration, potentially mineralizable N, microbial biomass) parameters (Doran and Parkin 1994; Wander and Drinkwater 2000; Schindelbeck et al. 2008).

Due to population density and expectations for cultural ecosystem services, aesthetic quality and recreational use values are more common in urban than agricultural landscapes. As a result, the relevant variables for urban soil quality may be quite different than those used to characterize the quality of agricultural soils. A bridge between agricultural and urban interpretations needs further development. For example, in a study that included characterization of urban vacant land soil quality, Schindelbeck et al. (2008) identified inputs of tillage, traffic, vegetative cover, and C-nutrient dynamics-transformations as important influences of soil quality. Also, relevant to urban soils is that their management may not be as central to the management of urban ecosystems as it would be to the management of agricultural ecosystems. Smagin et al. (2006), for example, found that plant selection and planting of vegetation in Moscow, Russia, was typically given priority over soil preparation. The result is landscape plantings that often fail (e.g., plants die from improper drainage or fertility or become outcompeted by weeds) soon after their establishment. The researchers go on to recommend the use of various soil measurements to inform proper management of soils. In this way, a landscape plan is front-loaded with appropriate management cues that can be acted upon such that soils can provide the requisite structural support for roots and provide fertility required by plants to maintain an aesthetically motivated landscaping. Developing an understanding of urban soils centered on their support for aesthetic landscapes is too limited given the greater range of desirable functions urban soils could provide. For example, measurements of water infiltration and redistribution of soil moisture, which are key soil functions for sustaining plant growth as well as providing stormwater volume management, are not typically represented in current MDS parameter sets.

Therefore, modification of the MDS is required to include characterization of factors that may limit not only plant vigor, but also hydrologic roles such as water infiltration and soil moisture redistribution. Workers studying urban soils agree that it is important to build soil organic matter (SOM) to increase soil resilience by achieving greater fertility and tilth (Jim 1998; Smagin et al. 2006). These attributes may furthermore protect soils from degradation via enhanced soil aggregation (Abiven et al. 2009) and could act on soils to improve water infiltration characteristics and thereby support the provisioning-regulating ecosystem service of stormwater detention. Much of the soil quality literature compares measurements with preestablished ranges in parameters to judge if a soil possesses a certain capacity to function. Current work shows how desirable soil functions and associated ecosystem services may be identified from field data. Although there is a dearth of these observations on urban soils, there is now at least foundational work that may inform on urban soil quality vis-à-vis ecosystem services (e.g., Rhea et al. 2014; Shuster et al. 2014a, 2014b; Herrmann et al. 2016). Rhea et al. (2014) explored how existing land cover may be used as a proxy to understand the extent to which vacant urban land (Cleveland, OH) might be used to provide stormwater detention. Results showed that type and proportion of vegetative cover are correlated to percent sand, drainage rate (which was measured *in situ*), and basic nutrient availability (N, P, K, Ca, Mg, C). In essence, urban soils are generally supportive of volunteer plant communities, especially in the lower disturbance perimeter areas of vacant lots which are usually spared demolition activity. Yet, this is one set of soil conditions, and as Rhea et al. (2014) point out, additional studies in different cities will be needed to identify or confirm the groupings of variables necessary for inclusion in an MSD.

A potentially fruitful alternative to the MSD is ordination techniques that can find associations between multiple soil parameters and other class variables, such as city or land use. Due to the nature of anthropogenic activities like cutting and filling, cities can be similar in terms of soil properties and function, compared to a rural-natural area (Groffman et al. 2014). Using ordination, Herrmann et al. (2016) investigated if there were relationships within and between the cities of Detroit and Cleveland for a set of vacant lot soil properties selected for their relevance to ecosystem services. In some cities, vacant or excavated areas are filled (e.g., Cleveland), whereas in others (e.g., Detroit), less fill material is utilized. The findings of Herrmann et al. (2016) suggest that soil-specific differences across cities outweigh the capacity for anthropogenic activities to blend these soils into providing common ecosystem services. The extent of stormwater retention ecosystem services was measured as the frequency of precipitation events (hourly time resolution) that exceeded the measured infiltration capacity. This metric was exceeded three times more frequently in Detroit than Cleveland, for both pre-existing and fill soils, based on regional differences in soil conditions that were developed through urbanization (Herrmann et al. 2016). Generally, in terms of support for plant growth, there are concerns about urban soil fertility due to (de)construction activities impacting the topsoil layer, or in some cases removing it completely (Craul 1992). Some urbanized soils are generally well developed with few fertility concerns (Pouyat et al. 2010). In fact, in residential lawn landscapes, soil N levels can be higher than in agricultural or natural landscapes (Golubiewski 2006; Raciti et al. 2011). Herrmann et al. (2016) found that these generalizations are affected by the linked histories of a city and its soil. For example, Cleveland vacant land had thinner topsoil layers but greater soil N levels compared to Detroit. There also was greater soil N levels in pre-existing soils compared to the fill soils. The differences between pre-existing and fill soils may be due to the nature of the initial fill material, having a low level of SOM and thus affecting soil N. Although fill material depth and type (i.e., texture) can be highly variable, other studies have shown that fill soils are lower in organic matter (Pouyat et al. 2002).

Taken as a group, these studies show that vacant lots can support plant growth and provide stormwater detention, and furthermore as multiuse passive recreation areas, green infrastructure for stormwater management, urban agriculture farms, or for recreational uses. In order to determine the suitability for these vacant lots, a site-specific evaluation is required. This is especially true in urban agriculture as there is a known history of metal contamination in these areas, necessitating the use of soil tests to measure lead levels before any long-term planning can commence (Schwarz et al. 2016). The criteria used in the assessment of soil quality (topsoil depth and fertility) is agronomic-specific (e.g., Andrews et al. 2002). The needs within a neighborhood are more varied and go beyond the agronomic capacity of soils. For passive recreation, it is sometimes necessary to reduce nutrient availability to control invasive species (Pavao-Zuckerman 2008). From a broad perspective, the assumption that urban soils are thin and have poor fertility is false; these soils often have moderate fertility and can be improved for desired outcomes. Whereas it was originally thought that urban soils were a limiting factor for providing ecosystem services from vacant lots, it is likely that neighborhood engagement and a long-term operation and maintenance strategy are more critical factors (Kremer and Hamstead 2015).

12.5 CONCLUSIONS

Ecosystem services are benefits to human populations derived from natural capitals like soil. When a soil is urbanized during infrastructure and superstructure development, the related processes modulate the state and quality of natural resources, along with the form and function of the resultant soil. The functional aspects of the soil in turn create bounds on the presence, absence, and overall continuity of supporting, provisioning, regulating, and cultural types of ecosystem services. A unique, field-based perspective is offered on the formation of urban soils, and illustrative examples of how form and formation under disturbance may lead to the rendering of certain ecosystem services, or the cessation of others.

ACKNOWLEDGMENTS

The authors thank Dr. Dustin Herrmann, Dr. Laura Schifman, and Jessica Mullane, MS, for their valuable comments and suggestions on both the technical content and organization of this chapter.

REFERENCES

Abiven, S., S. Menasseri, and C. Chenu. 2009. The effects of organic inputs over time on soil aggregate stability—A literature analysis. *Soil Biol Biochem* 41: 1–12.

Adhikari, K. and A.E. Hartemink. 2016. Linking soils to ecosystem services—A global review. *Geoderma* 262: 101–111.

Andrews, S.S., D.L. Karlen, and J.P. Mitchell. 2002. A comparison of soil quality indexing methods for vegetable production systems in Northern California. *Agric Ecosyst Environ* 90: 25–45. doi:10.1016/S0167-8809(01)00174-8.

Barthel, S. and C. Isendahl. 2013. Urban gardens, agriculture, and water management: Sources of resilience for long-term food security in cities. *Ecol Econ* 86: 224–234.

Bouma, J. and A. McBratney. 2013. Framing soils as an actor when dealing with wicked environmental problems. *Geoderma* 200–201: 130–139. doi:10.1016/j.geoderma.2013.02.011.

Bowman, A.O. and M.A. Pagano. 2004. *Terra Incognito: Vacant Land and Urban Strategies*. Washington, DC: Georgetown University Press.

Capra, G.F., S. Vacca, E. Cabula, E. Grilli, and A. Buondonno. 2013. Through the decades: Taxonomic proposals for human-altered and human-transported soil classification. *Soil Horiz* 54: 1–9. doi:10.2136/sh12-12-0033.

Chan, K.M.A., A.D. Guerry, P. Balvanera, S. Klain, T. Satterfield, X. Basurto, A. Bostrom, R. Chuenpagdee, R. Gould, B.S. Halpern, N. Hannahs, J. Levine, B. Norton, M. Ruckelshaus, R. Russell, J. Tam, and U. Woodside. 2012. Where are 'cultural' and 'social' in ecosystem services? A framework for constructive engagement. *Bioscience* 62(8): 744–756.

Craul, P.J. 1992. *Urban Soil in Landscape Design*. New York, NY: John Wiley and Sons.

Dominati, E., M. Patterson, and A. Mackay. 2010. A framework for classifying and quantifying the natural capital and ecosystem services of soils. *Eco Econ* 69: 1858–1868.

Doran, L.E. and T.B. Parkin. 1994. Defining and assessing soil quality. In *Defining Soil Quality for a Sustainable Environment*, ed. J.W. Doran et al., 3–32. SSSA Spec. Publ. 25. Madison, WI: SSSA and ASA.

Gallo, E.L., P.D. Brooks, K.A. Lohse, and J.E.T. McLain. 2013. Land cover controls on summer discharge and runoff solution chemistry of semi-arid urban catchments. *J Hydrol* 485: 37–53.

Golubiewski, N.E. 2006. Urbanization increases grassland carbon pools: Effects of landscaping in Colorado's front range. *Ecol Appl* 16: 555–571.

Gordon, C. 2003. Blighting the way: Urban renewal, economic development, and the elusive definition of blight. *Fordham Urban Law J* 31: 305–337.

Green O.O., A.S. Garmestani, S. Albro, N.C. Ban, A. Berland, C.E. Burkman, M.M. Gardiner, L. Gunderson, M.E. Hopton, M.L. Schoon, and W.D. Shuster. 2016. Adaptive governance to promote ecosystem services in urban green spaces. *Urban Ecosyst* 19: 77–93. doi:10.1007/s11252-015-0476-2.

Groffman, P.M., J. Cavender-Bares, N.D. Bettez, P.M. Groffman, J. Cavender-Bares, N.D. Bettez, J.M. Grove, S.J. Hall, J.B. Heffernan, S.E. Hobbie, K.L. Larson, J.L. Morse, C. Neill, K. Nelson, J. O'Neil-Dunne, L. Ogden, D.E. Pataki, C. Polsky, R.R. Chowdhury, and M.K. Steele. 2014. Ecological homogenization of urban USA. *Front Ecol Environ* 12: 74–81. doi:10.1890/120374.

Haase, D. 2008. Urban ecology of shrinking cities: An unrecognized opportunity? *Nat Cult* 3: 1–8. doi:10.3167/nc.2008.030101.

Haase, D., A. Haase, D. Rink. 2014b. Conceptualizing the nexus between urban shrinkage and ecosystem services. *Landsc Urban Plan* 132: 159–169.

Haase, A., D. Rink, K. Grossmann, M. Bernt, and V. Mykhnenko. 2014a. Conceptualizing urban shrinkage. *Environ Plan A* 46: 1519–1534. doi:10.1068/a46269.

Herrmann, D.L., W.D. Shuster, and A.S. Garmestani. 2016. Vacant urban lot soils and their potential to support ecosystem services. *Plant Soil* 179: 1–13. doi:10.1007/s11104-016-2874-5.

Jenny, H. 1941. *Factors of Soil Formation—A System of Quantitative Pedology*. New York, NY: McGraw Hill.

Jim, C.Y. 1998. Urban soil characteristics and limitations for landscape planting in Hong Kong. *Landsc Urban Plan* 40: 235–249.

Jones, C.G., J.H. Lawton, and M. Shachak. 1994. *Organisms as Ecosystem Engineers*. New York, NY: Springer, Oikos, Vol. 69, No. 3: 373–386.

Kelly, J. 2005. *Converting Turf to a Xeriscape Landscape: How to Eliminate a Bermudagrass Lawn Using Glyphosate*. Tucson, AZ: The University of Arizona Cooperative Extension: Publ. AZ1371. The University of Arizona, College of Agriculture and Life Sciences.

Kremer, P. and Z.A. Hamstead. 2015. Transformation of urban vacant lots for the common good: An introduction to the special issue. *Cities Environ* 8(2):Article 1.

Larson, E.K. and N.B. Grimm. 2012. Small-scale and extensive hydrogeomorphic modification and water redistribution in a desert city and implications for regional nitrogen removal. *Urban Ecosyst* 15: 71–85. doi:10.1007/s11252-011-0208-1.

Leguédois, S., G. Séré, A. Auclerc, J. Cortet, H. Huot, S. Ouvrard, F. Watteau, C. Schwartz, and J.L. Morel. 2016. Modelling pedogenesis of Technosols. *Geoderma* 262: 199–212.

Leitsinger, M. 2011. Foreclosed homes, empty lots are next 'Occupy' targets. http://usnews.msnbc.msn.com/_news/2011/12/02/9166035-foreclosed-homes-empty-lots-are-next-occupy-targets (Accessed February 5, 2012).

Lin, H., J. Bouma, Y. Pachepsky, A. Western, J. Thompson, R. van Genuchten, H.-J. Vogel, and A. Lilly. 2006. Hydropedology: Synergistic integration of pedology and hydrology. *Water Resour. Res.* 42: W05301. doi:10.1029/2005WR004085.

Lohmann, R., G.L. Northcott, and K.C. Jones. 2000. Assessing the contribution of diffuse domestic burning as a source of PCDD/Fs, PCBs, and PAHs to the UK atmosphere. *Environ Sci Technol* 34: 2892–2899.

Martinez-Fernandez, C., I. Audirac, S. Fol, and E. Cunningham-Sabot. 2012. Shrinking cities: Urban challenges of globalization. *Int J Urban Reg Res* 36:213–225. doi:10.1111/j.1468-2427.2011.01092.x.

Millennium Ecosystem Assessment. 2005. Ecosystems and human well-being: Synthesis. http://millenniumassessment.org/documents/document.356.aspx.pdf.

Neff, J.C., A.P. Ballantyne, G.L. Farmer, N.M. Mahowald, J.L. Conroy, C.C. Landry, J.T. Overpeck, T.H. Painter, C.R. Lawrence, and R.L. Reynolds. 2008. Increasing eolian dust deposition in the western United States linked to human activity. *Nat Geosci* 1: 189–195.

Pavao-Zuckerman, M.A. 2008. The nature of urban soils and their role in ecological restoration in cities. *Restor Ecol* 16: 642–649.

Pavao-Zuckerman, M.A. and L.B. Byrne. 2009. Scratching the surface and digging deeper: Exploring ecological theories in urban soils. *Urban Ecosyst* 12: 9–20.

Pouyat, R., P. Groffman, I. Yesilonis, and L. Hernandez. 2002. Soil carbon pools and fluxes in urban ecosystems. *Environ Pollut* 116: S107–S118.

Pouyat, R.V., K. Szlavecz, I.D. Yesilonis, P.M. Groffman, and K. Schwarz. 2010. Chemical, physical, and biological characteristics of urban soils. In *Urban Ecosystem Ecology, Agronomy M*, ed. J. Aitkenhead-Peterson and A. Volder, 119–152. Madison, WI: American Society of Agronomy, Crop Science Society of America, Soil Science Society of America.

Pouyat, R.V., I.D. Yesilonis, J. Russell-Anelli, and N.K. Neerchal. 2007. Soil chemical and physical properties that differentiate urban land-use and cover types. *Soil Sci Soc Am J*. 71: 1010–1019.

Raciti, S.M., P.M. Groffman, J.C. Jenkins, R.V. Pouyat, T.J. Fahey, S.T.A. Pickett, and M.L. Cadenasso. 2011. Accumulation of carbon and nitrogen in residential soils with different land-use histories. *Ecosystems* 14: 287–297. doi:10.1007/s10021-010-9409-3.

Reheis, M.C. 2006. A 16-year record of eolian dust in Southern Nevada and California, USA: Controls on dust generation and accumulation. *J Arid Environ* 67: 487–520.

Reynolds, R.L., M. Reheis, J. Yount, and P. Lamothe. 2006. Composition of aeolian dust in natural traps on isolated surfaces of the central Mojave Desert—Insights to mixing, sources, and nutrient inputs. *J Arid Environ* 66: 42–61.

Rhea, L., W. Shuster, J. Shaffer, and R. Losco. 2014. Data proxies for assessment of urban soil suitability to support green infrastructure. *J Soil Water Conserv* 69: 254–265. doi:10.2489/jswc.69.3.254.

Robinson, D.A., I. Fraser, E.J. Dominati, B. Davídsdóttir, J.O.G. Jónsson, L. Jones, S.B. Jones, M. Tuller, I. Lebron, K.L. Bristow, D.M. Souza, S. Banwart, and B.E. Clothier. 2014. On the value of soil resources in the context of natural capital and ecosystemservice delivery. *Soil Sci Soc Am J* 78(3): 685–700. doi:10.2136/sssaj2014.01.0017.

Schifman, L.A. and T.B. Boving. 2015. Spatial and seasonal atmospheric PAH deposition patterns and sources in Rhode Island. *Atmos Environ* 120: 253–261.

Schilling, J. and J. Logan. 2008. Greening the rust belt. *J Am Plan Assoc* 74: 451–466.

Schindelbeck, R.R, H.M. van Es, G.S. Abawi, D.W. Wolfe, T.L. Whitlow, B.K. Gugino, O.J. Idowu, and B.N. Moebius-Clune. 2008. Comprehensive assessment of soil quality for landscape and urban management. *Landsc Urban Plan* 88: 73–80.

Schoeneberger, P.J. and D.A. Wysocki. 2005. Hydrology of soils and deep regolith: A nexus between soil geography, ecosystems and land management. *Geoderma* 126: 117–128.

Schwarz, K., B.B. Cutts, J.K. London, and M.L. Cadenasso. 2016. Growing gardens in shrinking cities: A solution to the soil lead problem? *Sustainability* 8: 141.

Shuster, W.D., S.D. Dadio, C.E. Burkman, S.R. Earl, and S.J. Hall. 2014b. Hydropedological assessments of parcel-level infiltration in an arid urban ecosystem. *Soil Sci Soc Am J* 79(2): 398. doi:10.2136/sssaj2014.05.0200.

Shuster, W.D., S. Dadio, P. Drohan, R. Losco, and J. Scaeffer. 2014a. Residential demolition and its impact on vacant lot hydrology: Implications for the management of stormwater and sewer system overflows. *Landsc Urban Plan* 125: 48–56. doi:10.1016/j.landurbplan.2014.02.003.

Smagin, A.V., N.A. Azovtseva, M.V. Smagina, A.L. Stepanov, A.D. Myagkova, and A.S. Kurbatova. 2006. Criteria and methods to assess the ecological status of soils in relation to the landscaping of urban territories. *Eurasian Soil Sci* 39: 539–551.

USEPA. 2013. On the road to reuse: Residential demolition bid specification development tool. https://www.epa.gov/sites/production/files/2013-09/documents/road-to-reuse-residential-demolition-bid-specification-201309.pdf.

Wander, M.M. and L.E. Drinkwater. 2000. Fostering soil stewardship through soil quality assessment. *Appl Soil Ecol* 15: 61–73.

13 Biogeochemistry of Rooftop Farm Soils

Yoshiki Harada, Thomas H. Whitlow,
Nina L Bassuk, and Jonathan Russell-Anelli

CONTENTS

13.1 APPROACH

Rooftop farming draws on many specialized fields, each with its own terminology as well as different definitions for the same term. As a prime example, one can argue that there is no soil on a roof, so why should there be a chapter on rooftop farming in a volume devoted to soil science? A simple definition of soil that is universally accepted by agriculturalists, geologists, ecologists, and engineers is a perennial challenge, yet by calling the material placed on a roof in order to grow plants a *soil*, it is appropriate to follow Jenny's (1941) wise choice of leaving the definition open and inclusive.

Among the many subject areas informing rooftop farming are soil science, biogeochemistry, horticultural science, green roof design and management, and potting soils (Figure 13.1). In the biophysical sciences, *soil science* provides the foundation for understanding the physics and chemistry of water and nutrient movement in a rooftop soil while *biogeochemistry* treats these fluxes at the ecosystem scale. In the realm of applied science, production *horticulture* has historically dealt primarily with crops grown in ground in native soils. In contrast, roofs lack native soil, are disconnected from the underlying subsoil and parent material, and are disconnected from upland watersheds. A rooftop farm resembles a *green roof* yet would probably require deeper soil and supplemental irrigation to achieve acceptable yield and quality. A rooftop farm devoted to vegetable production also draws on the technologies developed for *greenhouse production*, including supplemental irrigation and nutrients and the use of artificial soilless or potting mixes. Since the 1950s, an extensive literature has dealt with synthetic soil mixes for greenhouse and nursery production, which provides information that could be used to develop soil for extensive outdoor landscapes like green roofs (Baker 1957; Dasberg 1999; Jozwik 2000; Newman 2008). Finally, *urban planning and design* lends understanding of how rooftop farming both drives and responds to the complex, coupled human and natural systems of modern cities. This literature is large and expanding rapidly. A recent search of bibliographic databases found over 1,000 peer-reviewed papers relevant to growing plants on roofs, while the intersection of the disciplines contains roughly 100 papers.

This chapter is intended for a wide audience, not just for soil scientists. After a review of the context for rooftop farming, the focus is on the influence of soil composition and depth on the water and nutrient budgets of rooftop farms and concludes with a case study from New York City (Figures 13.1 and 13.2).

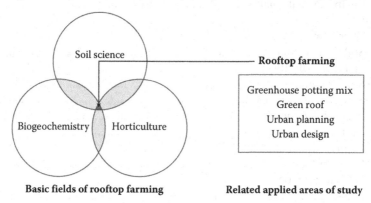

FIGURE 13.1   Related fields of rooftop farming.

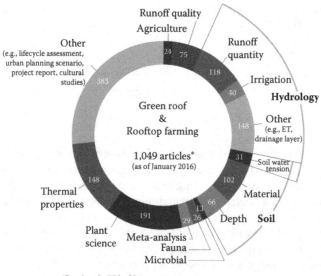

FIGURE 13.2   Topic groups in green roof and rooftop farming research.

13.2 BACKGROUND

Cities are hotspots for biogeochemical cycles, making them ideal locations for developing and testing novel ecosystems to enhance sustainability (Palmer et al. 2004; Grimm et al. 2008). Among these are a wide variety of green infrastructure projects intended to manage stormwater, save energy, and manage waste. Social dimensions to these practices include environmental education, investment, green job employment, eco-justice, food security, and building more cohesive communities.

Urban rooftop farms could potentially integrate many ecosystem services. These perceived services necessarily involve regulatory and investment sectors as well as public preferences (Plakias 2016). In this regard, urban rooftop farming could be viewed as one of the most creative components of twenty-first century planning for sustainable cities. The "combining" and "stacking" of multiple ecosystem services presents a unique opportunity for cross-disciplinary research (Felson and Pickett 2005; Lovell 2010; Robertson et al. 2014).

In New York City, green infrastructure projects are supported by a 20-year $1.5-billion capital initiative to fund community-based projects across the city (New York City Department of Environmental Protection 2010; Bloomberg 2011; De Blasio 2015). In April 2012, a new zoning code allowed retrofitting rooftops to include vegetable farms (New York City Department of City Planning 2012), prompting an influx of public and private funds into rooftop farms. A prominent example is the Brooklyn Grange, a 70,000-square-foot (0.65 ha) commercial rooftop farm (Figure 13.3), constructed in 2012 with a $592,730 grant from the Community-Based Green Infrastructure Program of the New York City Department of Environmental Protection (2011). This grant was partly based on the expectation that the farm would reduce stormwater runoff and the resulting N pollution caused by combined sewer discharges into surface waters.

Since its inception, the Grange has linked to the local community through businesses, schools, nonprofit organizations, and underrepresented populations by providing organic vegetables, collecting food wastes for composting, and offering educational and green job training programs (Plakias 2016). However, the functional environmental performance of rooftop farms has received little attention from the scientific community and there is little quantitative information on the design and operation of rooftop farms from a resource subsidies perspective. Understanding the water and nutrient budgets could lay the foundation for optimizing the environmental, yield, and economic return of the farm.

13.2.1 OPPORTUNITIES

A rooftop is a simple watershed system analogous to the Hubbard Brook experimental watershed. Inputs and outflows of water and nutrients are easiest to measure in watersheds with clear boundaries and topographic gradients, shallow soil overlying impervious rock, and a centralized stream network (Likens 2013). This approach provides a foundation for science to inform policy

FIGURE 13.3 The Brooklyn Grange Farm at Brooklyn Navy Yard, Summer 2015.

and practice across cities, thereby allowing comparisons of water and nutrient budgets across geographic locations and with varying degrees of human influence (Howarth et al. 1996). The extended community of scientists involved in the Long-Term Ecological Research (LTER) program supported by the US National Science Foundation (NSF) affords an ideal opportunity for generalizing the state of knowledge of urban biogeochemistry and applying this to practice (Fahey et al. 2015).

13.2.2 CHALLENGES

Although the biogeochemical processes of rooftop farms resemble those of forested and agricultural ecosystems, because many components of in-ground ecosystems are either *absent* (e.g., ground water recharge), *simplified* (e.g., soil horizonation), or *replaced with artificial materials* (e.g., soilless media), it is difficult to apply knowledge from other systems directly to rooftops.

Even among the horticultural disciplines that contribute to the emerging practice of rooftop farming, the existing scientific understanding of system performance may not translate directly. For example, extensive green roofs typically use soil media that are designed to be lightweight and drain rapidly to minimize roof loads. Nutrient supply after establishment is also relatively unimportant. While these soils may supply adequate water for slow growing, drought-tolerant *Sedum* species typically used on extensive green roofs, they are not optimal for a vegetable production system where yield and quality are important. In order to increase the nutrient and water holding capacity for vegetable cropping, green roof soil is amended with peat, coconut coir, biochar, and spent mushroom media, which are typical components used in greenhouse potting mixes managed for a single crop cycles. In contrast, rooftop farms use the same soil indefinitely under outdoor conditions with diurnal and seasonal cycles of temperature and moisture. Soils in rooftop farms are further amended with organic fertilizers for vegetable production (kelp meal, blood meal) and food-waste compost, which is low in many essential nutrients found in the native soil.

Ideally, each ingredient of a rooftop soil would contribute to optimizing the water and nutrient budgets. Optimum management of existing farms could aim at the steady state of nutrient budget in the traditional sense of biogeochemistry (Likens 2013), while "enhanced" steady state could be achieved by applying scientific understanding of site characteristics (e.g., substrates, irrigation systems) to improve the performance (Fahey et al. 2015). Beyond rooftop farms, similar challenges confront all green infrastructure projects, including green roofs, bioswales, rain gardens, and reclaimed urban parcels.

13.2.3 ENVIRONMENTAL QUALITY

New York City alone has about 8,600 ha of flat rooftop surface (Acks 2006), about 25 times the size of the Central Park. Converting even a small fraction of this space to agriculture raises concern about increased N load to surface water bodies draining the city. The appearance of "dead zones" resulting from low oxygen concentration in the Chesapeake Bay and Gulf of Mexico are the legacy of fertilizer runoff from farming practices intended to maximize crop yield (Rabalais et al. 2002; Kemp et al. 2005). In the United States, the passage of the Clean Water Act in 1970 and its subsequent amendments reflects a growing recognition that management practices can have a profound influence on downstream water quality. The establishment of total maximum daily loads (TMDLs) for each governmental jurisdiction in a watershed is intended to eliminate downstream pollution (New York State Department of Environmental Conservation 2000). The need to comply with increasingly stringent regulations has led to the development of best management practices (BMPs) to reduce both concentration and runoff volume from all land uses (Meals et al. 2010). Because urban rooftop farming is in its infancy, there exists an unprecedented opportunity to develop and implement BMPs before problems arise. Soil science is central to engineering soils that satisfy both the concerns of roof bearing capacity and nutrient and water retention. Among the most important soil properties affecting these are depth, composition, and pore-size distribution.

13.3 SOIL DESIGN

13.3.1 OVERVIEW

Over the past decade, reviews of the green roof literature report a wide range of water retention and nutrient loss (Mentens et al. 2006; Berndtsson 2010; Rowe 2011; Li and Babcock 2014; Driscoll et al. 2015). Reviews of green roof performance include both soilless (Ampim et al. 2010) and soil-based media (Best et al. 2015) and irrigation (Van Mechelen et al. 2015). The physical, chemical, and biological properties of the soil are the key factors, which relate the design and management to water retention and nutrient leaching from green roofs (Berndtsson 2010; Rowe 2011; Buffam and Mitchell 2015). It has proven difficult to develop standard specifications for rooftop soil that meet the competing demands of vegetable yield and quality, water retention, and nutrient leaching while keeping weight to a minimum. Green roofs with similar design and management vary widely in performance, yet the key information needed to explain such difference is often unknown or unreported. Studies often do not include statistics for both runoff quantity and quality, while coupled studies often lack details of soil composition. Insufficient detail stems in part from the difficulties of defining each component of green roof soil mixes. Even the standardized industrial-grade products like expanded shale would require laboratory testing to quantify their physical characteristics (e.g., pore-size distribution), and there are many components that are too variable for exact specification (e.g., composts), especially as their properties change over time in response to field conditions and the management history. Those factors present a challenge to synthesize the knowledge from different studies across disparate climactic zones and cultural contexts.

13.3.2 SOIL COMPOSITION

Rooftop farms use a variety of soils, including commercial potting mixes (e.g., planter mixes, garden mixes), commercial green roof mixes, experimental mixes using common/novel ingredients for a green roof, and commercial rooftop farming mixes (Table 13.1). Both commercial potting and green roof mixes use organic (e.g., compost, peat) and mineral (e.g., vermiculite, perlite) soilless materials, and both have been used in the studies on rooftop farming, while expanded shale, clay, and slate (ESCS) is the main mineral component typically used in commercial green roof mixes in order to meet the drainage guidelines and weight limitations (Ampim et al. 2010). Naturally sourced soils (e.g., sand, loam) are sometimes used for green roofs with or without being blended with soilless materials (Best et al. 2015). The review of available literature indicated only one commercial formulation specified for the rooftop farming, Rooflite®, which is reported in the feasibility review of rooftop farming by the New York State Energy Research and Development Authority (2013). It is a blend of heat-treated shale, spent mushroom media, and composts (Kong et al. 2015), and is used in the Brooklyn Grange, a rooftop farm in New York City, that is the subject of the "Case Study" later in this chapter. Given the variety of soils by rooftop farms, it is important to define the components in each study in order to obtain generalizable interpretations of the results.

In a broad sense, interest in rooftop farming is an outgrowth of ecological awareness, including adaptive reuse of waste products. Of 24 rooftop farming studies identified in this review, Grard et al. (2015) studied a rooftop farm in Paris, France that used locally sourced yard waste compost, crushed wood, and coffee grounds in its soil, growing lettuce (*Lactuca sativa*) and tomatoes

TABLE 13.1

Soil Type, Depth, and Plant Growth in Rooftop Farming Research

Soil Type	Soil Detail	Crops	Yield[a]	Irrigation[b]	Soil Depth (mm)	Location	References
Commercial potting mix	Sunshine Mix #4, Sun-Gro Horticulture (55%–65% peat, 35%–45% perlite)	Lettuce chicory[c]	S	Y	50, 100, 200	Korea (rooftop)	Cho (2008)
Commercial potting mix	Sunshine Mix #4, Sun-Gro Horticulture (55%–65% peat, 35%–45% perlite)	Lettuce chicory[d]	S	Y	150	Korea (rooftop)	Cho et al. (2010)
Commercial potting mix	Sunshine Mix Fisons (composition unspecified)	Kale[e]	S	Y	102	VA, USA (rooftop)	Elstein et al. (2006)
Commercial potting mix	1. Terre à planter, Brun brand (topsoil, blond sphagnum peat moss, composted bark, brown peat, horse manure and composted seaweed, ratio unspecified)						
Experimental mix	2. 100% yard waste compost, underlain by 100% crushed wood	Lettuce tomatoes	S	Y	300	Paris, France (rooftop)	Grard et al. (2015)
Experimental mix	3. 100% yard waste compost, underlain by 100% coffee ground layer, 100% crushed wood layer						
Experimental mix	4. 50% yard waste compost, 50% crushed wood						
Commercial green roof mix	Renewed Earth (50% expanded shale, 35% sand, 15% leaf compost)	Tomatoes[f] beans[g] cucumbers[h] peppers chives[i] basil[j]	S (except pepper)	Y	105	MI, USA (rooftop)	Whittinghill et al. (2013)
Experimental mix	Expanded shale, sand + 0, 20, 40, 60, 80, 100% yard waste compost	Cucumbers[h] peppers	S	Y	125	MI, USA (rooftop)	Eksi et al. (2015)

(Continued)

TABLE 13.1 (*Continued*)
Soil Type, Depth, and Plant Growth in Rooftop Farming Research

Soil Type	Soil Detail	Crops	Yield[a]	Irrigation[b]	Soil Depth (mm)	Location	References
Commercial rooftop farming mix	Rooflite, Skyland (lightweight mineral aggregates, mushroom compost, unspecified organic composted component, ratio unspecified) + (1) yard waste compost, (2) composted poultry manure, (3) vermicompost, (4) controlled-release fertilizer	Swiss chard	S	Y	110	NY, USA (greenhouse)	Kong et al. (2015)

[a]S, satisfactory yield.
[b]Y, irrigated.
[c]*Cichorium intybus* var. *folisum.*
[d]*Cichorium endivia* var. *endivia.*
[e]*Brassica oleracea* var. *acephala.*
[f]*Solanum lycopersicum.*
[g]*Phaseolus vulgaris.*
[h]*Cucumis sativus.*
[i]*Allium schoenoprasum.*
[j]*Ocimum basilicum.*

(*Lycopersicum esculentum* var. *cherry*) with irrigation. This study reports satisfactory yields of both crops and heavy metal levels lower than the European standard. However, the original soil depth of 300 mm soil decreased to 100–150 mm after the first growing season due to settling and decomposition of organic matter (OM). The volume reduction and consumption of urban wastes is an important ecosystem service, yet the changes in soil depth complicate management and point to the need to replenish OM frequently.

In green roof research, Ampim et al. (2010) reviewed the physical and chemical characteristics of recycled soil ingredients and emphasize the need of combining soil material research with observations of plant response, and runoff quantity and quality. More recently, an increasing number of studies have reported satisfactory growth of grass, sedum, and wildflowers in soils made from recycled construction materials (e.g., bricks, tiles) (Bates et al. 2015; Molineux et al. 2015), paper ash, and bark (Young et al. 2014; Molineux et al. 2015).

Biochar (Cao et al. 2014) and hydrogel products (Olszewski et al. 2010; Savi et al. 2014) have also been tested for their ability to increase the plant available water. Biochar could also improve the water and nutrient retention of soil (Beck et al. 2011), while the effects of hydrogel were species-dependent (Farrell et al. 2013). Unlike typical green roofs, which do not produce food, recycled materials used for rooftop farms would need to be tested for toxic residues to ensure public health.

13.3.3 Plant Growth Effects

Across a variety of soils, all seven studies of rooftop farms (Table 13.1) report satisfactory yield in all species except pepper (*Capsicum annuum*) (Whittinghill et al. 2013). All studies used irrigation yet none reported the effect of soil composition on irrigation requirements, although Kong et al. (2015) reported that the addition of composted yard waste to Rooflite® resulted in higher yields of Swiss chard (*Beta vulgaris*) and reduced leaching loss of nitrogen.

13.3.4 Soil Depth

Among the studies on rooftop farming summarized in Table 13.1, soil depth varied between 50–300 mm. Only Cho (2008) specifically tested different soil depths and reported positive but nonsignificant growth response to deeper soil. It is noteworthy that the manufacturer of Rooflite® specifies a minimum depth of 8 inches (≈200 mm) (Skyland USA LLC 2016), yet six of seven studies report satisfactory yields with soil less than 200 mm deep.

In addition to the studies of rooftop farms, green roof research includes an additional 60+ studies addressing the effect of soil depth both with and without irrigation. Standard extensive and intensive green roofs typically use drought-tolerant plants, which require much less water and nutrients in comparison to vegetable crops; hence, soils designed for green roofs may not be optimal for vegetable production. However, these studies still report important information on soil properties which could reduce the evaporative loss while maintaining plant available water (see "Soil Moisture and Evapotranspiration"). Ten studies using soil depths ranging between 20 and 400 mm reported that increasing depth increased available water and biomass across a wide range of species including *succulents* (VanWoert et al. 2005; Getter and Rowe 2009), *dry grassland species* (Dunnett et al. 2008), *turf grass species* (Nektarios et al. 2010; Ntoulas et al. 2012; Ntoulas et al. 2013), *drought-adapted shrubs* (Nektarios et al. 2011; Kotsiris et al. 2012; Savi et al. 2015), and *olive trees* (Kotsiris et al. 2013).

13.3.5 Soil Moisture and Evapotranspiration

In terms of the water budget of a rooftop farm, the ideal soil reduces runoff volume and minimizes the irrigation demand while achieving commercially viable yield and quality of crops. To this goal, it is important to balance evapotranspiration (ET) demand and soil water holding capacity. In a Mediterranean climate, Nektarios et al. (2011) grew *Dianthus fruticosus* in either 750 mm or 150 mm of pumice-based soil with irrigation. In this field experiment, shallow soil had higher evaporative losses presumably because the zone of capillary rise was closer to the soil surface, therefore both the diffusive resistance of the soil was less and the temperature gradient between the soil surface and capillary water was steeper. Pore-size distribution and soil depth are the key to controlling evaporative losses.

In a controlled greenhouse simulation of spring and summer conditions (Sheffield UK, average temperature 7.1 and 16.7°C, respectively; Poë et al. 2015), ET was positively correlated with the water holding capacity of the soil and was greatest in the soil with the highest OM and fine mineral fractions whose volumetric water content (VWC) was 25% at 33 kPa.

In a greenhouse simulation of summer conditions (Subtropical, New Zealand, average air temperature 22°C), Voyde et al. (2010) monitored ET of *Sedum mexicanum* and *Disphyma australe*, two drought hardy succulents, grown in 70 mm of pumice, zeolite, and compost. Over the first 9 days of the experiment, ET from planted treatments was consistently higher than the evaporative loss from bare soil. After day 17 when ET essentially ceased due to the drought stress, both treatments lost equal amounts of water. In a rooftop farm, transpiration in the absence of water stress could be much higher because of the large leaf area and could decline more rapidly as vegetables close their stomates in response to drought.

13.3.6 Water Retention

Soil water tension (SWT) plays important roles in the water retention characteristics of a soil, including the field capacity, plant available water, and the effects of porosity and self-mulching. Compared to greenhouse cropping systems, management of water is more important outdoors because of diurnal and seasonal variation in water supply and demand experienced in the field. Therefore, SWT of green roof and rooftop faming mixes are much less articulated or understood, containing large amount of soilless mixes typical to greenhouse cropping system. In addition, the low range of tension (<10 kPa) is particularly important for rooftop systems, because the soil depth is much shallower (<500 mm) than the in-ground systems (>1000 mm). Also, meta-analysis on SWT-based irrigation criteria reports 6–10 kPa for mustard greens, collard, and leaf lettuce in sandy loam (Shock and Wang 2011), and such leaf greens are important crops for rooftop farming. Therefore, for the effective management of water in rooftop farming, it is important to understand the water of soilless mixes in low range of tension (<10 kPa).

SWT is sensitive to the composition of soil mixes, and it is difficult to characterize the SWT of soil mixes solely based on SWT of each ingredient, yet the key effect of each material could be generalized to some extent. Within the tension range between 2 and 10 kPa, mixes of pumice and thermally treated clay showed 2%–7% VWC, in comparison to the mixes of crushed tile and pumice, which showed VWC between 15% and 20% (Ntoulas et al. 2015). Organic soilless materials and naturally sourced soil could also change VWC. Below 10 kPa, peat often holds more water than compost at the volumetric addition of 15% and 20% (Ntoulas et al. 2012, 2013, 2015), yet compost was more effective at the increased addition rates of 30% (Kotsiris et al. 2012, 2013). The type of compost also has an influence on the moisture contents (Ntoulas et al. 2015). In the mixture of pumice, compost, and zeolite, the 15% volumetric addition of sandy loam increased the VWC above 6.5 kPa tension, while mixes without sandy loam had higher VWC below 2 kPa (Nektarios et al. 2011). Adding 30% sandy loam to peat or compost treatments reduced the VWC above 3 kPa tension (Ntoulas et al. 2013). In order to design the SWT, it is important to establish the critical composition of each material.

13.4 RUNOFF VOLUME

The effect of particle size distribution on water retention and drainage is well understood and relatively easy to take for granted in native mineral soils, as are their effects on plant growth. Soils retain the maximum amount of available water in the texture class known as silt loam, where particle size classes, and hence pore sizes, are more or less evenly distributed. However, soils intended for rooftop farms often lack the familiar sand, silt, and clay size fractions and when organic materials (e.g., compost, biochar, highly modified skeletal components like expanded shale) are substituted for these mineral components, designing an optimal soil for rooftop farming becomes quite challenging.

In rooftop farms, surface runoff would not occur under normal conditions, and "runoff" as used in this context indicates the export of drainage water following the lateral flow through the soil and underlying drainage layer and along the impervious roof membrane. Rooftop farms are irrigated and it is important to differentiate "volumetric runoff reduction" from "water budget," which includes all fluxes of water including irrigation, while runoff reduction only compares precipitation to runoff. Irrigation can produce discharge. In green roofs and rooftop farms, irrigation could produce base flow if it is frequent and maintains water content near field capacity. Elstein et al. (2006) report the volumetric runoff reduction of a rooftop farm of 69.2%, while Whittinghill et al. (2015) report a reduction over 85% for 0–10 mm of precipitation, and below 60% for precipitation events over 10 mm. The study does not report the overall volumetric reduction based on the cumulative precipitation and drainage during the entire study period.

In studies of nonproduction green roofs, over 100 studies report volumetric runoff reduction, and five reviews show large variation across individual studies between 12%–85% (Figure 13.4).

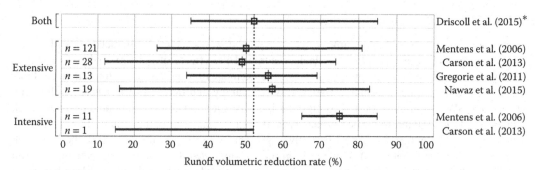

*Interquartile range and mean of site means from international stormwater best management practice database.
*Number and media depths of studies not reported.
*Dashed line indicates the mean of site means.

FIGURE 13.4   Runoff volumetric reduction of extensive and intensive green roofs.

(Carson et al. 2013; Gregoir and Clausen 2011; Mentens et al. 2006; Nawaz et al. 2015; Driscoll et al. 2015) This wide variation defies simple generalizations applicable across climate zones, soils, and plant cover types, but the average reduction is around 50%.

13.5   RUNOFF QUALITY

13.5.1   Overview

Through industry, agriculture, and urbanization, human activity is a major driver of the global N cycle, affecting the form and function of ecosystems across diverse scales (Howarth et al. 1996; Vitousek et al. 1997). It is useful to put N leaching from urban rooftop farming in the context of other land uses (Figure 13.5). Among 11 studies reporting N loss, including the NSF LTER projects in agricultural, forested, and urban watersheds (Berndtsson et al. 2006; Cameron et al. 2013; Campbell et al. 2004; Fenn et al. 1998; Gregoir and Clausen 2011; Groffman et al. 2004; Goulding et al. 2013; Likens 2013; Min et al. 2012; Pärn et al. 2012; Syswerda et al. 2012), N losses vary from <0.1 kg N ha^{-1} y^{-1} from a forested watershed (Fenn et al. 1998), to 277 kg N ha^{-1} y^{-1} from a vegetable farm (Min et al. 2012). Green roof losses average between 4 and 5 kg N ha^{-1} y^{-1} (Berndtsson et al. 2006; Gregoire and Clausen 2011). Although two studies report the N concentration in leachate from rooftop farms, as of January 2016, there have been no field observations on mass N leaching. Rooftop vegetable production could result in substantial N loss due to rapid soil drainage rates and fertility. Both urban design and policy need to consider N losses in order to quantify the environmental costs and benefits from urban rooftop farms.

13.5.2   N Concentration

N concentration is a key metric used by environmental regulators to gauge surface water quality (Groffman et al. 2004); hence, N concentration in runoff is a useful indicator of N loading from rooftop farms and green roofs. Two studies report the N concentration relevant to the runoff from rooftop farming. Whittinghill et al. (2015) reported average runoff nitrate concentration of 0.22 mg L^{-1} from a field experiment while in a greenhouse experiment, Kong et al. (2015) reported maximum nitrate N concentrations of 12.8 mg L^{-1} from vermicompost and 165.8 mg L^{-1} using a controlled-release fertilizer. Different nitrate N concentrations between Whittinghill et al. (2015) and Kong et al. (2015) reflect nitrate levels present in the soil products (65 vs. 118 mg L^{-1}), and N addition rates (35 vs. 126–189 kg N ha^{-1}).

Figure 13.6 includes the runoff nitrate N concentrations from these studies as well as the preliminary results from the Brooklyn Grange farm at the Brooklyn Navy Yard in New York City that uses the same soil used by Kong et al. (2015). The results from the Brooklyn Grange farm (2–12 mg L^{-1}) fall between the values from other two studies. With the oversight of the United States Environmental Protection Agency (2016), New York State sets the standard for nitrate N in rivers

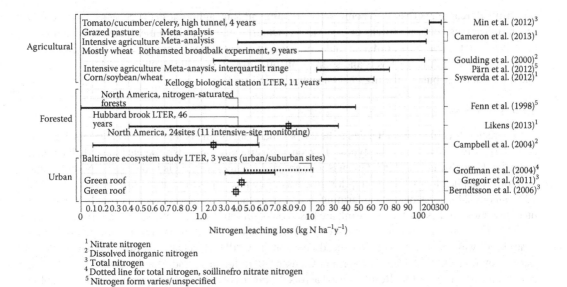

FIGURE 13.5 Nitrogen leaching loss from forested, agricultural, and urban watershed.

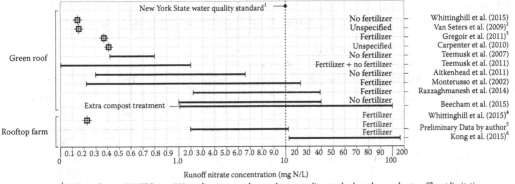

FIGURE 13.6 Runoff nitrate nitrogen concentration of green roofs and rooftop farms.

and streams at 10 mg L[-1] (6 NYCRR Part 703) (New York State Department of Environmental Conservation 2016). Note that roof drains are not directly subject to regulation.

Among 10 studies of nonproduction green roofs summarized in Figure 13.6 (Aitkenhead et al. 2011; Beecham and Razzaghmanesh 2015; Carpenter and Kaluvakolanu 2010; Gregoir and Clausen 2011; Monterusso et al. 2002; Razzaghmanesh et al. 2014; Teemusk and Mander 2007, 2011; Van Seters et al. 2009; Whittinghill et al. 2015), nitrate N concentration in runoff ranges from below 0.2 (Van Seters et al. 2009; Whittinghill et al. 2015) to 100 mg L[-1] (Beecham et al. 2015). Teemusk et al. (2011) report higher N concentration in runoff from their fertilized green roof yet this is not reflected in the ranking of maximum nitrate concentrations across all of the studies. The study conducted by Gregoire and Clausen (2011) reported low N concentration in runoff despite using twice the rate of N application used in studies that yielded higher concentrations (Monterusso et al. 2002; Razzaghmanesh et al. 2014).

Composts are another potential source of N leached from green roofs (Hathaway et al. 2008; Toland et al. 2012). In Figure 13.6, the two highest nitrate concentrations were observed in unfertilized treatments that used compost (Beecham and Razzaghmanesh 2015). Most commercial green roof mixes contain sources of organic N, including manure, blood meal, biosolids, and kelp even if it is not reported. In order to make meaningful comparisons among studies it is important for each study to define the levels and sources of soil nutrient, in order to understand how fertilizer and compost contribute to the high N concentration in runoff.

13.5.3 N Sink

Urban green infrastructure is intended to be the N sink, reducing the N load to surface waters. Driscoll et al. (2015) reviewed the seven types of green infrastructure projects (Bioretention, Media Filter, Detention Pond, Swale, Wetland, Green Roof), and report only green roofs behave as an N source due to fertilizer application, while rooftop farming was not within the scope of the review. As of January 2016, among 24 studies of rooftop farming, none report the N source/sink relationship. In green roof literature, most studies on N source/sink are based on observations of runoff N concentrations alone.

Table 13.2 compares N concentration of green roof runoff to that of precipitation or runoff from a control roof without vegetation (Aitkenhead-Peterson et al. 2011; Beecham and Razzaghmanesh 2015; Berndtsson et al.2006, 2009; Gregoire and Clausen 2011; Teemusk and Mander 2007; Toland et al. 2012; Van Seters et al. 2009). Results varied across the ranges of soil depths, vegetation, and geographic locations. The first four studies reported lower nitrate and ammonia N concentrations in green roof

TABLE 13.2
Comparison of N Concentration of Green Roof Runoff and Reference Stormwater

Green Roof Runoff[a]		Reference Stormwater[b]	Runoff Volume Reported	Soil Depth (mm)	Plants	Location	References
NO3	NH4						
Low	Low	Rain	N	30	Sedum	Sweden	Berndtsson et al.
Low	Low	Rain	N	400	Shrub, tree	Japan	(2009)
Low	Low	Rain	Y	30–40	Sedum	Sweden	Berndtsson et al. (2006)
Low	Low	Roof runoff	Y	140	Wildflower	Toronto	Van Seters et al. (2009)
Similar	Low	Rain	Y	102	Sedum	CT, USA	Gregoire and Clausen (2011)
Similar	Similar	Roof runoff	N	Unspecified	Sedum, moss	AR, USA	Toland et al. (2012)
Similar	High	Rain	Y[c]	100–300	Brachyscome, Chrysocephalum, Disphyma spp	South Australia	Beecham and Razzaghmanesh (2015)
Varies	Similar	Rain	N	71	Sedum, Delosperma, Talinum spp	TX, USA	Aitkenhead-Peterson et al. (2011)
High	High	Rain	Y[d]	100	Sedum, Thymus, Dianthus, Cerastium spp	Estonia	Teemusk and Mander (2007)

[a]Comparison of green roof runoff N concentration to reference stormwater (precipitation or runoff from unvegetated roof).
[b]Precipitation is used as a reference stormwater if runoff from unvegetated roof is also reported.
[c]Only retention rates reported.
[d]Rainfall events not specified in concentration.

runoff compared to precipitation or unvegetated roofs, hence green roofs are most likely an N sink. Among the remaining five studies, only Gregoire and Clausen (2011) compared mass N loading, reporting that the green roof was a net sink for N. The other four studies do not report runoff volume in terms that can be used to calculate loading rates. Because rooftop farms are likely sources of N due to the high soil infiltration and the application of irrigation and N fertilizer, it is necessary to measure both runoff volume and N concentration in order to calculate total load to downstream water bodies.

13.6 CASE STUDY

The literature review presented herein reveals that the water and N budgets of green roofs and rooftop farms varies widely in relation to soil composition, depth, vegetation, management regime, and regional climate. In an effort to disentangle these variables, a multiyear study was initiated at the Brooklyn Grange, a 70,000-square-foot (0.65 ha) commercial rooftop farm in New York City. The Grange uses Rooflite®, presently the only commercial soil available for rooftop farming in the United States. Observation included monitoring irrigation, ET, VWC, and runoff volume and quality and experimentation with a variety of soil mixes in an effort to optimize the use of water and nutrient subsidies. Even though the farm is irrigated multiple times each day using a combination of drip tape and overhead sprinklers in order to maintain VWC between 25%–35%, during the dry periods, VWC drops to 15%–25%. Irrigation consistently produces base flow even below 25% VWC, suggesting that opportunities exist for optimizing the water and nutrient budgets of the farm by modifying the soil mix currently in use.

13.6.1 Soil Water Retention

Biogeochemical performance is not the only criterion for rooftop soils. They must also meet construction material guidelines for weight, wind, and fire set by the American National Standards Institute (ANSI). Also, decisions about including reused waste products in soil mixes would ideally be informed by comprehensive life-cycle assessments, including the energy and carbon (C) emission by the transportation and manufacturing processes. The approach adopted in this study is to first narrow down the biogeochemically superior soil design through a controlled laboratory experiment, followed by field experiments, then consider regulatory and life-cycle perspectives.

Figure 13.7 compares the VWC of 26 potential rooftop soils and amendments obtained from replicated tension table experiments. Among 26 mixes, including eight Rooflite® and nine potting mixes, VWC at 0.93 kPa varied from <30% for expanded shale to 80% for coconut coir. The lowest VWC of both unused and spent Rooflite® is ca 35%, exceeding field observations during rainless summer periods. This could be due to nonuniform field conditions caused by preferential flows and hysteresis or by variation in irrigation. Amending Rooflite® with biochar, coir and compost increased maximum VWC up to 60%–70%, while six of nine potting mix treatments had maximum VWM > 70%, which indicates the potential for designing lightweight rooftop soil by substituting organic amendments for expanded shale. Based on these lab results, we are field testing new soils that include coir and biochar.

13.6.2 Yield

Despite the anticipated advantages afforded by local food production systems, if crop yield does equal or exceed conventional agriculture, rooftop farming may do little to advance urban sustainability. A comparison of yields for four of the many vegetable crops grown at the Grange with California, New York, and New Jersey shows that the Grange outperforms statewide averages for conventional farms. On a per hectare basis, the Grange yielded 1.2 times more lettuce than California in 2015 and five times more tomatoes than NY in 2014. Note that California was experiencing a multiyear drought during this period, which could have depressed yields and skewed the data in favor of the Grange. In any case, it appears that rooftop vegetable production can produce high yields with appropriate inputs of water and nutrients.

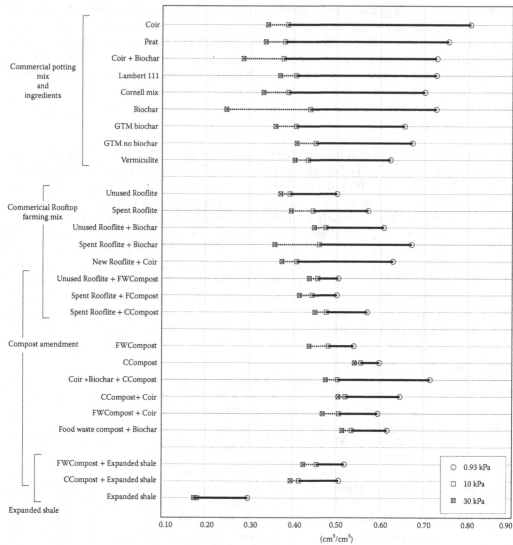

FIGURE 13.7 Volumetric water content of soilless mixes and ingredients at 0.93, 10, and 30 kPa.

13.6.3 WATER AND N BUDGET

Field monitoring of the farm's water budget indicates that 38% of the water supplied is lost to drainage (Table 13.4). Because irrigation relies on potable water from Upstate New York reservoirs, it would be desirable to reduce drainage losses. This could be accomplished by varying the soil mix and depth to maximize water retention in the range of plant available soil moisture.

Mass N input to the Grange through the fertilizer application and atmospheric N deposition (Table 13.5) is about 120 kg N ha⁻¹ y⁻¹, while mass N contained in vegetables leaving the farm as food crops is over 60% of N input. However, the sampled N loss by soil leachate (Table 13.5) is 10 times the initial estimation by this N balance model (Fertilizer application + Atmospheric deposition − Vegetable harvest ≈ 40 kg N ha⁻¹ y⁻¹). Organic N in soil imported in the original soil mix could

TABLE 13.3
Vegetable Yield of the Brooklyn Grange and In-Ground Agriculture

		Yield(metric ton/ha)			
Crop	Year	Brooklyn Grange	New York[b]	New Jersey[b]	California[b]
Snap beans	2014	20.40	7.07	3.70	12.35
	2015	19.79	7.30	3.59	13.47
Tomatoes	2014	68.58	13.47	24.14	35.36
	2015	47.62	14.59	25.26	34.80
Leaf lettuce	2014	34.22[a]	NA	NA	26.94
	2015	35.52[a]	NA	NA	29.19
Bell peppers	2014	128.26	NA	38.17	54.45
	2015	70.78	NA	34.24	51.08

[a]Based on the yield of leafy greens mix.
[b]Based on the National Agricultural Statistics Service, USDA.

TABLE 13.4
Water Budget of the Brooklyn Grange

Water Flux Type	Water Flux (10^6 L y^{-1})	Method
Irrigation input	≈1.5	Flow meter
Precipitation input	>5	Rain gauge
ET loss	>4	Penman–Monteith equation
Drainage loss	≈2.5	Water Balance Model (Irrigation + Precipitation – ET)

Note: Preliminary data on the water budget of the Brooklyn Grange, a rooftop farm at Brooklyn Navy Yard, NYC 2014–2015.

TABLE 13.5
Nitrogen Budget of the Brooklyn Grange

N Flux Type	N Flux (kg N ha^{-1} y^{-1})	N Form	Method
Fertilizer application	>110	Total N	Inventory analysis (farming record)
Atmospheric deposition	<10	Dissolved inorganic N	Bulk collector (with ion-exchange resin)
Soil leachate	>400	Dissolved inorganic N	Soil mesh bag (with ion-exchange resin)
Vegetable harvest	>80	Total N	Inventory analysis (farming record)

Note: Preliminary data on the N budget of the Brooklyn Grange, a rooftop farm at Brooklyn Navy Yard, NYC 2014–2015.

account for these high leaching losses. Further research will examine the accuracy of the sampling method, and denitrification potential of soil leachate in the drainage layer of the Grange.

13.7 CONCLUSION

Rooftop farming requires the synthesis of knowledge from many fields, including soil science, biogeochemistry, horticulture, and urban planning and design, among which soil science is central to both understanding and improving practices. This chapter reviewed the intersection of those fields with an emphasis on their relevance to rooftop farming. The perspective of soil science is central to both understanding and improving rooftop farming.

13.8   LIMITATIONS OF CURRENT RESEARCH

As of January 2016, more than 1,000 papers relevant to green roof design and rooftop ecosystems have been published. Of these studies, the bulk are focused on hydrologic responses, media and plant performance as well as climate effects. Green roof practices have been driven by the perceived need to reduce weight, drain rapidly, and contribute to building insulation. Few even attempt comprehensive integration of the many topics shown in Figure 13.2. As of yet, only 24 studies address rooftop agriculture. This review has identified these key gaps in knowledge:

- Commercial/custom potting mixes, conventional roofing ballast, and mineral soil could be important functional components roof infrastructure, but are not systematically studied in green roof or rooftop vegetable cropping systems.
- While there is increasing interest in soil composition, depth, and moisture, key factors are often unreported, preventing understanding of what is actually driving soil water and nutrient dynamics.
- Even when studies include both soil performance and plant growth, they do not include inputs and losses, or runoff volume, quality, and variation necessary to calculate loading rates and other ecosystem-level responses.

13.9   FUTURE WORK

Rooftop farms are ideal for investigating urban biogeochemical processes because they are simple enough to be studied in detail but complex enough to yield insights into the way cities function in the global context for food security, environmental quality, and waste management. Studies like the one at the Brooklyn Grange reveal how variables like soil composition, physical characteristics, depth, and application of fertilizer and irrigation subsidies affect the ability of a roof to deliver ecosystem services while at the same time actively informing daily management practices to improve the rapid development of BMPs. Future research should include:

- Detailed studies of engineered soils to conserve water, reduce leachate, and optimize partitioning of water into transpiration
- Systematic analysis of novel/different soil components, including repurposed waste and native soil
- Optimizing plant growth and quality along with ecosystem-level responses
- Applying the small science approach to studying individual farms and expanding to a research network including cities in different climate zones in order to develop a comprehensive framework of BMPs

REFERENCES

Acks, K. (2006). A framework for cost-benefit analysis of green roofs: Initial estimates. Green Roofs in the Metropolitan Region: Research Report. Columbia University Center for Climate Systems Research.

Aitkenhead-Peterson, J. A., B. D. Dvorak, A. Volder and N. C. Stanley (2011). Chemistry of growth medium and leachate from green roof systems in south-central Texas. *Urban Ecosystems* 14(1): 17–33.

Ampim, P. A., J. J. Sloan, R. I. Cabrera, D. A. Harp and F. H. Jaber (2010). Green roof growing substrates: Types, ingredients, composition and properties. *Journal of Environmental Horticulture* 28(4): 244.

Baker, K. F. 1957 (ed). *The U. C. System for Producing Healthy Container-Grown Plants*. California, CA: University of California Division of Agricultural Sciences.

Bates, A. J., J. P. Sadler, R. B. Greswell and R. Mackay (2015). Effects of recycled aggregate growth substrate on green roof vegetation development: A six year experiment. *Landscape and Urban Planning* 135: 22–31.

Beck, D. A., G. R. Johnson and G. A. Spolek (2011). Amending greenroof soil with biochar to affect runoff water quantity and quality. *Environmental Pollution* 159(8–9): 2111–2118.

Beecham, S. and M. Razzaghmanesh (2015). Water quality and quantity investigation of green roofs in a dry climate. *Water Research* 70: 370–384.

Berndtsson, J. C. (2010). Green roof performance towards management of runoff water quantity and quality: A review. *Ecological Engineering* 36(4): 351–360.

Berndtsson, J. C., L. Bengtsson and K. Jinno (2009). Runoff water quality from intensive and extensive vegetated roofs. *Ecological Engineering* 35(3): 369–380.

Berndtsson, J. C., T. Emilsson and L. Bengtsson (2006). The influence of extensive vegetated roofs on runoff water quality. *Science of the Total Environment* 355(1): 48–63.

Best, B. B., K. R. Swadek and L. T. Burgess (2015). *Soil-based green roofs*. In K. R. Sutton (ed) *Green Roof Ecosystems*. Cham, Switzerland: Springer International Publishing, 139–174.

Bloomberg, M. (2011). *PlaNYC: A Greener, Greater New York*. New York, NY: City of New York.

Buffam, A. P. D. I. and M. E. Mitchell (2015). Nutrient cycling in green roof ecosystems. *Green Roof Ecosystems*. Cham, Switzerland: Springer International Publishing, 107–137.

Cameron, K., H. Di and J. Moir (2013). Nitrogen losses from the soil/plant system: A review. *Annals of Applied Biology* 162(2): 145–173.

Campbell, J. L., J. W. Hornbeck, M. J. Mitchell, M. B. Adams, M. S. Castro, C. T. Driscoll, J. S. Kahl, J. N. Kochenderfer, G. E. Likens, J. A. Lynch, P. S. Murdoch, S. J. Nelson and J. B. Shanley (2004). Input-output budgets of inorganic nitrogen for 24 forest watersheds in the northeastern United States: A review. *Water Air and Soil Pollution* 151(1–4): 373–396.

Cao, C. T., C. Farrell, P. E. Kristiansen and J. P. Rayner (2014). Biochar makes green roof substrates lighter and improves water supply to plants. *Ecological Engineering* 71: 368–374.

Carpenter, D. D. and P. Kaluvakolanu (2010). Effect of roof surface type on storm-water runoff from full-scale roofs in a temperate climate. *Journal of Irrigation and Drainage Engineering* 137(3): 161–169.

Carson, T., D. Marasco, P. Culligan and W. McGillis (2013). Hydrological performance of extensive green roofs in New York City: Observations and multi-year modeling of three full-scale systems. *Environmental Research Letters* 8(2): 024036.

Cho, Y. Y. (2008). Soil depth affects the growth of lettuce and chicory in extensive green roofs. *Horticulture Environment and Biotechnology* 49(6): 434–438.

Cho, Y. Y., K. Y. Choi and Y.-B. Lee (2010). Effects of irrigation methods on the growth, water holding capacity of substrate and nutrient uptake of lettuce, chicory and endive grown in an extensive green roof system. *Horticulture Environment and Biotechnology* 51(4): 348–354.

Dasberg, S. (1999). The root medium. In Stanhill, G. and H. Z. Enoch (eds) *Greenhouse Ecosystems*. Netherlands: Elsevier Science Publishers.

De Blasio, W. W. (2015). One New York: The Plan for a Strong and Just City [Online]. New York City, NY. Accessed on 3 April 2016. <http://www.nyc.gov/html/onenyc/downloads/pdf/publications/OneNYC .pdf>.

Driscoll, C. T., C. G. Eger, D. G. Chandler, C. I. Davidson, B. K. Roodsari, C. D. Flynn, K. F. Lambert, N. D. Bettez and P. M. Groffman (2015). Green Infrastructure: Lessons From Science and Practice. A publication of the Science Policy Exchange. 32. Accessed on 3 April 2016. <https://projects.iq.harvard.edu/ science-policy/gi-report>.

Dunnett, N., A. Nagase and A. Hallam (2008). The dynamics of planted and colonising species on a green roof over six growing seasons 2001–2006: Influence of substrate depth. *Urban Ecosystems* 11(4): 373–384.

Eksi, M., D. B. Rowe, R. Fernández-Cañero and B. M. Cregg (2015). Effect of substrate compost percentage on green roof vegetable production. *Urban Forestry and Urban Greening* 14(2): 315–322.

Elstein, J., G. E. Welbaum, D. A. Stewart and D. R. Borys (2006). Evaluating growing media for a shallow-rooted vegetable crop production system on a green roof. IV International Symposium on Seed, Transplant and Stand Establishment of Horticultural Crops; Translating Seed and Seedling 782.

Fahey, T. J., P. H. Templer, B. T. Anderson, J. J. Battles, J. L. Campbell, C. T. Driscoll, A. R. Fusco, M. B. Green, K.-A. S. Kassam and N. L. Rodenhouse (2015). The promise and peril of intensive site based ecological research: Insights from the Hubbard Brook ecosystem study. *Ecology* 96(4): 885–901.

Farrell, C., X. Q. Ang and J. P. Rayner (2013). Water-retention additives increase plant available water in green roof substrates. *Ecological Engineering* 52: 112–118.

Felson, A. J. and S. T. Pickett (2005). Designed experiments: New approaches to studying urban ecosystems. *Frontiers in Ecology and the Environment* 3(10): 549–556.

Fenn, M. E., M. A. Poth, J. D. Aber, J. S. Baron, B. T. Bormann, D. W. Johnson, A. D. Lemly, S. G. McNulty, D. F. Ryan and R. Stottlemyer (1998). Nitrogen excess in North American ecosystems: Predisposing factors, ecosystem responses, and management strategies. *Ecological Applications* 8(3): 706–733.

Getter, K. L. and D. B. Rowe (2009). Substrate depth influences sedum plant community on a green roof. *HortScience* 44(2): 401–407.

Goulding, K. W. T., Poulton, P. R., Webster, C. P., and Howe, M. T. (2000). Nitrate leaching from the Broadbalk Wheat Experiment, Rothamsted, UK, as influenced by fertilizer and manure inputs and the weather. *Soil Use and Management* 16(4): 244–250.

Grard, B. J.-P., N. Bel, N. Marchal, F. Madre, J.-F. Castell, P. Cambier, S. Houot, N. Manouchehri, S. Besancon and J.-C. Michel (2015). Recycling urban waste as possible use for rooftop vegetable garden. *Future of Food: Journal on Food, Agriculture and Society* 3(1): 21–34.

Gregoire, B. G. and J. C. Clausen (2011). Effect of a modular extensive green roof on stormwater runoff and water quality. *Ecological Engineering* 37(6): 963–969.

Grimm, N. B., S. H. Faeth, N. E. Golubiewski, C. L. Redman, J. Wu, X. Bai and J. M. Briggs (2008). Global change and the ecology of cities. *Science* 319(5864): 756–760.

Groffman, P. M., N. L. Law, K. T. Belt, L. E. Band and G. T. Fisher (2004). Nitrogen fluxes and retention in urban watershed ecosystems. *Ecosystems* 7(4): 393–403.

Hathaway, A. M., W. F. Hunt and G. D. Jennings (2008). A field study of green roof hydrologic and water quality performance. *Transactions of the ASABE* 51(1): 37–44.

Howarth, R. W., G. Billen, D. Swaney, A. Townsend, N. Jaworski, K. Lajtha, J. Downing, R. Elmgren, N. Caraco and T. Jordan (1996). Regional nitrogen budgets and riverine N & P fluxes for the drainages to the North Atlantic Ocean: Natural and human influences. *Nitrogen Cycling in the North Atlantic Ocean and its Watersheds*. Netherlands: Springer, 75–139.

Jenny, H. (1941). *Factors of Soil Formationm*. New York, NY: McGraw-Hil.

Jozwik, F. X. (2000). *The Greenhouse and Nursery Handbook*. Mill, Wyoming: Andmar Press.

Kemp, W. M., W. R. Boynton, J. E. Adolf, D. F. Boesch, W. C. Boicourt, G. Brush, J. C. Cornwell, T. R. Fisher, P. M. Glibert and J. D. Hagy (2005). Eutrophication of Chesapeake Bay: Historical trends and ecological interactions. *Marine Ecology Progress Series* 303(21): 1–29.

Kong, A. Y., C. Rosenzweig and J. Arky (2015). Nitrogen dynamics associated with organic and inorganic inputs to substrate commonly used on rooftop farms. *HortScience* 50(6): 806–813.

Kotsiris, G., P. A. Nektarios and A. T. Paraskevopoulou (2012). Lavandula angustifolia growth and physiology is affected by substrate type and depth when grown under Mediterranean semi-intensive green roof conditions. *HortScience* 47(2): 311–317.

Kotsiris, G., P. A. Nektarios, N. Ntoulas and G. Kargas (2013). An adaptive approach to intensive green roofs in the Mediterranean climatic region. *Urban Forestry and Urban Greening* 12(3): 380–392.

Li, Y. and R. W. Babcock Jr (2014). Green roofs against pollution and climate change. A review. *Agronomy for Sustainable Development* 34(4): 695–705.

Likens, G. E. (2013). *Biogeochemistry of a Forested Ecosystem*. New York: Springer Science & Business Media.

Lovell, S. T. (2010). Multifunctional urban agriculture for sustainable land use planning in the United States. *Sustainability* 2(8): 2499–2522.

Meals, D. W., S. A. Dressing and T. E. Davenport (2010). Lag time in water quality response to best management practices: A review. *Journal of Environmental Quality* 39(1): 85–96.

Mentens, J., D. Raes and M. Hermy (2006). Green roofs as a tool for solving the rainwater runoff problem in the urbanized 21st century? *Landscape and Urban Planning* 77(3): 217–226.

Min, J., H. Zhang and W. Shi (2012). Optimizing nitrogen input to reduce nitrate leaching loss in greenhouse vegetable production. *Agricultural Water Management* 111: 53–59.

Molineux, C. J., A. C. Gange, S. P. Connop and D. J. Newport (2015). Using recycled aggregates in green roof substrates for plant diversity. *Ecological Engineering* 82: 596–604.

Monterusso, M., D. Rowe, C. Rugh and D. Russell (2002). Runoff water quantity and quality from green roof systems. XXVI International Horticultural Congress: Expanding Roles for Horticulture in Improving Human Well-Being and Life Quality 639.

Nawaz, R., A. McDonald and S. Postoyko (2015). Hydrological performance of a full-scale extensive green roof located in a temperate climate. *Ecological Engineering* 82: 66–80.

Nektarios, P. A., I. Amountzias, I. Kokkinou and N. Ntoulas (2011). Green roof substrate type and depth affect the growth of the native species Dianthus fruticosus under reduced irrigation regimens. *HortScience* 46(8): 1208–1216.

Nektarios, P., N. Ntoulas and E. Nydrioti (2010). Turfgrass use on intensive and extensive green roofs. XXVIII International Horticultural Congress on Science and Horticulture for People (IHC2010): International Symposium on 938.

Newman, J. 2008. Protecting water quality. In Newman, J. (ed) *Greenhouse and Nursery Management Practices to Protect Water Quality*. California, CA: University of California Publication, 3508.

New York City Department of City Planning (2012). Zone Green Text Amendment [Online]. New York City, NY. Accessed on 3 April 2016. <http://www1.nyc.gov/assets/planning/download/pdf/plans/zone-green/zone_green.pdf>.

New York City Department of Environmental Protection (2010). NYC Green Infrastructure Plan: A Sustainable Strategy for Clean Waterways [Online]. New York City, NY. Accessed on 3 April 2016. <http://www.nyc.gov/html/dep/html/stormwater/nyc_green_infrastructure_plan.shtml>.

New York City Department of Environmental Protection (2011). DEP Awards $3.8 Million in Grants for Community-Based Green Infrastructure Program Projects [Online]. New York City, NY. Accessed on 3 April 2016. <http://www.nyc.gov/html/dep/html/press_releases/11-46pr.shtml#.VvIn6fkrKUk>.

New York State Department of Environmental Conservation, Connecticut Department of Environmental Protection (2000). A total maximum daily load analysis to achieve water quality standards for dissolved oxygen in Long Island Sound. Connecticut Department of Environmental Protection, Hartford.

New York State Department of Environmental Conservation (2016). Water Quality Standards and Classifications [Online]. Accessed on 25 June 2016. <http://www.dec.ny.gov/chemical/23853.html>.

New York State Energy Research and Development Authority (2013). Sustainable Urban Agriculture: Confirming Viable Scenarios for Production, Final Report [Online]. Accessed on 3 April 2016. <http://urbandesignlab.columbia.edu/files/2015/04/2_Sustainable-Urban-Agriculture_NYSERDA.pdf>.

Ntoulas, N., P. A. Nektarios, E. Charalambous and A. Psaroulis (2013). Zoysia matrella cover rate and drought tolerance in adaptive extensive green roof systems. *Urban Forestry and Urban Greening* 12(4): 522–531.

Ntoulas, N., P. A. Nektarios, T.-E. Kapsali, M.-P. Kaltsidi, L. Han and S. Yin (2015). Determination of the physical, chemical, and hydraulic characteristics of locally available materials for formulating extensive green roof substrates. *HortTechnology* 25(6): 774–784.

Ntoulas, N., P. A. Nektarios and E. Nydrioti (2013). Performance of Zoysia matrella 'Zeon' in shallow green roof substrates under moisture deficit conditions. *HortScience* 48(7): 929–937.

Ntoulas, N., P. A. Nektarios, K. Spaneas and N. Kadoglou (2012). Semi-extensive green roof substrate type and depth effects on Zoysia matrella 'Zeon' growth and drought tolerance under different irrigation regimes. *Acta Agriculturae Scandinavica, Section B-Soil and Plant Science* 62(Suppl. 1): 165–173.

Olszewski, M. W., M. H. Holmes and C. A. Young (2010). Assessment of physical properties and stonecrop growth in green roof substrates amended with compost and hydrogel. *HortTechnology* 20(2): 438–444.

Palmer, M., E. Bernhardt, E. Chornesky, S. Collins, A. Dobson, C. Duke, B. Gold, R. Jacobson, S. Kingsland, R. Kranz, M. Mappin, M. L. Martinez, F. Micheli, J. Morse, M. Pace, M. Pascual, S. Palumbi, O. J. Reichman, A. Simons, A. Townsend and M. Turner (2004). Ecology for a crowded planet. *Science* 304(5675): 1251–1252.

Pärn, J., G. Pinay and Ü. Mander (2012). Indicators of nutrients transport from agricultural catchments under temperate climate: A review. *Ecological Indicators* 22: 4–15.

Plakias, A. C. (2016). *The Farm on the Roof: What Brooklyn Grange Taught Us about Entrepreneurship, Community, and Growing a Sustainable Business*. New York: Penguin Publishing Group.

Poë, S., V. Stovin and C. Berretta (2015). Parameters influencing the regeneration of a green roof's retention capacity via evapotranspiration. *Journal of Hydrology* 523: 356–367.

Rabalais, N. N., R. E. Turner and W. J. Wiseman Jr (2002). Gulf of Mexico hypoxia, AKA the dead zone. *Annual Review of Ecology and Systematics* 33: 235–263.

Razzaghmanesh, M., S. Beecham and F. Kazemi (2014). Impact of green roofs on stormwater quality in a South Australian urban environment. *Science of the Total Environment* 470: 651–659.

Robertson, M., T. K. BenDor, R. Lave, A. Riggsbee, J. Ruhl and M. Doyle (2014). Stacking ecosystem services. *Frontiers in Ecology and the Environment* 12(3): 186–193.

Rowe, D. B. (2011). Green roofs as a means of pollution abatement. *Environmental Pollution* 159(8): 2100–2110.

Savi, T., D. Boldrin, M. Marin, V. L. Love, S. Andri, M. Tretiach and A. Nardini (2015). Does shallow substrate improve water status of plants growing on green roofs? Testing the paradox in two sub-Mediterranean shrubs. *Ecological Engineering* 84: 292–300.

Savi, T., M. Marin, D. Boldrin, G. Incerti, S. Andri and A. Nardini (2014). Green roofs for a drier world: Effects of hydrogel amendment on substrate and plant water status. *Science of the Total Environment* 490: 467–476.

Shock, C. C. and F.-X. Wang (2011). Soil water tension, a powerful measurement for productivity and stewardship. *HortScience* 46(2): 178–185.

Skyland USA LLC (2016). Rooflite Intensive Ag [Online]. Accessed on 3 April 2016. <http://www.rooflitesoil.com/products/intensive-ag>.

Syswerda, S., B. Basso, S. Hamilton, J. Tausig and G. Robertson (2012). Long-term nitrate loss along an agricultural intensity gradient in the Upper Midwest USA. *Agriculture, Ecosystems and Environment* 149: 10–19.

Teemusk, A. and Ü. Mander (2007). Rainwater runoff quantity and quality performance from a greenroof: The effects of short-term events. *Ecological Engineering* 30(3): 271–277.

Teemusk, A. and Ü. Mander (2011). The influence of green roofs on runoff water quality: A case study from Estonia. *Water Resources Management* 25(14): 3699–3713.

Toland, D., B. Haggard and M. Boyer (2012). Evaluation of nutrient concentrations in runoff water from green roofs, conventional roofs, and urban streams. *Transactions of the ASABE* 55(1): 99–106.

United States Environmental Protection Agency (2016). Clean Water Act (CWA) Compliance Monitoring [Online]. Accessed on 3 April 2016. <https://www.epa.gov/compliance/clean-water-act-cwa-compliance-monitoring>.

Van Mechelen, C., T. Dutoit and M. Hermy (2015). Vegetation development on different extensive green roof types in a Mediterranean and temperate maritime climate. *Ecological Engineering* 82: 571–582.

Van Seters, T., L. Rocha, D. Smith and G. MacMillan (2009). Evaluation of green roofs for runoff retention, runoff quality, and leachability. *Water Quality Research Journal of Canada* 44(1): 33.

VanWoert, N. D., D. B. Rowe, J. A. Andresen, C. L. Rugh and L. Xiao (2005). Watering regime and green roof substrate design affect Sedum plant growth. *HortScience* 40(3): 659–664.

Vitousek, P. M., J. D. Aber, R. W. Howarth, G. E. Likens, P. A. Matson, D. W. Schindler, W. H. Schlesinger and D. G. Tilman (1997). Human alteration of the global nitrogen cycle: Sources and consequences. *Ecological Applications* 7(3): 737–750.

Voyde, E., E. Fassman, R. Simcock and J. Wells (2010). Quantifying evapotranspiration rates for New Zealand green roofs. *Journal of Hydrologic Engineering* 15(6): 395–403.

Whittinghill, L. J., D. B. Rowe, J. A. Andresen and B. M. Cregg (2015). Comparison of stormwater runoff from sedum, native prairie, and vegetable producing green roofs. *Urban Ecosystems* 18(1): 13–29.

Whittinghill, L. J., D. B. Rowe and B. M. Cregg (2013). Evaluation of vegetable production on extensive green roofs. *Agroecology and Sustainable Food Systems* 37(4): 465–484.

Young, T., D. D. Cameron, J. Sorrill, T. Edwards and G. K. Phoenix (2014). Importance of different components of green roof substrate on plant growth and physiological performance. *Urban Forestry and Urban Greening* 13(3): 507–516.

14 Managing Urban Soils for Food Production

Klaus Lorenz

CONTENTS

14.1 INTRODUCTION

Urban areas are expanding at an unprecedented rate and projected to increase by about 1.2 million km^2 by 2030, with the strongest expansion of cities and urban regions to occur in Africa and Asia (Seto et al. 2012). The global population living in urban areas is also increasing and this trend is set to continue with about 70% of the global population estimated to inhabit urban ecosystems by the year 2050 (UN DESA 2014). A major challenge is how to meet the needs of the increasing number of urban dwellers for housing, energy, water, and food while safeguarding the environment. Previously, reducing the environmental impact of urban areas has focused on minimizing fossil fuel intensive transport, reducing the energy consumption of buildings, and using an increasing share of renewable energy sources (Victor et al. 2014). However, a large environmental pressure is generated by food consumption in urban areas. Food security often focuses on how to sustainably provide for a world of 9–10 billion people in 2050 but much less attention has been paid to the fact that ~6.5 billion of these people will live in urban areas and that this will have many consequences for food systems (Seto and Ramankutty 2016). Supplying food to cities is one of the key contributors to greenhouse gas (GHG) emissions, biodiversity loss, water pollution, land-use change, nonrenewable resource exhaustion, and other pressing global environmental challenges (Goldstein et al. 2016). Urban agriculture (UA) has the potential to provide some of the food consumed by urban dwellers with up to 20% of food produced globally in urban areas (FAO 2014). Food supply from UA is an important provisioning service of urban ecosystems (EEA 2016). In January 2016, agriculture ministers of 65 nations underlined the importance of UA to contribute to the supply of urban agglomerations not only with food but also with a wide range of public goods and services (Global Forum for Food and Agriculture 2016). After a brief introduction, challenges for growing food crops in urban soils are discussed, and how soil biological, chemical, and physical properties can be improved to enhance crop production. This chapter concludes with an outlook on how soil-based agriculture for growing food crops in urban areas can be strengthened.

Credible assessments of the importance and extent of UA are challenging as numerous and disparate definitions of UA exist. Globally, about 800 million people may be engaged actively in UA practices (Smit et al. 2001), with 266 million households engaged in urban crop production in developing countries. However, the uncertainty and accuracy of global estimates and those for developed countries are not well known (Hamilton et al. 2014). The main UA activity is horticulture such as allotments for self-consumption, large-scale commercial farms, community gardens, and edible landscapes (Eigenbrod and Gruda 2015). Figure 14.1 shows an example for a community garden in Lombard, IL. Practices may also include animal husbandry, aquaculture, and arboriculture (Mok et al. 2014). Thus, some UA practices depend on urban soil resources in open areas while others occur on and in buildings, and some crops are cultivated without soil (e.g., hydroponics; Thomaier et al. 2015). Globally, UA occurs at any scale from rooftop gardens to larger cultivated open spaces. Thebo et al. (2014) estimated that in the year 2000, the global area of urban cropland was 67.4 Mha (5.9% of all cropland), with 23.6 Mha irrigated (11.0% of global irrigated cropland) and 43.8 Mha rainfed (4.7% of global rainfed cropland). Only one-third of the global urban area would be required to meet the global vegetable consumption of urban dwellers, but UA has only a limited potential to contribute to global cereal production as the global annual harvested area for cereals is 10 times larger than the global urban area (Martellozzo et al. 2014). How much urban area (i.e., soil) may actually be suitable and available for UA was not considered in this study. For example, the scientific community is only starting to recognize the potential of UA in the United States (Brown et al. 2016). Globally, there is a lack of credible data on the potential of UA to contribute to food security. In terms of urban land availability to grow the basic daily vegetable intake for the urban poor, UA can probably only make a limited contribution in achieving urban food security in low-income countries (Badami and Ramankutty 2015).

UA has many benefits and, in particular, contributes to urban food security in high-income countries by providing fruits and vegetables (Badami and Ramankutty 2015; Martellozzo et al. 2014). Integrating agriculture in urban areas will increase ecosystem services (ESs; Cooley and Emery 2016). Co-benefits of UA include, for example, reduced food transportation distance, carbon (C) sequestration, potentially reduced urban heat island effect, improved aesthetics, community building, employment opportunities, improved local land prices, shortened supply chains, provision of habitat for wildlife, waste recycling, and, in particular, improved physical and mental health supporting numerous health benefits (Mok et al. 2014).

FIGURE 14.1   Madison Meadow Community gardens, Lombard, IL. (Courtesy of Flickr.)

By terrestrial C sequestration (i.e., the increase in C stock in an area relative to the atmospheric C stock), UA can mitigate some of the loss of terrestrial C, including soil organic carbon (SOC) associated with the conversion of land to urban uses (Xiong et al. 2014). Specifically, for urban expansion about 0.05 Pg C (1 Pg = 1 petagram = 10^{15} g) may be lost annually from vegetation biomass in the pantropics by deforestation and forest degradation (Seto et al. 2012). Previously, the C sink in plant species and urban green systems has received attention (Muñoz-Vallés et al. 2013). However, soil is often the largest contributor to urban C storage (Lorenz and Lal 2015). For example, urban soils of China and the conterminous United States stored about 56% and 64% of urban C, respectively (Churkina et al. 2010; Zhao et al. 2013). UA contributes to C sequestration, particularly in polyculture systems that incorporate fruit or nut trees and other perennials (Grewal and Grewal 2012). Further, UA soils amended with composts from urban feedstocks will reduce C emissions by diverting wastes from landfills and restoring SOC stocks (Cooley and Emery 2016). For example, SOC stocks in 0–25 cm depth at old vegetable gardens in the city of Revda, Russia, were 205 Mg C ha^{-1} while only 136 Mg C ha^{-1} were stored to 50 cm depth in park soils under forest, which was 1.7 times more than SOC stocks in scalpless soils of the technogenic desert (Meshcheryakov et al. 2005). In addition to SOC stocks, UA may also enhance soil inorganic carbon (SIC) stocks. For example, 0–30 and 30–100 cm depth allotment gardens in Stuttgart, Germany, stored 37 and 186 Mg SIC ha^{-1}, respectively (Stahr et al. 2003). This was more than in soils under roads (35 and 135 Mg SIC ha^{-1}) and rural agricultural soils (7 and 2 Mg SIC ha^{-1}) in 0–30 and 30–100 cm depth, respectively (Stahr et al. 2003). In summary, UA soils can contribute to C sequestration in urban areas.

Previously, research on UA has focused on the social sciences (Lorenz 2015), but growing food in urban ecosystems and, especially, on degraded urban soils is challenging as soil properties can be far from those of prime agricultural land (Meharg 2016). Figure 14.2 shows the soil surface of a vacant lot remaining after demolition of houses in Youngstown, OH. Improving this soil for food production by UA is a major challenge.

Urban soil properties differentiate the type and value of food produced by UA (EEA 2016). However, the quality of urban soils can be improved for crop production, which is among the urban ecosystem goods and services. Improving soil fertility may be achieved, in particular, through

FIGURE 14.2 Soil surface of a vacant lot after houses have been removed in Youngstown, OH. (Courtesy of Klaus Lorenz.)

application of organic amendments such as abundant agricultural residues or postconsumer wastes, the use of cover crops, and judicious digging and tillage (Cogger and Brown 2016b). Recycled water, yard wastes, food scraps, and municipal biosolids may be treated before being used as amendments to improve UA soils (Brown and Goldstein 2016). For closing the resource loop, urban waste streams can be recycled to enhance soil fertility, while vacant unproductive lands can be put to a productive use (Smit et al. 2001). For example, "rustbelt" cities in the North Central United States contain >10,000 ha of vacant land which may be improved for UA (Beniston and Lal 2012; Pearsall and Lucas 2014). The abandoned, underused, and vacant 400,000 Brownfields in the United States offer opportunities for UA, but extent and nature of contamination must be understood before those are used for food production (Carroll 2016). Thus, best management practices (BMPs) are needed, and the knowledge must be disseminated among soil managers, policy makers, and local communities for improving soil-based UA. A key element of this knowledge exchange is raising awareness of ESs delivered by urban soils (Morel et al. 2015), and, thus, the natural capital they provide (Robinson et al. 2013) which will also help urban communities in improving human health and well-being.

However, UA can also have negative ecological effects or create ecosystem disservices as urban gardeners may rely heavily on external inputs, including seeds, plants, water, organic matter (OM), and synthetic fertilizers and pesticides (Taylor and Lovell 2014, 2015). For example, nitrogen (N), phosphorus (P), and potassium (K) may accumulate in high concentrations in UA soils. As only some garden waste is composted on site, much of it may enter the municipal waste stream where it contributes to filling of regional landfills and leaves open nutrient cycles within the garden. Further, gardeners import nutrient-rich compost and fertilizers from outside the garden, and a lack of careful management of nutrients may contribute to urban stormwater pollution. A major issue is the unmitigated contamination of UA soils with trace metals and organic contaminants as those are a potential threat to the health of gardeners and their families (Taylor and Lovell 2015). Examples of nonsite-related sources of urban soil contamination are bedrock/parent material and dust deposition, especially along traffic routes, whereas site-related sources may be agricultural/horticultural practices and deposits (Meuser 2010). Thus, pollution is a key issue for UA (Meharg 2016).

Industrial activity, dense and dirty transport infrastructure, burning of fossil fuels, and the plethora of chemicals released into domestic waste streams all contaminate urban soils (Meharg 2016). In addition to contaminants from suspended soil dust, urban air already has elevated levels of nitrogen oxides, sulfur oxides, hydrocarbons, and particulates from car exhausts. Air pollution is known to reduce urban crop yields (Thomaier et al. 2015), but the consequences of ingesting foodstuffs covered in these pollutants are not well understood (Meharg 2016). Only recently has urban science received increased attention especially with regard to urban health and human well-being (Editorial Nature 2016). In the following section, challenges for growing food crops in urban soils are discussed while other forms of UA such as open rooftop farms, rooftop greenhouses, and indoor farming including UA practices without soil are presented elsewhere (e.g., Thomaier et al. 2015).

14.2 CHALLENGES FOR GROWING FOOD CROPS IN URBAN SOILS

Investigations of anthropogenic soils in urban areas became popular since the 1960s and include mainly studies in Europe and North America (Capra et al. 2015). Thus, compared to nonurban soils, scientific interest in urban soils and research on their functions are relatively new developments (Burghardt et al. 2015). Urban soils occur on a continuum ranging from soils that are relatively undisturbed and functional (Edmondson et al. 2014a) to those that are altered by environmental change (e.g., temperature or water regimes) to anthropogenic, disturbed soils like those at vacant lots, old industrial sites, building demolition sites, and landfills (Lehman and Stahr 2007).

Urban soils are distinguished from natural soils by a range of anthropogenic factors such as the presence of human artifacts and contaminants, alkaline pH, high black carbon (BC) content, high bulk density, low soil moisture, warmer soil temperatures, and a relatively young stage of pedogenetic development (Table 14.1; Wessolek et al. 2011). The heterogeneity of urban substrates and their

TABLE 14.1

Comparison of Natural vs. Urban Soils

Parameter	Similarities	Differences in Urban Soils
Genesis	Lithogenesis, pedogenesis	Relatively young stage of genesis; anthropedogenesis imposes very rapid transformation cycles
Properties	Result of soil-forming processes and environmental factors	Modified soil parent material (human artifacts including contaminants, black carbon [BC]), bulk density, soil moisture and temperature, biotic community, vegetation cover, climate, and relief
Gaseous emissions		Modified climate
		Modified convective and diffusive gas transport
		Emissions reduced by impervious surfaces
Carbon dioxide	Sources: Decomposition of animal, microbial- and plant-derived organic matter (OM), and soil organic matter (SOM); root respiration	Decomposition altered by modified quantity, quality, and depth distribution of OM, and modified quantity, composition, and depth distribution of decomposer communities
Methane	Sources: Anaerobic decomposition of animal, microbial- and plant-derived OM, and SOM	Additional source: Leaking belowground natural gas infrastructure
	Aerobic soils are net sink	Smaller net sink due to nitrogen (N) fertilization, cultivation, excessively high or low soil moisture, incorporation of OM
Nitrous oxide	Sources: Microbial nitrification and denitrification	Modified quantity, quality, and activity of microbial communities
		Higher emissions due to fertilization and irrigation
Carbon (C) sequestration	Soil organic carbon (SOC) stocks determined by long-term balance between plant-derived soil inputs of C fixed from the atmosphere, losses by microbial decomposition, and losses associated with fire, erosion, and leaching	Modified input, losses, and depth distribution
		Modified SOM composition (e.g., BC)
		Addition of calcareous material
	Soil inorganic carbon (SIC) stock comprised of both primary carbonates inherited from the parent material or deposited as dusts, and secondary carbonates which form by precipitation of carbonate ions derived from root, microbial and faunal respiration, and calcium (Ca) and magnesium (Mg) ions from weathering	Irrigation with water saturated with carbonates
Contamination	Sources: Parent material, airborne pollutants	Additional sources: Human artifacts, pesticides, trace metal uses
		Enrichment of combustion-derived pollutants

Source: Modified from Lorenz, K. and R. Lal., *Environment International*, 35, 1–8, 2009.

constituents in anthropogenic urban soils, and of disturbances such as compaction and mixing of horizons, are much greater than those in natural soils (Howard and Shuster 2015). Soil horizons are often irregularly developed, with anthropogenic layers and a high degree of heterogeneity resulting from human disturbance and infill. Soil structure is often degraded due to artifacts and technogenic substrata, mechanical compaction, and human trampling (Yang and Zhang 2015). The physical properties of urban soils show a high degree of horizontal and vertical heterogeneity (Baumgartl 1998). UA soils, in particular, may be degraded as the use of urban soils to address essential needs such as housing, jobs, and social services takes precedence over UA unless strong land use and zoning laws are in place (Campbell 2016; Eriksen-Hamel and Danso 2010). Further, as land access for lower income urban farmers is difficult especially in developing countries, UA often occurs on marginal lands, that is, on soils of low fertility on steep slopes, valley bottoms, or in areas adjacent to polluting industries or roads (Orsini et al. 2013).

Urban soils are often less fertile than nonurban soils and their biological, chemical, and physical properties may be degraded. Thus, urban soil properties can make it difficult to grow a crop (Bartens et al. 2012). Specifically, urban soils may have poor physical properties including poor soil structure, high contents of coarse fragments including technogenic materials such as construction waste, high compaction levels, and impeded water infiltration rates. Some urban soils are also low in soil organic matter (SOM), high in inorganic and organic contaminants, and are characterized by altered soil moisture characteristics (Lehmann and Stahr 2007). The moisture characteristics of urban soils may be altered by physical disturbance, removal or desurfacing, burial or coverage of soil by fill material or impervious surfaces, and soil, water, and vegetation management practices (e.g., fertilization, irrigation, mowing, drainage; Lorenz and Lal 2012). Rubble from military conflicts or century-long accumulations of debris due to the cyclic demolition and reconstruction of houses often makes up a large component of the urban soil hidden beneath parks and private gardens, backyards, sidewalks, and other public places (Wessolek et al. 2011). Thus, to bring more urban soils into productive use for UA, it is essential to understand the problems inherent to degraded urban soils (Kumar and Hundal 2016).

Enrichment of waste materials associated with human activities, including nutrient elements (i.e., phosphorus, nitrate, ammonia), trace metals, and organic compounds, is the major problem of the urban soil environment (Burghardt et al. 2015; Yang and Zhang 2015). Specifically, contamination of urban soils by trace metals is a major concern because of the risk these elements pose to the environment and to human health (Kumar and Hundal 2016). The cumulative impact of multiple trace metals may be a particular health risk as trace metals do not exist in isolation at the majority of sites (Wijayawardena et al. 2016). Due to the long residence time of trace metals in soils, urban soils may act as both a sink and a source for these pollutants. Many of the trace metals may be present in parent material from which soils have developed and some urban soils may be inherently high in some of these metals. Urban soils may also be contaminated with organic pollutants, for example, petroleum hydrocarbons, polycyclic aromatic hydrocarbons (PAHs), and polychlorinated biphenyls (PCBs). In many cities, anthropogenic activities such as (1) unregulated waste management; (2) disposal of combustion residues of heating processes, feces, and refuse; (3) leakages and accidents associated with manufacturing and production processes at industrial and commercial sites; and (4) atmospheric emissions and depositions have resulted in the substantial contamination of urban soils (Meuser 2010; Wessolek et al. 2011). Further, anthropogenic activities such as fertilization and application of sewage sludge, wastewater, and pesticides may also contribute to the contamination of urban soils and restrict their use for UA. Thus, soil quality assessment is a necessary first step for designing UA sites (Montgomery et al. 2016). Some examples of approaches for improving soils for UA are discussed in the following section.

14.3 MANAGEMENT OF URBAN SOILS FOR GROWING FOOD CROPS

There is a paucity of published data on BMPs for managing and improving urban soils for UA crop production. Due to often naturally low soil fertility and other soil degradation, vegetable yields in urban areas may be lower than rural yields. On the other hand, vegetable yields in UA may also be considerably higher than those on rural farms due to the use of irrigation, relatively high input levels, and other BMPs. For example, achievable yields of up to 50 kg m^{-2} y^{-1} have been reported for fruit and vegetable cultivation in urban areas (Eigenbrod and Gruda 2015), but estimates of yields achieved by UA practices are generally not well known (CoDyre et al. 2015). Since very few empirical studies focusing on UA measured crop yields in urban areas, Martellozzo et al. (2014) assumed that, for each crop in each country, the UA yield equals the average national yield for conventional farming as reported by Food and Agriculture Organization Corporate Statistical Database (FAOSTAT). This assumption is uncertain as, for example, backyard tomato (*Solanum lycopersicum* L.) production in Guelph, ON, Canada, yielded 2.85 kg m^{-2}, while conventional tomato production yielded 7.68 kg m^{-2} in the tomato fields of an important tomato-producing region in southwestern Ontario (CoDyre et al. 2015). In another example, cucumbers (*Cucumis sativus* L.) in the backyard yielded 3.63 kg m^{-2} while conventional cucumber fields in Ontario only yielded 2.41 kg m^{-2}.

Yields for a range of crops from an urban garden are shown in Table 14.2. In conclusion, more science-based information on improving urban soils for UA is needed and knowledge about BMPs must be distributed among urban farmers to enhance the cultivation of crops by UA (Wortman and Lovell 2013).

TABLE 14.2

Range in Yields of Crop Cultivars or Varieties in a Vegetable Garden Plot Located in a Public Park in Seattle, WA

Crop	Yield (kg m^{-2})
Bush beans (*Phaseolus vulgaris*)	0.16–0.18
Beet (*Beta vulgaris*)	0.12–0.61
Bell pepper (*Capsicum annuum var. annuum*)	0.12
Broccoli (*Brassica oleracea*, Italica group)	0.08
Brussels sprouts (*Brassica oleracea*, Gemmifera group)	0.25
Carrots (*Daucus carota subspecies sativa*)	0.14–0.73
Cucumber	0.66
Eggplant (*Solanum melongena*)	0.19–0.23
Kale (*Brassica oleracea*, Acephala group)	0.28
Kohlrabi (*Brassica oleracea*, Gongylodes group)	0.10
Leeks (*Allium ampeloprasum*, Porrum group)	0.13–0.14
Lettuce (*Lactuca sativa*)	0.24–0.55
Onion (*Allium cepa*)	0.17–0.43
Pak choi (*Brassica rapa var. chinensis*)	0.31–0.40
Peas (*Pisum sativum var. macrocarpon*)	0.10–0.19
Potato (*Solanum tuberosum*)	0.30
Pumpkin (*Cucurbita pepo*)	1.26
Red radish (*Raphanus sativus*)	0.34
Spinach (*Spinacia oleracea*)	0.01–0.18
Swiss chard (*Beta vulgaris*, Cicla group)	0.45
Tomato (*Solanum lycopersicum*)	0.52
Turnip (*Brassica rapa*)	1.36
Zucchini (*Cucurbita pepo* L.)	0.57–1.08

Source: With kind permission from Springer Science+Business Media: *Sowing Seeds in the City*, How much can you grow? Quantifying yield in a community garden plot—One family's experience, 2016, 245–267, McGoodwin, M. et al.

14.3.1 MANAGEMENT OF SOC STOCKS TO IMPROVE URBAN SOIL QUALITY

The properties of urban soils can be maintained or improved for UA by managing SOC stocks (Lorenz and Lal 2015). Examples for urban SOC stocks under different land use and land cover are shown in Table 14.3. SOC and its turnover have well-known benefits as SOC has a fundamental role in the function and fertility of terrestrial ecosystems (Janzen 2006, 2015). The SOC stock supports critically important soil-derived ESs directly affecting food production and soil fertility,

TABLE 14.3
Organic Carbon Stocks of Urban Soils

Depth (cm)	Organic Carbon (Mg C ha^{-1})	Land Use	Land Cover	City/Region	References
0–10	4.2	Ornamental	Herbaceous	Victoria, Australia	Livesley et al. (2010)
	44.4	Ornamental	Herbaceous/ trees	Richmond, VA	Gough and Elliott (2012)
	28.1–70.7	Builtup/ ornamental	Herbaceous/ trees	Moscow; Serebryanye Prudy, Russia	Vasenev et al. (2013)
0–15	12.6	Ornamental	Herbaceous	Hong Kong, China	Kong et al. (2014)
	125.5	Ornamental	Herbaceous	Minneapolis, MN	Selhorst and Lal (2012)
	30.3–80.5	Builtup/forestry	Herbaceous/ trees	Louisville, KY	Trammell et al. (2011)
0–20	12.8–156.5	Builtup/ ornamental	Herbaceous/ trees	Chongqing Municipality, China	Liu et al. (2013)
0–25	96.5	Ornamental	Herbaceous/ trees	Vancouver, BC, Canada	Kellett et al. (2013)
	137.8	Ornamental	Herbaceous	Revda, Russia	Meshcheryakov et al. (2005)
0–30	48.5	Ornamental	Herbaceous/ trees	Village landscapes, China	Jiao et al. (2010)
	91.7	Ornamental	Herbaceous/ trees	Tianjin Binhai New Area	Hao et al. (2013)
0–50	44.8; 79.1	Forestry; ornamental	Trees; herbaceous/ trees	Revda, Russia	Meshcheryakov et al. (2005)
0–90	107; 159	Ornamental; forestry	Herbaceous/ trees	Apalachicola, FL	Nagy et al. (2013)
0–100	47.5; 184.6	Forestry/ ornamental	Herbaceous/ trees	Xinjiang, Northwest China; Hubei, Central-South China	Zhao et al. (2013)
	63.9	Ornamental	Herbaceous/ trees	Kaifeng city, China	Sun et al. (2010)
	199	Ornamental	Herbaceous/ trees	Leicester, UK	Edmondson et al. (2012)

Source: Modified from Lorenz, K. and R. Lal. *Carbon Management*, 6, 35–50, 2015.

including water filtration, erosion control, soil strength and stability, nutrient conservation, pollutant denaturing and immobilization, habitat and energy for soil organisms, and pest and disease regulation (Adhikari and Hartemink 2016; Franzluebbers 2010; Lal et al. 2012). The SOC stock on UA sites can be protected by minimizing losses resulting from erosion, for example, by reducing tillage and other soil-disturbing activities. To enhance SOC stocks, the vegetation should be maintained to optimize net primary production (NPP) by appropriate crop rotation, reduced physical disturbance, fertilization, and irrigation as NPP is the primary source of soil C inputs such as plant litter and exudates and dissolved organic carbon (DOC; Rumpel and Kögel-Knabner 2011). Specifically, cover cropping and reduced tillage are intensive management strategies to increase SOC stocks for increasing UA production (Beniston and Lal 2012).

The SOC-enhancing management practices should particularly optimize root growth and turnover as SOC is primarily originating from root-derived C inputs (Rasse et al. 2005). Additional storage of SOC at deeper soil depths for improving urban soil quality may also be possible by cultivating plants with deep and prolific roots systems (Kell 2012). The strategy for enhancing SOC is to select plant species with enhanced transfer of root-derived C into protected mineral-associated SOC fractions. Soil quality at deeper depths may also be improved by burial of OM during construction of UA sites (Chaopricha and Marín-Spiotta 2014). Further, adding BC compounds such as biochar or charcoal to UA soils has the potential to enhance SOC stocks and improve soil quality (Ghosh et al. 2012). However, little thought has been given to biochar application to UA soils (Renforth et al. 2011).

UA soils may be optimized for SOC accumulation by adding organic amendments such as biosolids, yard, and food wastes. Garden management practices such as the addition of peat, composts, and mulches, and cultivation of trees and shrubs may also contribute to greater SOC stocks in UA soils (Edmondson et al. 2014b). Recovered resources, such as municipal solid waste compost, biosolids, and harvested nutrients from wastewater, are available in urban areas and could be utilized for improving SOC stocks, and, thus, fertility and ecosystem functions of UA soils. Household composting and municipal-level solid waste composting are examples where urban waste is diverted from landfills and turned into compost, which is then used to amend UA soil to improve crop productivity as well as other ESs (Kumar and Hundal 2016).

14.3.2 IMPROVING SOIL HYDROLOGICAL AND PHYSICAL PROPERTIES FOR UA

Compaction is the most serious form of physical degradation in urban soils (Yang and Zhang 2015). Soil compaction can be reduced by the addition of biosolids and other OM amendments, which will also increase water infiltration and retention, reduce bulk density, improve soil structure, and reduce penetration resistance for plant roots (Brown et al. 2011). Intensive cover cropping with deep-rooted cover crops or crops having an extensive root profile are other appropriate management strategies to reduce soil compaction (Beniston and Lal 2012). Alleviation of compaction during construction of UA sites may also be achieved by subsoiling or deep tillage practices. For poorly drained urban soils, both surface and subsurface drainage are management options to improve the soil moisture characteristics for UA use (Craul 1999). On the other hand, urban soils may also exhibit elevated water deficit for growing food crops. The previously discussed addition of OM to soil in the form of compost, mulch, and other amendments can both improve infiltration and increase the water holding capacity of UA soils. In cases where it is not economically feasible to reduce soil compaction, soils for UA may be created on top of the compacted layers. For example, large quantities of compost, wood chips, and other forms of OM can be used to create raised beds over compacted urban soil layers. Figure 14.3 shows raised beds in a community garden in Mukilteo, WA. Growing crops in raised beds is also an option to overcome the limitations to crop growth in gravelly urban soils as the presence of coarse fragments in urban soils is ubiquitous (Yang and Zhang 2015). However, knowledge about UA practices using raised beds is scanty (Beniston and Lal 2012).

FIGURE 14.3 Raised beds at Mukilteo Community Garden, Snohomish County, WA. (Courtesy of Flickr.)

14.3.3 IMPROVING SOIL BIOLOGICAL PROPERTIES FOR UA

Creating favorable soil chemical and physical conditions for soil biotic activity can enhance soil fertility and UA production. BMPs include the addition of OM as a food resource for soil organisms and for improving habitat conditions, and the construction of raised beds at sites where chemical and physical constraints cannot be alleviated. Focus should be on ecosystem engineers comprising diverse soil biota (e.g., earthworms, termites, and ants) as those play key roles in creating habitats for other organisms (microfauna and mesofauna, microorganisms), and controlling their activities will drive soil processes through specific pathways (Lavelle et al. 2016). For example, garden soils under long-term UA management in Stuttgart, Germany, were enriched in SOC and N, contained substantial amounts of microbial biomass, and were metabolically active (Lorenz and Kandeler 2005). However, effects of UA and other urban soil uses on soil microorganism communities are less well known (Silva et al. 2010).

14.3.4 REDUCING SOIL CHEMICAL CONSTRAINTS FOR UA

Urban soils can have higher or lower pH than agricultural soils in rural areas depending on site history (Cogger and Brown 2016b). The alkaline soil pH in urban areas which is attributed to filling materials contaminated with alkaline building wastes, such as concrete and cement, may impact crop growth by nutrient imbalances (Yang and Zhang 2015). Adding OM to UA soils can lower the soil pH and alleviate detrimental effects of high pH values on crop growth (Beniston et al. 2016). Elemental sulfur may also be added to reduce alkalinity and lower soil pH. On the other hand, soil acidity may be corrected by addition of ground limestone (Cogger and Brown 2016b). In lieu of commercial limestone, residuals with a high calcium carbonate equivalence can be added to neutralize soil acidity (Brown and Goldstein 2016).

Typically N, P, and K are added to soils to support UA production but rarely other nutrients (Cogger and Brown 2016b). The full suite of required plant nutrients may be added with organic or residual-derived amendments such as compost, biosolids, and manures. Composts and biosolids, in particular, have been shown to result in increases in soil fertility. However, issues such as pathogens and contaminants (i.e., heavy metals) have to be addressed before compost- or biosolids-based soil amendments can be applied (Cogger and Brown 2016b).

The contamination of urban soils with trace metals and organic compounds such as petroleum hydrocarbons, PAHs, and PCBs restricts their use for UA (Meuser 2010). For decontamination, urban soils may be excavated and treated *ex situ* by using technical devices for soil washing, bioremediation, and thermal treatment (Meuser 2013). However, this is not a cost-effective management practice for urban gardeners in contrast to *in situ* stabilization methods. For example, the so-called "urban elements" such as copper (Cu), zinc (Zn), lead (Pb), and mercury (Hg) (Yang and Zhang 2015) may be stabilized *in situ* in UA soils by applying liming compounds or OM. Further, soil stabilization may also be possible for cadmium (Cd), nickel (Ni), chromium (Cr), and some organic pollutants.

Organic pollutants soluble to some degree in water may be treated *in situ* by bioremediation, a process depending on the activity of soil microorganisms (Meuser 2013). Thus, adding OM as a food resource for soil microorganisms may aid in soil decontamination. However, bioremediation of PAH-contaminated soils may cause the formation of more toxic transformation products (Chibwe et al. 2015), and, thus, other decontamination methods may be required. For example, decontamination of organic soil pollutants before establishing UA plots may be achieved by phytodegradation leading to microbial metabolism in the plant rhizosphere and subsequent degradation of organic pollutants. Poplar trees (*Populus* spp.) are resilient against many soil contaminants and adverse growing conditions and enhance the microbial degradation of hydrocarbons (Fisk 2014). Organic pollutants may also be removed from soil by phytovolatilization, a process leading to gaseous losses of organic contaminants after metabolism in plant tissue (Koptsik 2014).

Stabilized contaminants in UA soils may not result in quality reduction of fruits and vegetables, but may be a human health risk during gardening activities, in particular for infants, who may directly ingest soil. For example, inhalation and nondietary ingestion of PAH-contaminated house dust and soil likely are the most important routes of human exposure (Mahler et al. 2012). Thus, when UA practices are disturbing the soil structure (e.g., plant growing, tillage, soil amendment), increased volatilization/mobility of PAHs may occur. Otherwise, adding BC to UA soils may sequester carcinogenic PAHs and limit their oral bioavailability (Ruby et al. 2016). However, little is known about the extent and distribution of PAHs in UA soils (Marquez-Bravo et al. 2016).

If budgets are limited but sufficient time available to clean up soil before UA sites are established, natural attenuation may be used to reduce the concentration of organic pollutants in soil (Meuser 2013). However, when urban soils cannot be decontaminated to levels for safe production of fruits and vegetables, establishing UA on imported clean soil or raised beds on top of the contaminated soil may be an option. Alternatively, growing nonfood crops such as textile fiber plants, biomass crops, and timber would make use of urban waste land, green the city, recycle wastewater and biosolids, and produce crops that currently take up rural land that is ideal for food production (Meharg 2016).

After soil contamination levels have been reduced, surface deposition of contaminants with urban dust occurs, and soil contaminant levels may increase. Thus, monitoring contaminant concentrations in "clean" soil, imported soil, and raised beds at UA sites is needed and, if necessary, soil decontamination should be undertaken to reduce health risks associated with UA (Säumel et al. 2012). In conclusion, the properties of urban soils can be improved for growing crops but the challenge is to develop BMPs by interdisciplinary, multisectoral, and participatory (i.e., transdisciplinary) approaches which will be discussed in the following section.

14.4   THE FUTURE OF GROWING FOOD CROPS BY SOIL-BASED UA

The potential of UA to contribute to local food systems and the food supply to urban areas is receiving increasing attention from the public, policy makers, and private business (FAO 2014; GFFA 2016). For example, one of the world's biggest hotel chains has announced it will plant vegetable gardens at 1,000 of its hotels (http://www.theguardian.com/travel/2016/apr/13/accorhotels-to-grow-its-own-vegetables-in-push-to-cut-food-waste). Since the 1970s, New York City, NY, has had one of the largest and most robust community garden programs in the world (Campbell 2016). Currently, there are more than 700 farms and gardens throughout the five boroughs of New York City that grow food in schoolyards, grounds of public housing developments, community gardens, and public parks (http://www.fiveboroughfarm.org/). Vancouver, BC, Canada and Portland, OR, are focusing on UA research and policies that support small-scale urban farming and have identified urban lands where food could be grown (Worrel 2009). Similarly, urban planning reports in cities such as Detroit, MI; Berlin, Germany; and Leeds, United Kingdom; recommend the introduction or support of productive urban landscapes which includes UA (Bohn and Viljoen 2014). Further, residents of Los Angeles, CA, are allowed to grow fruits and vegetables in a strip of city-owned land between the sidewalk and the street (http://www.scpr.org/news/2015/03/04/50192/la-city-council-approves-curbside-planting-of-frui/). Some curbside areas can be suitable for growing food (Cogger and Brown 2016a).

In many urban areas, a paradigm shift in urban planning and zoning policy is needed to encourage food production by soil-based UA (FAO 2014). For example, in Cleveland, OH, the City of Cleveland and Ohio Department of Agriculture, in cooperation with the United States Department of Agriculture (USDA) and local community groups, have created the 26-acre (10.5 ha) Urban Agriculture Innovation Zone (BBC 2013). UA zoning districts should be equally integrated into urban planning similar to creating green zones for parks, botanical gardens, and golf courses within city boundaries (FAO/WB 2008). Specifically, the private sector must be directly involved in planning decisions, and experiences should be shared with local decision makers and actors from the public and private sectors, including nongovernmental organizations (NGOs) and growers' representatives (FAO/WB 2008). Rezoning of other urban land uses for UA may have a large potential to contribute to food production in urban areas. For example, converting a portion of turf grasses to UA in the United States may be sufficient to meet the actual and recommended vegetable consumption by urban dwellers (Martellozzo et al. 2014). However, more data are needed for urban areas of different sizes in the United States and worldwide to assess the potential of urban crop production.

As discussed previously, growing crops on often degraded urban soils is difficult. However, even in Europe with its long history in soil research the knowledge base on urban soils that would allow ecological soil valuation is largely missing or, at best, fragmented (EEA 2016). Further, there is a need to adapt soil function and soil-based service descriptions to an urbanized context, and to develop geospatial maps of soil characteristics and soil functions that are relevant to urban demand. More research is needed on BMPs to alleviate urban soil challenges for UA. Awareness must be raised among stakeholders toward the recognition of urban soils, their properties and ESs, and the spatial distribution of soils of different qualities within urban areas (Yigini et al. 2012). Enhanced recognition of the importance of urban soils is one of the prerequisites for developing BMPs. Urban soils may be contaminated, in particular, with trace metals, and risk assessments for single metals and mixtures are needed to establish safety of a site prior to its use for UA (Sharma et al. 2015; Wijayawardena et al. 2016). To improve degraded soils, huge quantities of organic waste products are available as soil amendments in urban areas. Effects of the application of organic wastes, such as compost or biochar, on improving biological, chemical, and physical properties of soils for UA must be studied in many urban areas worldwide (Edmondson et al. 2014a). This should also be accompanied by an assessment of closing nutrient cycles by urban waste recycling as a contribution to the sustainable development of urban areas (Taylor and Lovell 2015).

Science-based knowledge on the risks associated with soil-based UA and on BMPs for improving soils must be distributed among UA producers. A critical assessment is needed whether rural agriculture or UA is more ecologically, economically, and socially sustainable by making the best use of soils and other principal natural resources. Most urban producers belong to lower income groups often with no secured tenure status, which precludes any substantial investment in terms of soil fertility or infrastructure. However, access to land is the most critical factor for UA

(Badami and Ramankutty 2015). Temporary occupancy permits for urban producers should be tested as a strategy to get (temporary) access to land, especially for poor urban producers (Mougeot 2006). Further, UA practices are often not supervised and exposed to the "innocent" use of pesticides and polluted water. Thus, the lack of agricultural extension services provided to the urban producers must be properly addressed (FAO/WB 2008).

14.5 CONCLUSIONS

Food crops can be grown by improvements in biological, chemical, and physical properties of urban soils but more studies in urban areas of different sizes and, especially, in rapidly growing urban areas in Africa and Asia are needed to realize the full potential of soil-based UA. Also, more soil science involvement is needed to develop BMPs in close collaboration with urban gardeners, communities, waste and water departments, urban planners, health authorities, private companies, and other stakeholders. If all potential users recognize the long-term value of soil-based UA, they may mutually agree to collective use of the common property resource urban soil for UA.

REFERENCES

Adhikari, K. and A. E. Hartemink. 2016. Linking soils to ecosystem services—A global review. *Geoderma* 262:101–111.

Badami, M. G. and N. Ramankutty. 2015. Urban agriculture and food security: A critique based on an assessment of urban land constraints. *Global Food Security* 4:8–15.

Bartens, J., N. Basta, S. Brown et al. 2012. Soils in the city—A look at soils in urban areas. *CSA News* August 2012:4–13.

Baumgartl, Th. 1998. Physical soil properties in specific fields of application especially in anthropogenic soils. *Soil & Tillage Research* 47:51–59.

BBC. 2013. Urban agriculture innovation zone in the Kinsman neighborhood. Burten, Bell, Carr Development. http://www.bbcdevelopment.org/development/social-enterprise/urban-agricultural-innovation-zone/

Beniston, J. and R. Lal. 2012. Improving soil quality for urban agriculture in the North Central U.S. In *Carbon Sequestration in Urban Ecosystems*, ed. R. Lal, and B. Augustin, 279–314. Dordrecht: Springer.

Beniston, J. W., R. Lal, and K. L. Mercer. 2016. Assessing and managing soil quality for urban agriculture in a degraded vacant lot soil. *Land Degradation & Development* 27:996–1006.

Bohn, K. and A. Viljoen. 2014. Urban agriculture on the map: Growth and challenges since 2005. In *Second Nature Urban Agriculture: Designing Productive Cities*, ed. K. Bohn and A. Viljoen, 6–11. Florence: Routledge.

Brown, S. and N. Goldstein. 2016. The role of organic residuals in urban agriculture. In *Sowing Seeds in the City*, ed. S. Brown, K. McIvor, and E.H. Snyder, 93–106. Dordrecht: Springer.

Brown, S., K. Kurtz, A. Bary et al. 2011. Quantifying benefits associated with land application of organic residuals in Washington State. *Environmental Science & Technology* 45:7451–7458.

Brown, S., K. McIver, and E. H. Snyder (ed.). 2016. *Sowing Seeds in the City*. Dordrecht: Springer.

Burghardt, W., J. L. Morel, and G. L. Zhang. 2015. Development of the soil research about urban, industrial, traffic, mining and military areas (SUITMA). *Soil Science and Plant Nutrition* 61:3–21.

Campbell, L. K. 2016. Getting farming on the agenda: Planning, policymaking, and governance practices of urban agriculture in New York City. *Urban Forestry & Urban Greening* 19:295–305.

Capra, G. F., A. Ganga, E. Grilli et al. 2015. A review on anthropogenic soils from a worldwide perspective. *Journal of Soils Sediments* 15:1602–1618.

Carroll, A. 2016. Brownfields as sites for urban farms. In *Sowing Seeds in the City*, ed. S. Brown, K. McIvor, and E. H. Snyder, 339–349. Dordrecht: Springer.

Chaopricha, N. T. and E. Marín-Spiotta. 2014. Soil burial contributes to deep soil organic carbon storage. *Soil Biology & Biochemistry* 69:251–264.

Chibwe, L., M. C. Geier, J. Nakamura et al. 2015. Aerobic bioremediation of PAH contaminated soil results in increased genotoxicity and developmental toxicity. *Environmental Science & Technology* 49:13889–13898.

Churkina, G., D. Brown, and G. Keoleian. 2010. Carbon stored in human settlements: The conterminous US. *Global Change Biology* 16:135–143.

CoDyre, M., E. D. G. Fraser, and K. Landman. 2015. How does your garden grow? An empirical evaluation of the costs and potential of urban gardening. *Urban Forestry & Urban Greening* 14:72–79.

Cogger, C. and S. Brown. 2016a. Curbside gardens. In *Sowing Seeds in the City*, ed. S. Brown, K. McIvor, and E. H. Snyder, 351–360. Dordrecht: Springer.

Cogger, C. and S. Brown. 2016b. Soil formation and nutrient cycling. In *Sowing Seeds in the City*, ed. S. Brown, K. McIvor, and E. H. Snyder, 25–52. Dordrecht: Springer.

Cooley, C. and I. Emery. 2016. Ecosystem services from urban agriculture in the city of the future. In *Sowing Seeds in the City*, ed. S. BRown, K. McIvor, and E. H. Snyder, 1–22. Dordrecht: Springer.

Craul, P. J. 1999. *Urban Soil in Landscape Design*. New York, NY: John Wiley.

Editorial. 2016 Metropolis now. *Nature* 531:275–276.

Edmondson, J. L., Z. G. Davies, K. J. Gaston et al. 2014a. Urban cultivation in allotments maintains soil qualities adversely affected by conventional agriculture. *Journal of Applied Ecology* 51:880–889.

Edmondson, J. L., Z. G. Davies, S. A., McCormack et al. 2014b. Land-cover effects on soil organic carbon stocks in a European city. *Science of the Total Environment* 472:444–453.

Edmondson, J. L., Z. G. Davies, N. McHugh et al. 2012. Organic carbon hidden in urban ecosystems. *Nature Scientific Reports* 2:963.

EEA. 2016. *Soil Resource Efficiency in Urbanized Areas—Analytical Framework and Implications for Governance*. EEA Report No7/2016. Copenhagen: European Environment Agency.

Eigenbrod, C. and N. Gruda. 2015. Urban vegetable for food security in cities. A review. *Agronomy for Sustainable Development* 35:483–498.

Eriksen-Hamel, N. and G. Danso. 2010. Agronomic considerations for urban agriculture in southern cities. *International Journal of Agricultural Sustainability* 8:86–93.

FAO. 2014. *Growing Greener Cities in Latin America and the Caribbean*. Rome: Food and Agriculture Organization of the United Nations.

FAO/WB. 2008. *Urban Agriculture for Sustainable Poverty Alleviation and Food Security*. http://www.fao.org/fileadmin/templates/FCIT/PDF/UPA_-WBpaper-Final_October_2008.pdf

Fisk, S. 2014. Soil microbes and plants work together to clean up contaminated soils. *Soil Horizons* 55, doi:10.2136/sh2014-55-5-f

Franzluebbers, A. J. 2010. Will we allow soil carbon to feed our needs? *Carbon Management* 1:237–251.

Ghosh, S., D. Yeo, B. Wilson et al. 2012. Application of char products improves urban soil quality. *Soil Use and Management* 28:329–336.

Global Forum for Food and Agriculture (GFFA). 2016. *GFFA Communiqué 2016. How to feed our cities?—Agriculture and rural areas in an era of urbanization*. http://www.gffa-berlin.de/en/

Goldstein, B., M. Hauschild, J. Fernández et al. 2016. Urban versus conventional agriculture, taxonomy of resource profiles: A review. *Agronomy for Sustainable Development* 36:9.

Gough, C. M. and H. L. Elliott. 2012. Lawn soil carbon storage in abandoned residential properties: An examination of ecosystem structure and function following partial human-natural decoupling. *Journal of Environmental Management* 98:155–162.

Grewal, S. S. and P. S. Grewal. 2012. Can cities become self-reliant in food? *Cities* 29:1–11.

Hamilton, A. J., K. Burry, H. F. Mok et al. 2014. Give peas a chance? Urban agriculture in developing countries. A review. *Agronomy for Sustainable Development* 34:45–73.

Hao, C., J. Smith, J. Zhang et al. 2013. Simulation of soil carbon changes due to land use change in urban areas in China. *Frontiers of Environmental Science & Engineering* 7:255–266.

Howard, J. L. and W. D. Shuster. 2015. Experimental Order 1 soil survey of vacant urban land, Detroit, Michigan, USA. *Catena* 126:220–230.

Janzen, H. H. 2006. The soil carbon dilemma: Shall we hoard it or use it? *Soil Biology & Biochemistry* 38:419–424.

Janzen, H. H. 2015. Beyond carbon sequestration: Soil as conduit of solar energy. *European Journal of Soil Science* 66:19–32.

Jiao, J. G., L. Z. Yang, and J. X. Wu. 2010. Land use and soil organic carbon in China's village landscapes. *Pedosphere* 20:1–14.

Kell, D. B. 2012. Large-scale sequestration of atmospheric carbon via plant roots in natural and agricultural ecosystems: Why and how. *Philosophical Transactions of the Royal Society B-Biological Sciences* 367:1589–1597.

Kellett, R., A. Christen, N. C. Coops et al. 2013. A systems approach to carbon cycling and emissions modeling at an urban neighborhood scale. *Landscape & Urban Planning* 110:48–58.

Kong, L., Z. Shi, and L. M. Chua. 2014. Carbon emission and sequestration of urban turfgrass systems in Hong Kong. *Science of the Total Environment* 473–474:132–138.

Koptsik, G. N. 2014. Problems and prospects concerning the phytoremediation of heavy metal polluted soils: A review. *Eurasian Soil Science* 47:923–939.

Kumar, K. and L. S. Hundal. 2016. Soil in the city: Sustainably improving urban soils. *Journal of Environmental Quality* 45:2–8.

Lal, R., K. Lorenz, R. Hüttl et al. (ed.). 2012. *Recarbonization of the Biosphere*. Dordrecht: Springer.

Lavelle, P., A. Spain, M. Blouin et al. 2016. Ecosystem engineers in a self-organized soil: A review of concepts and future research questions. *Soil Science* 181:91–109.

Lehmann, A. and K. Stahr. 2007. Nature and significance of anthropogenic urban soils. *Journal of Soils and Sediments* 7:247–296.

Liu, Y., C. Wang, W. Yue et al. 2013. Storage and density of soil organic carbon in urban topsoil of hilly cities: A case study of Chongqing Municipality of China. *Chinese Geographical Science* 23:26–34.

Livesley, S. L., B. J. Dougherty, A. J. Smith et al. 2010. Soil-atmosphere exchange of carbon dioxide, methane and nitrous oxide in urban garden systems: Impact of irrigation, fertiliser and mulch. *Urban Ecosystems* 13:273–293.

Lorenz, K. 2015. Organic urban agriculture. *Soil Science* 180:146–153.

Lorenz, K. and E. Kandeler. 2005. Biochemical characterization of urban soil profiles from Stuttgart, Germany. *Soil Biology & Biochemistry* 37:1373–1385.

Lorenz, K. and R. Lal. 2009. Biogeochemical C and N cycles in urban soils. *Environment International* 35:1–8.

Lorenz, K. and R. Lal. 2012. Terrestrial carbon management in urban ecosystems and water quality. In *Carbon Sequestration in Urban Ecosystems*, ed. R. Lal and B. Augustin, 73–102. Dordrecht: Springer.

Lorenz, K. and R. Lal. 2015. Managing soil carbon stocks to enhance the resilience of urban ecosystems. *Carbon Management* 6:35–50.

Mahler, B. J., P. C. Van Metre, J. L. Crane et al. 2012. Coal-tar-based pavement sealcoat and PAHs: Implications for the environment, human health, and stormwater management. *Environmental Science & Technology* 46:3039–3045.

Marquez-Bravo, L. G., D. Briggs, H. Shayler et al. 2016. Concentrations of polycyclic aromatic hydrocarbons in New York City community garden soils: Potential sources and influential factors. *Environmental Toxicology and Chemistry* 35:357–367.

Martellozzo, F., J. S. Landry, D. Plouffe et al. 2014. Urban agriculture: A global analysis of the space constraint to meet urban vegetable demand. *Environmental Research Letters* 9:064025.

McGoodwin, M., R. McGoodwin, and W. McGoodwin. 2016. How much can you grow? Quantifying yield in a community garden plot—One family's experience. In *Sowing Seeds in the City*, ed. S. Brown, K. McIvor, and E. H. Snyder, 245–267. Dordrecht: Springer.

Meharg, A. A. 2016. City farming needs monitoring. *Nature* 531:S60.

Meshcheryakov, P. V., E. V. Prokopovich, and I. N. Korkina. 2005. Transformation of ecological conditions of soil and humus substance formation in the urban environment. *Russian Journal of Ecology* 1:8–15.

Meuser, H. 2010. *Contaminated Urban Soils. Environmental Pollution*, Vol. 18. Dordrecht: Springer.

Meuser, H. 2013. *Soil Remediation and Rehabilitation—Treatment of Contaminated and Disturbed Land. Environmental Pollution*, Vol. 23. Dordrecht: Springer.

Mok, H. F., V. G. Williamson, J. R. Grove et al. 2014. Strawberry fields forever? A review of urban agriculture in developed countries. *Agronomy for Sustainable Development* 34:21–34.

Montgomery, J. A., C. A. Klimas, J. Arcus et al. 2016. Soil quality assessment is a necessary first step for designing urban green infrastructure. *Journal of Environmental Quality* 45:18–25.

Morel, J. L., C. Chenu, and K. Lorenz. 2015. Ecosystem services provided by soils of urban, industrial, traffic, mining, and military areas (SUITMAs). *Journal of Soils and Sediments* 15:1659–1666.

Mougeot, L. (ed.). 2006. *in_focus: Growing Better Cities—Urban Agriculture for Sustainable Development*. Ottawa: International Development Research Centre.

Muñoz-Vallés, S., J. Cambrollé, E. Figueroa-Luque et al. 2013. An approach to the evaluation and management of natural carbon sinks: From plant species to urban green systems. *Urban Forestry & Urban Greening* 12:450–453.

Nagy, R. C., B. G. Lockaby, W. C. Zipperer et al. 2014. A comparison of carbon and nitrogen stocks among land uses/covers in coastal Florida. *Urban Ecosystems* 17:255–276.

Orsini, F., R. Kahane, R. Nono-Womdim et al. 2013. Urban agriculture in the developing world: A review. *Agronomy for Sustainable Development* 33:695–720.

Pearsall, H. and S. Lucas. 2014. Vacant land: The new urban green? *Cities* 40:121–123.

Rasse, D. P., C. Rumpel, and M. F. Dignac. 2005. Is soil carbon mostly root carbon? Mechanisms for a specific stabilisation. *Plant and Soil* 269:341–356.

Renforth, P., J. Edmondson, J. R. Leake et al. 2011. Designing a carbon capture function into urban soils. *Proceedings of the Institution of Civil Engineers-Urban Design and Planning* 164:121–128.

Robinson, D. A., N. Hockley, D. M. Cooper et al. 2013. Natural capital and ecosystem services, developing an appropriate soils framework as a basis for valuation. *Soil Biology & Biochemistry* 57:1023–1033.

Ruby, M. V., Y. W. Lowney, A. L. Bunge et al. 2016. Oral bioavailability, bioaccessibility, and dermal absorption of PAHs from soil—State of the science. *Environmental Science & Technology* 50:2151–2164.

Rumpel, C. and I. Kögel-Knabner. 2011. Deep soil organic matter—A key but poorly understood component of terrestrial C cycle. *Plant and Soil* 338:143–158.

Säumel, I., I. Kotsyuk, M. Hölscher et al. 2012. How healthy is urban horticulture in high traffic areas? Trace metal concentrations in vegetable crops from plantings within inner city neighborhoods in Berlin, Germany. *Environmental Pollution* 165:124–132.

Selhorst, A. and R. Lal. 2012. Effects of climate and soil properties on U.S. home lawn soil organic carbon concentration and pool. *Environmental Management* 50:1177–1192.

Seto, K. C., B. Güneralp, and L. R. Hutyra. 2012. Global forecasts of urban expansion to 2030 and direct impacts on biodiversity and carbon pools. *Proceedings of the National Academy of Sciences of the United States of America* 109:16083–16088.

Seto, K. C. and N. Ramankutty. 2016. Hidden linkages between urbanization and food systems. *Science* 352:943–945.

Sharma, K., N. T. Basta, and P. S. Grewal. 2015. Soil heavy metal contamination in residential neighborhoods in post-industrial cities and its potential human exposure risk. *Urban Ecosystems* 18:115–132.

Silva, L., A. Cachada, A. C. Freitas et al. 2010. Assessment of fatty acid as a differentiator of usages of urban soils. *Chemosphere* 81:968–975.

Smit, J., A. Ratta, and J. Nasr. 2001. *Urban Agriculture: Food, Jobs and Sustainable Cities* (2001 edition, published with permission from the United Nations Development Programme). New York, NY: The Urban Agriculture Network.

Stahr, K., D. Stasch, and O. Beck. 2003. *Entwicklung von Bewertungssystemen für Bodenressourcen in Ballungsräumen. BWPLUS-Projekt BWC 99001.* http://www.fachdokumente.lubw.baden-wuerttemberg.de/servlet/is/40148/?COMMAND=DisplayBericht&FIS=203&OBJECT=40148&MODE=METADATA

Sun, Y., J. Ma, and C. Li. 2010. Content and densities of soil organic carbon in urban soil in different function districts of Kaifeng. *Journal of Geographical Sciences* 20:148–156.

Taylor, J. R. and S. T. Lovell. 2014. Urban home food gardens in the Global North: Research traditions and future directions. *Agriculture and Human Values* 31:285–305.

Taylor, J. R. and S. T. Lovell. 2015. Urban home gardens in the Global North: A mixed methods study of ethnic and migrant home gardens in Chicago, IL. *Renewable Agriculture and Food Systems* 30:22–32.

Thebo, A. L., P. Drechsel, and E. F. Lambin. 2014. Global assessment of urban and peri-urban agriculture: Irrigated and rainfed croplands. *Environmental Research Letters* 9:114002.

Thomaier, S., K. Specht, D. Henckel et al. 2015. Farming in and on urban buildings: Present practice and specific novelties of Zero-Acreage Farming (ZFarming). *Renewable Agriculture and Food Systems* 30:43–54.

Trammell, T. L. E., B. P. Schneid, and M. M. Carreiro. 2011. Forest soils adjacent to urban interstates: Soil physical and chemical properties, heavy metals, disturbance legacies, and relationships with woody vegetation. *Urban Ecosystems* 14:525–552.

UN DESA. 2014. *2014 Revision of the World Urbanization Prospects.* New York, NY: Population Division, Department of Economic and Social Affairs, United Nations.

Vasenev, V. I., T. V. Prokof'eva, and O. A. Makarov. 2013. The development of approaches to assess the soil organic carbon pools in megapolises and small settlements. *Eurasian Soil Science* 46:685–696.

Victor, D. G., D. Zhou, E. H. M. Ahmed et al. 2014. Introductory chapter. In *Climate Change 2014: Mitigation of Climate Change. Contribution of Working Group III to the Fifth Assessment Report of the Intergovernmental Panel on Climate Change*, ed. O. Edenhofer, R. Pichs-Madruga, Y. Sokona et al., 111–150. Cambridge: Cambridge University Press.

Wessolek, G., B. Kluge, A. Toland et al. 2011. Urban soils in the vadose zone. In *Perspectives in Urban Ecology*, ed. W. Endlicher, P. Hostert, I. Kowarik et al., 89–132. Berlin: Springer.

Wijayawardena, M. A. A., M. Megharaj, and R. Naidu. 2016. Exposure, toxicity, health impacts, and bioavailability of heavy metal mixtures. *Advances in Agronomy* 138:175–234.

Worrel, G. 2009. Lawn be gone—The time has come for edible front yards. *Planning* August/September 2009:20–25.

Wortman, S. E. and S. T. Lovell. 2013. Environmental challenges threatening the growth of urban agriculture in the United States. *Journal of Environmental Quality* 42:1283–1294.

Xiong, X., S. Grunwald, D. B. Myers et al. 2014. Interaction effects of climate and land use/land cover change on soil organic carbon sequestration. *Science of the Total Environment* 493:974–982.

Yang, J. L. and G. L. Zhang. 2015. Formation, characteristics and eco-environmental implications of urban soils—A review. *Soil Science and Plant Nutrition* 61:30–46.

Yigini, Y., E. Aksoy, and K. Lorenz. 2012. *Urbanization: Challenges to Soil Management. Global Soil Week 2012—Rapporteurs' Reports, Global Soil Forum*, 53–59. http://globalsoilweek.org/wp-content/uploads/2015/01/Rapporteurs-Reports_GSW-2012.pdf

Zhao, S., C. Zhu, D. Zhou et al. 2013. Organic carbon storage in China's urban areas. *PLOS ONE* 8(8): e71975. doi:10.1371/journal.pone.0071975

15 Vertical Farming Using Hydroponics and Aeroponics

Dickson Despommier

CONTENTS

15.1 INTRODUCTION

As rapid climate change (RCC) continues to accelerate, conditions that favor traditional farming in soil continue to deteriorate (Lobel et al., 2011). RCC has resulted in an increased rate of crop failures throughout the world that are most frequently caused by floods and droughts. New solutions for producing commercial levels of edible plants and some domestic animals are needed, if RCC continues to worsen over the next 50 years. Fifty percent of us now live in cities. Urban agriculture, if adopted by the majority of the world's cities to feed at least 10% of their inhabitants, has the possibility of replacing significant amounts of farmland. Ground level in-bed and rooftop soil-based gardens can function to produce food crops wherever favorable conditions prevail, while enclosed rooftop greenhouses and vertical farms employing hydroponics and aeroponics allow for year-round production anywhere in the world. Indoor farms located in cities reduce dependency on outdoor farming. Urban indoor farming could help in restoring the carbon balance between the atmosphere and terrestrial carbon sinks by making way for the regrowth of hardwood forests on abandoned farmland. Another positive consequence of reducing our reliance on traditional agriculture is the eventual repair of soils that were degraded by years of commercial agricultural practices. This chapter details what is known regarding indoor growing strategies and how they are applied inside multilevel greenhouses (i.e., vertical farms). Advantages and disadvantages of farming vertically within the urban landscape are also discussed.

Advances in soil science focus on describing the current status of various soil types in selected eco-regions throughout the world, many of which are in various states of degradation (FAO, 2011). How curious it must seem then to include an entire chapter describing agricultural technologies that require no soil whatsoever, namely hydroponics and aeroponics. On further reflection, the rationale for doing so becomes clear enough. Removing land from the agricultural landscape by growing crops indoors would allow some portion of farmland to be returned to its original ecological function. In most cases, this means either the reestablishment of native grasslands or the regrowth of hardwood forests. That is because vertical farms are more efficient at producing certain crops, particularly leafy green vegetables, than traditional soil-based agriculture. For example, in temperate climate zones, outdoor lettuce farms, under the most ideal of growing conditions, can only produce two to three crops per year, whereas indoor facilities routinely produce one crop for every 35 days, year round (Brechner and Both. http://www.cornellcea.com/attachments/Cornell CEA Lettuce Handbook.pdf). Since the most productive farmland is found adjacent to rivers that once meandered through dense forest, reestablishment of a protective buffer zone of trees between existing farmland and large rivers would play a significant role in helping to reduce agricultural runoff by discouraging soil erosion. Sequestering atmospheric carbon by restoration of large swaths of hardwood forest would slow the rate of climate change (Paris Climate Conference: COP21.gouv.fr/en/: http://www.cop21.gouv.fr/en/). In addition, abandoning farmland is the first step in the eventual restoration of ecosystems that were fragmented by agriculture (Leopold, 1949; Hobbs and Cramer, 2007).

Globally, urban farming, and in particular controlled environment agriculture (CEA; Controlled Environment Agriculture. https://en.wikipedia.org/wiki/Controlled-environment_agriculture), has been gaining in popularity over the past 10 years, driven by a variety of factors, depending upon location. For example, in Japan, the rate of inclusion of vertical farms into the urban landscape was dramatically accelerated by the earthquake and resulting tsunami of March 11, 2011. In less than 2 hours, Japan lost approximately 23,600 ha of farmland in the northeast Tohoku region due to the tidal wave (Kingshuk et al., 2011). In addition, the release of radioactive material into the general environment adjacent to the damaged Fukushima Daiichi nuclear reactor found its way into some crops that then managed to show up in a few super markets (Gibney. http://www.nature.com/news/fukushima-data-show-rise-and-fall-in-food-radioactivity-1.17016). These related events served as a warning regarding food safety to a consumer public that is now wary of ingesting any vegetables grown outdoors anywhere in Japan. As a result, Japan has risen to a position of world leadership for urban agriculture by supporting the development of indoor farming as a practical means of providing locally grown food crops that are both safe to eat and sustainable to produce. As of 2016, there were over 150 plant factories (i.e., vertical farms) scattered throughout Japan (Kozai et al., 2016), with many more under construction or in the planning stage. Toshiba (QZ.com. http://qz.com/295936/toshibas-high-tech-grow-rooms-are-churning-out-lettuce-that-never-needs-washing/) and Panasonic (Panasonic.com. http://www.panasonic.com/sg/corporate/news/article/singaporeindoorfarm.html) have been encouraged by the government to develop commercial-level vertical farms. Taiwan and the American Midwest, particularly the states of Illinois, Michigan, and Indiana, are places where numerous commercial vertical farms have been constructed (Association for Vertical Farming. https://vertical-farming.net/).

CEA and its shift from soil-based to nonsoil-based technologies had its origins hundreds of years earlier in Europe. Portuguese and Italian explorers were among the first to sail out beyond the horizon, discovering new continents, several of which were predominantly tropical. Several centuries of venturing into the unknown generated a steady flow of foreign goods, enriching the lives of the stay-at-homes. The bounty included an astonishing variety of tropical and subtropical plants together with their seeds. Getting them to grow in nontropical climates, however, proved challenging. In Italy, for example, environmental conditions in most parts of the Etruscan peninsula did not favor the year-round cultivation of most of those exotics, so an old strategy was resurrected and improved upon to address that issue. Special buildings were constructed to allow for continuous cultivation throughout the year, though obviously their designs could not incorporate much in the way of transparent windows or air conditioning. Nonetheless, protecting plants from harsh conditions simply by growing them indoors took advantage of the older Roman practice of moving certain sensitive plant species indoors each night so as to avoid exposing them to the often-lethal diurnal outdoor temperature variations associated with many regions of Italy. Despite its limitations, raising plants inside buildings proved so successful that most of Europe was soon building similar structures. The majority of that activity was focused on raising decorative flowering plants. In 1832, William Chance perfected the industrial process of manufacturing transparent window glass (Chance, 1919). Following that breakthrough, elaborate conservatories of various sizes were soon to be constructed (e.g., Palm House at Kew Gardens, London, 1848; Enid A. Haupt Conservatory, New York Botanical Gardens, 1902). Many of them were built to house private collections of exotic plants, while the largest examples served as status symbols for those imperialistic nations engaged in founding colonies, many of which were in the tropics. As the result, more and more exotic plant species went on display in what would otherwise be inhospitable environments.

Constructing new conservatories and enclosed portions of botanical gardens continues to this day. The apex of this activity is embodied in several spectacular iterations: the Eden Project in Cornwall, England (Eden Project.com. https://www.edenproject.com/), the remarkable indoor Cloud Forest in Singapore (Gardens By The Bay.com. http://www.gardensbythebay.com.sg/en.html), and the geodesic-domed Climatron at the Missouri Botanical Gardens in St. Louis (Missouri Botanical Gardens.com. http://www.missouribotanicalgarden.org/). The Eden Project has as its main objective saving endangered plant species, giving them a safe haven until such time as it becomes feasible to rehabilitate the damaged ecosystems from which they were rescued. At that moment, they can then be reintroduced back into their native habitats.

Botanical gardens have thrived because of their wide appeal to a general public that appreciates being able to observe live exotic plants up close, especially orchids, without the need of searching them out in the dense thicket of a cloud forest, or scaling a steep cliff next to a waterfall in some remote tropical wilderness. For example, Kew Gardens in London averages 1.35 million visitors per year (Kew Gardens.com. http://www.kew.org/). These artificial habitats are living proof that we can replicate indoors the optimal growing environments for virtually any plant species by applying technologies developed over the past 50 years, regardless of how fastidious their biological requirements.

Indoor growing strategies applied to most edible plants have been slower to develop. The general lack of appreciation for the need to grow food indoors can be traced back to several factors. Until the advent of the industrial revolution in the 1800s, the human population numbered around 1 billion (Census.gov. https://www.census.gov/popclock/world). While food shortages were common, traditional farming was up to the challenge, and the industrialization of agriculture catalyzed the shift in our population growth from an arithmetic to a geometric rate of increase (ibid.). By 1950, there were two and a half billion people, and as of 2016, we had increased to 7.4 billion (ibid.). Throughout this period, farming, particularly soil-based agriculture, continued to expand, providing more and more food for more and more people. Farming was deemed such an essential human activity, it was allowed, and often times strongly encouraged, to encroach into more and more grassland and hardwood ecosystems (National Aeronautics and Space Administration. NASA.gov. https://www.nasa.gov/) with devastating consequences to ecological process. Traditional soil-based farming has short-circuited the carbon cycle (3) on a global scale by eliminating enormous tracts of hardwood forests (Tillman, 1999). Hardwood forests sequester carbon, whereas farms do not. Farmland occupies ~17.6 million km^2 (6.8 million square miles) (21) while grazing land occupies an additional 36.3 million km^2 (14 million square miles).

15.2 EVIDENCE-BASED DOCUMENTATION OF RCC DERIVED FROM GLOBAL AGRICULTURAL DATA

Before beginning a detailed discussion of hydroponic and aeroponic indoor farming technologies and the commercial vertical farms that employ them, a brief discussion follows as to the reasons why, at least in urban settings, switching to these agricultural methods is both advantageous and necessary (Parry and Rosenzweig, 2004). By the 1960s, it was obvious that our climate was changing at an accelerating rate (National Aeronautics and Space Administration. NASA.gov. http://climate.nasa.gov/evidence). The first clues that something other than normal was occurring on a global scale to the environment came from a host of independent observations on the rates of melting of glaciers (ibid.), and the rise of sea surface temperatures (ibid.). Around the globe, the frequency of crop failures triggered by massive floods and protracted droughts was also on the increase (ibid.). These trends have continued up to the present (nca2014.globalchange.gov/ http://nca2014.globalchange

.gov/report/sectors/agriculture; United States Department of Agriculture. USDA.gov/ http://www .usda.gov/oce/climate_change/effects_2012/effects_agriculture.htm; United States Department of Agriculture. USDA.gov/ http://www.usda.gov/oce/climate_change/effects_2012/CC%2020and%20 20Agriculture%2020Report%2020%2802-2004-2013%2029b.pdf). Some specific examples will serve to reinforce these observations.

Data quantifying total annual precipitation from 2010–2016 clearly show that California has experienced its worst drought in recorded history, and there is no end in sight (United States Geological Survey. USGS.gov. http://droughtmonitor.unl.edu/; Swain et al., 2013/2014). The drought has been maintained by a shift in the jet stream, bringing mostly dry air over the American West, while enhancing the hydrological cycle for the eastern portion of the United States (Climate.gov. https:// www.climate.gov/news-features/event-tracker/us-temperature-extremes-and-polar-jet-stream). As a direct consequence of California's protracted drought, food prices have steadily risen (United States Department of Agriculture. USDA.gov. http://www.ers.usda.gov/topics/in-the-news/california-drought-farm-and-food-impacts/california-drought-food-prices-and-consumers.aspx), in contrast to global food prices, which continue to fall (Food and Agriculture Organization.org. http://www .fao.org/worldfoodsituation/foodpricesindex/en/). In the past, in developed countries, increases in the food price index were attributable to increases in the cost of a barrel of crude oil (The Energy Collective.com. http://www.theenergycollective.com/gsitty/233751/driving-food-prices). This linkage to fossil fuel use is no longer valid, given the glut on the market of oil and natural gas selling for a fraction of their peak prices in the 1990s and 2000s (United States Department of Agriculture. USDA.gov. http://www.ers.usda.gov/data-products/ag-and-food-statistics-charting-the-essentials.aspx). In the United States, the beginning of the decline in the cost of oil and natural gas did cause a drop in the consumer food price index, but it quickly rose again, now independent of the cost of fossil fuels, as climate change issues began seriously impacting food production (Intergovernmental Panel on Climate Change. IPCC.ch. https://www.ipcc.ch/publications_and_ data/ar4/syr/en/spms5.html). The changes in our environment that dictate what can and cannot be grown and where the best growing conditions are will most likely be the norm for some time to come (Bevitori et al., 2014).

There are other consequences of drought that impact how much of a given crop can be produced. Droughts encourage the spread of rice blast, a fungal pathogen that kills rice plants before they mature (Watson et al., 2001). In past years, most countries in Southeast Asia harvested at least two crops of rice per year, and in good years, three. Today, due to sporadic outbreaks of rice blast facilitated by dried weather regimes in that region, a single crop per year has become the norm. Within the last 10 years, China has experienced a similar reduction in rice production that has triggered a mass migration of farmers from central China to the coastal cities. This unanticipated degree of urbanization began in the early 2000s and continues to the present (The Economist. com. http://www.economist.com/news/china/21640396-how-fix-chinese-cities-great-sprawl-china), creating enormous logistical problems that have yet to be resolved (Slideshare.net. http://www .slideshare.net/FGInstitute/urbanization-china-vs-india).

Thailand once exported more rice each year than any other country:

> For a long time Thailand was the world's largest rice exporter. It slipped to No. 3 behind Vietnam and India in 2012. In the late 1980s, Thailand became a leading exporter of rice. It exported over 10 million tons in 2008, compared to 6.8 million tons in 2001 and 4.5 million tons in 1994. The world's top export-ers of rice in 1991 were 1) Thailand, 2) the U.S., 3) Pakistan, 4) Vietnam, 5) China, 6) Australia, 7) Italy, 8) India, 9) Uruguay, 10) Spain (Facts and Details.com. http://factsanddetails.com/southeast-asia/ Thailand/sub5_8h/entry-3321.html).

Since 2011, the year it experienced of one of the most devastating floods in its recent history, Thailand has been forced to import rice on several occasions, due to the increased frequency and severity of flooding (Asia Foundation.org. http://asiafoundation.org/in-asia/2011/2012/2014/

thailands-flood-disaster) and droughts (Reuters.com. http://www.reuters.com/article/thailand-rice-idUSL3N0ZH30M20150701) in that part of Southeast Asia.

In summary, a crop that used to grow to abundance at a given latitude now struggles to survive there, as annual precipitation and temperature profiles continue to shift. Several hundred kilometers to the north in the northern hemisphere and to the south in the southern hemisphere that same crop may be doing much better in its new location than at its original latitude. A good example of the shift in latitude for optimal growing conditions can be gleaned from examining what has happened over the last 20 years to the global wine industry (Hannah et al., 2013). The University of California, Davis campus has a world-class oenology department and has published a brief overview on the current state of affairs regarding the adverse effects of RCC on wine production (University of California, Davis Campus.edu., UCDavis.edu. http://aic.ucdavis.edu/publications/posters/wine_climate6.pdf). Often called the "canary in the cage," the *pinot noir* grape is also referred to as the "polar bear" of wine grape varieties (Slate.com. http://www.slate.com/articles/technology/future_tense/2014/2012/wine_and_climate_change_pinot_noir_is_the_vintner_s_polar_bear.html). Its climate tolerance limits of latitude for maximum yields are defined by less than five degrees. In the northern hemisphere, the ideal conditions for producing quality *pinot noir* wines have been steadily moving north, from California's Sonoma Valley and Monterey peninsula to the cooler Willamette Valley in Oregon (Ashenfelter and Storchmann, 2016), while conditions favoring their growth in the southern hemisphere have been slowly migrating down the cone of Patagonia (ibid.). One of the regional winners created by RCC has been the southwestern portion of England (Yale University.edu. http://e360.yale.edu/feature/what_global_warming_may_mean_for_worlds_wine_industry/2478/), where once difficult to grow champagne grapes now do quite well there. According to Frazer Thompson, chief executive of Chapel Downs Wines in England: "For every degree centigrade of annual global temperature increase, growing conditions for most grape varieties shift some 270 km" (The Atlantic.com. http://www.theatlantic.com/international/archive/2012/2009/in-the-future-your-champagne-will-come-from-england/262784). These are a small sample of the effects of climate change on the agricultural landscape. In other instances, more serious consequences arise when essential food crops are adversely affected (World Food Programme.org. https://www.wfp.org/content/2015-food-insecurity-and-climate-change-map).

Novel agricultural strategies are needed to compensate for the loss of crops if the world is to avoid massive food riots and starvation over to coming years. This chapter details the application of CEA methods to urban populations, providing them with a fresh, safe, and sustainable food supply. The future of urban farming is part of a larger construct, in which cities are reengineered to emulate the best features of ecosystems (biodiverse, sustainable, and resilient), using technology to do so. Primary productivity is a constant feature of every terrestrial ecosystem (de Groot et al., 2002). Cities that produce a portion of their own food will have taken the first step toward self-sufficiency and long-term sustainability.

By the year 2050, nearly 80% of us will be living in cities. Urbanization is an obvious expression of what it means to be human. Yet, even though we have lived in cities for centuries, we still need to learn more about how to achieve balance between the natural world and the one we have created. Cities are by their very nature parasitic on the landscape (energy, water, and food), and this behavior needs to change or the world will become less and less nurturing for humans over the following millennium. Currently, uncontrolled urban growth is creating chaos in most underdeveloped countries. But there is hope for change in the right direction. As CEA becomes more accepted and eventually is insinuated into the very fabric of all cities worldwide, perhaps then some of the more pressing problems associated with urban sprawl (e.g., food shortages, out-of-control unemployment, high infant mortality rates, and chronic illnesses) will be greatly reduced by the creation of more employment opportunities, healthier food choices with the availability of year-round fresh produce, and thus less malnutrition among the poorest members of the global urban community.

What are the challenging issues that need to be addressed to enable this scenario to unfold over the next 50 years? Vertical farming has come of age only within the last few years. The popularity of any new technology needs some time for the word to spread to all sectors. There are now some 500 vertical farms in operation worldwide. Predicting the future is always a slippery slope, but given the trend for more and more urban agricultural production, it is not unrealistic to expect that over the next 50 years, every city might be producing at least 20%–40% of its food from *in situ* farming operations.

Government support at the academic level for encouraging the development of more crop-diverse applications for vertical farming (e.g., grains and root vegetables), increased manufacturer buy-in for making available more innovative, cheaper indoor growing systems, more consumer awareness regarding the food safety and security issues of outdoor versus indoor farming, and greater political support for establishing vertical farming inside the urban landscape, if properly implemented, would jump-start a movement toward establishment of the eco-city, the "holy grail" of urban cultural evolution.

15.3 CONTROLLED ENVIRONMENT AGRICULTURE

The term CEA (University of Arizona.edu. http://cals.arizona.edu/ceac) refers to growing plants indoors in a protected enclosure (e.g., greenhouse) under conditions in which all of the factors affecting the growth of the plant—temperature, humidity, nutritional requirements, and so on—can be adjusted to mimic their ideal biological and physical requirements. As of 2016, a wide variety of soilless technologies have become available for the production of edible plants and some animals in a controlled environment (Lakkireddy et al., 2012; University of Arizona.edu. http://ag.arizona.edu/ceac/sites/ag.arizona.edu.ceac/files/jensen%20Taiwan%20World%20Review%20of%20CEA.pdf). Included into the mix of applied technologies for food production are hydroponics (Sardare and Admane, 2013) and aeroponics (Stoner, 1983). Each of these methods uses significantly less water— hydroponics = 70%; aeroponics = 95%—than traditional outdoor farming, which uses over 70% of the available liquid freshwater on Earth (United States Geological Survey.gov. http://water.usgs.gov/edu/wuir.html). Hydroponics and aeroponics are the two most adaptable technologies for producing commercial levels of crops indoors. Due to their frugal water requirements, mass production facilities can be highly successful regardless of geographic location. Because of these advantages, growing edible plants indoors has become a major global industry in just over 30 years of application (Cuestraroble.com. http://cuestaroble.com/statistics.htm), with over 0.4 million ha (1 million acres) of greenhouses in production as of 2016. The greenhouse industry employs both soil-based and soilless methods for year-round production. The same is true for vertical farms—multistory greenhouses. Before discussing the advantages and disadvantages of vertical farming, it is instructive to first give an overview of the various growing technologies that, when used in combination, were immediately adopted by the fledgling vertical farm industry, and that were primarily responsible for its current level of economic success and popularity.

15.4 INDOOR GROWING METHODS

15.4.1 HYDROPONICS

Two growing strategies have become widely adopted for the indoor commercial production of edible plants: *hydroponics* and *aeroponics*. A third hybrid method, *aquaponics*, in which fish production is also essential, will also be described. The application of hydroponics for growing edible plants probably had its origins in ancient times (Historia. http://www.hydor.eng.br/HISTORIA/C1-I.pdf). In Europe, by the middle of the eighteenth century—The Age of Enlightenment—the emergence of the scientific method began to replace outdated religious dogma, serving as the basis for creating a more rational and just future. Those individuals driven by an insatiable curiosity and a passion to know more about the natural world were now free to explore any question, as science began to change how people lived. Nothing was assumed to be true until physical evidence was obtained by observation and experimentation. The scientific method was rapidly adopted by institutions of higher learning throughout the western world and applied to all subjects dealing with life on Earth. In that spirit, two academic botanists, Julius von Sachs and Wilhelm Knop, working independently in Germany between 1860 and 1875, described many of the physiological parameters that defined the conditions for optimal growth of most green plants, and laid the foundation for the development of modern hydroponics (Douglas, 1975). Sachs (von Sachs, 1868) focused on determining the essential elements that green plants require in order to complete their life cycles. He grew plants suspended in a series of aqueous solutions, each composed of a defined amount of dissolved purified chemical compounds, to determine which plant diets were sufficient for optimum growth (von Sachs, 1887). Knop also concentrated his studies on aquatically grown plants and their nutritional requirements. As a direct result of their studies and a few more that followed shortly thereafter, it was firmly established that all green plants require just 16 elements (Knop, 1868). Sachs and Knop further demonstrated that an inorganic source of nitrogen in the form of potassium nitrate was sufficient for maximum growth, and that the addition of manure, a traditional farming practice thought to be required as fertilizer, while aiding in making soil more tillable, did not contribute in a major way to plant nutrition.

It was not until 1929, when most of the basic biochemical and physiological parameters of plant growth had been described, that someone took agriculture to the next level, namely growing commercial levels of edible plants in a defined nutrient-enriched aqueous solution. Dr. William F. Gericke, working at the Berkeley campus of the University of California, pioneered the science of modern hydroponics, a name that he coined to describe that agricultural method. His book, *The Complete Guide to Soilless Gardening* (Gericke, 1949), became the primary reference for a new generation of urban farmers who came to appreciate the many advantages such a radical approach offered, including being able to grow edible plants virtually anywhere, including inside the urban landscape, independent of soil, as long as there was enough water and nutrients available. Gericke's research amply demonstrated that almost any crop could be grown that way, including vine, tuber, and root vegetables (e.g., tomatoes, potatoes, carrots, and beets). But his seminal work was not without its critics, especially some plant scientists from his home institution (Hoagland and Arnon, 1938). After thoroughly reviewing Gericke's work, Hoagland and Arnon concluded that the water-culture method was not any more effective when it came to plant growth than raising the same plants traditionally in soil. But instead of serving as a criticism of the new method, their paper had the opposite effect, paving the way for the large-scale use of hydroponics.

Commercial-level hydroponic-based crop production is currently employed in a large number of greenhouses and vertical farms throughout the world (Resh, 2013; Kozai et al., 2016). Several growing system configurations define each of the several methods employed in hydroponic greenhouses (Table 15.1). A short review of the principles and applications of hydroponics can be found at https://en.wikipedia.org/wiki/Hydroponics. Many commercial growers employ some variation of either nutrient film technology (NFT) or continuous-flow solution culture (CFS). Each system has its advantages and disadvantages. NFT systems are typically configured with regularly spaced holes to accommodate plants (e.g., Boston bib lettuce) in a shallow PVC pipe that is angled in such a way as to allow the passive flow of a nutrient solution to move from one pipe rack to another by gravity, such as ones that might be found in a vertical farm with multiple vertical layers of growing beds. The seedlings grow through evenly spaced holes in the pipe in specially fitted inert fiber-filled (e.g., shredded coconut shell or finely crushed volcanic rock) containers that snug tightly into each hole. The roots barely touch the bottom of the pipe, encountering the nutrient solution, and continue to grow downstream as the plants mature. The film of nutrient solution is shallow enough to allow the roots direct exposure to the air as well as to the growth medium, achieving a high level of oxygen that encourages healthy root systems. NFT systems, due to their simplicity of design, are easy to maintain and many commercial versions have built-in automated monitoring equipment for real-time tracking of nutrient concentrations, oxygen levels, pH and Eh, temperature, and so on. While NFT is a highly efficient method with respect to energy and water consumption, the density of plants is limited by the configuration of the growing system and requires labor-intensive harvesting strategies.

TABLE 15.1

Hydroponic and Aeroponic Growing Systems for Vertical Farms

Hydroponic

Nutrient film technology (NFT)

Single plants grow in individual wells along with a PVC pipe. Each pipe is slightly tilted. Root tips rest on the bottom of the pipe and are continuously in touch with a thin layer of nutrient solution, which flows past each plant aided by gravity. Roots are well-oxygenated, since the greater portion of each root on all plants is exposed to the air.

Static solution culture

Individual plants grow in a container filled with nutrient solution, which must be aerated.

Passive subirrigation

Root systems grow into an inert medium. Nutrient solution reaches the roots via capillary action from a sub-root nutrient reservoir. Roots are well-aerated.

Ebb and flow (flood and drain) subirrigation

Nutrient solution floods the roots via a pump at regular intervals, then returns to reservoir. Roots are well-aerated.

Run to waste

Nutrient solution floods the surface of an inert material that supports a growing plant, then drains into a collection reservoir. The nutrient solution is used only once, then discarded or recharged with nutrients. Roots are well-aerated.

Deep-water culture

A tray is partially filled with a nutrient solution. A floating piece of Styrofoam in the tray holds the plants, and the roots grow into the tray. Usually, multiple trays are connected to one another and the solution moves from tray to tray by gravity.

Top-fed deep-water culture

Aerated nutrient solution floods the tray at regular intervals from above the surface of the growing rack of plants, then returns to the reservoir below the tray.

Rotary grow system (e.g., Omega Grow system®)

A specially constructed hollow barrel with plant holders on its inner surface and a grow light located in the middle of the barrel rotates the plants through a tray of nutrient solution located at the bottom of the barrel. The barrel rotates at a rate that does not allow the plants to dry out.

Hydro-stacker Vertical Hydroponic Growing System®

Planters, with four cups per planter, can be stacked one on top of the another by a central connector spike up to five layers high. The nutrient solution is introduced from above the highest planter via a central feeding tube. Nutrient solution saturates the soil around each plant until all layers of the array are fed. A timer delivers the nutrient solution at regular intervals.

Aeroponic

Flat-bed configuration

A flat piece of stiff material with regularly spaced holes such as "3/4" marine plywood is "planted" with seedlings that are fed nutrients from below in the form of a mist. Roots are well-aerated and grow longer than in hydroponic systems. Plants also grow more rapidly in this configuration compared with hydroponic systems.

Soft fiber matrix configuration (e.g., Aerofarms®)

A patented cloth-like material is seeded with a lawn of seedlings. Plants are sprayed from below with a nutrient solution. Harvesting is automated. Excellent system for producing a wide variety of microgreens.

Slant wall configuration

Same general approach as the flat-bed configuration, but the roots are more easily accessed for monitoring growth, and so on.

Vertical tube configuration (e.g., Tower Garden®)

A polyvinyl chloride plastic pipe is fitted with a series of evenly spaced holes to accommodate seedlings that are embedded in an inert fiber matrix. The PVC tube is inserted into a water-tight circular tub filled with nutrient solution and a submersed pump. The pump delivers a stream of the nutrient solution to the top of the enclosed pipe and the specially configured inner surface of the top converts it into a mist that slowly descends the tube, encountering all the roots. Any mist that does not get absorbed returns to the reservoir in the tub for recycling. There are Tower Garden systems for home and commercial use.

Stacker with multiple plant ports configuration (e.g., Hydro-Stacker®)

Plants in soil occupy individual wells in a molded grow unit that can be stacked on top of one another, facilitated by a spindle-in-hole system. A total of five units can be accommodated in this fashion, with four plants per stackable unit. The top of the upper most unit has a port, onto which the nutrient delivery system inserts. Each unit receives a steady stream of nutrient solution delivered by a pump. There are many commercial versions of this configuration, some of which accommodate more plants than the original Hydro-Stacker.

In contrast, CFS uses shallow trays of 91.4 cm × 121.9 cm × 15.2 cm (~3 ft × 4 ft × 0.5 ft) partially filled with a nutrient solution in which Styrofoam® rafts of densely packed plants are floated. The germinated seedlings, each nested in a fiber-filled bottomless container, send their roots into the solution below through regularly spaced holes in the plastic rafts. In this method, more water is used compared with NFT, and stronger pumps are required to move the solution from tray to tray, even though the trays are often configured to take advantage of gravity. Ultimately, the choice of growing method is dictated by the kind of crops selected for production and the desired yields. Some examples of commercial vertical farms are listed in Table 15.2, together with the crops grown.

15.4.2 Aeroponics

Aeroponics is a relatively new method for growing edible plants and was developed in 1983 by Dr. Richard J. Stoner II while working on contract from the National Aeronautics and Space Administration (Stoner and Clawson, 1999–2000). Aeroponics utilizes a fine mist of nutrient-laden water created by its passage through a pressurized nozzle that is then directed toward the enclosed root system of the plants (Aeroponics.com. http://www.aeroponics.com/aero2.htm). A short review of the principles and applications of aeroponics can be found at https://en.wikipedia.org/wiki/Aeroponics. Aeroponics uses approximately 70% less water than hydroponics, while delivering the same amount of nutrients to the roots. Its main advantage is its application to growing high densities of grains such as wheat (*Triticum aestivum*) (National Aeronautics and Space Administration. NASA.gov. http://ntrs.nasa.gov/archive/nasa/casi.ntrs.nasa.gov/19900009537.pdf) and microgreens (Aerofarms.com. http://aerofarms.com/). Recent advances in nozzle design have improved the reliability of the system for creating the spray by eliminating clogging, a major issue in earlier models (Stoner and Clawson, 1999–2000). As a result, more vertical farms are adopting aeroponics as their main growing strategy. Several configurations of aeroponic growing systems will serve to illustrate the creative process following Stoner's original aeroponic equipment.

Tower Garden (Tower Gardens.com. http://www.towergarden.com/), a patented vertical tubular aeroponic system designed and manufactured by Tim Blank (Tower Gardens.com. http://www.towergarden.com/content/towergarden/en-us/about/future-growing/tim-blank.html) has evolved into a scalable technology for both home and commercial application (Future Growing.com .http://www.futuregrowing.com/info.html). In brief, a large tub filled with a nutrient solution and a submersible pump is fitted with a PVC tube that has portals cut into it to accommodate holders of seedlings of various types (i.e., leafy greens, tomatoes, green beans, and zucchini). The pump delivers a stream of nutrient solution up into the hollow body of the pipe. When it reaches the inner surface of the sealed top, a specially engineered contour disperses the stream, converting it into a mist that then bringing the nutrients to the roots that are hanging inside the PVC cavity. The main disadvantage of this system appears to be the limited number of individual plants (28 plants per tube) that can be accommodated by the tube. Larger diameter tubes with double the capacity and larger tubs of nutrient solution are available for commercial vertical farms.

A second patented aeroponic growing system developed by Edward Harwood (Aerofarms.com. http://aerofarms.com/technology/) makes use of a cloth-like porous fiber matrix onto which a lawn of germinated seeds is evenly distributed. Nutrient solutions are sprayed from a reservoir below the cloth, while the plants are lit from above with high efficiency LED grow lights. This is the configuration employed by Aerofarms, and their technology is ideally suited for the large-scale production of a wide variety of microgreens. In the near future, other aeroponic growing systems will undoubtedly emerge, as this version of urban agriculture becomes more popular not only with commercial growers, but with restaurants (e.g., *Bidwell* in Washington, DC (Bell Book and Candle.com. http://bidwelldc.com/).

TABLE 15.2

Selected List of Commercial Vertical Farms and Their Main Crops

The United States
  Vertical Harvest
    Jackson, Wyoming
      Butterhead lettuce, red head lettuce, baby arugula, Bright Lights Swiss Chard, baby romaine, Vertical Harvest sugar pea, Vertical Harvest rock chive, Vertical Harvest kaiware, Vertical Harvest Shiso Jewel, Vertical Harvest Garden Gem, Vertical Harvest bumble bee tomato, Vertical Harvest rebel tomato
  Aerofarms
    Newark, New Jersey
      Leafy greens
  FarmedHere
    Bedford Park, Illinois
    New farm in Louisville, Kentucky
      Microgreens, basil, and basil vinaigrette
  Green Spirit Farms
    Portage, Indiana
    20 VFs for China.
      Butterhead lettuce (4 varieties), Basil, chives, dill, mint, rosemary, thyme, Sagee oregano, arugula, collard greens, kale, Swiss chard, mizuna, watercress, garnet mustard, green mustard, komatsuna, lime basil, red cabbage, Thai basil, amarath
  Urban Produce
    Irvine, California
      Arugula, sunflower, broccoli, cilantro, wasabi, bok choy, kale
  GreenFarms A&M
    Valparaizo, Indiana
      Leafy greens
  Caliber Biotherapeutics
    Byran, Texas
      Value-added products from GMO tobacco plants (e.g., influenza vaccine)
Japan (165 VFs)
  Spread (Nuvege)
  Kyoto
    Leafy greens
  Osaka: New automated farm to produce 10,000 heads of lettuce per day
Taiwan (56 VFs)
China (35 VFs)
Singapore
  Sky Greens
    Sky Nai Bai
    Sky Cai Xin
    Sky Xiao Bai Cai
    Sky Chinese Cabbage
Panasonic Industrial Vertical Farm
  Leafy greens
The Netherlands
  The Hague
  Urban Farming De Schilde
    Leafy greens, fish (aquaponics)
Panama
  Panama City
    Urban Vertical Farms

15.4.3 AQUAPONICS

Aquaponics employs an advanced, more complex form of indoor agriculture (The Plant.org. http://plantchicago.org/), in which bacteria-laden water, a byproduct of freshwater fish farming (e.g., tilapia), is circulated through an NFT system and serves as the nutrient solution for the edible plants. The removal of nutrients by the plants purifies the water, and it is then returned to the fish tanks for reuse. Ammonia, a volatile compound excreted by fish, must be passed through a biofilter colonized with specialized bacteria that enzymatically converts it into plant-friendly nitrates and nitrites. Aquaponics growers liken their method of agriculture to a kind of *closed loop* ecosystem approach to food production. To begin the circulation of nutrients, the fish are fed pellets of compressed vegetable material. The fish feces contain the essential nutrients for the plants in the form of bacteria. When the bacteria die and then lyse, they release their stored elements into the water, which are taken up by the plants. Since two different systems are required to complete the loop, one for fish and the other for the plants, monitoring equipment for each part of the system must be incorporated into the overall design of the indoor farm (ibid.). Aquaponics is more difficult to operate and maintain than hydroponics or aeroponics, and that is why many indoor crop production facilities have opted for the latter simpler modes of growing.

15.5 THE VERTICAL FARM (AKA PLANT FACTORY IN JAPAN)

While greenhouse agriculture has matured and flourished (University of Arizona.edu. https://ag.arizona.edu/ceac/sites/ag.arizona.edu.ceac/files/WorldGreenhouseStats.pdf), evolving into an extensive global industry within the past 20 years (Yilmaz and Ozkan, 2005; Resh, 2013), growing edible plants in multistory buildings only recently emerged as an alternative urban agricultural strategy (Despommier, 2010; Kozai et al., 2016). The reasons why vertical farming has been slow to develop include the lack of commercial manufacturers of hydroponic and aeroponic equipment designed specifically for tall buildings, efficient LED grow lights, and financial support for small-scale development of demonstration vertical farms. Starting in 2013, many of these issues had either been solved or were in the process of being addressed, encouraging a spurt of vertical farm construction throughout Japan, Singapore, Taiwan, and the United States (Table 15.2; Association for Vertical Farming.org. https://vertical-farming.net/). As of this writing, worldwide, there were an estimated 300 vertical farms, with many more in the planning stage or under construction. As pointed out, some countries got an earlier start than others due to catastrophic environmental events, such as the earthquake and tsunami that occurred in Japan. Singapore's vertical farm industry (ASEANUP.com. http://aseanup.com/vertical-farming-in-singapore/) began by considering food as a security and safety issue. In the United States, vertical farming began as an outgrowth of the locavore (Slow Food USA. org. https://www.slowfoodusa.org/) and slow food movements (ibid.), with food safety as a separate problem that also needed to be resolved (United States Food and Drug Administration. USFDA.gov. http://www.fda.gov/Food/FoodborneIllnessContaminants/FoodborneIllnessesNeedToKnow/default. htm). Vertical farming produces locally grown food without the use of herbicides or pesticides in a clean environment that permits harvesting and sales without the need to wash the plants before packaging and shipping (Despommier, 2010). Interestingly, nearly all of the commercial vertical farm companies in the United States are currently expanding their businesses and advising other countries as to how to establish and operate them. Nearly every vertical farm specializes in high-value leafy green vegetables (see Table 15.2). In the case of Aerofarms, nearly 40 different varieties are grown for a wide range of customers (wholesale grocery stores, restaurants, etc.).

15.5.1 DESIGNS

The first iteration of a vertical farm is credited to Andrew Kranis in 2000 (Sofa.aarome.org. http://sofa.aarome.org/andrew-kranis), who designed "Agrowanus" as his third year student project in the School of Architecture at Columbia University (Google.com. https://www.google.com/

search?q=vertical+farm+2016&safe=off&biw=1381&bih=1275&tbm=isch&tbo=u&source=univ&
sa=X&sqi=2012&ved=2010ahUKEwj2055tysraXLAhUCVT2014KHUjaDnUQsAQIGw). Within
several years thereafter, more vertical farm designs emerged from like-minded architects and
designers, and some renderings found their way on to the internet and into the hearts and minds
of millions of viewers (Work.ac. http://work.ac/locavore-fantasia/). New designs continue to be
uploaded, as interest in vertical farming remains strong among those with expertise in architecture and design. Unfortunately, while the vast majority of these often highly imaginative projects
are attractive to the eye (Tree Hugger.com. http://www.treehugger.com/sustainable-product-design/
vertical-farms-for-london-are-lovely-green-eye-candy.html.; Today.com. http://www.today.com/id/
21153990/ns/today-today_tech/t/could-these-be-farms-future/#.VtiE21153992xiOrMo), if actually
built, none appear to be practical with respect to crop yields. They remain visual expressions of
a larger creative force whose rationale is to envision a brighter, perhaps more ecologically balanced future for the built environment. A commercial farm, irrespective of location (i.e., indoors or
outside) has as its first consideration the production of a given crop. Dedicating space for nonfood
production (i.e., for purely aesthetic reasons) is never an option when it comes to agriculture. It is the
same basic principle at work when designing a nuclear submarine or the International Space Station.
Indeed, space is the "last frontier." So, while ultramodern designs are exciting to contemplate as
buildings, they fall far short on this one critical issue. The majority of traditional soil-based farmers plant their crops right down to the riverbank, and in fact, that is one of the major drawbacks of
traditional farming. In times of floods, the soil simply washes away, leaving less for farming when
the water retreats. As a direct consequence of this practice, agricultural runoff is widely acknowledged as the world's most prevalent form of nonpoint source pollution (Centers for Disease Control
and Prevention.gov. http://www.cdc.gov/healthywater/other/agricultural/contamination.html), since
farmers cannot control the weather, although most wish that they could. Indoor farming eliminates
agricultural runoff as an issue by design. While radically designed vertical farms may not succeed
in real life, the vision they represent will continue to serve an important function: to further stimulate the imagination of those more practical individuals who can take the idea of growing food in
tall buildings to the next phase. In fact, that is exactly what happened next.

The exact date of when the first vertical farm was built is not known, but starting in the early 2000s
a few versions of plant factories were established in Japan, preceding the Fukushima event. By 2012,
there were also vertical farms in Chicago (The Plant.org. http://plantchicago.org/) and in Suwon,
South Korea (Korea.net. http://www.korea.net/NewsFocus/Sci-Tech/view?articleId=103942). The
next few years witnessed an explosion of global activity that continues to the time of this writing.

The designs for functional vertical farms fall into two categories: new buildings and repurposed
abandoned buildings. The great majority of vertical farms are of the latter variety. Most typically,
an abandoned warehouse (e.g., FarmedHere in Bedford Park, Illinois) or factory (e.g., The Plant in
Chicago; Aerofarms in Newark, New Jersey) serves as the shell for repurposing the structure. Since
grow lights have become efficient enough to justify their use in buildings without windows, these
"big box" flat-roofed buildings have become prime candidates for start-up vertical farm operations. By using an existing building, often a vertical farm can begin operations within months after
closing the deal with the city of their choice (e.g., Green Spirit Farms, New Buffalo, Michigan).
The other obvious advantage is in reducing the start-up cost. Often, mayors are willing to cobble
together attractive economic incentive packages to entice prospective vertical farm companies to
locate in these buildings. The cost to a city of tearing down an abandoned building and to then have
to find a contractor to erect a replacement is frequently much more than the tax breaks the city is
willing to offer in the vast majority of deals. For example, in the United States, as of this writing,
there were hundreds of empty Walmart warehouses up for sale (Business Insider.com. http://www
.usinessinsider.com/list-of-walmart-stores-closing-2016-1). The situation is even more dire in
Detroit, where it is estimated that as of July 2016, there were over 70,000 abandoned buildings, the
vast majority of which could theoretically be converted to some form of indoor growing space, and
many of which would immediately qualify as excellent candidates for establishing a vertical farm.

Both hydroponics and aeroponics equipment is commercially available that can function in any size enclosure, and some of the newly growing systems can be adapted for multistory rack systems for NFT and CFS. A novel structure for a potential vertical farm has recently been identified: storage silos that were once used to keep surplus grains in (News Leader.com. http://www.news-leader.com/story/news/education/2016/2002/2017/msu-leases-downtown-silos-urban-farming-start-up/80524672/). Missouri State University, owner of several silos (grain elevators), has given permission to lease several silos to a start-up company to repurpose them into vertical farms. If successful, other silos in the general vicinity have been targeted for similar retrofits. Hydroponics will be the favored growing system for these iconic hollow cement cylinders. Ironically, if this project succeeds, grains could eventually emanate from silos rather than being stored in them.

Newly constructed vertical farms number far less than those that took advantage of already existing buildings. But new vertical farms (VFs) have the obvious advantage of designing with specific grow systems in mind. Custom-fitted NFT and CFS rack systems with integrated high efficiency LED lighting fixtures maximize the space dedicated to growing, thus making them far more cost-effective in that regard than retrofitted older buildings. In addition, heating, ventilation, and air conditioning (HVAC) systems can also be designed for maximum efficiency, adding even more savings to the bottom line. Examples of newly built vertical farms include *Vertical Harvest* in Jackson, Wyoming (Vertical Harvest.com. http://verticalharvestjackson.com/), *Nuvege*, in Kyoto, Japan (Spread.com. http://spread.co.jp/en/), and *Sky Greens* in Singapore (Sky Greens.com. http://www.skygreens.com/).

15.6   CONCLUSIONS

Nature's grandest design is the ecosystem. Municipalities must emulate these assemblages of diverse life forms in order to prevent further disruption of ecosystem services (Grunewald and Bastian, 2015). Failure to address this issue could well lead to the near-total collapse of some terrestrial ecosystems with catastrophic consequences for numerous species, including humans (World Health Organization.org. http://www.who.int/globalchange/ecosystems/biodiversity/en/). Agriculture is a human activity that has been seriously altered by changes in our climate. Empowering cities to produce a significant portion of its food through the construction of vertical farms speaks volumes regarding the direction forward in the creation of the eco-city, in which all municipal functions are eventually fully integrated, resilient, and sustainable (see *State of the World: Can a City Be Sustainable?* World Watch Institute meeting. May 10, 2016, Washington, DC). If the idea of living in the built environment while coexisting with the natural world around us eventually becomes the accepted norm, then perhaps most of the damage we have caused by encroachment can be reversed simply by adopting new ways of obtaining our food that does not threaten to eliminate the rest of the natural world. While learning how to farm in tall buildings requires a "hands-on" approach, learning to live in balance with the rest of nature demands a "hands-off" strategy. Urban agriculture, and vertical farming in particular, has the potential to significantly reduce our rate of encroachment into natural systems for the purpose of establishing more farms, while at the same time supplying urban dwellers with a varied, sustainable, and safe food supply.

REFERENCES

Aerofarms.com. Accessed March 2016. http://aerofarms.com/.
Aerofarms.com. Accessed March 2016. http://aerofarms.com/technology/.
Aeroponics.com. Accessed March 2016. http://www.aeroponics.com/aero2.htm.
ASEANUP.com. Accessed March 2016. http://aseanup.com/vertical-farming-in-singapore/.
Ashenfelter, O. and Storchmann, S. 2016. The economics of wine, weather, and climate change. *Review of Environmental Economics and Policy*, 10:25–46.

Asia Foundation.org. Accessed March 2016. http://asiafoundation.org/in-asia/2011/2012/2014/thailands-flood-disaster.

Association for Vertical Farming. Accessed July 2016. https://vertical-farming.net/.

Bell Book and Candle.com. Accessed March 2016. http://bbandcnyc.com/; http://bidwelldc.com/.

Bevitori, R.O.M., Grossi-de-Sá, M.F., Lanna, A.C., da Silveira, R.D., and Petrofeza, S. 2014. Selection of optimized candidate reference genes for qRT-PCR normalization in rice (*Oryza sativa* L.) during *Magnaporthe oryzae* infection and drought. *Genetics and Molecular Research*, 13:9795–9805.

Brechner, M and Both, A.J. http://www.cornellcea.com/attachments/Cornell CEA Lettuce Handbook.pdf.

Business Insider.com. Accessed July 2016. http://www.businessinsider.com/list-of-walmart-stores-closing-2016-1.

Census.gov. Accessed March 2016. https://www.cencus.gov/popclock/world.

Centers for Disease Control and Prevention.gov. Accessed March 2016. http://www.cdc.gov/healthywater/other/agricultural/contamination.html.

Chance, J.F. 1919. *A History of the Firm of Chance Brothers & Co: Glass and Alkali Manufacturers*. London: Spottiswood, Ballantyne & Co.

Climate.gov. Accessed March 2016. https://www.climate.gov/news-features/event-tracker/us-temperature-extremes-and-polar-jet-stream.

Controlled Environment Agriculture. Accessed March 2016. https://en.wikipedia.org/wiki/Controlled-environment_agriculture.

Cuestraroble.com. Accessed March 2016. http://cuestaroble.com/statistics.htm.

de Groot, R.S., Wilson, M.A., and Boumans, R.M.J. 2002. The dynamics and value of ecosystem services: Integrating economic and ecological perspectives: A typology for the classification, description and valuation of ecosystem functions, goods and services. *Ecological Economics*, 41:393–408.

Despommier, D.D. 2010. *The Vertical Farm: Feeding the World in the 21st Century and Beyond*. New York, NY: St. Martin's Press: p. 400.

Douglas, J.S. 1975. *Hydroponics*, 5th ed. Bombay: Oxford UP.

Eden Project.com. Accessed March 2016. https://www.edenproject.com/.

Facts and Details.com. Accessed March 2016. http://factsanddetails.com/southeast-asia/Thailand/sub5_8h/entry-3321.html.

Food and Agriculture Organization, 2011. SUMMARY REPORT: The State of the World's Land and Water Resources for Food and Agriculture. http://www.fao.org/docrep/015/i1688e/i1688e1600.pdf.

Food and Agriculture Organization.org. Accessed March 2016. http://www.fao.org/worldfoodsituation/foodpricesindex/en/.

Future Growing.com. Accessed March 2016. http://www.futuregrowing.com/info.html.

Gardens By The Bay.com. Accessed March 2016. http://www.gardensbythebay.com.sg/en.html.

Gericke, W.F. 1949. *The Complete Guide to Soilless Gardening*. Upper Saddle River, NJ: Prentice Hall: p. 304.

Gibney, E. Fukushima data show rise and fall in food radioactivity. Nature News, 27 February 2015, Accessed July 2016. http://www.nature.com/news/fukushima-data-show-rise-and-fall-in-food-radioactivity-1.17016.

Google.com. Accessed March 2016. https://www.google.com/search?q=vertical+farm+2016&safe=off&biw=1381&bih=1275&tbm=isch&tbo=u&source=univ&sa=X&sqi=2012&ved=2010ahUKEwj2055tysraXL AhUCVT2014KHUjaDnUQsAQIGw.

Grunewald, K. and Bastian, O., eds. 2015. *Ecosystem Services. Concepts, Methods and Case Studies*. Berlin, Heidelberg: Springer: p. 312.

Hannah, L., Roehrdanz, P.R., Ikegami, M., et al., 2013. Climate change, wine, and conservation. *Proceedings of the National Academy of Sciences of the USA*, 110:6907–6912.

Historia. Accessed March 2016. http://www.hydor.eng.br/HISTORIA/C1-I.pdf.

Hoagland, D.R. and Arnon, D.L. 1938. The water-culture method for growing plants without soil. Agricultural Station Berkeley, California Circular 347.

Hobbs, R.J. and Cramer, V.A., eds. 2007. *Old Fields*. Washington, DC: Island Press: p. 352.

Intergovernmental Panel on Climate Change. IPCC. ch. Accessed March 2016. https://www.ipcc.ch/publications_and_data/ar4/syr/en/spms5.html.

Kew Gardens.com. Accessed March 2016. http://www.kew.org/.

Kingshuk, R., Katsuhiro, S., and Kohno, E. 2011. Salinity status of the 2011 Tohuku-okitsunami affected agricultural lands in northeast Japan. *International Soil and Water Conservation Research*, 2(2):40–50.

Knop, W. 1868. Der Kreislauf des Stoffs. Lehrbuch der Agrikultur-Chemie 2 BdeLeipzig.

Korea.net. Accessed March 2016. http://www.korea.net/NewsFocus/Sci-Tech/view?articleId=103942.

Kozai, T., Nui, G., and Tagaki, M. 2016. *Plant Factory*. Cambridge, MA: Academic Press: p. 405.

Lakkireddy, K.K.R., Katsuri, K.R.S., and Sambasiva Rao, K.R.S. 2012. Role of hydroponics and aeroponics in soilless culture in commercial food production. *Journal of Agricultural Science and Technology*, 1:26–35.

Leopold, A. 1949. *A Sand County Almanac*. Oxford: Oxford University Press: p. 227.

Lobel, D.B., Schlenker, W., and Costa-Roberts. 2011. Climate trends and global crop production since 1980. *Science*, 333:616–620.

Missouri Botanical Gardens.com. Accessed March 2016. http://www.missouribotanicalgarden.org/.

National Aeronautics and Space Administration. NASA.gov. Accessed March 2016. https://www.nasa.gov/.

National Aeronautics and Space Administration. NASA.gov. Accessed March 2016. http://climate.nasa.gov/evidence.

National Aeronautics and Space Administration. NASA.gov. Accessed March 2016. http://ntrs.nasa.gov/archive/nasa/casi.ntrs.nasa.gov/19900009537.pdf.

National Geographic Society. Accessed July 2016. http://news.nationalgeographic.com/news/2005/2012/1209_051209_crops_map.html.

nca2014.globalchange.gov. Accessed March 2016. http://nca2014.globalchange.gov/report/sectors/agriculture.

News Leader.com. Accessed March 2015. http://www.news-leader.com/story/news/education/2016/2002/2017/msu-leases-downtown-silos-urban-farming-start-up/80524672/.

Paris Climate Conference: COP21.gouv.fr/en/: http://www.cop21.gouv.fr/en/.

Panasonic.com. Accessed March 2016. http://www.panasonic.com/sg/corporate/news/article/singaporeindoorfarm.html.

Parry, M.L. and Rosenzweig, C. 2004. Effects of climate change on global food production under SRES emissions and socio-economic scenarios. *Global Environmental Change*, 14:53–67.

QZ.com. Accessed March 2016. http://qz.com/295936/toshibas-high-tech-grow-rooms-are-churning-out-lettuce-that-never-needs-washing/.

Resh, H.M. 2013. *Hydroponic Food Production: A Definitive Guidebook for the Advanced Home Gardener and Commercial Hydroponic Grower*, 7th ed. Boca Raton, FL: CRC Press: p. 560.

Reuters.com. Accessed March 2016. http://www.reuters.com/article/thailand-rice-idUSL3N0ZH30M20150701.

Sardare, M.D. and Admane, S.V. 2013. A review on plant without soil—Hydroponics. *International Journal of Research in Engineering and Technology*, 2:299–304.

Sky Greens.com. Accessed March 2016. http://www.skygreens.com/.

Slate.com. Accessed March 2015. http://www.slate.com/articles/technology/future_tense/2014/2012/wine_and_climate_change_pinot_noir_is_the_vintner_s_polar_bear.html.

Slideshare.net. Accessed March 2016. http://www.slideshare.net/FGInstitute/urbanization-china-vs-india.

Slow Food USA.org. Accessed March 2016. https://www.slowfoodusa.org/.

Sofa.aarome.org. Accessed March 2016. http://sofa.aarome.org/andrew-kranis.

Spread.com. Accessed July 2016. http://spread.co.jp/en/.

Stoner, R. 1983. Growing in Air. *Greenhouse Grower*, Vol. 1.

Stoner, R.J. and Clawson, J.M. (1999–2000). Low-mass, Inflatable Aeroponic System for High Performance Food Production. NASA SBIR NAS10-00017.

Swain, D.L.T., Haugen, M., Singh, D., et al. 2014. The Extraordinary California Drought of 2013/2014: character, context, and the role of climate change. *Bulletin of the American Meteorological Society*, suppl Explaining Extreme Events of 2013, 95(9):S3–S7.

The Atlantic.com. Accessed March 2016. http://www.theatlantic.com/international/archive/2012/2009/in-the-future-your-champagne-will-come-from-england/262784.

The Economist.com. Accessed March 2016. http://www.economist.com/news/china/21640396-how-fix-chinese-cities-great-sprawl-china.

The Energy Collective.com. Accessed March 2016. http://www.theenergycollective.com/gsitty/233751/driving-food-prices.

The Plant.org. Accessed March 2016. http://plantchicago.org/.

Tillman, D. 1999. Global environmental impacts of agricultural expansion: The need for sustainable and efficient practices. *Proceeding of the National Academy of Sciences of the USA*, 96:5995–6000.

Today.com. Accessed March 2016. http://www.today.com/id/21153990/ns/today-today_tech/t/could-these-be-farms-future/#.VtiE21153992xiOrMo.

Tower Gardens.com. Accessed March 2016. http://www.towergarden.com/.

Tower Gardens.com. Accessed March 2016. http://www.towergarden.com/content/towergarden/en-us/about/future-growing/tim-blank.html.

Tree Hugger.com. Accessed March 2016. http://www.treehugger.com/sustainable-product-design/vertical-farms-for-london-are-lovely-green-eye-candy.html.

United States Department of Agriculture.USDA.gov. Accessed July 2016. http://www.usda.gov/oce/climate_change/effects_2012/effects_agriculture.htm.

United States Department of Agriculture. USDA.gov. Accessed March 2016. http://www.usda.gov/oce/climate_change/effects_2012/CC%2020and%2020Agriculture%2020Report%2020%2802-2004-2013%2029b.pdf.

United States Department of Agriculture. USDA.gov. Accessed March 2016. http://www.ers.usda.gov/topics/in-the-news/california-drought-farm-and-food-impacts/california-drought-food-prices-and-consumers.aspx.

United States Department of Agriculture. USDA.gov. Accessed March 2016. http://www.ers.usda.gov/data-products/ag-and-food-statistics-charting-the-essentials.aspx.

United States Food and Drug Administration. USFDA.gov. Accessed March 2016. http://www.fda.gov/Food/FoodborneIllnessContaminants/FoodborneIllnessesNeedToKnow/default.htm.

United States Geological Survey. USGS.Gov. Accessed March 2016. http://droughtmonitor.unl.edu/.

United States Geological Survey.gov. Accessed March 2016. http://water.usgs.gov/edu/wuir.html.

University of Arizona.edu. Accessed March 2016. http://cals.arizona.edu/ceac.

University of Arizona.edu. Accessed March 2016. https://ag.arizona.edu/ceac/sites/ag.arizona.edu.ceac/files/WorldGreenhouseStats.pdf.

University of Arizona.edu. Accessed March 2016. http://ag.arizona.edu/ceac/sites/ag.arizona.edu.ceac/files/jensen%20Taiwan%20World%20Review%20of%20CEA.pdf.

University of California, Davis Campus. UCDavis.edu. Accessed March 2106. http://aic.ucdavis.edu/publications/posters/wine_climate6.pdf.

Vertical Harvest.com. Accessed March 2016. http://verticalharvestjackson.com/.

von Sachs, J. 1886. *Lehrbuch der Botanik*. Leipzig, Germany: Verlag von Wilhelm Englemann: p. 916.

von Sachs, J. 1887. *Lectures on the Physiology of Plants*. Oxford: Clarendon Press: p. 836.

Watson, R.T. and Core Writing Team, 2001. *Climate Change 2001: Synthesis Report: Third Assessment Report of The Intergovernmental Panel on Climate Change*. Cambridge: Cambridge University Press: p. 396.

Wikipedia.com. Accessed March 2016. https://en.wikipedia.org/wiki/Aeroponics.

Wikipedia.com. Accessed March 2016. https://en.wikipedia.org/wiki/Hydroponics.

Work.ac. Accessed March 2016. http://work.ac/locavore-fantasia/.

World Food Programme.org. Accessed March 2016. https://www.wfp.org/content/2015-food-insecurity-and-climate-change-map.

World Health Organization.org. Accessed March 2016. http://www.who.int/globalchange/ecosystems/biodiversity/en/.

Yale University.edu. Accessed March, 2016. http://e360.yale.edu/feature/what_global_warming_may_mean_for_worlds_wine_industry/2478/.

Yilmaz, I.S.C. and Ozkan, B. 2005. Trukish greenhouse industry: Past, present, your-champagne-will-come-from-england/262784. *New Zealand Journal of Crop and Horticultural Sciences*, 33:233–240.

16 Soils and Waste Management in Urban India

Simi Mehta, Arjun Kumar, and Rattan Lal

CONTENTS

16.1 INTRODUCTION

Soil is a dynamic natural body at the interphase of the atmosphere and lithosphere, which performs several functions in the ecosystem including recycling of dead and decaying organic matter into plant nutrients, denaturing of pollutants and filtering of water, moderating climate and sequestering carbon, storing germplasm and enhancing biodiversity, and providing the basis of all terrestrial life (Lal 2016), along with air and water (USDA 2005). Thus, urban soil by definition includes materials that have been manipulated, disturbed, and distributed by activities of human beings in the process of urbanization (Craul 1991). Urban centers are concentrated on soils in watersheds that provide drinking water, food, waste utilization and recycling, and natural resources to communities. These are also located within cities in areas of recreation, community gardens, green belts of roads and highways, lawns, septic absorption fields, sediment basins, and so on.

One of the principal uses of land, with increasing mechanization and industrialization in India, has been appropriation toward urbanization. Urbanization has been understood to be the most significant economic transformation for any society. Cities of any nation determine the trends they pursue toward development. While cities have been offering better avenues of employment and hence decent standards of living since the British era, these "pull-factors" of migration became more prominent when agriculture in the hinterlands increasingly became an unprofitable profession since the late 1990s. This phenomenon has been preceded and simultaneously occurred through the drastic transformation of ecosystems. The major protagonist in all this has been human beings, who through their rapidly growing population have created demand pressures for their needs and civic amenities, at the expense of the agricultural lands.

The various facets of this phenomenon generate vibrant debates on present trends along with their challenges and prospects, and become especially true for a developing country like India. The total land area of India is about 330 million ha (Mha) and the total area under urban settlements (where around one-third of the population lives) was only around 8 Mha (around 2.4% of total) in 2001. Apart from the obvious increase in the population, the number of urban households has increased due to the addition of new urban areas (Census Towns) and densification since Census 2001 (Kumar 2015a). Strikingly, the urban population of 377 million in 2010 is predicted to be 590 million by 2030 (Sankhe et al. 2010). One of the attributes of urban areas is waste generation that is beyond the assimilative capacity of the local ecosystem, and hence requires careful management. Municipal solid waste (MSW), which gets generated primarily by households and commercial establishments, is a heterogeneous mix of combustibles, organic matter, inert materials, and moisture. The urban solid waste generation in 1995 was 41 Tg (Teragram = 10^{12} g = million metric ton)/year (Hoornweg and Thomas 1999), which increased to 62 Tg/year in 2011, as revealed by the data of the Census of India, 2011, and is expected to increase by nearly five times to 260 Tg/year by 2047 (Infrastructure Development Finance Company Limited 2006). For judicious management of the waste, India must invest $1.1 trillion including $100 billion for proper sewage treatment (Sankhe et al. 2010)—an important addendum to managing soil and human health in urban ecosystems.

Accompanying this shift in residence of human beings is their adverse impacts on environmental resources like air, water, soils, and ecoregions. Being at the frontline of land use conversion, soils are subjected to a number of pollutants by the process of urbanization (Facchinelli et al. 2001). The encroachment of cities on agricultural lands and sensitive ecosystems poses a formidable challenge as they industrialize and introduce pollutants into the environment. With an increase in the quantity of waste production, and its open disposal, there are overflowing landfills, drainage of surface water becomes disrupted as the small natural channels and low-lying areas become filled with municipal waste (Roychoudhury 2014). Thus, the levels of soil pollution in urban areas warrant urgent action. A major reason for the urgency is that most cities depend upon river waters and groundwater to meet their water needs and, ironically, also as the disposal yards for most of their industrial effluents and untreated solid residues.

There has also been an inevitable increase in waste production. For instance, the amount of waste generated in New Delhi will increase from 10,000 Mg (Megagram = 10^6 g = metric ton) per day in 2011 to 18,000 Mg/day by 2021 (Kumar and Mehta 2014). Statistics from the Ministry of Environment, Forest, and Climate Change (MoEF&CC) revealed that the per capita generation of waste in Indian cities is about 200–600 g/day (Press Information Bureau 2016a). Thus, Indian cities presently generate MSW at the rate of 0.15 Tg/day (India Energy Security Scenarios 2015).

River Ganges, which supports >600 million lives over its stretch of 2,500 km, like all other rivers, is plagued with the ill-effects of urbanization. Prominent cities situated along its banks are Haridwar, Varanasi, Allahabad, Kanpur, Patna, and Kolkata. Rising population, coupled with inadequately planned urbanization and industrialization, has had severe effects on water recharge capacity and quality, soil salinity, and urban infrastructures in the basin of river Ganges (Misra 2011).

Its waters are polluted by the incessant deluge of sewage, by the large quantities of solid and industrial waste, and exacerbated by human and economic activities along its banks (World Bank 2015). Drainage of surface water has been disrupted as the small natural channels and low-lying areas have been clogged, often with municipal waste.

While heavy metals naturally occur in the earth's crust and are essential for marine life, anthropogenic activities like rapid industrialization and poor management of industrial effluent, mining, smeltering, combustion, and so on have exacerbated heavy metal pollution of urban soils. For instance, in many cities in India (e.g., New Delhi, Patancheru industrial development area of Andhra Pradesh, Lucknow, Kanpur, Agra, and Surat), soils are contaminated by the excessive presence of heavy metals including iron, barium, copper, zinc, arsenic, and others (Govil et al. 2001; Singh et al. 2002). Soil contamination has further reduced the fertility of urban soils and threatened the food chain thereby seriously compromising human health as well as the environment (Kumar 2011; Sharma et al. 2009), along with negative effects on soil microorganisms and crops (Rajindiran et al. 2015).

Hence, it is not a surprise that the erstwhile Planning Commission of India in its report[*] stated that the pace of urbanization presented unprecedented managerial and policy challenges, which if continued unabated would impose massive stress upon the ecosystem. This fosters a "messy and hidden" urbanization, constraining the prospects of economic prosperity and sustainable development (Ellis and Roberts 2016). Therefore, in order to overcome potentially stressful conditions of urban soils, it is important to objectively and effectively address the challenges posed by solid wastes in the urban areas of India.

16.2 MUNICIPAL SOLID WASTE MANAGEMENT: POLICY FRAMEWORK

The massive scale of urbanization in the country has enhanced the demand for solid waste services. As per the Constitution of India, solid waste management (SWM) is a state responsibility. MSW is defined to include refuse from households, nonhazardous solid waste from industrial, commercial and institutional establishments (including hospitals), market waste, yard waste, and street sweepings. Municipal solid waste management (MSWM) encompasses the functions of collection, transfer, treatment, recycling, resource recovery, and disposal of MSW. Failure to properly carry out these functions results in serious local, regional, and global public and environmental health problems, like air pollution, soil and groundwater contamination, and greenhouse (GHG) emissions. The United Nations Framework Convention on Climate Change reported estimates of total emission of GHGs (carbon dioxide equivalent as CO_2 eq. in India during the year 2000) at 10.3 Tg by solid waste disposal on land or unmanaged waste disposal sites, 42.3 Tg by wastewater handling as domestic and commercial wastewater, and 52.6 Tg by waste incineration and other systems.

With growing population in the urban areas, significant increases in waste generation have raised both the demand for waste collection and the urgency of finding innovative treatment options. The urban local bodies (ULBs) are responsible for the tasks of solid waste service delivery, with their own staff, equipment, and funds. In most cases now, part of the work is contracted out to private enterprises.

The most preferred hierarchy of MSWM involves waste avoidance (like refusing polybags for shopping), reduction in the use of unnecessary packaging, using reusable and durable products, recycling degraded or worn-out products, and recovering benefits from waste (e.g., energy recovery through incineration, etc.) ensuring zero waste and circular economy. Treatment and disposal are among the less preferred methods of MSWM because all safeguards must be in place to ensure that presence of MSW does not pose environmental risks, pest problems, and social, health, and safety issues.

The three main MSW treatment options are landfilling, composting, and incineration. Each of these is discussed in the following sections.

---

[*] Available at: http://12thplan.gov.in/12fyp_docs/17.pdf

16.2.1 Landfills

Disposal of waste in a landfill involves burying the waste, and this remains a common practice in urban India. Landfills and garbage dumps are examples of intentional burial sites. Construction debris, garbage, and other items are often placed in dumps or low-lying areas and then covered by soil or dredge materials. Operators of most landfills are required to cover each layer of garbage with a layer of soil on a daily basis. Landfills have often been established in abandoned or unused quarries, mining voids, or burrow pits. A properly designed and well-managed landfill can be a hygienic and relatively inexpensive method of disposing of waste materials. Older, poorly designed, or poorly managed landfills can create a number of adverse environmental impacts such as wind-blown litter, attraction of vermin, and generation of liquid leachate. Another common byproduct of landfills is gas (mostly composed of methane—CH_4, carbon dioxide—CO_2, and nitrous oxide—N_2O), which is produced as organic waste breaks down anaerobically. These are GHGs and contribute to global warming, besides creating the nuisance of bad odor. Many landfills have landfill gas extraction systems installed to extract the landfill gas. Gas is pumped out of the landfill using perforated pipes and flared off or burnt in a gas engine to generate electricity.

Problems concerning landfills are CO_2 and CH_4 emission, leachate contamination of underground water, aesthetic degradation from increased dust, litter, noise, vermin, increased traffic congestion, and property devaluation.

16.2.2 Composting

Composting is a form of recycling and saves a lot of landfill sites (Press Information Bureau 2016b). It involves the bacterial conversion of the organic substances present in MSW in the presence of air under hot and moist conditions. The final product obtained after bacterial activity is called compost (humus), which has high agricultural value when added to the soil as an amendment, as it increases the organic matter and key nutrients in soils. Soil organic matter (SOM) interacts with trace metals, and reduces the levels of their toxicity to plants, and also improves soil structure, increases its water-holding capacity, enhances soil aggregation, and reduces the risks of soil erosion. The application of MSW compost (which is free of odor and pathogens) to agricultural soil as a fertilizer can be a method to return organic matter to agricultural soil and help in the reduction of the cost of MSW disposal. Composting methods may use either manual or mechanical processes and waste volume is reduced by 50%–85%. Application of compost on soils of low nutrient concentration (e.g., nitrogen) can increase productivity (US EPA 2006). It was estimated that by 2010, India had the potential to produce over 6 Tg of compost per year worth about US$600 million (Asian Development Bank [ADB] 2011). Alternatively, if the entire 27.3 Tg of organic waste were used to produce electricity from biogas, the potential was estimated at 1,365 million m^3 of biogas and 2,730 million kilowatt-hours (kWh)/year of electricity (ADB 2011).

Some other benefits of composting include the presence of both macro- and micronutrients; gradual or slow release of nutrients, improved soil structure and aggregation, favorable air, and moisture regimes; better root growth and proliferation, and better drainage and aeration; low soil-erodibility; increase in soil biota that support healthy plant growth; and increase in soil nutrient-holding capacity, thereby reducing the need for commercial fertilizers, and so on (Haering and Evanylo 2005).

However, the key issue for promoting the use of compost in agricultural soils is the methods of production, which must ensure that the compost is of suitable quality for safe and beneficial reuse and reliable performance in agricultural production (ADB 2011).

16.2.3 Incineration

Incineration is a waste disposal method that involves combustion of waste material and is sometimes described as "thermal treatment." Incinerators convert waste materials into heat, gas, steam, and ash. It is used to dispose of solid, liquid, and gaseous waste. It is recognized as a practical method of disposing of certain hazardous waste materials (such as biological and medical waste). Nevertheless, controversies surround over method because of emission of gaseous pollutants.

16.2.4 Policy Implications of MSWM

16.2.4.1 MSW (Management and Handling) Rules, 2000

These rules were set forth by the Ministry of Environment and Forests under the Environment (Protection) Act of 1986. These were based on the study of the previous recommendations of the various committees and subcommittees, namely, the Committee on Urban Waste 1972, the Sivaraman Committee 1975, the Bajaj Committee 1995, and the Burman Committee 1999. The most important points that these committees highlighted included the various aspects of collecting and transportating solid wastes, setting up composting plants, and improving SWM practices through cost-effective technologies, among others.

These rules were considered to be the first comprehensive guidelines on SWM that designated the Central Pollution Control Board (CPCB) as the nodal agency to monitor its implementation in the union territories (UTs) and the State Pollution Control Boards in the states. The rules gave municipal authorities responsibility for the implementation of the provisions and for development of infrastructure for collection, storage, segregation, transportation, processing, and disposal of MSWs. It delineated the various parameters of waste management under four "schedules," ranging from authorization of grants for establishing waste processing and disposal facilities including landfills to rules for their implementation, including collection, segregation, storage, transportation, processing, and disposal of MSW. While the various procedures of MSW Rules 2000 were holistic, it failed to achieve the objectives of following the "norms" it set for the municipal bodies, as a result of which the environmental impact, social acceptance, and functional sustainability severely undermined pollution prevention goals.

16.2.4.2 SWM Rules, 2016

Under this new set of rules, the MoEF&CC has greatly expanded the scope of the coverage beyond municipal areas, extending to urban agglomerations, census towns, notified industrial townships, areas under the control of Indian Railways, airports, airbase, port and harbor, defense establishments, special economic zones, State and Central government organizations, pilgrim locations, and religious and areas of historical importance (Gazette of India 2016).

Some of the major highlights of the new SWM Rules, 2016 include (1) segregating at source by households and institutions to channelize the "waste to wealth" by recovery, reuse, and recycling, (2) handing over recyclable material either to authorized waste-pickers and recyclers or to the ULBs, (3) implementing the "polluter pays principle" with strict enforcement, and (4) integrating waste-pickers into the formal system through State Governments and Self-Help Groups. The new rules have empowered local authorities across India to decide user fees. Municipal authorities will levy user fees for collection, disposal, and processing from bulk generators.

The new rules provide for the processing, treatment, and disposal of biodegradable waste through composting or biomethanation within the premises as far as possible and the residual waste shall be given to the waste collectors or agency as directed by the local authority. The developers of special economic zones, industrial estates, and industrial parks have been required to earmark at least 5% of the total area of the plot for a recovery and recycling facility.

The MSW Rules stipulate that waste processing facilities will have to be set up by all local bodies having a population of 1 million or more within 2 years. For census towns with a population below 1 million or for all local bodies having a population of 0.5 million or more, common or stand-alone sanitary landfills will have to be set up within 3 years. Also, common or regional sanitary landfills are required to be set up by all local bodies and census towns with a population under 0.5 million in 3 years (Press Information Bureau 2016a,b).

Also, the rules have mandated investigation and analyses of all old open dumpsites and existing operational dumpsites for their potential for biomining and bioremediation. Horticulture waste and garden waste generated from its premises should be disposed as per the directions of local authority. The use of compost has been encouraged with plans to set up laboratories to test the quality of compost produced by local authorities or their authorized agencies. MSW Rules 2016 has set up a Central Monitoring Committee under the chairmanship of the Secretary, MoEF&CC to annually monitor the overall implementation of the rules (Sambyal 2016).

16.2.4.3   Swachh Bharat (Clean India) Mission—Urban 2014

Launched on Gandhi Jayanti (October 2), 2014, one of the primary objectives of SBM-Urban is to achieve 100% collection and scientific processing/disposal reuse/recycle of MSW and liquid waste, for which the ULBs have been entrusted with the task to prepare detailed project reports having a viable financial model for management of solid and liquid waste of their respective cities with the aid and advice of the state governments. It also opened the scope of smaller cities to jointly become clusters to attract private investment, and also incorporated street sweeping and litter control interventions as being essential for a clean city, with the liberty of the states to choose the technology for SWM projects and street sweeping. The funding for the SBM-Urban SWM projects would be in the ratio of 75:25 (for union government and state government, respectively) and 10% in the case of North East States and special category states (Government of India 2014).

Under the Swachh Bharat Mission (SBM), 80% of the urban population (allowing for a 2% increase year-on-year) has to be covered for SWM services. It is therefore imperative that environmentally sound, socially, and techno-economically viable SWM practices are adopted to ensure sustainability in the long run. The concept of partnership in Swachh Bharat has been introduced in the MSW Rules 2016. Bulk and institutional generators, market associations, event organizers, and hotels and restaurants have been made directly responsible for segregation and sorting of waste and management in partnership with local bodies. All hotels and restaurants should segregate biodegradable waste and set up a system of collection or follow the system of collection set up by the local body to ensure that such food waste is utilized for composting/biomethanation (Gazette of India 2016).

16.2.4.4   Waste to Energy 2015

It cannot be denied that waste management is a key priority for the government and hence effective waste management needs to incorporate the feasibility of turning waste into an energy resource. Waste management provides an opportunity to tap potential energy to meet part of India's energy demands, simultaneously addressing environmental concerns. As a corollary to the objectives of SBM, the proposed waste-to-energy program is directed toward effective waste management, with energy recovery procedures employed after the maximum possible reuse and recycling efforts. According to the Ministry of New and Renewable Energy (MNRE), there exists a power potential of about 1,700 MW from urban waste (1,500 from solid waste and 225 MW from sewage) and about 1,300 MW from industrial waste (Chinwan and Pant 2014; Business Standard 2015).

Given that MSW is a heterogeneous mix of combustibles, organic matter, inerts, and moisture, energy generation through biochemical conversion or combustion depends on the levels of segregation and collection efficiency of MSW. This is a key focus area of Ministry of Urban Development (MoUD) as well as ULBs across the country and hence it is assumed that by 2047, approximately 30% of total "waste to electricity generation" potential would be realized, resulting in the generation of around 5,850 MW power. As much as 63% of segregated urban combustibles would be used as fuel, which would yield 0.012 Mt OE (million ton of oil equivalent) of thermal energy (India Energy Security Scenarios 2015). It must be ensured that waste to energy (W to E) prioritizes biomethanation over incineration, given the latter's potential to pollute the atmosphere. Biogas production from waste in landfills would be helpful.

This program also implies that the energy thus generated from waste would be eligible for government subsidies as are available to other renewable sources of power.

According to the report of the Task Force on W to E of the erstwhile Planning Commission of India, untapped waste has a potential of generating 439 MW of power from 32,890 Mg/day of combustible wastes including refuse-derived fuel (RDF), 1.3 million m^3 of biogas per day or 72 MW of electricity from biogas, and 5.4 Tg of compost annually to support agriculture (Planning Commission 2014).

Strategies like zero waste[†] should be adopted that could eliminate the need for landfills in the long run by implementing a cradle-to-cradle approach, that is, to segregate waste at the source, which enables scientific processing and recycling of waste. India's adoption of zero waste as a strategy for SWM has largely been at the local level, driven by local governments such as Pune and Mysore, at the state-government level by states such as Gujarat, or by nongovernment organizations (NGOs) such as Chintan and others. The city of Pune has efficiently reduced the amount of waste going to landfills from 10 Mg/day to 2 Mg/day (MoUD 2015). Zero waste strategies would also look at an existing landfill as a source of materials and energy and try to ensure that as much as is environmentally, financially, and technically feasible, the resources are mined from the landfill before it is capped and made redundant. Existing waste in landfills needs to be secured in an environmentally safe manner (Center for Policy Research 2016).

The SWM Rules, 2016 mandated all industrial units using fuel and located within 100 km from a solid waste-based RDF plant to make arrangements by October 2016 to replace at least 5% of their fuel requirement with RDF.

The rules also direct that nonrecyclable waste having energy value of 1,500 Kcal/kg or more shall be utilized for generating energy either through RDF not disposed of in landfills and be utilized for the generation of electrical energy either or through RDF or by using as feed stock for preparing RDF. High calorific wastes shall be used for coprocessing in cement or thermal power plants.

As per the rules, MNRE sources should facilitate infrastructure creation for W to E plants and provide appropriate subsidies or incentives for such W to E plants (TERI 2015). The Ministry of Power would fix tariffs or charges for the power generated from the W to E plants based on solid waste and ensure compulsory purchase of power generated from such W to E plants by discoms (Sambyal 2016). According to the latest data available from the MoUD, current W to E production is 88.4 MW as against the planned production of 493.7 MW. Further, the waste to compost production as of September 2016 was 0.065 Tg.

16.2.4.5 Sustainable Development Goals of the UN and Their Agenda for Sustainable Soil and Waste Management

With the aim to transform the world for the better and building upon the legacy of the Millennium Development Goals (MDGs), the UN General Assembly (UNGA) adopted the Sustainable Development Goals (SDGs) in September 2015, also called Agenda 2030. While there are 17 Goals for action in priorities that are of "critical importance for humanity and the planet," those having direct and overlapping consequences upon the sustainable soil and waste management are discussed here.

To decisively address climate change and environmental degradation, conserve biodiversity, tackle water scarcity and pollution and soil degradation, and to reduce the negative impacts of urban activities and of hazardous chemicals through the environmentally sound management and safe use of chemicals, and the reduction and recycling of waste in order to ensure sustainable urban development and management for enhancing the quality of life of the people, the UNGA member-states reaffirmed that "Mother Earth" and her ecosystems are a common home, and it is for its maintenance that the roadmap to sustainable development was outlined (UNGA 2015).

Several SDGs that have direct and indirect relationships to sustainable soil and waste management and inclusive and sustainable urbanization are depicted in Table 16.1 (UNGA 2015).

[†] The term Zero-Waste was coined by chemist Paul Palmer in 1973, in Oakland, California.

TABLE 16.1

The Sustainable Development Goals of the UN in the Context of Rapid Urbanization in India

Goal 3	Target	Ensure healthy lives and promote well-being for all at all ages
	3.9	By 2030, substantially reduce the number of deaths and illnesses from hazardous chemicals and air, water, and soil pollution and contamination
Goal 6		**Ensure availability and sustainable management of water and sanitation for all**
	6.2	By 2030, achieve access to adequate and equitable sanitation and hygiene for all and end open defecation, paying special attention to the needs of women and girls and those in vulnerable situations
	6.3	By 2030, improve water quality by reducing pollution, eliminating dumping and minimizing release of hazardous chemicals and materials, halving the proportion of untreated wastewater and substantially increasing recycling and safe reuse globally
	6.3a	By 2030, expand international cooperation and capacity-building support to developing countries in water- and sanitation-related activities and programs, including wastewater treatment, recycling and reuse technologies
	6.3b	Support and strengthen the participation of local communities in improving water and sanitation management
Goal 11		**Make cities and human settlements inclusive, safe, resilient, and sustainable**
	11.1	By 2030, ensure access for all to adequate, safe, and affordable housing and basic services and upgrade slums
	11.3	By 2030, enhance inclusive and sustainable urbanization
	11.6	By 2030, reduce the adverse per capita environmental impact of cities, including by paying special attention to air quality and municipal and other waste management
Goal 12		**Ensure sustainable consumption and production patterns**
	12.4	By 2020, achieve the environmentally sound management of chemicals and all wastes throughout their life cycle, in accordance with agreed international frameworks, and significantly reduce their release to air, water, and soil in order to minimize their adverse impacts on human health and the environment
	12.5	By 2030, substantially reduce waste generation through prevention, reduction, recycling, and reuse
Goal 14		**Conserve and sustainably use the oceans, seas, and marine resources for sustainable development**
	14.1	By 2025, prevent and significantly reduce marine pollution of all kinds, in particular from land-based activities, including marine debris and nutrient pollution
Goal 15		**Protect, restore, and promote sustainable use of terrestrial ecosystems, sustainably manage forests, combat desertification, and halt and reverse land degradation and halt biodiversity loss**
	15.3	By 2030, combat desertification, restore degraded land and soil, including land affected by desertification, drought and floods, and strive to achieve a land degradation-neutral world

Source: Adapted from UNGA 2015. Resolution adopted by the General Assembly on 25 September 2015, A/RES/70/1 Transforming our world: The 2030 Agenda for Sustainable Development. New York.

Addressing the UN Summit on the adoption of SDGs, the Prime Minister of India Narendra Modi was appreciative of the lofty but comprehensive goals set forth by the UNGA, and reiterated that much of India's development agenda were mirrored in the SDGs (NITI Aayog 2015).

It is widely underscored that the SDGs provide promising prospects for India to renew and integrate its domestic efforts in order to meet the objectives within the prescribed time frame. There could, however, be some challenges posed by the demand for financing the commitments made at the UN. A study by the Technology and Action for Rural Advancement group estimated that implementing SDGs in India by 2030 would cost around US$14.4 billion, and that for realizing

TABLE 16.2
The Financial Implication of Advancing the Sustainable Development Goals of the UN

Goal	Money Required (in Billions of USD)
Goal 3	880
Goal 6	199 + 123 (for universal sanitation coverage in the country and cleaning of the Ganga River) = 322
Goal 11	2.067
Goal 12	1.344
Goals 14 and 15	489
Total funds required	5.102

Note: These funds are required over the period 2015–2030.

the goals listed in Table 16.1 it would need a total of $5.102 billion. Table 16.2 is a tabloid presentation of the amount of funds that would be required by India (Kapur 2015; Technology and Action for Rural Advancement 2015).

In essence, it is important to understand that while the SDGs provide expansive goals and targets, it remains at the disposal of the union and the state governments to identify the priorities and decide locally suitable relevant policies, in order to harness innovation and ensure that a proper implementation and monitoring plan is executed (Kapur 2015).

16.3 LINKAGES BETWEEN SOIL HEALTH AND WASTE MANAGEMENT IN URBAN INDIA

Estimates of the Government of India to feed an expected total population of 1.4 billion by 2025 would need about 300 million Mg (Tg) of food grains by 2025 as against the current 257 Tg being produced at present. For achieving this target, around 35 Tg of chemical fertilizers and about 10 Tg of organic manure and biofertilizers would be needed per year to provide the correct ratio of nitrogen, phosphorus, and potassium (NPK) nutrients (Government of India 2005). Thus, there is an enormous potential for increased production of compost from municipal organic waste that would serve as an important soil amendment in order to support better soil health, agricultural productivity, and ensure food security (UNEP 2005).

16.3.1 MANAGING SOILS OF URBAN ECOSYSTEMS

Having understood that India is urbanizing at a very fast pace, and so is the consequent generation of solid and liquid wastes, this section describes how the soils in the urban ecosystems are increasingly been degraded, running the risk of increased challenges for the urban poor in the short term and for the affluent in the medium to longer term. The massive scale of urbanization in the country has enhanced the demand for solid waste services. It is argued that in all circumstances, if all anthropogenic activities are continued unabated in the same manner, India would be far from achieving the SDGs by 2030.

Human beings use land for reasons ranging from habitation, vegetation, setting up factories and industries, recreation, augmenting the forest cover and clearing the forests for each of the previous activities, and several others. It is from soil that these temporary or permanent activities are appropriated. Thus, with soils being the basis from which humans appropriate ecosystem services, land use under urban ecosystems has increased manifold, primarily to accommodate the endogenous needs of the growing population as well as to cater to the demands posed by increasing rural to urban migration.

On the one hand, for India to be at par with developed countries, urbanization is a prerequisite, as is also an important constituent to strengthen its economy. Yet on the other hand, one must not shy away with the problems that this phenomenon has brought about and the consequences that unregulated urban expansion could bring in the future, especially in unplanned, unauthorized, informal, squatter, and slum settlements.

Land use in urban areas adversely impacts upon soils in the environment and increases their erodibility and exacerbates the risks of degradation by a range of processes including contamination and pollution. For instance, there is a sharp rise in the amount of soil erosion during and after construction activities in cities. Sediments are either blown or washed away with air and rainfall or deposited on any low-lying area. Intensified construction leading to concretization of any particular land induces soil denudation and desertification. Further, increase in the rate of mining of sand and gravel for construction activities mostly through illegal or no sanctions and irresponsible disposal of construction waste has intervened with the environment and degraded soil health. Continued human activities to disturb the sensitive soil for their needs is hostile to the natural strength of the soil, which increases the possibilities of landslides and other natural disasters. This is more conspicuous in areas of high densities of people and supporting structures such as roads, homes, and buildings (Einstein 1999). Continued land use in the urban areas has also increased the magnitude and frequency of floods and has become a cause of concern for the low-lying areas amid the phenomenon of climate change. The observed and projected warming of the planet Earth along with simultaneously expansion of urbanization for human use has increased the likelihood of flash floods and inundation.

This trend of flash floods was imminent in the summer of 2013, when cataclysmic cloud bursts in the state of Uttarakhand in India led to massive landslides not only in the pilgrimages of Kedarnath, but also in the towns of Srinagar, Rishikesh, and Haridwar, causing large-scale loss of lives of humans and livestock, destruction of habitat, and overflooding of the river Ganges (Rao 2013; Sampath 2013). The insensitivity with which places of tourism were developed with a total disregard for the ecology, rampant and unauthorized construction of hotels off the mountains in close proximity to the river, the completed and ongoing hydro-electric projects bypassing environmental clearances, faulty engineering techniques to build them, the eco-insensitivity of politics that allowed it, and polluting the water by disposal of dead and innumerable solid wastes all added to the misery of the people and caused an unbearable strain on the Himalayan ecosystems (Rahman 2013; Sampath 2013). The natural habitat of living organisms and species have also been threatened as a result of failing to understand the fragility of ecology and focusing on uncontrolled development of the region (Grumbine and Pandit 2013).

The previous section detailed the various policies and programs of the governance of the MSWM, but the majority of the objectives remain unfulfilled. The several challenges to the smooth functioning of waste management that can be identified are as follows:

- The lackadaisical approach of governments, municipal authorities, and government health centers to implement the policies and indulgence in blame games each time a crisis occurs have put citizens' lives at risk.
- The weak performance in source collection, segregation, and disposal of waste throughout the region are the primary constraints to the sector.
- An unequal playing field for organic products to compete with highly subsidized chemical fertilizers and nonrenewable energy sources.
- Municipalities' poor financial standing to invest in waste recycling.
- Low municipal capacity to operate and maintain such facilities and to engage private sector partners.
- Low community awareness and participation.

- Coordination deficit between stakeholders such as governments at all levels, parastatals, various development boards and authorities, cooperatives, civil society, and private and informal sectors.
- Weak regulatory and enforcement systems, coupled with poor monitoring capabilities to ensure high standards of compost.

Figure 16.1 exemplifies the ways in which rapid and unplanned urbanization in India continues to impact the quality of soils and water by the mismanagement of MSW and sewage disposal.

While the above nexus of soil denudation, soil degradation as a result of deforestation for urbanization has been a consistent phenomenon in the developed countries, with elaborate mechanisms to restore the soil health and achieve the balance, India has been unable to successfully implement the restoration processes. Added to this has been the disposal of municipal and industrial solid and liquid wastes directly into the water bodies leading to water pollution and accumulation of wastes therein (Sharholy et al. 2008). These activities result in a gradual increase in the levels of the river beds, and even a slightly heavy downpour leads to overflowing of the contaminated water, most of the time entering unplanned settlements. As the river water becomes extremely unfit for consumption, its continued use for drinking, cooking, washing, bathing, and so on by the slum dwellers compromises their health and general well-being. Research as well as studies from the World Health Organization (WHO) have demonstrated that accompanied by lack of basic amenities the normalcy of life is uprooted when these flood or flood-like situations and gradual recession in the water levels thereafter lead to a spread of contagious diseases, especially vector-borne diseases like malaria, dengue fever, and diarrhea (Chandra et al. 2010).

Absence of proper sanitation and sewage facilities in India, where around 10 million urban households practiced open defecation as per the Census of India 2011, directs drain water, wastewater, and fecal sludge to the surface, which during the rainy season diffuses to the surface water and other aquifers, creating conditions for the growth of microbial pathogens, and inevitably giving way to the rise of numerous viral and bacterial diseases. Disposal of industrial hazardous waste along with municipal waste exposes people to chemical and radioactive hazards, directly

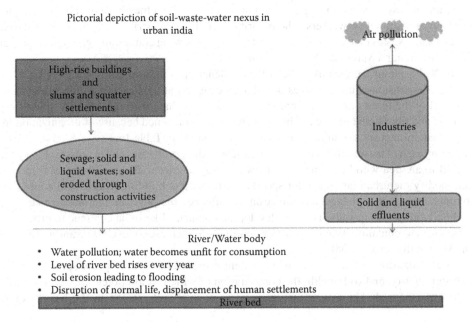

FIGURE 16.1 Causes of degradation by erosion and contamination of soil, water, and air resources of urban ecosystems in India.

affecting the health of those living in the slums, especially children, rag pickers, and those working in facilities producing toxic and infectious material (Raman and Narayanan 2008).

There have been numerous studies that have highlighted the needs for housing, especially for the lower strata of the urban population. The fact that more than 65.5 million people live in urban slums and sprawls as per the Census of India 2011 brings to the fore several challenges for the people, the communities in which they live, and the wider ecology. The latest Census of India 2011 offers a glimpse of the situation regarding access to basic amenities and is explained below. Twenty-nine percent of the households did not have available drinking water facilities within their houses. In addition, around 62% of households received treated tap water, and 15 million of the 79 million households did not have latrine facilities within their premises. Only 32.7% of the households were connected to a piped sewer system and 38.2% to septic tanks. Only 45% of the households had closed drainage in their houses for wastewater outlet and 37.3% and 18.2% had open and no drainage connectivity, respectively.[‡]

The glaring example is New Delhi, a megacity of the world, home to over 17 million people, reporting malaria, dengue, and chikungunya deaths within the first 15 days of September 2016, becomes significant as it underscores the shortcomings associated with a shortsighted and unplanned approach toward urbanization. En masse spread of fevers among people of all age groups with symptoms of these diseases, especially among those residing in the outskirts, slums, and in unauthorized colonies of Delhi, adversely affected their livelihood, with most of ill being daily wage earners (The Hindu 2016).

Further, the increase in the construction activities in the urban spaces generates sediments and debris, the improper disposal of which blocks the drains with silt and other phosphorus and nitrogen carrying materials which contaminate the soil and the water bodies wherever these are eventually deposited (UNEP 2005). The lives around the water bodies are affected by their contamination, with phenomenon like algal blooms on the rise.

These recent cases raise wider concerns about the inadequacy and inefficacy of the various welfare schemes of the state and union governments. Added to this unfortunate situation is the power politics at the level of the ULBs, and their unwillingness to act in times of such adversities (Press Trust of India 2015; Singh 2015). Citing their own resource crunch and passing on the blame to the state and union governments, there have been instances in which they have refrained from paying the salaries to the sanitation workers, who in turn resorted to mass strikes, which in turn led to open disposal of solid wastes on the roads, pavements, and overflowing community garbage bins (if any!) (NDTV 2016; Indian Express 2016; Millennium Post 2016).

Figure 16.2 explains the vicious circle of the challenges posed by urbanization in India.

The rapid expansion of urban spaces in India is occurring at the expense of the invaluable soils that are part of the forests or agricultural lands. These soils that otherwise are best suited for providing food and fiber to the people are being increasingly consumed because of urbanization in the country. Predominant soils of urban ecosystems are depicted in Table 16.3.

Managing each of these soils for different uses affects their properties and processes. For instance, adding fill to an area with huge quantities of waste may aggravate the undesirable soil properties or even modify the urban landscape for specific purposes like building recreational zones, lawns, arboretums, and so on. Urbanization can strongly influence the physical and biochemical properties and pollution levels in soils. Thus, as developing countries like India continue to urbanize and industrialize, soil contamination in the cities continues to increase to levels warranting immediate action (Marcotullio et al. 2008).

Rapid urbanization of the Ganga basin has demonstrated harmful impacts upon the environment, water quality, and soil health, the continuation of which would potentially create a drastic environmental hazard. There is an urgent need to control soil erosion in the basin, conserve

---

[‡] Available at: Houses, Household Amenities and Assets, Figures at a Glance, Census of India, 2011. URL: http://censusindia.gov.in/2011census/hlo/hlo_highlights.html

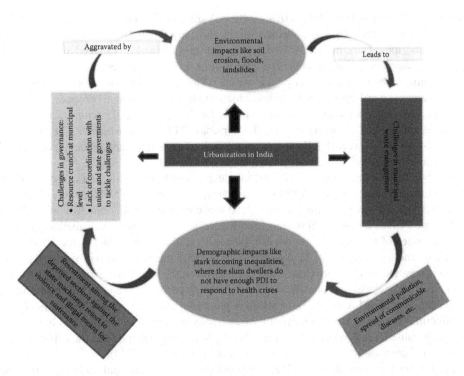

FIGURE 16.2 Environmental ramifications of urbanization and the challenges to effective governance.

TABLE 16.3
Two Predominant Types of Soils in Urban Ecosystems

Natural Soils	Anthropogenic Soils
The relatively undisturbed soils which are formed by materials naturally mixed or transformed by the action of water, wind, gravity, and so on.	Those soils are formed by human activity, such as addition of human-deposited fill. These may include natural soils moved around by humans, construction debris, coal ash,
These provide living spaces and supply air, water, and nutrients for microorganisms, plants, animals, and human beings.	municipal solid waste, materials collected from the waterways, and so on. These soils are called "Anthrosols."

valuable water and nutrients, minimize pollution and contamination, and improve and restore productivity and environment quality of both urban and the surrounding ecosystems. These methods would contribute in the upkeep of soil health and also enhance farm productivity (Misra 2011).

16.4 EMPIRICAL EVIDENCE—RECENT TRENDS AND PATTERNS

Per capita waste generation of MSW ranged between 0.2 kg and 0.6 kg per day in Indian cities amounting to about 0.115 Tg of waste per day and 42 Tg/year in 1995. Also, as a city expands, average per capita waste generation increases (IDFC 2006). The Energy and Resources Institute (TERI) has estimated that waste generation will exceed 260 Tg/year by 2047—more than five times the 1995 level. Cities with 100,000 plus population contribute 72.5% of the waste generated in the country as compared to the other 3,955 urban centers that produced only 27.5% of the total waste in 2005 (IDFC 2006).

As per a CPCB report from 2012–2013, municipal areas in the country generated 0.134 Tg/day of MSW, of which 0.091 Tg was collected (68%) and only 0.026 Gt (28%) was treated by the municipal authorities. Thus, merely 19% of the total waste generated was treated and the remaining disposed of at dumpsites/landfill sites. The composition of MSW was organic waste (51%), inert and nonorganic waste (32%), and recyclable waste (17%). The MSW, therefore, dumped in low-lying urban areas is a whopping 0.108 Tg/day, which needs 0.213 million m^3 of space every day and 776 ha of precious land per year (Planning Commission 2014). In its 2015 report, it was estimated that waste generation in the country was 143,449 tons/day (TPD) out of which 32,871 TPD was processed. It also added a caveat that there has been a lack of systematic and authentic data on MSW generation at the national, state, and city/town levels (CPCB 2015).

As per the 2011 census, there are 377 million people living in 7,935 urban centers (with 4,041 statutory municipal authorities and 3,894 Census towns with more than 5,000 people of which 75% are males involved in nonagricultural activity) and this constitutes more than 30% of total population. There also exists a big-city bias, unacknowledged and exclusionary urbanization, disparities among class sizes of towns/cities, regional imbalances, and challenges facing the ULBs (Kumar 2015b). Urban India generates 0.17 Tg/day and 62 Tg of MSW per year which is based on an average per capita generation of 450 g per person per day. However, the 62 Tg of waste generation reported annually does not include wastes picked up by kabadiwalas from households and from the streets by rag pickers. As per MoUD, 70 Tg of waste is generated currently in urban centers. There are thus conflicting data about the quantum of waste actually generated in urban areas in the country, principally because there is no system of periodically collecting and updating the national database on quantity and composition of urban waste (Planning Commission 2014).

As per information available for 2012, compiled by CPCB, municipal authorities have so far only set up 279 compost plants, 172 biomethanation plants, 29 RDF plants, and 8 W to E plants in the country. The Task Force on Waste and Energy constituted by the Planning Commission in 2014 cited WHO's observation that around 22 types of diseases including dysentery, diarrhea, malaria, dengue, and others could be prevented/controlled in India by improving the MSWM system. Scientific management of MSW will save huge financial resources currently spent on medical services and the health of the young population. Principal reasons for the prevailing unhygienic conditions in Indian cities is the casual attitude of the citizens as well as the municipal authorities toward managing solid waste, lack of priority to this essential service, inadequate and inappropriate institutional structure, lack of technical know-how, and paucity of financial resources (Planning Commission 2014). Some notable properties of the MSW in India are given below:

1. **Composting:** In urban areas, the major fractions of MSW are (1) compostable materials (40%–60%) and (2) inert (30%–50%) components. In general, the percentage of recyclables (e.g., paper, glass, plastic, and metals) is rather low, because of rag pickers who segregate and collect the materials at generation sources, collection points, and disposal sites. Relatively more per capita waste is generated in bigger cities, and its composition changes because of decrease in compost and biomethanation waste (NIUA 2015).

2. **Processing:** In 2016, the total daily solid waste generated in urban India was 0.147 Tg, of which only 18% was processed. About 44% of wards (34,259 of total 78,633) had 100% door to door waste collection as per MoUD, SBM-Urban MIS database as of 31 December 2016 (NSSO 2016).

3. **Collecting:** Arrangements made for collecting garbage from household was by municipality/corporations—51.9%, by resident/group of residents—21.5%, others—2.4%, and no arrangement—24.2% during 2012 as per NSSO (NSSO 2013).

4. **Depositing:** The sites where garbage is deposited after removal from the household were 2.4% to biogas plant or manure pit, 50% to a community dumping spot, 9.7% to a household's individual dumping spot(s), 19.3% to others, and 18.6% unaccounted for, during 2012 as per NSSO.

5. **Disposing:** The frequency of garbage disposal/clearance from community dumping spots as reported by households was daily—57.8%, not daily but at least once in a week—30.2%, not even once in a week—5.6%, and not known—6.2%.

6. **Draining:** The drainage arrangements by urban households were 45.2%, 14.9%, 22.4%, 5.0%, and 12.5% for underground, covered cemented, open cemented, open uncemented, and no drainage, respectively.

7. **Treating:** The disposal of household wastewater was safe for reuse after treatment—0.0%; disposed of without treatment to open low land areas—15.9%, ponds—1.7%, nearby river—0.9%, drainage system—73.3%; and disposed of with or without treatment to other places—7.0%, not known—1.1% the during 2012 (NSSO 2013).

8. **Managing:** In case of urban SWM, around 70%–90% of the solid waste is collected and transported. But issues remain in the proper disposal, segregation, and treatment of the solid wastes. "Waste to Wealth" programs modelled on public–private partnerships (PPPs) have mixed results in different cities (Rajkot was successful while Lucknow was not) (MoUD 2011).

ADB (2011) reported that a large fraction of the total solid waste generated in the urban centers in India was biodegradable and had high moisture content. These statistics indicate the need for their proper disposal, in the absence of which this organic waste has the potential to emit harmful GHGs like methane and add to global warming, putting the public and environmental health at risk. Methods of waste treatment (e.g., composting and generating RDF) offer sustainable alternatives vis-à-vis the business-as-usual (BAU) methods of open dumping and unhygienic landfilling. These methods also result in some environmental and economic benefits (e.g., production of renewable energy and organic fertilizers). Soil amendments or conditioners increase the porosity of the soil and provide amicable environments for plants, maximizing the nutrient content available to them (United Nations Human Settlements Program [UN-HABITAT] 2010).

Managing compost and organic waste is also important to improving soil health and plant nutrition, biodegradation, and denaturing of toxins and pollutants in soil. Two key ways to manage waste are filtering and buffering. Waste is filtered when it flows through the soil and is slowly trapped and bound to soil particles. Soil buffering traps waste particles and transforms and denatures them. Composting and using septic systems are examples of waste management in urban soils. Organic substances (partially or completely decomposed organic materials and their microbial by-products) are needed to hold water and nutrients in the soil for plant growth. In community parks and urban gardens, composting can recycle most of the leaves and grass and other biomass.

Further, households, especially those in the slums and squatter settlements, can harness nitrogen and phosphorus for use by the community lawns or parks, thereby returning some nutrients to the soil, which if otherwise left unattended, may contaminate nearby water bodies leading to eutrophication and contamination.

16.5 RESEARCH NEEDS AND PRIORITIES

India needs to adopt a holistic approach toward the management of urban waste that would be asymptomatic to improving soil health and sustainable development. These strategies would have positive implications upon the well-being of all, including the environment. Toward this goal, urgent and concerted policy responses are needed as the human dimensions, ecological transformation, and physical impacts of urbanization are intricately tied to the soil waste environment nexus and its management. Holistic research for robust development strategies and comprehensive policy interventions are required to develop better land-use plans for rural and urban areas and manage the finite but essential natural and man-made resources with efficiency.

16.5.1 BUILDING INCLUSIVE AND SUSTAINABLE CITIES

In pursuance of this approach, which is also listed as one of the SDGs of the UN, inclusive and sustainable cities (ICS) would not only ensure human settlement (safe, affordable, and adequate) for all, but through participatory involvement of all groups of people, ICS would also reduce the per capita environmental impact of the urban built environment with a special focus on municipal waste management ensuring zero waste and a circular economy. Techniques like aerobic composting of waste like vegetables and food scraps, slaughterhouse waste, fallen leaves, and so on would turn these wastes into organic manure, which when applied to the soil would enhance their fertility and productivity. In addition, recovering energy from non-biodegradable waste through integrated waste management, waste treatment, and waste disposal would avoid landfill overflow and safeguard the health of the public while improving the environment.

16.5.2 FINANCIAL EMPOWERMENT OF ULBs

Empowerment of ULBs would improve city administration and enhance the efficacy of service delivery (Kumar 2015b). To enable ULBs to invest in awareness and execution, possibilities must be explored for implementing best practices and innovation in the waste sector, and allocating substantial financial resources for operations and management (Water and Sanitation Program-South Asia 2007). Some of these goals could be realized through private investments, international collaborations, strict implementation of the "polluter pays principle," and others (TERI 2015).

16.5.3 IMPLEMENTING BEST PRACTICES IN SOIL AND WASTE MANAGEMENT

Soil management strategies require holistic sustainable development policy frameworks wherein the task of disseminating knowledge about soils is significant. The challenges to soil management vis-à-vis human well-being have been discussed in the preceding sections. Drainage and sewerage facilities must be upgraded to avoid stagnation of water and blockage of drains especially during rainy seasons where epidemics may endanger the health of people and other biota. This is because urban runoff can carry polluted water from cities to streams and oceans, polluting the environments even outside of the city (Einstein 1999). Also, new facilities for sewage waste treatment should be created so that these are not allowed to be dumped in the rivers without primary treatment. The wastewater must be reused after proper treatment for agricultural and nonpotable purposes. This would reduce the demand on groundwater sources (Raychaudhuri et al. 2014). NGOs, the private sector, and other stakeholders must be encouraged to establish community-based segregation at the source, separate collection, and resource recovery from wastes with particular focus on composting (ADB 2011). New policies for sustainable development of water resources, conservation education, and pollution prevention should be formulated to provide effective and timely information on water management (Misra 2011). Storm water management is needed in highly developed areas to prevent flooding and emergency discharge of untreated sewage into rivers. Some former landfills could be converted into parks and gardens. A soil-based site management plan could include ways to create a favorable hydrological balance that would reduce surface runoff. Use of the composted biomass would enhance SOM and thus increase the supply of water and nutrients to plants. The attendant improvement in soil structure would enhance total and macroporosity and thus reduce the risks of compaction and storm runoff. The recent initiative of Soil Health Card Scheme by the Ministry of Agriculture, Government of India, launched in 2015, needs to be extended to other nonagriculture sectors in urban areas, and needs to maintain a real-time database that would suggest measures to restore and sustainably manage urban soils.

16.5.4 Adopting the Public–Private–People Partnership Approach

While the major investments for creating wealth out of waste would come through effective governance and political willpower, it would also require orchestration of the expertise of the private sector and most importantly the contribution of the well-informed citizenry, who would undertake unto themselves the task of efficiently managing the wastes from their houses and workplaces (NIUA 2015). Governments need to make a greater commitment to enhance the reach of community awareness programs in order to encourage segregation of waste at the source and to provide an enabling environment and timely incentive and resources (TERI 2015; Kumar 2015b). Public awareness programs should be coupled with strict enforcement of existing laws. Quality standards and promotion of composting by the public sector are vital for augmenting the effort. Private sector participation can significantly reduce costs and enhance service delivery.

16.5.5 Planning Urban/Land Use

One of the biggest problems faced by most of the urban areas of the Ganga Basin is unregulated urban expansions and land use patterns, which have triggered conflicts between urban and periurban interests. By using technologies like remote sensing and GIS, cities need to be planned for sustainable habitat and consumption. The use of open and abandoned areas for plantation purposes and establishment of water recharge structures should be emphasized. Areas under natural vegetation must not be appropriated for any construction activities.

16.5.6 Strengthening Urban Agriculture

India has a long tradition of urban agriculture, which also received prominent attention during the food scarcity era of 1960s. While grain production in India has been a success story, rapid urbanization has created a new urgency for identifying innovative ways to enhance food production. It is widely argued that megacities (e.g., Delhi, Mumbai, Kolkata, Chennai, Bengaluru, and Hyderabad among others) must produce fresh foods and vegetables within their limits by recycling MSW and gray/black water. The goal should be to produce ~20% of the food consumed in a city from within the city through innovative agriculture. Sky farming (Despomier, this volume) is an innovative option. As was also highlighted by PM Modi during his address to the nation via All India Radio, using compost from urban waste and promoting community gardens and greenhouse farming are other options which must be given a high priority (Modi 2016).

16.6 CONCLUSION: POLICY IMPLICATIONS AND THE WAY FORWARD

Soil being the basic foundation of human civilization and the environment, the total world land area of about 14.9 billion ha must be used and restored to meet basic needs of human well-being and nature conservancy. While the progress of humanity has been based on the use of soils, one of the features of this progress—urbanization—has been increasingly reaping the benefits without any reciprocation. Waste (mis)-management and scarcity of soil and sand/gravels are some such challenges that have added tremendous stress on the system (Development Alternatives 2016). The process of urbanization is virtually irreversible, and human activities lead to unintended consequences due to ignorance and neglect. The effects can be far reaching and potentially damaging (Merrifield and Swyngedouw 1997). Excessive sand mining to cater to the demand of urban construction has led to the degradation of rivers, and enlargement of river mouths and coastal inlets and could become a threat to adjacent infrastructure as well (The Ojos Negros Research Group 2004). This uncoordinated and improperly planned approach has led to the depletion of groundwater, lesser availability of water for industrial and drinking purposes, destruction of agricultural land, threat to livelihoods, and damage to roads and bridges (Saviour 2012). Excavation of topsoil for the purpose of producing bricks has negatively affected the fertility of agricultural land and has added to food security concerns.

An assessment of MSW in India shows that diverting waste into resource recovery systems offers an enormous untapped potential for creating economic and environmental benefits, and ultimately reducing the pressure on municipalities to manage their waste. It will take at least 4–5 years to witness changes in the waste management regimes in India. The SWM Rules, 2016 need an extensive awareness campaign in association with communities, NGOs, students, and other stakeholders for timely and effective implementation along with proper scrutiny, monitoring, and evaluation.

An enormous opportunity exists to improving upon the BAU rut of uncollected waste and open dumping and converting the waste into value-added resources, such as alternative fuels and agricultural amendments. As around half of the country's municipal waste is currently organic (biodegradable) waste, methods such as composting, anaerobic digestion, and conversion to RDFs present a sustainable course of action (ADB 2011).

With a projected population of around 1.7 billion by 2050 and an estimated increase in urbanization, India must understand that adhering to the BAU approach would jeopardize the natural resources already under great stress. The losses must be minimized and efficiency enhanced by 3Rs: reduce, reuse, and recycle (Lal 2013).

Key recommendations for scaling up organic waste management in the region are as follows: Composting is an option for greater use in urban and periurban agriculture, and should be promoted vis-à-vis the chemical fertilizers as an affordable option for the farmers. Adopting appropriate quality standards is essential to making the compost facilities operational, with close coordination between urban, agriculture, and energy departments (Manna et al. 2015). Strictly enforcing the "polluter pays" principle along with recognition of those households that segregate the waste at the source are steps in the right direction. Training of the staff is critical to the entire design and construction of the waste management system. It is also important to leverage the expertise of the private sector along with zealous participation of the communities.

It is of urgent necessity to explore methods such as cultivating phyto-remediation plants and using metal-contaminated lands for production of biofuel crops like sugarcane, maize, sunflower, jatropha, and so on for the bioremediation of soils so as to have a safe and healthy environment, essential for sustainable development (Indian Institute of Soil Science 2011). Restored soils, with the use of compost and soil amendments from MSW, must be used to promote urban agriculture to meet 20% of the food demand of the megacities of India.

REFERENCES

Asian Development Bank (ADB). 2011. *Toward Sustainable Municipal Organic Waste Management in South Asia: A Guidebook for Policy Makers and Practitioners.* Mandaluyong City, Philippines: Asian Development Bank.

Business Standard. 2015. India's waste to wealth potential lies in reusing resources. *Business Standard*, June 2, 2015. http://www.business-standard.com/article/sponsored-content/india-s-waste-to-wealth-potential-lies-in-reusing-resources-115060200857_1.html

Center for Policy Research. 2016. Conceptualising Zero-Waste in India under Swachh Bharat: Possibilities and Challenges, NITI Aayog—CPR Open Seminar Series, Room 122, NITI Aayog, Sansad Marg, New Delhi 110001, 29th June 2015.

Central Pollution Control Board (CPCB). 2015. Action Plan for Management of Municipal Solid Waste. Ministry of Environment, Forests and Climate Change, Government of India.

Chandra, S. et al. 2010. Dengue syndrome: An emerging zoonotic disease. *North-East Veterinarian* 9(4): 21–22.

Chinwan, D. and S. Pant. 2014. Waste to energy in India and its management. *Journal of Basic and Applied Engineering Research* 1(10): 89–94.

Craul, P.J. 1991. Urban Soil: Problems and Promise. *Urban Soils* 51(1): 23–32.

Development Alternatives. 2016. *Achieving Resource Synergies for a Rapidly Urbanizing India.* New Delhi. http://www.devalt.org/images/L2_ProjectPdfs/Resources_and_Urbanisation_Background_Paper_HBF.pdf

Einstein, D.E. 1999. *Urbanization and Its Human Influence. Seminar in Global Sustainability.* Irvine, USA: University of California.

Ellis, P. and M. Roberts. 2016. *Leveraging Urbanization in South Asia: Managing Spatial Transformation for Prosperity and Livability.* Washington, DC: World Bank Group.

Facchinelli, A. et al. 2001. Multivariate statistical and GIS-based approach to identify heavy metal sources in soils, *Environmental Pollution* 114: 313–324.

Gazette of India. 2016. Ministry of Environment and Forests and Climate Change. Government of India, April 8, 2016.

Government of India. 2005. Report of the Inter-Ministerial Task Force on Integrated Plant Nutrient Management Using City Compost. Delhi.

Government of India. 2014. Guidelines for Swachh Bharat Mission (SBM), Ministry of Urban Development. http://swachhbharaturban.gov.in/writereaddata/SBM_Guideline.pdf.

Govil, P.K. et al. 2001. Contamination of soil due to heavy metals in the Patancheru industrial development area, Andhra Pradesh, India, *Environmental Geology* 41: 461–469.

Grumbine, R.E. and M.K. Pandit. 2013. Threats from India's Himalaya Dams. *Science* 339(6115): 36–37.

Haering, K. and G. Evanylo. 2005. *Composting and Compost Used for Water Quality. Department of Crop and Soil Environmental Sciences.* Blacksburg, VA: Virginia Tech.

Hoornweg, D. and L. Thomas. 1999. *What a Waste: Solid Waste Management in Asia.* Urban and local government working paper series; no. UWP 1. Washington, D.C.: The World Bank. http://documents.worldbank.org/curated/en/694561468770664233/What-a-waste-solid-waste-management-in-Asia

India Energy Security Scenarios. 2015. User Guide for India's 2047 Energy Calculator Municipal Waste to Energy, NITI Aayog, Government of India.

Indian Express. 2016. Not our fault: BJP-led MCDs, Delhi govt's PWD blame each other for waterlogging. *Indian Express*, July 18, 2016. http://indianexpress.com/article/cities/delhi/delhi-rain-water-logging-traffic-bjp-mcd-delhi-government-kejriwal-blame-game-2918690/.

Indian Institute of Soil Science. 2011. *Vision 2030.* Bhopal: Indian Institute of Soil Science (Indian Council of Agricultural Research).

Infrastructure Development Finance Company Limited. 2006. *India Infrastructure Report, Urban Infrastructure 2006.* New Delhi: Oxford University Press.

Kapur, A. 2015. Four challenges that India faces in achieving sustainable development goals, *Business Standard*, October 26, 2015. http://www.business-standard.com/article/punditry/four-challenges-that-india-faces-in-achieving-sustainable-development-goals-115102600232_1.html

Kumar, A. 2015a. Housing shortages in urban India and socio-economic facets. *Journal of Infrastructure Development* 7(1): 19–34.

Kumar, A. 2015b. Housing amenities in urban India-issues and challenges. *Shelter* 16(1):7–14.

Kumar, A. and S. Mehta. 2014. An appraisal of municipal solid waste management in India: With special reference to Delhi. *Educator-The FIMT Journal* 6:10–32.

Kumar, S. 2011. Status of Soil Contamination in India, Water and Mega Cities, December 14. http://www.waterandmegacities.org/status-of-soil-contamination-in-india/.

Lal, R. 2013. The Nexus Approach to Managing Water, Soil and Waste, Introductory Talk at the International Kick-Off Workshop, Dresden, Germany, November 11–12, 2013.

Lal, R. 2016. Soil Carbon Sequestration: Science & Implementation of the "4 per Thousand Initiative" on U.S. Croplands and Grasslands. C-AGG Meeting, Denver, CO, July 12–13, 2016.

Manna, M.C. et al. 2015. *Rapid Composting Technique: Ways to Enhance Soil Organic Carbon, Productivity and Soil Health.* Bhopal: ICAR—Indian Institute of Soil Science Technology Folder.

Marcotullio, P.J. et al. 2008. The impact of urbanization on soils. In *Land Use and Soil Resources*, ed. Braimoh and Vlek, 201–250, the Netherlands: Springer.

Merrifield, A. and E. Swygedouw. 1997. *The Urbanization of Injustice.* New York, NY: New York University Press.

Milennium Post. 2016. BJP Municipal Bodies Not Working: Satyendar Jain. August 31, 2016. http://www.millenniumpost.in/NewsContent.aspx?NID=322956.

Ministry of Urban Development. 2011. Report on Indian Urban Infrastructure and Services. The High Powered Expert Committee (HPEC) for Estimating the Investment Requirements for Urban Infrastructure Services.

Ministry of Urban Development. 2015. Zero Garbage and SWaCH Model Solid Waste Management in Pune, Best Case Studies, Swachh Bharat-Urban.

Misra, A.K. 2011. Impact of urbanization on the hydrology of Ganga Basin (India). *Water Resource Management* 25: 705–719.

Modi, N. 2016. Mann Ki Baat. *All India Radio*, September 25, 2016. https://www.youtube.com/watch?v=ijoR5HRRbaU

NDTV. 2016. No Money To Pay Salaries, North Delhi Civic Body Tells High Court, February 3, 2016. http://www.ndtv.com/delhi-news/no-money-to-pay-salaries-north-delhi-civic-body-tells-high-court-1273006.

NITI Aayog. 2015. PM's Statement at the UN Summit for the Adoption of Post-2015 Development Agenda. http://niti.gov.in/content/pm%E2%80%99s-statement-un-summit-adoption-post-2015-development-agenda.

NIUA. 2015. *Urban Solid Waste Management in Indian Cities, Compendium of Good Practices, PEARL Initiative*. New Delhi: National Institute of Urban Affairs, Ministry of Urban Development.

NSSO. 2013. Key Indicators of Drinking Water, Sanitation, Hygiene and Housing Condition in India, NSS 69th Round (July 2012–December 2012), National Sample Survey Organization, Ministry of Statistics and Programme Implementation, Government of India.

NSSO. 2016. *Swachhta Status Report*. National Sample Survey Office, Ministry of Statistics and Programme Implementation. Governmnet of India.

Planning Commission. 2014. Report of the Task Force on Waste to Energy (Volume I) (In the context of Integrated MSW Management). Government of India.

Press Information Bureau. 2016a. Solid Waste Management Rules Revised after 16 Years; Rules Now Extend to Urban and Industrial Areas: Javadekar. Ministry of Environment and Forests. Government of India, April 5, 2016. http://pib.nic.in/newsite/PrintRelease.aspx?relid=138591.

Press Information Bureau. 2016b. Cabinet Approves Policy on Promotion of City Compost, Government of India, January 20, 2016.

Press Trust of India. 2015. Delhi reels under waterlogging; AAP government, municipal bodies spar. *The Economic Times*, July 12, 2015. http://articles.economictimes.indiatimes.com/2015-07-12/news/64334001_1_aap-government-rains-yamuna-pushta.

Rahman, M., P. 2013. Indians question how far flash-flooding disaster was manmade. *The Guardian*, June 28, 2013. https://www.theguardian.com/world/2013/jun/28/indians-flash-flooding-disaster-manmade

Rajindiran, S. et al. 2015. Heavy metal polluted Soils in India: Status and countermeasures, *JNKVV Research Journal* 49(3): 320–337.

Raman, N. and D.S. Narayanan. 2008. Impact of solid waste effect on ground water and soil quality nearer to Pallavaram solid waste landfill site in Chennai. *Rasayan Journal of Chemistry* 1(4): 828–836.

Rao, K. 2013. Were India's floods caused by reckless human greed? *The Guardian*, June 24, 2013. https://www.theguardian.com/environment/terra-india/2013/jun/24/india-floods-himalayas

Raychaudhuri, S. et al. 2014. *Impact of Urban Wastewater Irrigation on Soil and Crop. DWM Bulletin No. 64*. Bhubaneswar, Odisha: Directorate of Water Management.

Roychoudhury, S. 2014. For Clean India to work, country needs to solve its waste disposal problem. *Scroll*, November 1, 2014. http://scroll.in/article/682335/forcleanindiatoworkcountryneedstosolveitswaste disposalproblem.

Sambyal, S.S. 2016. Government notifies new solid waste management rules. *Down To Earth*, April 5, 2016. http://www.downtoearth.org.in/news/solid-waste-management-rules-2016-53443.

Sampath, G. 2013. Don't blame nature for the Uttarakhand flood disaster. *Livemint*, June 27, 2013. http://www.livemint.com/Opinion/hzKmWekwYOOtYKv8N6dZlN/Dont-blame-nature-for-the-Uttarakhand-flood-disaster.html

Sankhe, S. et al. 2010. *India's Urban Awakening: Building Inclusive Cities, Sustaining Economic Growth*. McKinsey Global Institute, McKinsey & Company, India.

Saviour, N. 2012. Environmental impact of sand and soil mining: A review, *International Journal of Science, Environment and Technology* 1(3): 125–134.

Sharholy, M. et al. 2008. Municipal solid waste management in Indian cities–A review, *Waste Management* 28: 459–467.

Sharma, R.K. et al. 2009. Heavy metals in vegetables collected from production and market sites of a tropical urban area of India. *Food and Chemical Toxicology* 47: 583–591.

Singh, D. 2015. Experts say it could take three years to upgrade Delhi's drainage system. *Daily Mail*, July 14, 2015. http://www.dailymail.co.uk/indiahome/indianews/article-3159863/Experts-say-THREE-YEARS-upgrade-Delhi-s-drainage-system.html

Singh, M. et al. 2002. Heavy metals in freshly deposited stream sediments of rivers associated with urbanization of the Ganga Plain, India, *Water Air and Soil Pollution* 141: 35–54.

Technology and Action for Rural Advancement. 2015. Achieving the Sustainable Development Goals in India: A Study of Financial Requirements and Gaps, Submitted to the Ministry of Environment, Forest and Climate Change, New Delhi. http://www.devalt.org/images/L3_ProjectPdfs/AchievingSDGsinIndia_DA_21Sept.pdf?mid=6&sid=28.

TERI. 2015. *Industrial and Urban Waste Management in India*. New Delhi: The Energy and Resources Institute.

The Hindu. 2016. Blame game as chikungunya claims five lives in Delhi. *The Hindu*, September 13, 2016. http://www.thehindu.com/news/cities/Delhi/Blame-game-as-chikungunya-claims-five-lives-in-Delhi/article14636761.ece

The Ojos Negros Research Group. 2004. Impact of Sand Mining. http://ponce.sdsu.edu/three_issues_sandminingfacts01.html.

UNEP. 2005. *Solid Waste Management* (Volume I). United Nations Environment Programme, Kenya, http://www.unep.or.jp/ietc/publications/spc/solid_waste_management/Vol_I/Binder1.pdf.

UNGA. 2015. Resolution adopted by the General Assembly on 25 September 2015, A/RES/70/1 Transforming our world: The 2030 Agenda for Sustainable Development. New York.

United Nations Human Settlements Program (UN-HABITAT). 2010. *Solid Waste Management in the World's Cities, Water and Sanitation in the World's Cities 2010*, London, Washington, DC: Routledge.

United States Environmental Protection Agency (US EPA). 2006. *Solid Waste Management and Greenhouse Gases: A Life-Cycle Assessment of Emissions and Sinks*. 3rd Edition. Washington, DC: USEPA.

USDA. 2005. *Urban Soil Primer*, Washington, DC: Natural Resources Conservation Service Soils.

Water and Sanitation Program-South Asia. 2007. *Implementing Integrated Solid Waste Management Systems in India, Moving Towards the Regional Approach*. New Delhi: World Bank.

World Bank. 2015. The National Ganga River Basin Project. http://www.worldbank.org/en/news/feature/2015/03/23/india-the-national-ganga-river-basin-project.

17 Enhancing Awareness about the Importance of Urban Soils

Christina Siebe, Silke Cram, and Lucy Mora Palomino

CONTENTS

"In the end we will conserve only what we love; we will love only what we understand; and we will understand only what we are taught."

—Baba Dioum (1968)

17.1 INTRODUCTION

The year 2015 was declared International Year of Soil by the Food and Agriculture Organization (FAO) of the United Nations (http://www.fao.org/soils-2015/about/en/). As a result, a large variety of actions were undertaken by research institutions, government agencies, and civil associations to promote awareness about the importance of soil for sustaining life on Earth. Central topics were the agricultural use of soil to produce food, fiber, and fuel and the soil's regulating functions of the water, carbon, and nutrient cycles together with the increasing recognition of the huge and yet widely unknown biodiversity hosted within soils. All the activities received positive feedback, yet soil degradation is ongoing, and much larger and longer lasting efforts need to be undertaken to stop it. Soil formation processes have millennial time scales, and it is to be expected that soil rehabilitation will need at least decades to centuries. In this sense, at the end of 2015, the president of the International Union of Soil Science (IUSS) promoted an International Decade of Soil (2015–2024) (http://www.iuss.org/index.php?article_id=588).

In the years to come, efforts will continue to enhance awareness and to promote soil conservation strategies. This is an opportunity to focus on soil aspects that have received less attention in the past.

Among the less acknowledged topics are soils in urban environments, despite the fact that more than half of the world's population lives currently in cities and the prospects for the next 30 years indicate that this figure will reach 75% by the year 2050 (United Nations 2014). However, only a small proportion of the world's soil research is conducted on urban soils, and important stakeholders within cities lack essential knowledge about soil distribution patterns and soil processes in cities. A wide range of soil attributes need to be considered before decisions are made on the use and management of urban soils and to achieve city sustainability. Additionally, among all human inhabitants on Earth, urban dwellers are particularly at loss of awareness of the soil ecosystem, which hinders the achievement of soil protective actions.

Therefore, the focus of this chapter is on how to enhance soil awareness, particularly among city inhabitants. The first section of the chapter describes the causes of the alienation that city inhabitants experience with respect to the importance of soil to sustaining life. It includes a brief discussion on how the welfare of people living in cities depends on soil management in periurban and rural regions within the same watersheds. Further, there is a discussion on the interests of different groups within an urban society on soil, its attributes, and functioning. Another section reviews the awareness of different stakeholders regarding the importance of urban soils, and there is a description of the strategies being followed in different countries to communicate existing knowledge. Finally, there is a review of activities that enhance soil valuation by fostering emotional bonds with soil in urban locations.

17.2 AWARENESS ON THE IMPORTANCE OF SOIL FOR HUMAN WELL-BEING IS DECREASING

Awareness on the importance of healthy or clean soil for human well-being is much less developed than that of clean water or clean air. The attributes of clean water and air are universally known and accepted, readily perceivable, and easy to grasp by everyone. Soil is hidden, much more heterogeneous, complex, and costly to study, and its dark color, muddy consistency, and earthy smell remind people rather of dirt than of the life-supporting resource (Müller 2015). Also soil organisms, such as earthworms, millipedes, and all bacteria, cause fear and rejection among many people, particularly those living in the cities.

The concept of a "healthy" soil is predominantly associated with its productivity, and traditional farmers are the main group in human societies who are aware of it. They are not only aware of the yield potential of their fields, but also judge the soil quality on behalf of the consistency of the soil, its color, its water infiltration capacity, its aggregation, and the strength of the aggregates, among many other soil attributes (Barrera-Bassols and Zinck 2003; Table 17.1). They do not only acknowledge the quality of the soil in their own fields, but also have a perspective of the surrounding landscape (Toledo 1994). The latter is perceivable in rural areas, since neither buildings nor tall and dense vegetation obstruct the view. Another important fact when considering the awareness of soil among traditional farmers is that they are generally emotionally bound to the soil. Most of them inherited the land from their ancestors; they grew up in the region, played during their childhood with soil, and almost have daily contact with soil: they carry soil underneath their fingernails.

Several ancient rural societies developed a strong relation to soil and natural resources: their perception and knowledge evolved and adapted over hundreds and even thousands of years and was transferred from one generation to the next (Hillel 2005; Barrera-Bassols and Zinck 2003; Szulczewski 2015). Traditional farmers preserve and further develop the perception and knowledge they inherited from their ancestors. They clearly conceive soil as the sustaining resource of agriculture and also recognize its regulatory role in the hydrologic cycle. Nature is conceived as sacred, and the natural elements have a life force (Gottlieb 2004). Soil has a strategic role and is honored

TABLE 17.1
Soil and Site Quality Attributes Acknowledged by Ancient Rural Societies/Traditional Farmers, Modern Farmers, and Urban Inhabitants

Traditional Farmers/Ancient Rural Societies	Modern Farmers	Urban Inhabitants
Sacred "Mother Earth"	Productivity (yield)	Size and price of the ground
Part of nature (holistic view)	Slope	Ownership
Living force	Workability	Assigned land use category by authorities
Substrate on which food, fiber, and fuel are produced	Stone content	Connectivity
Water "factory" (water infiltration upslope and discharge downslope in springs)	Road access	Surrounding infrastructure
Color, consistency (Barrera-Bassols and Zinck 2003)	Nutrient contents	Access to potable water supply
Moisture/drainage conditions (Barrera-Bassols and Zinck 2003)	Access to irrigation	Access to electricity
Organic matter (Barrera-Bassols and Zinck 2003)	Access to market	Connection to natural gas pipeline, access to internet, cable TV
Stone content, soil depth (Barrera-Bassols and Zinck 2003)	Water infiltration and storage capacity	Soil mechanical characteristics (carrying capacity, compressive strength, deformability, plastic limits) (Bell 2013)
Slope (Barrera-Bassols and Zinck 2003)	Drainage	Vulnerability to landslides, subsidence, soil creeping, etc.
Structure/aggregation (Barrera-Bassols and Zinck 2003)		Concentrations of pollutants

Source: Barrera-Bassols, N. and J. A. Zinck., *Geoderma*, 111, 171–195, 2003; *Engineering Properties of Soils & Rocks*, 3rd edn, Bell, F. G. (2013), Elsevier.

as the deity of fertility with ceremonies, oblations, and sacrifices. Ritual activities are preformed to reinforce the engagement humans have in restoring the damage they have committed by perturbing the soil through cultivation (Barrera-Bassols and Zinck 2003). Traditional farmers have a tight emotive and mystic bond with the soil, and also a complex cognition of the different components of their land, including rocks and minerals, water, meteorological and astronomical phenomena, plants, animals, and their interactions. This knowledge has been acquired through trial and error experiences and innovation, both constantly evolving and being transmitted from generation to generation over millennia (Toledo and Barrera-Bassols 2008).

Modern agriculture based on scientific knowledge and technical tools has changed the mystic and emotive relation to soil among farmers; they perceive it increasingly as a production factor, rather than as a sacred life force. Land attributes as workability, road connectivity, and access to markets are more appreciated than the physical, chemical, or biological soil properties (Table 17.1). But even more radical is the shift in valuation of soil in urban societies. Worldwide people are migrating from rural areas into cities; currently more than half (54%) of the world's population (United Nations 2016) lives in cities where people seek better education and health services, better paying jobs, and more diverse entertainment opportunities. In some counties, particularly in Africa, migration is triggered by decreasing land productivity, wars, and religious conflicts, among many other factors (Cohen 2006; Achankeng 2003). City dwellers have much less contact with soil, and buildings and pavements obscure visibility of the surrounding landscape. Only those who maintain a garden keep in touch with soil. But as population density increases and ground prices rise, less people can afford to maintain a garden and more people live in apartments. People perceive soil increasingly as dirt or filth and associate it with a disease-fostering substrate (Johnson and Catley 2009). Modern urban children play less in open spaces and are entertained dominantly at home by electronic devices. Their parents are less fond of allowing them to get dirty, bringing home clothes and shoes covered with soil. They are afraid of soil being plagued by pathogenic organisms, so gardens or plant beds are replaced by stone or gravel covered yards and by well-contained plant pots in order to avoid family members getting "contaminated" with soil. To decrease maintenance costs, schoolyards are preferably no longer covered with grass, but rather paved, and grass football courts are replaced by artificial plastic and rubble materials. The valuation of soil shifts from sacred (ancient societies) and life-supporting natural resource (traditional farmers) to production factor (modern agriculture), and support for infrastructure and waste receptor (urbanizations). The soil attributes which are valued are also distinctly different in cities: the size of the plot, its ownership, its location with respect to communicating roads and important infrastructure, the monetary value of the land, its carrying capacity, and other mechanical soil properties as well as the possible presence of contaminants are important attributes to decide its use (Table 17.1). Ecological soil functions as a medium for plants, particularly allotments, ornamental trees and plants, as well as grass are of minor relevance, and regulating soil functions such as water infiltration, habitat of organisms, and carbon storage are seldom appreciated.

In cities, the price of the ground m^{-2} is much higher than in rural locations, which makes it affordable to adapt unfavorable soil attributes that limit the potential of soils to support infrastructure. Soils are compacted, drained, removed, and sealed in order to habilitate their use for habitation, industry, traffic, and waste disposal. City planners are seldom aware of ecological soil functions. An example is the new international airport of Mexico City, which is currently constructed on a 60 km^2 large area owned by the federal government located less than 10 km from the city center. Size, ownership, and proximity to the city center were the most valued land attributes to decide its location. Civil engineers consider that the nonfavorable soil attributes such as the very small carrying capacity (bulk density of the soil is <1.0 g cm^{-3}, porosity >70%), a water table near the surface (0.4–2 m), and high salt content (EC$_{sat}$: 2–410 dS m^{-1}) (Jazcilevich et al. 2015; Fernández-Buces et al. 2006) can be overcome through adequate, but costly, construction techniques. Soil functions such as water cycle regulation, soda extraction, habitat of endemic flora and fauna, and the value of these soils as documents of the landscape history (Siebe et al. 2004) as well as the potential for local climate regulation, solar energy production, recreation, and conservation are considered of minor relevance (Jazcilevich et al. 2015).

There are also several examples in which city governments have established regulations to protect urban soils from surface sealing, or to restrict urban growth in aquifer recharge areas or urbanization of sites prone to inundation, landslides, or subsidence, but the low perception of these hazards by city dwellers hinder their understanding. Again an example from Mexico City helps to illustrate this: despite the fact that the government issued, in 1996, a law to hinder surface sealing at aquifer recharge areas in the southern part of the city (Ley de Desarrollo Urbano del Distrito Federal, GDF 1996), there are continuously invasion reports by migrants and neighbors (Molla Ruiz-Gómez 2006; Aguilar 2008).

17.3 IMPORTANCE OF SOIL CONSERVATION PRACTICES IN RURAL AREAS TO PRESERVE LIFE QUALITY IN CITIES

In order to guarantee food, water, and air quality for urban locations, soils need to be preserved in rural and periurban areas. Many cities are located in flat or gently sloping areas within valleys. Urban sprawl is thus displacing agricultural activities from high quality soils (Hasse and Lathrop 2003; Setälä et al. 2014), which then have to migrate to rather unfavorably sloping land that is prone to erosion-induced degradation. Erosion affects surface water quality, since it pollutes rivers and lakes with suspended particles and nutrient-laden sediments. The supply of clean water in the cities, either from groundwater or surface water, depends on the surrounding watersheds (Postel and Thompson 2005), and it can be costly for water suppliers not to take care of the watershed. The costs of water purification are either charged to consumers, or need to be covered by government subsidies. Erosion by wind also impacts life in cities, since it deteriorates the air quality (Wiggs et al. 2003). During the dry season, the air quality index of the metropolitan area of Mexico City, for instance, is affected by suspended soil particles, which originate in fields of the piedmont of the Sierra Nevada, in the western part of the city, and from the former lake Texcoco area, where saline soils with sparse vegetation cover prevail (Jazcilevich et al. 2015).

Thus, there is a need to enhance the awareness of urban inhabitants about soil ecological functioning, not only within the city, but also in periurban regions and even in rural locations within the same watershed. Maintenance of catchment areas results in economic benefits for the cities (Postel and Thompson 2005), and city inhabitants should therefore be willing to support soil conservation practices through payments for ecosystem services (Wunder 2005; Pagiola et al. 2007).

17.4 IDENTIFICATION OF LAND-USE CATEGORIES AND STAKEHOLDERS NEEDING SOIL INFORMATION IN URBAN AREAS

Land-use categories assigned in urban areas differ from those assigned in rural areas. Rural land may be used for agriculture, forestry, or pasture, or assigned to conservation (Johnston and Swallow 2010), and the policy makers and stakeholders deciding on the land-use category and giving advice to land owners on how to use, manage, and conserve the soil are dominantly agricultural engineers, foresters, or landscape planners. In urban areas, the land-use categories are habitation, industry, waste disposal, and public or private green areas. The latter can be assigned to recreation, conservation, or sport, for example. The professional and training backgrounds of decision makers differ from those in rural areas (Table 17.2).

An important group of stakeholders in urban areas is developers and builders. The group is primarily interested in distinct soil attributes, such as the soil particle size distribution, soil strength, carrying capacity, plastic limits, and other mechanical soil properties (Bell 2013). Also, the depth of the water table, salt, and contaminant concentrations matter when decisions on the drainage, cementation, and risk are to be taken into account. Bachelor programs in civil engineering and architecture offer training in these particular soil characteristics.

Also, urban planners need to have information about soils, not only of their mechanical properties, but also about the distribution of soils and landforms in the city and its surroundings. Important subjects for urban planners include, therefore, hydrology, particularly to regulate runoff and to foster water infiltration and aquifer recharge. Also, the analysis of land stability and vulnerability to landslides, subsidence, soil creeping, and earthquakes requires information about soil characteristics. Since this knowledge is generally not included in the training of urban planners, geoscientsists like geomorphologists, geologists, and geohydrologists become increasingly involved in city planning (Pavlovskaya 2002).

Waste management is another important subject that is tightly related to soil attributes and processes. Not only urban planners but also new disciplines, such as environmental science and management, even waste management, are currently part of higher education programs at universities and technical schools. Here watershed management is another important subject, since it gives a more regional perspective and allows the relation of cities with the surrounding rural areas.

Local authorities have the responsibility to control soil sealing and to ensure the living quality of citizens; accordingly, they need to be aware of the importance of soil functions (Artmann 2014).

Gardeners and urban farmers, who maintain either gardens for ornamental use or vegetable production (allotments), need similar information about their soils, namely nutrient contents, water

TABLE 17.2

Comparison of Land-Use Categories, Stakeholders, and Professional Training Skills in Urban and Rural Areas

	Land-Use Categories	Stakeholders	Professions
Rural areas	Agriculture	Farmers	Agricultural engineers and scientists
	Forestry	Foresters	Rural development geographers
	Pasture	Agricultural extension services	Landscape planners
	Conservation	Rural development agencies	Forest engineers
		Civil associations	Biologists
		Management agencies	Ecologists
		Indigenous communities	Economists
		Government agencies	Environmental lawyer
		Nature conservation foundations	
Urban areas	*Sealed areas*	Local developers	Architects
	Habitation	Builders	Civil engineers
	Industrial	Consultants	City planners
	Traffic	Owners	Waste managers
	Waste disposal	Resident associations	Economists
		Regulation authorities	Lawyers
		Urban planning department	
	Green areas	Children	Environmental scientists
	Public green areas	Runners, dog walkers	Urban farmers
	Private green areas (gardens)	General public	Gardeners
	Allotments	Public health department	City planners
	Graveyards	Environmental protection agency	Landscape planners
	Sport fields	Researchers	Teachers

infiltration rate, and water holding capacity. They also need to consider small scale heterogeneity, which is much larger in urban than in rural areas (Pickett and Cadenasso 2009; Greinert 2015) as well as pollutant contents (Golden 2013; Warming et al. 2015), particularly if their plots are on abandoned industrial sites or near traffic roads. Soil maps provide information about the spatial variability of urban soils, but the survey techniques need to consider land-use history as a highly relevant soil forming factor, which requires special training (Effland and Pouyat 1997). City gardeners are also interested in using organic amendments as compost, and want to learn about how to produce these. They are also often confronted with polluted grounds, which is not the case in rural areas.

Since many city dwellers cannot afford to have a garden, some might be interested in establishing green walls or green roofs, or at least some vertical garden strategies on balconies and the like. Some city governments are increasingly promoting green roofs and walls, since their positive effects on the temperature and relative humidity, the isolation of buildings optimizing heating costs have been demonstrated (Berardi et al. 2014).

Other stakeholders who need to perceive the importance of the urban soil are those who are responsible for the maintenance of green areas in public spaces (such as parks, garden squares, and tree-lined streets). The importance of urban green areas for health and quality of life is undeniable (Tzoulas et al. 2007; Bertram and Rehdanz 2014; Sandifer et al. 2015). These areas also maintain and increase biodiversity in the city. But people living in cities are increasingly disconnected from the natural world (Miller 2005) and often have a fixed expectation on the kind of plants which should be part of a city park, namely English grass, tall trees like oaks and firs, and exotic ornamental plants. The introduction of exotic species has an ecological impact on the native species (Goddard et al. 2010). Large efforts have been recently undertaken in some city parks to promote local species, not only in parks but also in green areas in between roads, and on roundabouts and sidewalks. This is of particular interest in arid and semiarid regions where it is unaffordable to maintain species that require large amounts of water; here cacti and other succulents are increasingly used for providing a distinct character to these cities. An example of these efforts is the city of Phoenix in Arizona (Davies and Hall 2010).

Special land-use categories are botanical gardens. Many of these were created by former landlords, kings, and emperors, who set up collections of plant species within the castle parks, which are now taken care of by botanical societies and research institutes. Also university campuses frequently have botanical gardens. Even conservation areas of special local plant and animal communities can be established within cities. One example is the Reserve of the Pedregal de San Ángel, in the southern part of Mexico City (Box 17.1).

People in charge of the maintenance of schoolyards and playgrounds are also interested in soil information. Here an important topic is how to deal with soil compaction produced by frequent trampling and mechanized grass cutting. These gardeners need to be additionally instructed in how excessive fertilization and irrigation affect the groundwater. This is particularly the case with the maintenance of golf courses, which are spreading in several cities and have caused environmental

BOX 17.1

The Pedregal de San Ángel is an area of 237 ha within the campus of the National Autonomous University of Mexico. This natural conservation area is located on the Xitle lava flow, on which a highly biodiverse (1800 species) xerophytic shrub community developed. The plant and animal community of this ecosystem is unique not only within the basin of Mexico, but also worldwide. It hosts a high number of endemic species, several of which are endangered. The maintenance of a conservation area surrounded by a dense population, crossed by a heavily trafficked road, poses special challenges (Lot et al. 2012). Photo courtesy of Armando Peralta.

damage through nitrate leaching and greenhouse gas (GHG) emissions (Brown et al. 2016). Also football fields need similarly careful consideration and, therefore, adequate management strategies (Puhalla et al. 1999).

Graveyards are another land-use category that has special management and strategic demands. Old traditional cemeteries are sited within cities and in periurban areas, while new ones are spreading. Special soil attributes such as moist conditions and anaerobic conditions have caused problems in the decomposition process (Fiedler et al. 2012) indicating a need to establish guidelines for the land attribute requirements that should be fulfilled prior to the assignment of this land-use category (Dent et al. 2004). Also, environmental concerns arise due to the increasing use of metal coffins and other artificial materials and groundwater pollution problems (Jonker and Oliver 2012). On the other hand, cemeteries also contribute to city biodiversity, but frequently by introducing exotic species rather than giving preference to local ones (Rahman et al. 2008; Löki et al. 2015).

17.5 STRATEGIES TO PROMOTE AWARENESS OF SOIL FUNCTIONS IN URBAN AREAS

In the following section, we review different strategies to promote awareness of the importance and functioning of urban soils among the different stakeholders mentioned before.

17.5.1 SCIENTIFIC RESEARCH

Scientific research is the motor of mind-oriented awareness about urban soil characteristics, distribution, functions, and processes. Although the first written research paper on urban soils was published in 1847 (Ferdinand Senft, cited in Lehmann and Stahr 2007), this subdiscipline of soil science is rather young and emerging. Relatively few soil scientists are currently interested in this topic: for example, from 6296 members of the Soil Science Society of America only 817 have registered in the Urban and Anthropogenic Soils Division (3%), and only 83 (1.4%) mentioned this topic as their primary interest (Andrea McDonald, SSSA, 2016, personal communication). The working group on Soils in Urban, Industrial, Transport, Mining, and Military Areas (SUITMA) of the IUSS, which has approximately 20,000 full members, has existed since 1998, and counts approximately with 600 soil scientists (3%) who have participated in at least one of the eight meetings organized to date. The oldest working group on urban soils within a scientific society was registered in 1987 in the German Soil Science Society, which also organized the first symposium on this topic in Berlin in 1982 (Burghardt 1994). From the 2182 members of the German Soil Science Society (DBG), 194 (9%) are registered in the urban soils working group (C. Schlüssing-Ahl, sectertariat DBG, personal communication), and around 30 (1.4%) are regularly participating (W. Burghardt, personal communication).

Research findings are primarily published in scientific journals. A search in the Scopus® database of the terms "urban soils" in either title or keywords yielded a total of 5183 documents (articles and review papers) published between 1990 and 2015. In the selected time frame, a clear increase of number of documents published per year can be observed after the Montpellier World Congress of Soil Science in 1998, and an even greater increase after 2007 (Figure 17.1). The sources that mostly publish articles related to urban soils are journals of disciplines related to Environmental Science (Science of the Total Environment) (232 documents), Environmental Pollution (170 documents), Environmental Monitoring and Assessment (159 documents), Chemosphere (147 documents), and Atmospheric Environment (126 documents). Among the soil science journals, the *Journal of Soils and Sediments* (49), *Eurasian Soil Science* (46), *Geoderma* (36), and *Catena* (18) published the most manuscripts on the topic in the searched period. It seems that the topic "urban soils" does not primarily interest the soil science community, but rather environmental scientists. Also, urban and landscape planners are apparently more interested than soil scientists (*Landscape and*

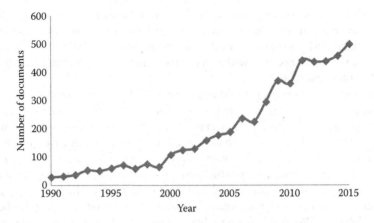

FIGURE 17.1   Number of articles and reviews published in scientific journals (Scopus® database) between 1990 and 2015.

Urban Planning, 69 documents). The fact that most research in urban soils is published in more environmental and urban planning–oriented journals rather than in soil science journals clearly indicates that information on urban soils is demanded by a different readership than the one soil science journals are focused on, namely researchers and professionals dealing with soils in rural and natural areas.

17.5.2 Strategies to Communicate Research Findings to Other Stakeholders

Particularly useful to communicate scientific research to specific stakeholders are review articles and books. These provide syntheses of the current knowledge of urban soils (Yang and Zhang 2015; Burghardt et al. 2016; Bellows and Hamm 2003; Burghardt 1994), or discuss specific topics related to soils in urban environments, for example, trace element contamination (Lou et al. 2012; Galušková et al. 2011; Wei and Yang 2010; Meuser 2010) contents of polycyclic aromatic hydrocarbons (Peng et al. 2011), soil carbon pools (Tong et al. 2007; Lal and Augustin 2012; Lou et al. 2012), tree root systems (Watson et al. 2014), compost benefits (Cogger 2005), soil bank seeds in urban domestic gardens (Thompson et al. 2005), or anthropogenic sealing (Scalenghe and Marsan 2009).

Effective communication of scientific findings occurs during congresses and symposia. Congresses are not only meeting points for scientists; government agencies and members of civil associations frequently assist these events, particularly the annual soil science meetings organized by scientific societies which promote the participation of different stakeholders offering training courses and exhibitions, and inviting directors of government agencies and ministers as keynote speakers. Communication between researchers and government agencies as well as representatives of civil associations occurs in regular intervals in these kinds of events. They are specially attractive for building links between the academic sector and the governmental and consultancy sectors since their organization occurs in regular intervals and the logistics are assumed by the organizing societies, so that little effort has to be undertaken by individuals to obtain high quality information at first hand. It is important, though, that government agencies and civil organizations assign regularly a small part of their budget to pay congress fees and travel expenses for their employees, and excuse them of their work duties for a few days to keep their knowledge updated.

Since the year 2000, an interdisciplinary soil science congress specifically on SUITMA is organized every 2 years in a different city around the planet (Burghardt et al. 2016). Eight international meetings have been organized until now including those in Germany in 2000, France in 2003, Egypt in 2005, China in 2007, the USA in 2009, Morocco in 2011, Poland in 2012, and Mexico in 2015. The SUITMA congresses aim to foster interdisciplinary research specifically among soil scientists, not only in urban sites, but also in mining and military areas and former industrial sites (environmental liabilities), and soils affected by traffic are included within the research objectives. Regular assistance to these meetings will foster collaboration and knowledge transfer between the academic and institutional sectors.

A more directed and specific topic-oriented communication occurs through workshops, which are meetings of a smaller number of participants, organized by either working groups of scientific societies or by government agencies, frequently by international agencies (e.g., FAO, UNESCO, UN-WATER, and UN-HABITAT). These workshops are especially effective in gathering together scientists, decision makers, and representatives of civil associations to foster governance and design public policy instruments through interinstitutional cooperation. They help in identifying common objectives and produce synergies among stakeholders to synchronize the work programs.

An international initiative to communicate experiences and promote synergies between stakeholders is the Global Soil Week (GSW), a multistakeholder platform that seeks to promote sustainable development in the areas of sustainable soil management and responsible land governance. Policy makers, scientists from different disciplines, stakeholders from civil society organizations, students, farmers, and artists, as well as stakeholders from international organizations and industry have met since 2012 every 2 years to find strategies to ensure the sustainable management of soil and land. However, the future of GSW is uncertain because of the change in priorities of the host institution (the Institute of Advanced Sustainability Studies). Among the specific thematic areas considered by GSW are urban soils (http://www.landgovernance.org/events/global-soil-week-2015/; globalsoilweek.org).

17.5.3 EDUCATIONAL PROGRAMS

17.5.3.1 University Study Plans and Strategies for Higher Educated Stakeholders

The inclusion of lectures on urban soil formation, characteristics, and processes linking them with other environmental compartments is another effective strategy to communicate knowledge and also to enhance awareness on urban soils. The direct pathway for achieving this is through lecture plans at universities, since a great majority of researchers are committed to lecture.

Soil science is dominantly part of the undergraduate curricula of agricultural and horticultural sciences, forestry, geography, biology, and so on (Figure 17.2). A global search of the universities that offer soil science training in their study plans indicates 167 universities throughout the world. (https://en.wikipedia.org/wiki/List_of_universities_with_soil_science_curriculum and https://www.universities.com/programs/soil-sciences-degrees). The study plans were grouped according to the offering faculties and major regions of the world (Figure 17.2). Although a wide range of soil science topics are included in the study plans of these careers, the curricula and practical training courses are focused on soils in rural environments. Curricula related specifically to urban soils are rare. The search indicated only seven universities, all within the United States, which specifically offer courses on urban soils. It might well be that there are other faculties or universities offering such courses, but they are not accessible through an internet search. Several topics concerning urban soils are part of the study plans of undergrad programs of environmental sciences. These topics include pollution, waste disposal, sustainable resource management, public policies, and governance of natural resources conservation. Nonetheless, most examples include water and air resources, while soil is clearly underrepresented in the courses being offered. A similar trend is observed when higher education programs are critically reviewed in urban planning, architecture, and so on. Programs in civil engineering offer lectures in soil mechanical behavior, but no specific lectures are related to soil functions such as the regulation of the water cycle. There seems to be a clear lack in solid soil science training of professionals, who make decisions on the soil use in urban environments. An important strategy for enhancing awareness on urban soils would be to include this topic in the study plans of

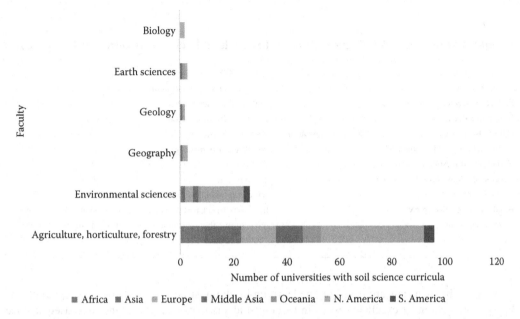

FIGURE 17.2 Number of universities in different regions of the world offering a soil science curriculum within a specific faculty the courses are offered. (From https://en.wikipedia.org/wiki/List_of_universities_with_soil_science_curriculum; https://www.universities.com/programs/soil-sciences-degrees.)

BS and MS programs of urban planners, architects, civil engineers, and environmental scientists. Study of properties, mapping, formation, and ecological functioning of urban soils should be a compulsory theme in the curricula of soil science in agriculture, forestry, horticulture, and other natural sciences. Therefore, researchers teaching soil science disciplines, predominantly within agriculture or geography faculties, should also actively participate in soil science training in urban planning, architecture, or civil engineering faculties.

17.5.3.2 Educational Programs for Children and Youths

Primary school books contain little information about soil. While the importance of water and air is present in most lesson plans, soil is only briefly mentioned in school books. In general, only soil fertility is mentioned in relation to agriculture and sustainability. Enhancing education on urban soil functions from preschool education throughout primary school is, therefore, a clear opportunity to increase awareness on urban soils. The latter would help overcome the prevailing prejudice that soil is synonymous with dirt or filth. Particularly, preschool education could well motivate toddlers to explore soil and understand its omnipresence. Simply put, children can access soil in any schoolyard or a park and describe its properties and salient features. They can further investigate the presence of living organisms by using a hand lens. Simple experiments can be done such as putting soil of different kinds into funnels, adding the same amount of water to each, and observing the results; these experiments can help children understand that water moves at different velocities through soil, or is retained in different quantities on a unit mass and volume basis (Müller 2015).

In this context, it is interesting to note the large number of internet web pages which offer all kinds of materials, from information, exercises, and experiment descriptions to videos and so on to allow teachers of all grades, from preschool to high school, to include soil science topics in science, biology, geography, and chemistry lessons. These web pages are maintained mainly by universities (university outreach programs) and civil associations, like national soil science societies, nature conservation foundations or environmental protection associations, as well as by government ministries and agencies. Some examples are listed in Table 17.3.

TABLE 17.3

Examples of Internet Web Pages with Ideas to Teach Soil Science at Different Grade Levels

	Web Pages
Nutrients for Life Foundation	www.thescienceofsoil.com
Soil Science Society of America	www.soils4teachers.org
Brainpop Educators	educators.brainpop.com
Natural Resources Conservation Service Soil Education	www.nrcs.usda.gov
Edible Schoolyard Foundation	edible.schoolyards.org
National Soil Resources Institute	Soil.net.org
Geoscience News and Information	geology.com
TES Global Limited	www.tesglobal.com
Bundesverband Boden eV	http://www.bodenwelten.de/navigation/boden-und-bildung
Umweltbundesamt	www.umweltbundesamt.at/leistungen/seminare_schulungen/ boden_schule/

Most of the materials are based on the learning-by-doing strategy (Field et al. 2011), which has been proven to render excellent results in helping students to experience an inquiry-based science learning approach (Johnson and Catley 2009).

Some schools maintain a horticultural garden, in which children can produce vegetables, and experiment with composting, water infiltration and retention, and nutrient cycling. A remarkable program is the "edible school yard project" promoted by a civil association (edibleschoolyard.org) originated in the United States. It currently registers 5503 locations among which 5208 maintain a "garden classroom" predominantly within the United States, but also starting in some other countries. Based on the saying "I am what I eat" children learn to grow, cook, and share food at the table.

Another very efficient strategy to promote knowledge about soils are the so-called Mobile Environmental Education Projects (MEEPs). These consist of vehicles equipped with the materials needed to explore the soil and its interaction with other environmental compartments in parks or schoolyards. They drive around cities promoting informal soil science outreach activities among children and youths. The web page www.umweltmobile.de lists more than 23 different countries in which MEEPs are currently operating, many sponsored by environmental agencies, but others being part of nongovernmental organizations (Box 17.2).

17.5.4 LOCAL GOVERNMENT INITIATIVES

Local city authorities and municipalities assign land-use categories and are also responsible for preserving green areas, organizing waste disposal, and maintaining the drainage and sewer systems. They have, therefore, excellent opportunities to preserve ecological soil functions in cities. They have the legal instruments to plan the spread of urbanization and to induce citizens to a more sustainable soil use in the city. In Europe, many cities have implemented programs to protect urban soils. There even exists a civil association (European Land and Soil Alliance [ELSA]), which aims to foster collaboration between European city governments to jointly pursue the aims of sustainable urban development—in particular, the promotion of a sustainable utilization of soils—in the context of Agenda 21 (http://www.bodenbuendnis.org). Among their specific objectives are to limit soil and land consumption, to direct settlement development inwards and promote the quality of the settlement management, to register and redevelop dangerous old waste deposits and prepare the land for an appropriate reuse, to take measures for the improvement of the microclimate and the water

BOX 17.2

"Terramóvil" is a mobile soil and geohazards education project sponsored by the Institute of Geology of the National University of Mexico (UNAM). Since 2013, geoscience bachelor students volunteer to provide nonformal education to increase awareness of the importance of soil and water conservation among children, young scholars, and the general public within the basin of Mexico City. Through interactive experiments brought to the schoolyard or to public places, participants experience how the sustainable management of natural resources improves their daily life. Specific emphasis is given to soil ecological functions within the city. Different geologic hazards, to which the inhabitants of the basin of Mexico are exposed, are also shown together with the corresponding protection strategies (www.terramovil-igl.unam.mx). Photos courtesy of C. Siebe.

balance, thus contributing to the enhancement of the quality of life in the settlement area, and to promote the marketing of regionally grown products.

There are many other examples of initiatives of local governments to improve soil use, conservation, and functioning in cities, many of these activities are related to motivation of food production in urban gardens and creating public green areas. The city of Andernach, Germany has, for example, promoted a program called "Edible city," which has received a lot of attention, not only in Germany, but in many other European cities (Box 17.3).

BOX 17.3

Andernach, the "edible city": In 2010, the small city of Andernach, which has 30,000 inhabitants, launched a project where 8000 m² of traditional park areas were converted into urban orchards. Around 20 permanently unemployed people were recruited to serve as gardeners to replace ornamental plants with vegetables, fruit trees and fruit shrubs, and vineyards among other edible plants. All citizens are invited to harvest the product for free during the growing period. With this program, the city contributes not only toward reducing unemployment, but also to promoting local food production and awareness on sustainable city management by fostering social participation. The maintenance costs of ornamental plant beds in parks are also significantly reduced with this initiative (agriculturers.com; March 2, 2015). Photos courtesy of the city of Andernach, Germany. ©90Grad Photography/Hilger & Schneider GbR.

17.5.5 INTERNATIONAL INITIATIVES

In 2012, the FAO of the United Nations established a program called "The Global Soil Partnership" (GSP) as a mechanism to develop a strong interactive partnership and enhance collaboration and synergism between all stakeholders, from land users through policy makers (www.fao.org/global-soil-partnership). The aim is to improve the governance and promote sustainable management of soils in general. The GSP is specifically focused on rural areas to assure sustainable food production in the world. Nevertheless, the five pillars of GSP include promotion of sustainable soil management, raising awareness, promoting research, enhancing soil data collection, and harmonizing methods and indicators of soil quality measurements. These pillars of GSP are also applicable to urban soils.

The GSP should ideally be replicated at a national scale, and even on a local scale, since local governments are the ones that are directly in contact with land owners and are responsible for taking decisions that impact urban land use. In this context, the rector of the National University of Mexico invited ministers and directors of all government agencies as well as all research institutions dealing with soil to join in a National Soil Partnership ("Alianza Nacional por el Suelo") in August 2015 (Box 17.4). The next step should then be to replicate this at the state level. Yet, these events should ideally be organized by a politically neutral entity, which may improve acceptance; these events also require financial support, which still has to be arranged.

BOX 17.4

National Soil Partnership organized by the National Autonomous University of Mexico in Mexico City on August 18th, 2015, involved ministers of Agriculture (SAGARPA), Environment (SEMARNAT), and Rural and Urban Planning (SEDATU) as well as directors of 13 government agencies, rectors and academics of universities, representatives of science academies, and representatives of nongovernment organizations to constitute a National Soil Partnership (Allianza Nacional por el Suelo). http://www.dgcs.unam.mx/boletin/bdboletin/2015_473.html. Photo courtesy of C. Siebe.

17.5.6 OUTREACH PROGRAMS AND STRATEGIES FOR THE GENERAL PUBLIC

Further initiatives to increase knowledge and perception about soils and the importance of their conservation are museums, exhibitions, and natural teaching trails. Among the soil museums around the world are the World Soil Museum in Wageningen, The Netherlands, the Vasily Dokuchaev Museum in Russia, the "Museum-am-Schölerberg" in Osnabrück, and next year the Dubai Museum of Soils. Many exhibits were presented all around the world during the International Year of Soils (2015): "Dig It!—The Secrets of Soils" created by the Smithsonian's National Museum of Natural History, Washington, DC in 2006–2008, is one example, but "Fertile Soils, Secret Lives" in Luxembourg, "Our Land, Your Land" in France, "Solos de Minas" in Brazil, "Art and Soil" in Germany, and "Soil-Walking on Life" in Spain are further examples of recent exhibits aimed to enhance the perception on the importance of soils for human life.

Another effective strategy to draw attention of people to soils is by hiking trails or bike tours passing by sites at which different soils can either be visualized in outcrops, or are performing specific ecological functions, like green walls, roof gardens, parks established on former environmental liabilities, and so on. Art exhibits can also be shown along these trails. One example of the latter is the initiative "Art in the Forest" or "Kunst im Wald" in the periurban area of Berlin, where former soil-aquifer wastewater treatment systems have been transformed into an afforested recreation area, to which artists have added additional attraction value by distributing within the area sculptures and other art monuments. Urban trials have a much larger possibility to attract visitors than trials in natural environments or within geoparks; they can be reached easily by any kind of transport and in a short time, the visit time can be flexible, since within a city there will be many interruption possibilities, and a large variety of potential users as walkers, runners, dog walkers, and people seeking a lunch break, but also short school class excursions, can be directed to them with less effort than that needed to visit a nature trail in rural or conservation areas.

Written media such as newspapers, journals, atlases (Menegat et al. 1998), and illustrated books are some other options of promoting information about soils in urban environments, although they will reach a relatively small and continuously decreasing proportion of the population, since they are already preoccupied by a range of digital media. The latter, however, have a large and increasing outreach, particularly among young people (<35 years). Social networks such as Facebook, Twitter, and YouTube should be increasingly used to reach out to the younger generations. But open access publication of research articles is also important, particularly considering the discrepancy between soil training faculties and their preferred publication journals and the disciplines requiring information about urban soils. The latter often do not have access to periodic publications from other disciplines.

Radio and TV programs are also options that transmit information to a large audience. In a short time (30 minutes to the maximum of 120 minutes), visual and audio impressions can achieve a long lasting impression to a large populace of all ages.

17.6 RESTORATION OF AFFECTIVE BONDS WITH THE SOIL

As stated before, urban societies have lost the close and mystic bonds that traditional rural communities have with soil. This estrangement is primarily because people are not in close contact with nature and do not touch, feel, and smell soil in their daily life. In megacities, a vast proportion of the urbanized area is sealed. In the megalopolis of Mexico City, for example, 80% of the surface of several districts is sealed (Cram et al. 2008). In order to reverse this alienation, the most successful actions to enhance awareness about soil should involve feelings and emotions. This requires enforcement of the social, cultural, and artistic connections to the soil (Landa and Feller 2009; Feller et al. 2015; Wessolek and Toland 2010). The restoration of the sense of place and the establishment of an effective bond with soil are a primary step in achieving this goal. Another important strategy is to reintroduce the concept of the sacredness and divine nature of the soil. Many traditional farmers and indigenous people believe in the divine and mystic aspects of soil, and it motivates them to protect agricultural soils and follow the stewardship (Tobasura-Acuña et al. 2015; Toland 2015).

The use of all senses (observe, listen, touch, taste, and smell) evokes feelings and creates emotional bridges. It detonates in people an interest to know more about soil and motivates them to care about it. Some examples given below illustrate how can such a task be realized while implementing some of the previously described strategies to enhance awareness about the soil.

Evoking the cult of sacred Mother Earth during a visit to the geopark Mixteca Alta in Mexico local inhabitants ask sacred Mother Earth for permission before they start cultivating the land. They remove firewood or simply walk over the land with great respect and bury small offerings of sweets and beverages so Mother Earth forgives the intrusion. (Palacio-Prieto et al. 2016).

Painting with soil is an art that has been practiced since ancient times by indigenous tribes, and many modern artists make artwork with the colors of the soil, such as Yusuke Asai from Japan and Jackie Yeomans of the United Kingdom. It is also an activity that fascinates people of all ages and it is a useful tool to transmit knowledge about the soil, especially to kids (Capeche 2010); it is used in many countries such as Brazil (Museu de Ciências da Terra Alexis Dorofeef, Muggler 2006, 2012), Mexico (Terramovil, www.terramovil-igl.unam.mx/), Austria (*Painting with the colors of the Earth,* Szlezak, 2009), and others.

Artists can be effective in restoring the relationship of humans with the soil and the landscape they inhabit, since they create sensual and aesthetic experiences which evoke personal and cultural associations, triggering emotional responses (Toland and Wessolek 2010). In recent years, a large number of publications deal with soil and art, and there is a lot of evidence indicating the interest that many artists have in soil (Toland and Wessolek 2014; van Breemen 2010). Exhibitions strongly benefit when they include or emphasize the aesthetic aspects of soil as colors and textures. For example, the exhibition "Dig It!" includes soil profile monoliths as artistic expressions (Drohan et al. 2010). The wonderful profiles painted in watercolors by Kubiena (1954) as teaching materials to illustrate the different morphologies of soils in Europe are first of all beautiful, and win the hearts of the general public, not just soil scientists. Elvira Wersche has put together different colored soils sampled from around the world (*Sammlung Weltensand*), and created a color palette using distinct hues and textures.

A remarkable collection is located at the altar of the Schneverdingen church in Lower Saxony in Germany. The altar is decorated with more than 5000 acrylic books filled with soil from a specific part of the world, which also tells a story reminding people the soil exists everywhere and is a form of connection.

The exhibition "Rooted in soils" at the DePaul Art Museum explores the underappreciated role of soil in human life and connects art with environmental science (http://resources.depaul.edu/newsroom/news/press-releases/Pages/rooted-in-soil.aspx-).

Other artistic expressions that promote the perception of soils are cinema (Landa 2010) and audiovisuals. Among the latter are "Soil Saturdays" and "Nature Is Speaking." Also pottery out of clays from a specific region and painted with the colors of the soil foster emotional bonds to soil (Box 17.5).

BOX 17.5

"The Hand of Maize" and "The Man of Birds" made by artist Manuel Reyes from Yanhuitlán, Oaxaca, Mexico. He uses different local clays depending on the qualities of the clay mix; he creates sculpture and masks, among other utilitarian pieces, which he paints with natural pigments and with a mix of different kinds of soils to create a broad range of paint colors, tones, and textures. The pieces reflect the color of the surrounding landscape. Photos courtesy of Manuel Reyes.

In 2015, the sister cities Mexico and Denver exchanged a 30-m drill core to symbolize the existence of one culture (implant) within the grounds of the other. The implant project was designed by Mexican artist Marcela Armas (implant.marcelaarmas.net) as a reflection of human intervention on Earth.

Activities like urban gardening, in schoolyards or as part of city programs to alleviate unemployment or food scarcity, have a collateral positive effect on psychic and physical health, when people get involved, or have a social benefit, when neighbors or classmates work together to produce something (Bellows and Hamm 2003; Wakefield et al. 2007).

These activities produce pleasure and fun, by simply touching the soil or feeling the warm and humid soil by walking on it with bare feet. Children enjoy being connected with nature; as Wilson (2011) from New Zealand says: "They need to be outside, they need to explore, get dirty, find stuff—they need to have fun."

Finally, it is important to identify the concept of "Terroir," which Stahr (2015) mentions as an additional opportunity to bring soil closer to the citizens. Terroir is a concept that describes a location with specific soil, topography, climate, landscape, and biodiversity attributes, which defines a geographic region and confers a distinctive feature to wine produced out of grapes that are grown on it. The terroir is used widely by winemakers to promote their wines. All those people who enjoy a good glass of wine get to "taste" the characteristic flavor of soil. The concept is increasingly applied to fruits, vegetables, cheese, olive oil, coffee, cacao, and other crops, creating a link between the uniqueness and quality of the product to the soil that produces it and giving the consumer a sense of place (Costantini et al. 2014).

17.7 CONCLUDING REMARKS

The urban population has lost the perception of soil as a life-sustaining natural resource. Strategies to promote awareness are needed for all stakeholders. This includes scientific researchers, since currently only less than 2% of the members of the soil science community are actively studying urban soil environments. Professionals making decisions in cities also have a clear lack of knowledge with respect to soil properties and functions, since they do not receive adequate training in basic soil concepts. This is due to the fact that the soil curricula are primarily offered by department and faculties that train decision makers working in rural areas. There is a strong need to foster not only interdisciplinarity, but also to include soil science curricula in urban planning, civil engineering, and architecture. The curricula in urban soil science offers numerous opportunities to enhance awareness about soils.

Communication of research findings through scientific papers and books has increased strongly since 2000. Also congresses, symposia, and workshops have been increasingly organized. Yet, government agencies should assign a permanent budget line so that their employees can periodically assist in these events. This is a cost-effective strategy to update personnel and to foster exchange between the academic and government sectors.

Schoolbooks do not include soil as a topic in primary schools; however, a large variety of "learning by doing" and "hands-on" strategies are available on the Internet for including soil-related topics in school's curricula at all levels from kindergarden (K) to the 12th grade (K–12). Digital media are very effective in spreading information to a broad range of audiences. Even researchers can take advantage of open access publications.

Local authorities can play an important role through direct contact with landowners and regulation and surveillance procedures. The GSP will strongly benefit if it fosters not only collaboration between national ministries, but also between local governments. Many existing enthusiastic activities are indicative of the success of soil awareness strategies implemented at the local level, including school teaching, local city governments, and initiatives of civil associations.

The most effective strategy to increase awareness about the importance of soil to human well-being must consider the following three aspects:

1. To increase the mental understanding about soil functioning and related soil ecosystem services, and help to enhance the need for "soil care" in the human **mind**
2. To foster the acquisition of understanding by providing experiences, that is, "learning by doing" and therefore including the participation of the **body**
3. To create emotional links to the soil (**soul**), that is, spiritual connections to soils

In the latter, artists, filmmakers, and writers play an important role. Knowledge will only be transferred if it is assimilated, not only by the brain, but also by the soul.

REFERENCES

Achankeng, E. 2003. Globalization, urbanization and municipal solid waste management in Africa. In *African Studies Association of Australasia and the Pacific AFSAAP 2003 26th Annual Conference*, 1–22. Adelaide, South Australia.

Aguilar, A. G. 2008. Peri-urbanization, illegal settlements and environmental impact in Mexico City. *Cities* 25(3): 133–145.

Artmann, M. 2014. Institutional efficiency of urban soil sealing management—From raising awareness to better implementation of sustainable development in Germany. *Landscape and Urban Planning* 131: 83–95.

Barrera-Bassols, N. and J. A. Zinck. 2003. Ethnopedology: A worldwide view on the soil knowledge of local people. *Geoderma* 111(3): 171–195.

Bell, F. G. 2013. *Engineering Properties of Soils & Rocks*, 3rd edn. Oxford, UK: Elsevier.

Bellows, A. C. and M. W. Hamm. 2003. International origins of community food security policies and practices in the U.S. *Critical Public Health, Special Issue: Food Policy* 13(2):107–123.

Berardi, U. G., A. H. GaffiaranHosseini, and A. GaffiaranHosseini. 2014. State-of-the-art analysis of the environmental benefits of green roofs. *Applied Energy* 115: 411–428.

Bertram, C. and K. Rehdanz. 2014. The role of urban green space for human well-being. *Ecological Economics* 120(1911): 1–39.

Blume, H. P. and E. Schlichting (Eds). 1982. Soil problems in urban areas. *Mitteilungen Deutsche Bodenkundliche Gesellschaft* 33: 1–280.

Brown, S., A. Coday, and I. Frame. 2016. Managing golf greens: Aligning golf green quality with resource inputs. *Urban Forestry and Urban Greening.* doi:10.1016/j.ufug.2016.06.012.

Burghardt, W. 1994. Soils in urban and industrial environments. *Zeitschrift für Pflanzenernährung und Bodenkunde* 157(3): 205–214.

Burghardt, W., J. L. Morel, S. A. Tahoum, G. L. Zhang, R. K. Shaw, A. Boularbah, P. Charsynski, C. Siebe, and K.-H. J. Kim. 2016. Activities of SUITMA: From origin to future. In: Levin, J.M., K.-H. J. Kim, J. L. Morel, W. Burghardt, P. Charzynski, R. K. Shaw (Eds)., *Soils within Cities*, pp. 15–18. Stuttgart Germany: CATENA, Schweizerbart Science Publishers.

Capeche, C. L. 2010. *Educação Ambiental Tendo O Solo Como Material Didático: Pintura Com Tinta de Solo E Colagem de Solo Sobre Superfícies. Embrapa Solos, 123*. Rio de Janeiro, Brasil.

Cogger, C. G. 2005. Potential compost benefits for restoration of soils disturbed by urban development. *Compost Science and Utilization* 13(4): 243–251.

Cohen, B. 2006. Urbanization in developing countries: Current trends, future projections, and key challenges for sustainability. *Technology in Society* 28: 63–80.

Costantini, E. V., G. Jones, S. Mocali, and A. Cerdà (Eds.). 2014. Geosciences and wine: The environmental processes that regulate the terroir effect in space and time. *SOIL* 2 (Special Issue). http://www.soil-journal .net/special_issue2.html.

Cram, S., H. Cotler, L. M. Morales, I. Sommer, and E. Carmona. 2008. Identificación de los servicios ambientales potenciales de los suelos. En: El paisaje urbano del Distrito Federal. *Investigaciones Geográficas, Boletín Del Instituto de Geografía, UNAM* 66: 81–104.

Davies, R. and S. Hall. 2010. Direct and indirect effects of urbanization on soil and plant nutrients in desert ecosystems of the Phoenix metropolitan area, Arizona (USA). *Urban Ecosystems* 13(3): 295–317.

Dent, B. B., S. L. Forbes, and B. H. Stuart. 2004. Review of human decomposition processes in soil. *Environmental Geology* 45(4): 576–585.

Drohan, P. J., J. L. Havlin, J. P. Megonigal, and H. H. Cheng. 2010. The "Dig It!" Smithsonian soils exhibition: Lessons learned and goals for the future. *Soil Science Society of America Journal* 74: 697–705.

Effland, W. R. and R. V. Pouyat. 1997. The genesis, classification, and mapping of soils in urban areas. *Urban Ecosystems* 1(4): 217–228.

Feller, C., E. R. Landa, A. R. Toland, and G. Wessolek. 2015. Case studies of soil in art. *SOIL* 1(2): 543–559.

Fernández-Buces, N., C. Siebe, S. Cram, and J. L. Palacio. 2006. Mapping soil salinity using a combined spectral response index for bare soil and vegetation: A case study in the former lake Texcoco, Mexico. *Journal of Arid Environments* 65(4): 644–667.

Fiedler, S., J. Breuer, C. M. Pusch, S. Holley, J. Wahl, J. Ingwersen, and M. Graw. 2012. Graveyards—Special landfills. *Science of the Total Environment* 419: 90–97.

Field, D. J., A. J. Koppi, L. E. Jarrett, L. K. Abbott, S. R. Cattle, C. D. Grant, A. B. McBratney, N. W. Menzies, and A. J. Weatherley. 2011. Soil science teaching principles. *Geoderma.* 167–168: 9–14

Galušková, I., L. Boruvka, and O. Drábek. 2011. Urban soil contamination by potentially risk elements. *Soil and Water Research* 6(2): 55–60.

GDF. 1996. Gobierno del Distrito Federal. Ley de Desarrollo Urbano del Distrito Federal. *Gaceta Oficial del Distrito Federal*, 29 de enero de 1996.

Goddard, M. A., A. J. Dougill, and T. G. Benton. 2010. Scaling up from gardens: Biodiversity conservation in urban environments. *Trends in Ecology and Evolution* 25(2): 90–98.

Golden S. 2013. Urban agriculture impacts: Social, health, and economic: A literature review. UC Sustainable Agriculture Research and Education Program Agricultural Sustainability Institute at UC Davis. http://asi.ucdavis.edu/programs/sarep/publications/food-and-society/ualitreview-2013.pdf (accessed October 20, 2016).

Gottlieb, R. S. 2004. *This Sacred Earth: Religion, Nature, Environment.* New York, NY: Routledge, 673 p.

Greinert, A. 2015. The heterogeneity of urban soils in the light of their properties. *Journal of Soils and Sediments* 15(8): 1725–1737.

Hasse, J. E. and R. G. Lathrop. 2003. Land resource impact indicators of urban sprawl. *Applied Geography* 23: 159–175.

Hillel, D. 2005. Civilization, role of soils. In: D. Hillel, J.H. Hatfield, D.S. Powlson, C. Rosenzweig, K.M. Scow, M.J. Singer, and D.L. Sparks (Eds.), *Encyclopedia of Soils in the Environment*, pp. 199–204, Elsevier/Academic Press.

Jazcilevich, A., E. Gacría-Chávez, C. Estrada, Aguillón, and C. Siebe. 2015. Retos y oportunidades para el aprovechamiento y manejo ambiental del ex lago de Texcoco. *Boletín de la Sociedad Geológica Mexicana* 67(2): 145–166.

Johnson, E. A. and K. M. Catley. 2009. Urban soil ecology as a focal point for environmental education. *Urban Ecosystems* 12: 79–93.

Johnston, R. J. and S. K. Swallow. 2010. *Economics and Contemporary Land Use Policy: Development and Conservation at the Rural-Urban Fringe.* Washington, DC: RFF Press.

Jonker, C. and J. Olivier. 2012. Mineral contamination from cemetery soils: Case study of Zandfontein Cemetery, South Africa. *International Journal of Environmental Research and Public Health* 9: 511–520.

Kubiena, W. L. 1954. *Atlas of Soil Profiles.* London: Thomas Murby and Company.

Lal, R. and B. Augustin. 2012. *Carbon Sequestration in Urban Ecosystems.* The Netherlands: Springer.

Landa, E. R. and Feller, C. (Eds). 2009. *Soil and Culture.* Dordrecht, Heidelberg, London, New York, Netherlands: Springer, 488 pp.

Landa, E. R. 2010. In a supporting role: Soil and the cinema. In: Landa, E. R. and C. Feller (Eds.), *Soil and Culture*, pp. 83–105. Doordrecht, The Netherlands: Springer, Science+Business Media B.V., 465 p.

Lehmann, A. and Stahr, K. 2007. Nature and significance of antrhopogenic urban soils. *Journal of Soils and Sediments* 7(4): 247–260.

Löki, V., J. Tökölyi, K. Süveges, Á. Lovas-Kiss, K. Hürkan, G. Sramkó, and V. A. Molnár. 2015. The orchid flora of Turkish graveyards: A comprehensive field survey. *Willdenowia* 45: 231–243.

Lot, A., M. Pérez-Escobedo, G. Gil-Alarcon, S. Rodríguez-Palacios, and P. Camarena. 2012. *La Reserva Ecológica Del Pedregal de San Ángel: Estudios Ecosisteémicos*, Ed. ICyTDF UNAM. México: Reserva Ecológica Del Pedregal de San Ángel.

Lou, Y., X. Minggang, C. Xianni, H. Xinhua, and Z. Kai. 2012. Stratification of soil organic C, N and C:N ratio as affected by conservation tillage in two maize fields of China. *Catena* 95: 124–130.

Menegat, R., M. L. Porto, C. C. Carraro, and L. A. Dávila Fernandez. 1998. *Atlas Ambiental de Porto Alegre.* Brasil: Universidade Federal do Rio Grande do Sul.

Meuser, H. 2010. *Contaminated Urban Soils.* Doordrecht, The Netherlands: Springer, Science+Business Media B.V., 333 p.

Miller, J. R. 2005. Biodiversity conservation and the extinction of experience. *Trends in Ecology and Evolution* 20(8): 430–434.

Molla Ruiz-Gómez, M. 2006. The irregular settlements in conservancy areas Tlapans´s municipality. *Investigaciones Geográficas* 60: 83–109.

Muggler, C. C. 2006. Educação em solos: Principios, teoria e métodos (1) seção vii -ensino da ciência do solo. *Revista Brasileira de Ciência do Solo* 30: 733–740.

Muggler, C. C. 2012. Raising awareness about soil diversity: The Education Programme of the Earth Sciences Museum Alexis Dorofeef, Minas Gerais, Brazil. *Geophysical Research Abstracts* 14. EGU2012-14428-1: Z-138.

Müller, K. 2015. Boden und Bildung—Was ist zu tun? In: Wessolek, G. (Ed.), *Von ganz unten. Warum wir unsere Böden Besser Schützen Müssen*, pp. 285–298. München: oekom Verlag.

Pagiola, S., E. Ramírez, J. Gobbi, C. de Haan, M. Ibrahim, E. Murgueitio, and J. P. Ruíz. 2007. Paying for the environmental services of silvopastoral practices in Nicaragua. *Ecological Economics* 64(2): 374–385.

Palacio-Prieto, J. L., E. Rosado-González, X. Ramírez-Miguel, O. Oropeza-Orozco, S. Cram-Heydrich, M. A. Ortiz-Pérez, J. M. Figueroa-Mah-Eng, and G. Fernández de Castro-Martínez. 2016. Erosion, culture and geoheritage: The case of Santo Domingo Yanhuitlán, Oaxaca, México. *Geoheritage* 8(4):359–369.

Pavlovskaya, M. 2002. Mapping urban change and changing GIS: Other views of economic restructuring. *Gender, Place and Culture* 9(3): 281–289.

Peng, C., C. Weiping, L. Xiaolan, W. Meie, O. Zhiyun, J. Wentao, and B. Yang. 2011. Polycyclic aromatic hydrocarbons in urban soils of Beijing: Status, sources, distribution and potential risk. *Environmental Pollution* 159(3): 802–808.

Pickett, S. T. A. and M. L. Cadenasso. 2009. Altered resources, disturbance, and heterogeneity: A framework for comparing urban and non-urban soils. *Urban Ecosystems* 12(1): 23–44.

Postel, S. L. and B. H. Thompson. 2005. Watershed protection: Capturing the benefits of nature's water supply services. *Natural Resources Forum* 29(2): 98–108.

Puhalla, J., J. Krans, and M. Goatley. 1999. *Sport Fields: A Manual for Design, Construction and Maintenance*. Chelsea, MA: Ann Arbor Press.

Rahman, A. H. M., M. Anisuzzaman, S. A. Haider, F. Ahmed, A. K. M. Rafiul Islam, and A. T. M. Naderuzzaman. 2008. Study of medicinal plants in the graveyards of Rajshahi city. *Research Journal of Agriculture and Biological Sciences* 4(1): 70–74.

Sandifer, P. A., A. E. Sutton-Grier, and B. P. Ward. 2015. Exploring connections among nature, biodiversity, ecosystem services, and human health and well-being: Opportunities to enhance health and biodiversity conservation. *Ecosystem Services* 12: 1–15

Scalenghe, R. and F. A. Marsan. 2009. The anthropogenic sealing of soils in urban areas. *Landscape and Urban Planning* 90(1): 1–10.

Setälä, H., R. D. Bardgett, K. Birkhofer, M. Brady, L. Byrne, P. C. Ruiter, F. T. de Vries, C. Gardi, K. Hedlund, L. Hemerik, S. Hotes, M. Lüri, S. R. Mortimer, M. Pavao-Zuckerman, R. Pouyat, M. Tsiafouli, and W. H. van der Putten. 2014. Urban and agricultural soils: Conflicts and trade-offs in the optimization of ecosystem services. *Urban Ecosystems* 17(1): 239–253.

Siebe, C., S. Cram, and N. Fernandez. 2004. Loss of soil functions and potentials by urbanization projects in Mexico-city: Its importance for land use planning and decision making. In *Eurosoil*. Freiburg, Germany: Institute of Soil Science and Forest Nutrition.

Stahr, A. 2015. Bodenbewusstsein oder: Die Wahrnehmung des Bodens. Ahabc.de. Das Magazin für Boden und Garten. http://www.ahabc.de/bodenbewusstsein-oder-die-wahrnehmung-des-bodens/ (consulted October 14, 2016).

Szulczewski, M. 2015. *Soil's Social and Cultural Connections*. Open Access. Soil Horizons. Soil Science Society of America (Eds.). Madison, WI: The American Society of Agronomy. doi:10.2136/sh2015-56-6-gc.

Tobasura Acuña, I., F. H. Obando Moncayo, F. A. Moreno Chávez, C. S. Morales Londoño, and H. Castaño. 2015. From soil conservation to land husbandry: An ethical-affective proposal of soil use. *Ambiente and Sociedade* 18(3): 121–136.

Toledo, V. M. 1994. La diversidad biológica de México. *Ciencias* 34: 43–57.

Toledo, V. M. and N. Barrera-Bassols. 2008. *La Memoria Biocultural: La Importancia Ecológica de Las Sabidurias Tradicionales*, Consejería de Agricultura y Pesca. Junta de Andalucía (Eds.). Barcelona: Icaria and Editorial.

Thompson, K., S. Colsell, J. Carpenter, R. M. Smith, P. H. Warren, and K. J. Gaston. 2005. Urban domestic gardens (VII): A preliminary survey of soil seed banks. *Seed Science Research* 15(2): 133–141.

Toland, A. 2015. Soil art. Transdisciplinary approaches to soil protection. Doctoral thesis at the TU Berlin. Department of Soil Protection. 206 p.

Toland, A. R. and G. Wessolek. 2010. Merging horizons—Soil science and soil art. In: Landa, E. R. and C. Feller (Eds.), *Soil and Culture*, pp. 45–66. Dordrecht, Heidelberg, London, New York, Netherlands: Springer Science+Business Media B.V.

Toland, A. and G. Wessolek. 2014. Picturing soil: Aesthetic approaches to raising soil awareness in contemporary art. In: Churchman, G. J. and E. R. Landa (Eds), *The Soil Underfoot. Infinite Possibilities for a Finite Resource*, Chap. 7, pp. 83–102. Boca Raton, FL: CRC Press, Taylor & Francis Group.

Tong, C. and Y. Dong. 2007. Characteristics of soil carbon pool in urban ecosystem. *Chinese Journal Ecology* 10: 1616–1621.

Tzoulas, K., K. Korpela, S. Venn, V. Yli-Pelkonen, A. Kaźmierczak, J. Niemela, and P. James. 2007. Promoting ecosystem and human health in urban areas using green infrastructure: A literature review. *Landscape and Urban Planning* 81(3): 167–178.

United Nations. 2014. *World Urbanization Prospects: The 2014 Revision, Highlights (ST/ESA/SER.A/352)*. New York.

United Nations. 2016. *World Economic Situation and Prospects*. United Nations publication, Sales No. E.16. II.C.2

van Breemen, N. (2010). Transcendental aspects of soil in contemporary visual arts. In: Landa, E. R. and C. Feller. (Eds.), *Soil and Culture*, pp. 37–43. The Netherlands: Springer.

Wakefield, S., F. Yeudall, C. Taron, J. Reynolds, and A. Skinner. 2007. Growing urban health: Community gardening in South-East Toronto. *Health Promotion International* 22(2): 92–101.

Warming, M., M. G. Hansen, P. E. Holm, J. Magid, T. H. Hansen, and S. Trapp. 2015. Does intake of trace elements through urban gardening in Copenhagen pose a risk to human health? *Environmental Pollution* 202: 17–23.

Watson, G. W, A. M. Hewitt, M. Custic, and M. Lo. 2014. The management of tree root systems in urban and suburban settings: A review of soil influence on root growth. *Arboriculture and Urban Forestry* 40: 193–217.

Wei, B. and L. Yang. 2010. A review of heavy metal contaminations in urban soils, urban road dusts and agricultural soils from China. *Microchemical Journal* 94(2): 99–107.

Wessolek, G. and A. Toland. 2010. Merging horizons—Soil science and soil art. In: Feller, C. and E. R. Landa. (Eds.), *Soil and Culture*. Dordrecht, Heidelberg, London, New York, Netherlands: Springer Science+Business Media B.V.

Wiggs, G. F. S., S. L. O'hara, J. Wegerdt, J. Van Der Meer, I. Small, and R. Hubbard. 2003. The dynamics and characteristics of aeolian dust in dryland Central Asia: Possible impacts on human exposure and respiratory health in the Aral Sea Basin. *Geographical Journal* 169(2): 142–157.

Wilson, C. 2011. *Effective Approaches to Connect Children with Nature; Principles for Effectively Engaging Children and Young People with Nature*. New Zealand: New Zealand Department of Conservation Government.

Wunder, S. 2005. *Payments for Environmental Services: Some Nuts and Bolts. CIFOR Occasional Paper*, Vol. 42. Bogor Barat, Indonesia: Center for International Forestry Research.

Yang, J.-L. and G.-L. Zhang. 2015. Formation, characteristics and eco-environmental implications of urban soils—A review. *Soil Science and Plant Nutrition* 61(1): 30–46.

INTERNET RESOURCES

https://en.wikipedia.org/wiki/List_of_universities_with_soil_science_curriculum. Accessed October 15, 2016
https://www.universities.com/programs/soil-sciences-degrees. Accessed October 15, 2016
www.thesciencedeofsoil.com
www.soils4teachers.org
www.nrcs.usda.gov
edible.schoolyards.org
Soil.net.org
http://www.bodenwelten.de/navigation/boden-und-bildung
www.umweltbundesamt.at/leistungen/seminare_schulungen/boden_schule/
www.umweltmobile.de
http://www.bodenbuendnis.org
http://www.dgcs.unam.mx/boletin/bdboletin/2015_473.html

http://www.terramovil-igl.unam.mx/
https://soilarts.wordpress.com/2013/03/06/painting-with-soil/
http://resources.depaul.edu/newsroom/news/press-releases/Pages/rooted-in-soil.aspx
Soil Saturdays—Shown at European Parliament Soil Conference https://vimeo.com/146634470
Nature Is Speaking—Edward Norton Is the Soil | Conservation International (CI). https://www.youtube.com/watch?v=Dor4XvjA8Wo

18 Feeding Megacities by Urban Agriculture

Rattan Lal

CONTENTS

18.1 INTRODUCTION

Until the Industrial Revolution (circa 1750), most of the population lived in rural communities. In 1500, there were only four cities in Europe (e.g., Paris, Florence, Rome, and Naples) with a population >100,000 (Burghardt et al. 2015). In contrast, the twenty-first century is an era of urban civilization. The urban population was 50% of the world population of 6.8 billion (bn) in 2008 and is projected to be 60% of the world population of 8.1 bn by 2030, and 70% of the world population of 9.7 bn by 2070 (Lal 2016). Future urban growth will be the greatest in Africa, Asia, and Latin America. The number of megacities (population >10 million) was 1 in 1950 and 28 in 2015, and will be 41 in 2030. It takes ~30,000 ha to provide accommodation and other amenities to 1 million people (Lal 2016). With global increase in annual population of ~75 million between 2015 and 2024, it will take approximately 2.25 million ha (Mha) of new land under urban ecosystems (UEs) to provide basic amenities to the ever growing population. The urban land area of 250–400 Mha in 2015 will increase at the rate of ~2.25 Mha/annum. Thus, urban soils presently represent and will be even a prominent landscape within the urban boundaries in the future. There is a strong and growing need to maintain soil functions and services within urban centers. The quality and amount of food produced depends on the quality and management of urban soil. Understanding properties, temporal and spatial variation, and management of urban soils are important to advancing food security and improving the environment. Therefore, the objective of this chapter is to describe the nature, properties, restoration, and sustainable management of urban soils for advancing food security and improving the environment.

18.2 URBAN ECOSYSTEMS

UEs encompass landscapes that account for exchanges of materials and influence between cities and the surrounding environment. However, UEs are not just expanding rapidly during the twenty-first century, but their inherent characteristics are also changing. Cities, especially those in North America, are no longer compact but rather sprawled and merging with wildlands, and include exurbs, edge cities, and houses interspersed within natural vegetation including forest, savanna, and desert (Pickett et al. 2001, 2011). Also, interspersed within exurbs and suburbs are industrial installations and civil structures (shopping malls, airports, utility companies, and academic and education institutions), along with community gardens, family farms, and recreational grounds. Therefore, research on urban soils and ecosystems must also be all encompassing and multidisciplinary, including soil science, restoration ecology, hydrology, biogeochemistry, climatology, forestry, agronomy and horticulture, pedology, anthropology, economics, health sciences, and others. Indeed, there is a strong need for an integrated approach to long-term and coordinated studies of urban ecological ecosystems (Grimm et al. 2000). Furthermore, study of modern ecosystems would be incomplete without a thorough understanding of ancient but deeply transformed man-made soils such as those of iron ages in Moscow region (Golyeva et al. 2015), the Mayan cities in the Yucatan Province of Mexico (Sweetwood et al. 2009), Aztec culture (Clendinnen 1991), Mesopotamia (Kriwaczek 2014), the Indus Valley (Bates and Mio 2015), and the Yangtze River valley (Keys 2003).

18.3 URBAN SOILS

Urban soils come under the overall category of Anthropic soils (FAO/UNESCO 1990). Drastic perturbations of urban soils hinder numerous ecosystem functions and services, because anthropogenic controls predominate over natural factors with regards to the pedogenic processes. Urbanization leads to strong changes in physical, chemical, and biological properties. Mechanical disturbance changes the substrata and horizonation. Soil horizons are irregular and contain mixture of soils from diverse sources. Substrata of urban soils are refilled or redeposited and contain technogenic materials including gravel, stones, and sandy substances (Yang and Zhang 2015). Substrata may also contain bricks, debris, rubble, and artifacts (e.g., glass, plastic wood, textiles, and coal).

Anthropogenic control of pedogenic processes leads to convergence (Pouyat et al. 2015). Yet, characteristics of urban soils vary widely depending on the degree of soil disturbance, intensity and type of input, the vehicular traffic, and climate. The convergence is strong in properties associated with biogenic processes such as soil organic carbon (SOC) and total nitrogen (TN). In comparison, other properties (e.g., P and K concentration) are controlled by local pedogenic factors such as the parent materials. However, some other properties (e.g., pH) are affected by interaction between anthropogenic and local pedogenic factors (Pouyat et al. 2015). Input of anthropogenic particles also impacts some chemical and geophysical properties of urban soils.

In general, urban soils may contain high amounts of phosphorus (P), because P brought in with food is not recycled back to agricultural lands. It is estimated that 250 Tg of P (Tg = teragram = 10^{12}g = 1 million metric ton), out of 1000 Tg of P that has been globally mined, is retained in urban environments (Yang and Zhang 2015). Urban soils may also be contaminated by trace metals such as Cr, Co, Ni, Cu, Zn, As, Cd, and Pb. A large proportion of home gardens are contaminated with Pb and As (Cheng et al. 2015). Parent materials must also be considered while assessing the concentration of trace elements in urban soils. Thomas and Lavkulich (2015) reported that metal concentrations are affected by geographical location and local management practices. Soil contamination on previously industrial sites may affect productivity and the health hazards of farm produce grown on these soils. Ectomycorrhizal fungi and other biota have a varying level of tolerance to trace metals in a metaliferous soil (Evans et al. 2015). Some urban soils derived from building rubble may release sulfates into the groundwater (Abel et al. 2015). The level of contamination, both of soil and water along with properties of air, may be high in soils derived from technogenic parent martials (Huot et al. 2015). Examples of such Technosols include those developed on industrial deposits (e.g., iron) which also exhibit a high spatial variability and temporal discontinuities. Some highly contaminated and now abandoned industrial lands are termed "brownfields." The term brownfields refers to "abandoned, idled or under-used commercial facilities where expansion or development is complicated by real or perceived environmental contamination" (EPA 2006; Triantafyllidis 2003). Thus, management of soils for urban agriculture (UA) must be based on understanding of site-specific attributes.

Physical properties of urban soils are drastically perturbed. These soils are compacted and have a high skeleton (gravel, stones) content. Soil particles may be derived from industrial materials (e.g., ash, slag, and broken materials). Similar to particles of inorganic matrix, total C may originate from ashes, root, coal, cement, and other sources. In general, urban soils may contain more inorganic carbon than natural soils, and thus have a high C:N ratio. The original structure of the soil is drastically altered; the structure is poor, with little or no aggregation. High bulk density and compaction are widespread problems (Yang and Zhang 2015), along with reduction in total and macroporosity. Surface sealing with asphalt is a common feature of urban soils. Surface sealing and pavements reduce water infiltration rate, and most of the precipitation water is discharged into a sewer system, without recharging the groundwater or wetting the soil profile (Burghardt et al. 2015). Installation of impervious surface in urban lands has a strong effect on microbial biomass C and functional diversity of biota, and decreases the overall SOC pool (Wei et al. 2013; Piotrowska-Dlugosz et al. 2015). Soil sealing with asphalt and cement pavement has numerous adverse effects on soil functions, and can increase the following properties of the soil beneath: pH, CaO and inorganic C contents, EC, and S contents due to the impact of chemicals in asphalt (Kida and Kawahigashi 2015). Thus, application of asphalt on soil leads to alkalization, calcification, and mixing with technogenic material. However, material cycling and exchange with the atmosphere does occur through the asphalt/concrete layer because of the presence of pores in the asphalt/concrete and cracks (Kida and Kawahigashi 2015). Nonetheless, removal of asphalt from an old road can reduce SOC concentration by increasing oxidation, and also affect formation of secondary carbonates through transport of Ca^{+2} into the subsoil. Urban-induced soil sealing can divert agricultural land to unproductive use. For example, soil sealing during 2010–2013 in the Nile Delta covered 1397 ha of highly productive soil. The urban spiral is projected to cover 12.4% of the land in the Nile Delta by 2020, equivalent to an additional 3400 ha of productive soils (Mohamed et al. 2015).

Urbanization has strong adverse impacts on biological properties of urban soils, and this impact is among the leading causes of species extinction because of replacement of natural vegetation by sealed and impervious surfaces, pavements, and buildings, as well as nonnative species. Change in species due to formation of technogenic soils also depends on the specific industry. For example, technogenic soils in close proximity to a soda industry in Poland were enriched in NaCl and $CaCl_2$. These soils provide a unique habitat for halophytic species (Piernik et al. 2015), which can also be used as a diagnostic feature. Bioavailability of pyrine and other contaminants may be affected by the concentration of humic substances (Kobayashi and Sumide 2015). Further, soil contamination (by heavy metals and other pollutants) can adversely impact soil biota (e.g., microbes, earthworms), reduce fungal population, impact base respiration and consumption of SOC, and affect enzyme activity (Yang and Zhang 2015). Halophytic vegetation can predominate in soils affected by the soda industry (Piernik et al. 2015).

Drastic perturbations of soil properties (physical. chemical, biological, and ecological) by anthropologic pressure impact several processes (Figure 18.1) with a strong impact on ecosystem functions and the environment. Temperature transfer to deeper layers can be faster because of high soil density and the presence of artifacts (Miyajima et al. 2015). Hazardous impacts on urban soils are caused by inorganic and organic pollutants from diverse industries and increase

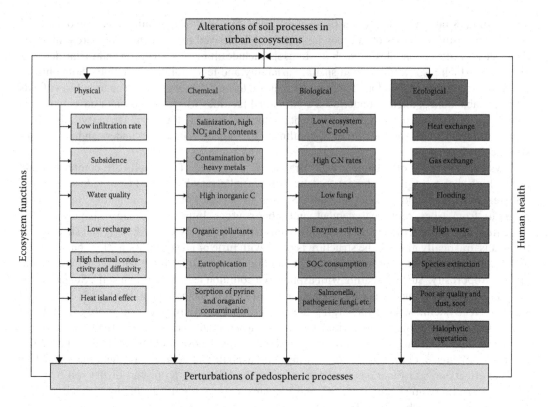

FIGURE 18.1   Adverse impacts of urbanization on processes of drastically disturbed soils.

in pests and pathogens. Environmental impacts are exacerbated by sealing through increase in floods, erosion and high temperature, and traffic and industry-related fumes. Airborne dust can carry heavy metals (Cr and Ni) to long distances (Kim et al. 2015). Despite drastic alterations through urbanization, geophysical properties have not been widely studied (Burghardt et al. 2015). Urbanization-induced alterations in soil processes can create some disservices, especially in relation to adverse effects on the environment and on health of human and other biota (Figure 18.1). These disservices and trade-offs are a major concern to UA and the environment, and must be effectively addressed.

18.4   HAZARDOUS SOIL PROCESSES CAUSED BY URBANIZATION

Specific activities responsible for disservices and trade-offs in urban soils and UEs are (1) complete vegetation removal and destruction of habitat, (2) disruption in waterways, (3) removal of topsoil and compaction of subsoil, (4) surface sealing by asphalt and concrete pavement, (5) high runoff, transport of pollutants into natural waters causing eutrophication, and accelerated erosion of the exposed soil surfaces and of stream banks, (6) air containing dust, soot, and other particulate fine materials (PM2.5, PM10), (7) application of materials containing lime, organic contaminants, and heavy metals, (8) subsidence of soils (peat) from drainage and excessive withdrawal of water on irrigation-induced leaching of carbonates/gypsum, and so on, (9) contamination of groundwater, and (10) extinction of species and loss of biodiversity. Adverse changes in soil quality/health also occur in man-made soils used in greenhouse agriculture. In general, soils used in greenhouse differ significantly from field soils in contents of SOC, total C, total N, NPK fertility, nonexchangeable K, water soluble Ca and SO_4^{-2}, and so on (Sano et al. 2015). Soils under greenhouse use are salinized over time. Prior to using these urbanized soils for food production (raising of crops, fruits, plants, and livestock) and ornamental plants (flowers and shrubs), precautionary measures must be taken to understand site-specific soil-related constraints (physical, chemical, biological, and ecological) and processes which impact amount and quality (heavy metals, pathogens) of produce, and the related water resources.

18.5 QUALITY OF URBAN SOILS

With a strong impact on the ecosystem and the environment, sustainable management of urban soil must be integral to urban planning. Thus, identification of key soil properties and their integration into a soil quality index (SQI) are important to improving UEs. Several SQIs have been developed (Table 18.1) and used for soils of UEs (Vrscaj et al. 2008; Schindelbeck et al. 2008). Characteristics of urban soils vary widely because of vast differences in soils of UEs, depending on land-use history and any industrial facilities in close proximity. Thus, identification of soil properties for use into a SQI must be site specific. Properties to be included in a SQI for urban soils must reflect hydrological processes (storm runoff, infiltration, and drought), gaseous emission and respiration rate (CO_2, CH_4, N_2O, NH_3, Hg, etc.), the heat island effect, low soil fertility, and reduced and variable agronomic productivity. While urban soils are used for many purposes, the specific objective for UA is to produce food within the urban limits. It is the use for food production on contaminated and polluted soils that has raised many concerns (Jim 1998). The SQI must also be pertinent to the choice of best management practices (BMPs) for sustainable intensification (SI) to produce food for rapidly expanding cities (Drechsel and Zimmerman 2005). For example, the SQI must be appropriate to the problem of managing soil fertility (Jin et al. 2011; Okello et al. 2013), restoring soils contaminated with heavy metals (Cheng et al. 2015), and reducing risks of eutrophication of water courses in vicinity of urban centers by use of animal manure (Jia et al. 2015). Similarly, assessment must also be made of overland flow from urban centers (Ferreira et al. 2015) exacerbated by surface sealing (Artmann 2014), and the attendant impacts on groundwater and its quality (Mohrlok and Schiedek 2007; Kramer et al. 2015). Anthropogenically transformed flood-plain soils of Moscow are characterized by the presence of newly formed phosphates of Fe and Ca in the lower horizon. This is a typical diagnostic feature of urban-illuvial pedogenesis (Prokojeva et al. 2010). Surface layers of urban soils in industrial areas are also enriched in magnetite (Fe_3O_4) (Vodyanitskii 2013)

TABLE 18.1
Some Examples of Soil Quality Indexes for Urban Soils

Location	Country	Objective	References
Ithaca	NY, USA	Soil quality assessment protocol, and selection of soil quality indicators (SQI) = Cornell Soil Health Test	Schindelbeck et al. (2008)
Ascona	Switzerland	SQI for urban land use to support public services and good environment quality management	Vrscaj et al. (2008)
British Colombia	Canada	Fire-induced loss of soil organic matter by 10% is a warning level and by 20% is critical. These are critical thresholds of soil organic matter for urban forests	Blanco et al. (2014)

which can be used in mapping such soils. Use of GIS and web-based systems (Hossack et al. 2004) may be useful.

18.6 URBAN WASTE AS A SOIL AMENDMENT

Using the power of soil and plants to strengthen the restorative processes of nature is the best strategy (Ingram 2014) to enhance ecological functionality of urban landscapes. Just as crop residue isn't waste (Lal 2007), so refuse isn't rubbish (Stocking and Albaladejo 1994). Urban waste can be a precious soil amendment for restoring degraded soils of urban and peri-UEs including the adjacent watersheds. Thus, urban waste—called by a range of terms including municipal solid waste (MSW), composted urban waste (CUW), or solid urban waste (SUW) among others—can be a major asset as an amendment to restore the quality of urban soils (Table 18.2). While enhancing SOC concentration, application of compost from MSW can improve physical,

TABLE 18.2
Impacts of Compost Made from Urban Waste on Soil Quality and Agronomic Yield

Region	Country	Composting Material	Experiment Duration (years)	Impact on Urban Soil	References
Mediterranean	Spain	MSW	17	Improvement in SOC, aggregation, AWC, microbial activity, and overall soil quality	Bastida et al. (2007)
Mediterranean	Central Spain	CUW	2	Reduced bulk density, improved soil structure, enhanced aggregate stability, improved growth of pine	De Sa et al. (2008)
Mediterranean	Spain	–	–	Increased AWC, improved nutrient supply, decreased runoff and restored soil	Stocking and Albaladejo (1994)
Europe	Italy	–	7	Increased macroporosity, reduced bulk density, increased enzyme activity	Giusquiani et al. (1995)
Europe	Italy	CUW	3	Compost increased bulk density due to inert fraction content, improved soil physical properties, reduced maize yield	Bazzoffi et al. (1998)
South America	Brazil	SUW	2	Improved P and K and increased soil pH	Ruppenthal and Castro (2000)
South America	Argentina	MSW	–	Increased SOC, TN, and available P, reduced bulk density, increased infiltration rate, and increased plant growth	Civeira (2010)
South America	Colombia	MSW	–	Increased SOC, control erosion on watershed level, increased biomass	Binder and Patzel (2001)
West Africa	Burkina Faso	SUW	10	Improved sorghum yield	Kiba et al. (2012)

AWC, available water capacity; CUW, composted urban waste; MSW, municipal urban waste; SUW, solid urban waste.

chemical, and biological properties of soils (Giusquiani et al. 1995). However, the main concern is the loading of soil with heavy metals and other pollutants (Hargreaves et al. 2008). Concerns about high concentrations of heavy metals in urban soils have been reported widely from Europe (Plyaskina and Ladonin 2009; Nikiforova and Koshelva 2007; Vodyanitski 2013; Romic and Romic 2003), India (Khurana et al. 2014; Rattan et al. 2005), China (Li et al. 2001, 2004; Wei and Yang 2010; Chen et al. 1997, 2005), Southeast Asia (Wilcke et al. 1998), West Asia (Moller et al. 2005), Australia (Markus and McBratney 1996; Tiller 1992), and elsewhere. The most commonly observed heavy/trace metals in urban areas include Pb, As, Cu, Zn, Cr, Cd, Ni, Ti, Mn, Al, and Co. Contamination of food grown in urban soils is a major health concern, especially for children (Mielke et al. 1999), and must be effectively addressed. Aerosols are also a cause of air quality, health hazard, agricultural productivity, and ecosystem issues (Brodie et al. 2007). Yet little research information exists regarding management of urban soils to reduce aerosols, especially in rapidly urbanizing countries, such as China (Chen 2007) and India, which have severe air quality issues.

While a considerable literature exits regarding the risks of heavy metals and how to manage urban soils to alleviate these problems, little research data are available on management of soil physical geophysical and hydrological properties. High compaction, low water infiltration rate (Lassabatere et al. 2010), and high storm water runoff from urban soils (Gregory et al. 2006) are serious environmental and productivity issues and are a researchable priority. Other soil physical properties of great relevance but without sufficient information include soil surface temperature, thermal conductivity, and temperature-induced mineralization and gaseous exchange (CO_2, CH_4, N_2O, Hg, NH_3, etc.). While research information on the "heat island effect" is available (Weng et al. 2004), information on its relation to soil temperature and the attendant biophysical processes is not. Even soils under urban forests do not oxidize CH_4 at the same rate as those under undisturbed forests (Goldman et al. 1995), and the cause-effect relation must be established.

18.7 DISTINGUISHING FEATURES OF URBAN SOILS FOR MAPPING

High soil compaction, surface seal, and heavy metals are among the distinguishing features of urban soil. Key distinguishing properties of urban soils (Scharenbroch et al. 2005; Lehmann and Stahr 2007; Pouyat et al. 2007) are soil bulk density, texture, microbial biomass and enzymatic activity, SOC and dynamics, infiltration rate, and heavy metals. Magnetic properties of the soil (Hu et al. 2007) are another distinguishing feature and can be used for mapping by using remote sensing techniques. Anthropogenic soils may be more alkaline with 1 or more units higher pH (Yang and Zhang 2015), 3–4 times more electrical conductivity (EC), and up to 20 times more magnetic susceptibility (MS) than undisturbed soils (Howard and Orlicki 2015). Thus, EC and MS can be used as parameters for remote mapping of urban soils. Classification and mapping of urban and industrial soils, following the protocol of the World Reference Base (WRB) (FAO 2015), must be based on some or all of these unique distinguishing features. Mapping is also essential, not only because UEs are expanding rapidly (Lal 2018), but also because this land use must be quantified and characterized for sustainable management of UA.

18.8 SOIL MANAGEMENT FOR URBAN LANDSCAPES

There are numerous concerns with regards to management of Anthropic soils and landscapes (Figure 18.2). While concerns about heavy metals have been widely known, those on physical, hydrological, and biological issues are equally important but not widely addressed. Yet, urban environments are strongly impacted by these interacting factors. Understanding of these concerns also has implications to planning for appropriate land use, and identification of strategies for restoration (Pavao-Zuckerman 2008).

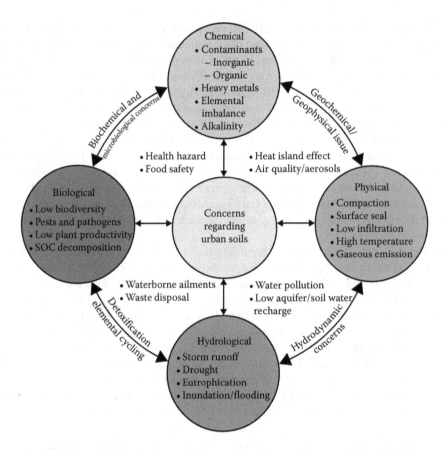

FIGURE 18.2 Concerns regarding use of urban soils for agriculture and horticultural land uses in terms of environment quality and health hazards.

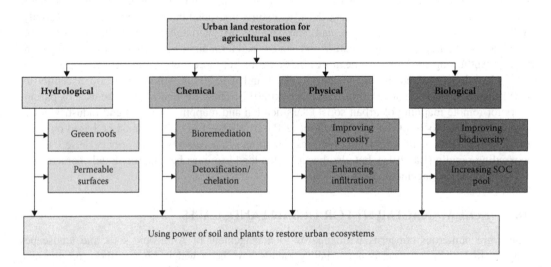

FIGURE 18.3 Principles of restoring urban soils.

On the basis of the concepts outlined shown in Figure 18.3, important options for hydrological restoration include green roofs or vegetative roofs which can mimic or enhance function of vegetated surfaces such as evapotranspiration, infiltration, surface detention, and so on (Palla et al. 2010), and those which moderate runoff. Choice of appropriate species of grass (Fischer et al. 2013) is pertinent for establishing grass cover on home lawns and recreational grounds. Ornamental plants (shrubs and grasses) can play a duel role of improving the aesthetical value and restoring the landscape (Larcher et al. 2012). Along with green roofs, increasing green areas within the city is an important component of bioengineering and ecological restoration (Orsini et al. 2010), and among the innovative concepts for reducing risks (Ho and Li 2003). Thus, turf grasses and their judicious management are important to enhancing urban landscapes (Beard and Green 1994; Selhorst and Lal 2012). Use of road salts for deicing is a major source of pollution of surface waters, leading to an increasing salinity trends in streams and affecting aquatic life (Cooper et al. 2014). Establishing vegetation cover is also pertinent to reducing/moderating gaseous emissions from urban lands (Lompo et al. 2012) and enhancing SOC sequestration (Zirkle et al. 2011).

18.9   SOIL MANAGEMENT FOR UA

Intensively managed soils for UA and forestry have not been widely studied (De Kimpe and Morel 2000). Highly heterogeneous and drastically disturbed Anthrosols are a growing concern because of environmental and health hazards. These soils differ from those in rural areas by the scale and intensity of anthropogenic perturbations (Bullock and Gregory 1991; Craul 1992).

Two considerations in judicious soil management for UA are productivity and quality or safety of food products. Growing urbanization can impact food production by taking agricultural land for nonagricultural uses (Lal 2018; Nizeyimana et al. 2001). Similar to agricultural lands, soils under UA also use water, nutrients, energy, and other resources with low efficiency. Thus, heavy and indiscriminate use of chemicals and uncontrolled water must also be avoided in UA (Sangare et al. 2012).

The efficient use of resources is also linked to soil compaction in UEs. Thus, use of heavy equipment and vehicular traffic must be objectively considered (Materechera 2009). Compaction-related decline in water infiltration rate can increase runoff, and negatively affect eco-environment of the city (Yang and Zhang 2011). Improving soil structure and aggregation is closely linked with increase in SOC concentration (Caravaca et al. 2001) and with population and density of earthworms (Smetak et al. 2007), which varies among different-aged UEs. High bulk density in young urban lands ($1.6 \, Mg \, m^{-3}$) can limit the activity of earthworms compared with that of soil under parks or old residential plots ($1.3 \, Mg \, m^{-3}$) (Smetak et al. 2007). However, not all biotic communities are affected by conversion to urban land use. In Naples, Italy, Santorufo et al. (2015) observed that *Collembolan* communities in urban soils were comparable to those observed in forest soils. Similar to croplands, there are also critical thresholds of soil organic matter (SOM) in soils of UEs. For soils under urban forest, the reduction of SOM by 10% is considered a warning level and 20% a critical level (Blanco et al. 2014). Any increase in SOM in urban soils under agricultural land use is related to increase in productivity of biomass and maintenance of the greenzone (management). On the contrary, decomposition may be reduced by input of organic pollutants and other contaminants (Vodyanitskii 2015). In the US Rust Belt, metalliferous soils have strong adverse effects on biotic activity, such as that of ectomycorrhizal fungi, and on plant productivity (Evans et al. 2015). Thus, an urban environment may increase or decrease SOM/SOC depending on the site-specific situations.

18.10   INNOVATIONS IN URBAN SOIL MANAGEMENT

There are some special features and distinct characteristics of urban soil that must be duly considered for their sustainable management. Specific constraints, outlined in the previous sections, must be alleviated through innovative options such as those listed in the following sections.

18.10.1   Green Roofs

Green roofs—also called eco-roofs, nature roofs, and vegetated roof covers—can provide hydrologic control over storm water runoff. Experiments conducted by Palla et al. (2010) in Genova, Italy, showed that green roofs also reduced storm water pollution. Green roofs are important to moderating flashfloods and improving the urban environments (Schrader and Böning 2006; Palla et al. 2010). Depending on the age of a green roof, new soil formation and improved niche occupancy of collembolans can stabilize the environment. Because of soil formation in the old roofs in Lower Saxony, Germany, Schrader and Böning (2006) observed that pH was lower and SOC and TN concentrations were higher in soil under old compared with young roofs.

18.10.2 PLANNING WATER-CENTRIC CITIES

The concept of ecocities involves restoration of their hydrological functionality while reducing gaseous emissions (Novotny and Novotny 2011). On the basis of this concept, ecosystems can be built even in harsh environments (e.g., arid climate) or decontaminated brownfields. However, cite-specific research is needed to limit recycling and reduce the buildup of contaminants.

18.10.3 SYNTHETIC SOILS

Because of the scarcity of good quality topsoil, the importance of making synthetic soils cannot be over-emphasized. Making synthetic soil is an emerging branch of engineering ecology, soil science, and agriculture (Smagin and Sadovnikova 2015). Synthetic soils have applications for agricultural lands, sports grounds, and construction of structures for geostabilization, soil reclamation, and salt protection. The design criteria, layering characteristics, and material used differ among diverse uses of synthetic soils (Smagin and Sadovnikova 2015).

18.10.4 ECOLOGICAL RESTORATION OF DEGRADED/POLLUTED URBAN SOILS

The strategy of restoration is to create novel soil types that enhance vegetative production even of nonnative species, denature pollutants, reduce salinization, and strengthen biogeochemical and biogeophysical cycling. The goal is to restore the functions of UEs (Pavao-Zuckerman 2008).

18.11 BUILDING UPON HISTORIC KNOWLEDGE

UA has been a pertinent feature of ancient civilizations, including those of the Classic Mayan of the first millennium AD (McNeil et al. 2010), Byzantine and Constantinople (Barthel and Isendahl 2013), the Harappan (Giosan et al. 2012), and the Mesopotamian (Paulette 2012; Charles et al. 2010). Modern civilization, with a population of 9.7 bn by 2050, can learn from the successes and failures of these ancient cultures. Specific attention should be paid to their strategies of addressing recurring drought/flooding syndromes and other uncertainties and risks (Paulette 2012). The water management strategies of the Mayan civilization—multicomponential, diverse across space, and shifting over time—were part of its success (Isendahl 2011). In most cases, UA has been a prominent feature of urban support systems over the long term. With unprecedented and rapid urbanization under a changing and uncertain climate, modern civilization must learn from these ancient cultures, especially now when humanity is faced with uncertainties about our collective future. The challenges facing the C civilization (Lal 2007) of the urban era are even direr than those of the long vanished but once thriving civilizations. The term C civilization refers to the era since 1750 and the dependence of human population on use of C from fossil fuel.

18.12 CONCLUSIONS

- The twenty-first century is the era of urban civilization. The urban land area of 250–400 Mha in 2015 may increase at the rate of 2.25 Mha/yr.
- UEs encompass complex landscapes, and drastically disturbed soils which come under the overall category of anthropic soils. Drastic disturbances hinder numerous ecosystem functions and services.

- Among distinguishing features of urban soils are high compaction, poor structure and aggregation, surface seal, high concentration of heavy metals, low infiltration rate, and high gaseous emissions. Anthropogenic soils may be more alkaline with 1 or more units higher pH, 3–4 times more EC, and up to 20 times more MS than undisturbed soils. These features can be used to map urban soils and develop SQIs.
- Restoration of urban soils is important to increasing productivity and food safety. In addition to increasing the vegetation cover, use of compost from MSW can improve soil structure and increase SOC concentration.
- Innovative options of UA may include bioremediation of soil contaminated by heavy metals, improvement of soil fertility and nutrient balance, and enhancement of water infiltration rate and storage in the root zone. Other innovative options include green roofs, planning of water-centric cities, manufacturing synthetic soils, and adopting principles of ecological restoration.
- UA is not a new concept; it has been practiced for millennia. The humanity of the modern urban era must learn from the successes and failures of ancient civilizations, especially now when faced with uncertainties of our collective future due to changing and uncertain climate, rapidly expanding urban population, growing affluence, and high expectations of living standards.

REFERENCES

Abel, S., A. Peters, T. Nehls, and G. Wessolek. 2015. Long-term release of sulfate from building rubble-composed soil: Lysimeter study and numerical modeling. *Soil Science* 180:175–181.

Artmann, M. 2014. Institutional efficiency of urban soil sealing management—From raising awareness to better implementation of sustainable development in Germany. *Landscape and Urban Planning* 131:83–95.

Barthel, S. and C. Isendahl. 2013. Urban gardens, agriculture, and water management: Sources of resilience for long-term food security in cities. *Ecological Economics* 86:224–234.

Bastida, F., J. Moreno, C. Garcia, and T. Hernandez. 2007. Addition of urban waste to semiarid degraded soil: Long-term effect. *Pedosphere* 17:557–567.

Bates, C. and M. Mio. 2015. *Cities in South Asia*. New York, NY: Routledge (ISBN:9781317565123)

Bazzoffi, P., S. Pellegrini, A. Rocchini, M. Morandi, and O. Grasselli. 1998. The effect of urban refuse compost and different tractors tyres on soil physical properties, soil erosion and maize yield. *Soil and Tillage Research* 48:275–286.

Beard, J. and R. Green. 1994. The role of turfgrasses in environmental-protection and their benefits to humans. *Journal of Environmental Quality* 23:452–460.

Binder, C. and N. Patzel. 2001. Preserving tropical soil organic matter at watershed level. A possible contribution of urban organic wastes. *Nutrient Cycling in Agroecosystems* 61:171–181.

Blanco, J., D. Dubois, D. Littlejohn, D. Flanders, P. Robinson, M. Moshofsky, and C. Welham. 2014. Soil organic matter: A sustainability indicator for wildfire control and bioenergy production in the urban/forest interface. *Soil Science Society of America Journal* 78:S105–S117.

Brodie, E., T. DeSantis, J. Parker, I. Zubietta, Y. Piceno, and G. Andersen. 2007. Urban aerosols harbor diverse and dynamic bacterial populations. *Proceedings of the National Academy of Sciences of the United States of America* 104:299–304.

Bullock, P. and P.J. Gregory. 1991. *Soils in the Urban Environment*. Oxford: Blackwell.

Burghardt, W., J. Morel, and G. Zhang. 2015. Development of the soil research about urban, industrial, traffic, mining and military areas (SUITMA). *Soil Science and Plant Nutrition* 61:3–21.

Caravaca, F., A. Lax, and J. Albaladejo. 2001. Soil aggregate stability and organic matter in clay and fine silt fractions in urban refuse-amended semiarid soils. *Soil Science Society of America Journal* 65:1235–1238.

Charles, M., H. Pessin, and M. Hald. 2010. Tolerating change at Late Chalcolithic Tell Brak: Responses of an early urban society to an uncertain climate. *Environmental Archaeology* 15:183–198.

Chen, J. 2007. Rapid urbanization in China: A real challenge to soil protection and food security. *Catena* 69:1–15.

Chen, T., J. Wong, H. Zhou, and M. Wong. 1997. Assessment of trace metal distribution and contamination in surface soils of Hong Kong. *Environmental Pollution* 96:61–68.

Chen, T., Y. Zheng, M. Lei, Z. Huang, H. Wu, H. Chen, K. Fan, K. Yu, X. Wu, and Q. Tian. 2005. Assessment of heavy metal pollution in surface soils of urban parks in Beijing, China. *Chemosphere* 60:542–551.

Cheng, Z., A. Paltseva, I. Li, T. Morin, H. Huot, S. Egendorf, Z. Su, R. Yolanda, K. Singh, L. Lee, M. Grinshtein, Y. Liu, K. Green, W. Wai, B. Wazed, and R. Shaw. 2015. Trace metal contamination in New York City garden soils. *Soil Science* 180:167–174.

Civeira, G. 2010. Influence of municipal solid waste compost on soil properties and plant reestablishment in peri-urban environments. *Chilean Journal of Agricultural Research* 70:446–453.

Clendinnen, I. 1991. *Aztecs: A History*. London: Macmillan (ISBN: 0-333-12404-9).

Cooper, C., P. Mayer, and B. Faulkner. 2014. Effects of road salts on groundwater and surface water dynamics of sodium and chloride in an urban restored stream. *Biogeochemistry* 121:149–166.

Craul, P.J. 1992. *Urban Soil in Landscape Design*. New York, NY: John Wiley.

De Kimpe, C. and J. Morel. 2000. Urban soil management: A growing concern. *Soil Science* 165:31–40.

De Sa, M., R. Roman, and C. Fortun. 2008. Use of composted urban waste in the reforestation of a degraded calcic regosol in Central Spain. *Agrochimica* 52:23–33.

Drechsel, P. and U. Zimmermann. 2005. Factors influencing the intensification of farming systems and soil-nutrient management in the rural-urban continuum of SW Ghana. *Journal of Plant Nutrition and Soil Science-Zeitschrift Fur Pflanzenernahrung Und Bodenkunde* 168:694–702.

EPA. 2006. Brownfields success story: Solid waste and emergency response (51051). EPA-560-F-06-267, October 2006 (www.epa.gov/brownfields/).

Evans, J., A. Parker, F. Gallagher, and J. Krumins. 2015. Plant productivity, ectomycorrhizae, and metal contamination in urban brownfield soils. *Soil Science* 180:198–206.

FAO. 2015. *World Reference Base for Soil Resources 2014*. Rome: FAO.

FAO/UNESCO. 1990. Soil map of the world. Revised legend with corrections and updates. World Soil Resource Report 60, FAO, Rome.

Ferreira, C., R. Walsh, T. Steenhuis, R. Shakesby, J. Nunes, C. Coelho, and A. Ferreira. 2015. Spatiotemporal variability of hydrologic soil properties and the implications for overland flow and land management in a peri-urban Mediterranean catchment. *Journal of Hydrology* 525:249–263.

Fischer, L., M. von der Lippe, and I. Kowarik. 2013. Urban grassland restoration: Which plant traits make desired species successful colonizers? *Applied Vegetation Science* 16:272–285.

Giosan, L., P. Clift, M. Macklin, D. Fuller, S. Constantinescu, J. Durcan, T. Stevens, G. Duller, A. Tabrez, K. Gangal, R. Adhikari, A. Alizai, F. Filip, S. VanLaningham, and J. Syvitski. 2012. Fluvial landscapes of the Harappan civilization. *Proceedings of the National Academy of Sciences of the United States of America* 109:E1688–E1694.

Giusquiani, P., M. Pagliai, G. Gigliotti, D. Businelli, and A. Benetti. 1995. Urban waste compost—Effects on physical, chemical, and biochemical soil properties. *Journal of Environmental Quality* 24:175–182.

Goldman, M., P. Groffman, R. Pouyat, M. Mcdonnell, and S. Pickett. 1995. CH_4 uptake and N availability in forest soils along an urban to rural gradient. *Soil Biology and Biochemistry* 27:281–286.

Golyeva, A., E. Zazovskaia, and I. Turova. 2016. Properties of ancient deeply transformed man-made soils (cultural layers) and their advances to classification by the example of early Iron Age sites in Moscow Region. *Catena* 137:605–610.

Gregory, J., M. Dukes, P. Jones, and G. Miller. 2006. Effect of urban soil compaction on infiltration rate. *Journal of Soil and Water Conservation* 61:117–124.

Grimm, N., J. Grove, S. Pickett, and C. Redman. 2000. Integrated approaches to long-term studies of urban ecological systems. *Bioscience* 50:571–584.

Hargreaves, J.C., M.S. Adl, and P.R. Warman. 2008. A review of the use of composted municipal solid waste in agriculture. *Agriculture Ecosystems and Environment* 123(1–3):1–14.

Hossack, I., D. Robertson, P. Tucker, A. Hursthouse, and C. Fyfe. 2004. A GIS and web-based decision support tool for the management of urban soils. *Cybernetics and Systems* 35:499–509.

Howard, J. and K. Orlicki. 2015. Effects of anthropogenic particles on the chemical and geophysical properties of urban soils, Detroit, Michigan. *Soil Science* 180:154–166.

Hu, X., Y. Su, R. Ye, X. Li, and G. Zhang. 2007. Magnetic properties of the urban soils in Shanghai and their environmental implications. *Catena* 70:428–436.

Huot, H., M.-O. Simonnot, and J.L. Morel. 2015. Pedogenetic trends in soils formed in technogenic parent materials. *Soil Science* 180(4–5):182–92.

Ingram, M. 2014. Washing urban water: Diplomacy in environmental art in the Bronx, New York City. *Gender Place and Culture* 21:105–122.

Isendahl, C. 2011. The weight of water: A new look at pre-Hispanic Puuc Maya water reservoirs. *Ancient Mesoamerica* 22:185–197.

Jia, W., Z. Yan, D. Chadwick, L. Kang, Z. Duan, Z. Bai, and Q. Chen. 2015. Integrating soil testing phosphorus into environmentally based manure management in pen-urban regions: A case study in the Beijing area. *Agriculture Ecosystems and Environment* 209:47–59.

Jim, C. 1998. Soil characteristics and management in an urban park in Hong Kong. *Environmental Management* 22:683–695.

Jin, J.W., H.C. Ye, Y.F. Xu, C.Y. Shen, and Y.F. Huang. 2011. Spatial and temporal patterns of soil fertility quality and analysis of related factors in urban-rural transition zone of Beijing. *African Journal of Biotechnology* 10(53):10948–10956.

Keys, D. 2003. Ancient cities discovered in Yangtze Valley. *Independent News* 2 August 2003.

Khurana, M., B. Kansal, and R. Setia. 2014. Long-term impact of irrigation with sewage water on cadmium concentration in soils and crops. *Agrochimica* 58:19–34.

Kida, K. and M. Kawahigashi. 2015. Influence of asphalt pavement construction processes on urban soil formation in Tokyo. *Soil Science and Plant Nutrition* 61:135–146.

Kiba, D., F. Lompo, E. Compaore, L. Randriamanantsoa, P. Sedogo, and E. Frossard. 2012. A decade of non-sorted solid urban wastes inputs safely increases sorghum yield in periurban areas of Burkina Faso. *Acta Agriculturae Scandinavica Section B-Soil and Plant Science* 62:59–69.

Kim, H., K. Kim, G. Lim, J. Kim, and K. Kim. 2015. Influence of airborne dust on the metal concentrations in crop plants cultivated in a rooftop garden in Seoul. *Soil Science and Plant Nutrition* 61:88–97.

Kobayashi, T. and H. Sumida. 2015. Effects of humic acids on the sorption and bioavailability of pyrene and 1,2-dihydroxynaphthalene. *Soil Science and Plant Nutrition* 61:113–122.

Kramer, S., B. Monninkhoff, S. Seifert, T. Koch, B. Pfutzner, and F. Reinstorf. 2015. A groundwater management concept against soil wetness in urban areas. *Wasserwirtschaft* 105:32–37.

Kriwaczek, P. 2014. *Babylon: Mesopotamia and the Birth of Civilization*. New York, NY: Thomas Dunne Book.

Lal, R. 2007. Soil science and the carbon civilization. *Soil Science Society of America Journal* 71:1095–1108.

Lal, R. 2018. Urban agriculture in the 21st century. In R. Lal and B.A. Stewart (eds.), *Urban Soils*, Chap. 1. Boca Raton, FL: Taylor & Francis.

Larcher, F., W. Gaino, M. Devecchi, F. Ajmone-Marsan, and G. Groening. 2012. Evaluation of ornamental plants as metals bioaccumulators in urban low contaminated soils. XXVIII International Horticultural Congress on Science and Horticulture For People (IHC2010): International Symposium on Advances in Ornamentals, Landscape and Urban Horticulture 937:1109–1114.

Lassabatere, L., R. Angulo-Jaramillo, D. Goutaland, L. Letellier, J. Gaudet, T. Winiarski, and C. Delolme. 2010. Effect of the settlement of sediments on water infiltration in two urban infiltration basins. *Geoderma* 156:316–325.

Lehmann, A. and K. Stahr. 2007. Nature and significance of anthropogenic urban soils. *Journal of Soils and Sediments* 7:247–260.

Li, X., C. Poon, and P. Liu. 2001. Heavy metal contamination of urban soils and street dusts in Hong Kong. *Applied Geochemistry* 16:1361–1368.

Li, X., S. Lee, S. Wong, W. Shi, and L. Thornton. 2004. The study of metal contamination in urban soils of Hong Kong using a GIS-based approach. *Environmental Pollution* 129:113–124.

Lompo, D., S. Sangare, E. Compaore, M. Sedogo, M. Predotova, E. Schlecht, and A. Buerkert. 2012. Gaseous emissions of nitrogen and carbon from urban vegetable gardens in Bobo-Dioulasso, Burkina Faso. *Journal of Plant Nutrition and Soil Science* 175:846–853.

Markus, J. and A. McBratney. 1996. An urban soil study: Heavy metals in Glebe, Australia. *Australian Journal of Soil Research* 34:453–465.

Materechera, S. 2009. Tillage and tractor traffic effects on soil compaction in horticultural fields used for periurban agriculture in a semi-arid environment of the North West Province, South Africa. *Soil and Tillage Research* 103:11–15.

McNeil, C., D. Burney, and L. Burney. 2010. Evidence disputing deforestation as the cause for the collapse of the ancient Maya polity of Copan, Honduras. *Proceedings of the National Academy of Sciences of the United States of America* 107:1017–1022.

Mielke, H., C. Gonzales, M. Smith, and P. Mielke. 1999. The urban environment and children's health: Soils as an integrator of lead, zinc, and cadmium in New Orleans, Louisiana, USA. *Environmental Research* 81:117–129.

Miyajima, S., N. Uoi, T. Murata, M. Takeda, W. Morishima, and M. Watanabe. 2015. Effect of structural modification on heat transfer through man-made soils in urban green areas. *Soil Science and Plant Nutrition* 61:70–87.

Mohamed, E., A. Belal, and A. Shalaby. 2015. Impacts of soil sealing on potential agriculture in Egypt using remote sensing and GIS techniques. *Eurasian Soil Science* 48:1159–1169.

Mohrlok, U. and T. Schiedek. 2007. Urban impact on soils and groundwater—From infiltration processes to integrated urban water management. *Journal of Soils and Sediments* 7:68–68.

Nikiforova, E. and N. Kosheleva. 2007. Dynamics of contamination of urban soils with lead in the eastern district of Moscow. *Eurasian Soil Science* 40:880–892.

Nizeyimana, E., G. Petersen, M. Imhoff, H. Sinclair, S. Waltman, D. Reed-Margetan, E. Levine, and J. Russo. 2001. Assessing the impact of land conversion to urban use on soils with different productivity levels in the USA. *Soil Science Society of America Journal* 65:391–402.

Novotny, V. and E.V. Novotny. 2011. Water centric cities of the future—Towards macro scale assessment of sustainability. In C. Howe and C. Mitchell (eds.), *Water Sensitive Communities*. London: IWA Publishing.

Okello, J., C. Largerkvist, M. Ngigi, and N. Karanja. 2014. Means-end chain analysis explains soil fertility management decisions by peri-urban vegetable growers in Kenya. *International Journal of Agricultural Sustainability* 12:183–199.

Orsini, F., M. Morbello, M. Fecondini, and G. Gianquinto. 2010. Hydroponic gardens: Undertaking malnutrition and poverty through vegetable production in the suburbs of Lima, Peru. II International Conference on Landscape and Urban Horticulture 881:173–177.

Palla, A., I. Gnecco, and L. Lanza. 2010. Hydrologic restoration in the urban environment using green roofs. *Water* 2:140–154.

Paulette, T. 2012. Domination and resilience in bronze age mesopotamia. In J. Cooper and P. Sheets (eds.), *Surviving Sudden Environmental Change: Understanding Hazards, Mitigating Impacts, Avoiding Disasters*, pp. 167–196. Boulder, Colorado: University Press of Colorado.

Pavao-Zuckerman, M. 2008. The nature of urban soils and their role in ecological restoration in cities. *Restoration Ecology* 16:642–649.

Pickett, S., M. Cadenasso, J. Grove, C. Boone, P. Groffman, E. Irwin, S. Kaushal, V. Marshall, B. McGrath, C. Nilon, R. Pouyat, K. Szlavecz, A. Troy, and P. Warren. 2011. Urban ecological systems: Scientific foundations and a decade of progress. *Journal of Environmental Management* 92:331–362.

Pickett, S., M. Cadenasso, J. Grove, C. Nilon, R. Pouyat, W. Zipperer, and R. Costanza. 2001. Urban ecological systems: Linking terrestrial ecological, physical, and socioeconomic components of metropolitan areas. *Annual Review of Ecology and Systematics* 32:127–157.

Piernik, A., P. Hulisz, and A. Rokicka. 2015. Micropattern of halophytic vegetation on technogenic soils affected by the soda industry. *Soil Science and Plant Nutrition* 61:98–112.

Piotrowska-Dlugosz, A. and P. Charzynski. 2015. The impact of the soil sealing degree on microbial biomass, enzymatic activity, and physicochemical properties in the Ekranic Technosols of Torun (Poland). *Journal of Soils and Sediments* 15:47–59.

Plyaskina, O. and D. Ladonin. 2009. Heavy metal pollution of urban soils. *Eurasian Soil Science* 42:816–823.

Pouyat, R., I. Yesilonis, M. Dombos, K. Szlavecz, H. Setala, S. Cilliers, E. Hornung, D. Kotze, and S. Yarwood. 2015. A global comparison of surface soil characteristics across five cities: A test of the urban ecosystem convergence hypothesis. *Soil Science* 180:136–145.

Pouyat, R., I. Yesilonis, J. Russell-Anelli, and N. Neerchal. 2007. Soil chemical and physical properties that differentiate urban land-use and cover types. *Soil Science Society of America Journal* 71:1010–1019.

Prokojeva, T., O. Varava, S. Sedov, and A. Kuznetsova. 2010. Morphological diagnostics of pedogenesis on the anthropogenically transformed floodplains in Moscow. *Eurasian Soil Science* 43:368–379.

Rattan, R., S. Datta, P. Chhonkar, K. Suribabu, and A. Singh. 2005. Long-term impact of irrigation with sewage effluents on heavy metal content in soils, crops and groundwater—A case study. *Agriculture Ecosystems and Environment* 109:310–322.

Romic, M. and D. Romic. 2003. Heavy metals distribution in agricultural topsoils in urban area. *Environmental Geology* 43:795–805.

Ruppenthal, V. and A. Castro. 2005. Effect of urban waste compost on nutrition and yield of gladiolus. *Revista Brasileira De Ciencia Do Solo* 29:145–150.

Sangare, S., E. Compaore, A. Buerkert, M. Vanclooster, M. Sedogo, and C. Bielders. 2012. Field-scale analysis of water and nutrient use efficiency for vegetable production in a West African urban agricultural system. *Nutrient Cycling in Agroecosystems* 92:207–224.

Sano, S., H. Kongo, and T. Uchiyama. 2015. Characteristics of man-made soils of greenhouse fields in urban areas, Osaka prefecture, Japan. *Soil Science and Plant Nutrition* 61:123–134.

Santorufo, L., J. Cortet, J. Nahmani, C. Pernin, S. Salmon, A. Pernot, J. Morel, and G. Maisto. 2015. Responses of functional and taxonomic collembolan community structure to site management in Mediterranean urban and surrounding areas. *European Journal of Soil Biology* 70:46–57.

Scharenbroch, B., J. Lloyd, and J. Johnson-Maynard. 2005. Distinguishing urban soils with physical, chemical, and biological properties. *Pedobiologia* 49:283–296.

Schindelbeck, R., H. van Es, G. Abawi, D. Wolfe, T. Whitlow, B. Gugino, O. Idowu, and B. Moebius-Clune. 2008. Comprehensive assessment of soil quality for landscape and urban management. *Landscape and Urban Planning* 88:73–80.

Schrader, S. and M. Boning. 2006. Soil formation on green roofs and its contribution to urban biodiversity with emphasis on Collembolans. *Pedobiologia* 50:347–356.

Selhorst, A. and R. Lal. 2012. Effects of climate and soil properties on U.S. homelawn soil organic carbon concentration and pool. *Environmental Management* 50:1177–1192.

Smagin, A. and N. Sadovnikova. 2015. Creation of soil-like constructions. *Eurasian Soil Science* 48:981–990.

Smetak, K., J. Johnsn-Maynard, and J. Lloyd. 2007. Earthworm population density and diversity in different-aged urban systems. *Applied Soil Ecology* 37:161–168.

Stocking, M. and J. Albaladejo. 1994. Refuse isnt rubbish. *Ambio* 23:229–232.

Sweetwood, R., R. Terry, T. Beach, B. Dahlin, and D. Hixson. 2009. The maya footprint: Soil resources of Chunchucmil, Yucatan, Mexico. *Soil Science Society of America Journal* 73:1209–1220.

Thomas, E. and L. Lavkulich. 2015. Anthropogenic effects on metal content in urban soil from different parent materials and geographical locations: A Vancouver, British Columbia, Canada, Study. *Soil Science* 180:193–197.

Tiller, K. 1992. Urban soil contamination in Australia. *Australian Journal of Soil Research* 30:937–957.

Triantafyllidis, T., K. Ho, and K. Li. 2003. Green brownfields and urban development—Innovative concepts, visions and risks. *Geotechnical Engineering Meeting Society's Needs* 3:149–160.

Vodyanitskii, Y. 2013. Biogeochemical role of magnetite in urban soils (review of publications). *Eurasian Soil Science* 46:317–324.

Vodyanitskii, Y. 2015. Organic matter of urban soils: A review. *Eurasian Soil Science* 48:802–811.

Vrscaj, B., L. Poggio, and F. Marsan. 2008. A method for soil environmental quality evaluation for management and planning in urban areas. *Landscape and Urban Planning* 88:81–94.

Wei, Z., S. Wu, S. Zhou, and C. Lin. 2013. Installation of impervious surface in urban areas affects microbial biomass, activity (potential C mineralisation), and functional diversity of the fine earth. *Soil Research* 51:59–67.

Wei, B. and L. Yang. 2010. A review of heavy metal contaminations in urban soils, urban road dusts and agricultural soils from China. *Microchemical Journal* 94:99–107.

Weng, Q., D. Lu, and J. Schubring. 2004. Estimation of land surface temperature-vegetation abundance relationship for urban heat island studies. *Remote Sensing of Environment* 89:467–483.

Wilcke, W., S. Muller, N. Kanchanakool, and W. Zech. 1998. Urban soil contamination in Bangkok: Heavy metal and aluminium partitioning in topsoils. *Geoderma* 86:211–228.

Yang, J. and G. Zhang. 2011. Water infiltration in urban soils and its effects on the quantity and quality of runoff. *Journal of Soils and Sediments* 11:751–761.

Yang, J. and G. Zhang. 2015. Formation, characteristics and eco-environmental implications of urban soils—A review. *Soil Science and Plant Nutrition* 61:30–46.

Zirkle, G., R. Lal, and B. Augustin. 2011. Modeling carbon sequestration in home lawns. *HortScience* 46:808–814.

Appendix

Studies and Data Used in Analysis of Urban SOC

Study	City	References
32	Baltimore, MD	Pouyat et al. (2007)
19	Fort Collins, CO	Kaye et al. (2005)
20	Phoenix, AZ	Kaye et al. (2008)
6	Montgomery County, VA	Chen et al. (2013b)
15	Village landscapes, China	Jia-Guo et al. (2010)
18	Chuncheon, Kangleung, and Seoul, Korea	Jo (2002)
16	Hong Kong, China	Jim (1998)
24	Chongqing Municipality, China	Liu et al. (2013)
22	Hong Kong, China	Kong et al. (2014)
21	Vancouver, BC Canada	Kellett et al. (2013)
42	Atlanta, GA	Selhorst and Lal (2012)
59	Irvine, CA	Townsend-Small and Czimczik (2010)
36	New York, NY	Raciti et al. (2012a)
30	Portland, ME	Selhorst and Lal (2012)
43	New York, NY	Pouyat et al. (2002)
11	Richmond, VA	Gough and Elliot (2012)
12	Phoenix, AZ	Green and Oleksyszyn (2002)
41	Phoenix, AZ	Selhorst and Lal (2012)
44	Bloomington, IN	Schmitt-Harsh et al. (2013)
45	Orlando, FL	Selhorst and Lal (2012)
46	Houston, TX	Selhorst and Lal (2012)
47	San Francisco, CA	Selhorst and Lal (2012)
48	Cheyenne, WY	Selhorst and Lal (2012)
49	Dallas, TX	Selhorst and Lal (2012)
58	Washington, DC	Short et al. (1986)
13	Tianjin Binhai New Area, China	Hao et al. (2013)
38	Moscow, ID	Scharenbroch et al. (2005)
60	Louisville, KY	Trammell et al. (2011)
50	Wichita, KA	Selhorst and Lal (2012)
10	Denver, CO	Golubiewski (2006)
14	Auburn, AL	Huyler et al. (2014)
51	Las Vegas, NV	Selhorst and Lal (2012)
52	Albuquerque, NM	Selhorst and Lal (2012)
53	Seattle, WA	Selhorst and Lal (2012)
54	Denver, CO	Selhorst and Lal (2012)
4	Montgomery County, VA	Campbell et al. (2014)
55	Duluth, MN	Selhorst and Lal (2012)

(Continued)

(Continued)

Study	City	References
35	Baltimore (Gwynns Falls), MD	Raciti et al. (2011)
56	Wooster, OH	Selhorst and Lal (2012)
5	Guangdong Province, China	Chen et al. (2013a)
23	Arlington County, VA	Layman et al. (2016)
25	Melbourne, Australia	Livesley et al. (2010)
2	Kiel, Germany	Beyer et al. (1995, 1996)
63	Kaifeng, China	Sun et al. (2010)
1	Liverpool, England	Beesley (2012)
61	New York, NY	Pouyat et al. (2006)
28	Revda, Russia	Meshcheryakov et al. (2005)
7	Leicester, England	Edmondson et al. (2014)
33	Denver, CO	Pouyat et al. (2009)
26	Melbourne, Australia	Livesley et al. (2016)
31	Baltimore, MD	Pouyat et al. (2006)
34	Baltimore, MD	Pouyat et al. (2009)
57	Minneapolis, MN	Selhorst and Lal (2012)
29	Apalachicola, FL	Nagy et al. (2014)
37	Boston, MA	Raciti et al. (2012b)
9	Leicester, England	Edmondson et al. (2012)
17	Chicago, IL	Jo and McPherson (1995)
39	Chicago, IL	Scharenbroch (unpublished)
62	Moscow, Russia	Vasenev et al. (2013)
8	North East England, England	Edmondson et al. (2015)
27	Stuttgart, Germany	Lorenz and Kandeler (2005)
3	Rostock, Germany	Beyer et al. (2001)
40	Eckernförde, Germany	Schleuß et al. (1998)

Study	n	Depth Range (cm)	Mean SOC (kg m^{-2})	SD SOC (kg m^{-2})
32	122	0–10	0.3	0.0
19	3	0–30	0.7	0.5
20	204	0–30	1.4	0.4
6	18	0–30	1.6	0.1
15	170	0–30	2.4	1.0
18	30	0–60	2.5	0.2
16	50	0–66	2.6	0.4
24	220	0–20	2.6	1.1
22	14	0–15	2.8	0.3
21	20	0–25	3.5	0.0
42	12	0–15	3.5	0.9
59	9	0–20	3.8	1.5
36	62	0–15	4.0	0.5
30	9	0–10	4.2	2.9
43	12	0–15	4.2	0.6
11	60	0–10	4.3	0.2
12	6	0–30	4.5	1.0
41	100	0–15	4.7	1.2
44	12	0–15	4.7	1.6
45	12	0–15	4.7	0.8
46	12	0–15	4.8	0.6

(Continued)

(Continued)

Study	n	Depth Range (cm)	Mean SOC (kg m^{-2})	SD SOC (kg m^{-2})
47	12	0–15	5.1	0.6
48	12	0–15	5.2	0.7
49	12	0–15	5.2	0.4
58	100	0–75	5.3	0.4
13	25	0–30	5.4	0.0
38	48	0–15	5.6	0.3
60	21	0–15	5.6	1.1
50	12	0–15	5.8	0.1
10	14	0–30	6.0	0.1
14	67	0–50	6.0	0.4
51	12	0–15	6.2	0.4
52	12	0–15	6.3	0.9
53	12	0–15	6.6	1.2
54	12	0–15	6.6	1.6
4	64	0–30	6.8	0.5
55	12	0–15	6.9	0.9
35	32	0–100	7.0	0.6
56	12	0–15	7.4	1.0
5	11	0–40	7.9	0.8
23	24	0–100	8.1	1.0
25	6	0–25	8.5	0.7
2	6	0–15	8.7	7.2
63	15	0–100	9.1	2.1
1	9	0–15	9.2	0.1
61	10	0–200	9.2	8.9
28	4	0–50	9.5	4.2
7	45	0–21	9.9	0.2
33	13	0–100	11.0	0.9
26	13	0–30	11.7	1.1
31	18	0–100	12.2	1.1
34	26	0–100	12.3	1.9
57	12	0–15	12.6	1.3
29	28	0–90	13.3	3.5
37	94	0–100	13.6	0.3
9	136	0–100	14.5	1.9
17	24	0–60	16.3	2.3
39	191	0–100	33.1	19.1
62	108	0–150	36.8	29.0
8	55	0–100	48.0	17.0
27	10	0–190	69.5	77.0
3	4	0–235	103.0	71.0
40	8	0–150	134.8	113.0

Study	MAT (°C)	MAP(mm yr^{-1})	Lithology	Stratigraphy	Ecoregion	Vegetation
32	14.5	1,075	Sedimentary	Cenozoic	TeBMF	Trees and grasses
19	8.9	385	Sedimentary	Upper Paleozoic	TeCF	Trees and grasses

(Continued)

(Continued)

Study	MAT (°C)	MAP(mm yr⁻¹)	Lithology	Stratigraphy	Ecoregion	Vegetation
20	23.0	193	Endogenous	Precambrian	DrXS	Trees and grasses
6	10.8	1,038	Sedimentary	Lower Paleozoic	TeBMF	Trees
15	16.7	1,209	Sedimentary	Precambrian	TrSMBF	Trees and grasses
18	12.5	1,450	Sedimentary	Precambrian	TeBMF	Trees and grasses
16	24.1	2,150	Sedimentary	Precambrian	TrSMBF	Trees
24	18.0	1,090	Sedimentary	Lower Paleozoic	TrSMBF	Trees and grasses
22	24.1	2,150	Sedimentary	Precambrian	TrSMBF	Grasses
21	11.1	1,058	Extrusive	Mesozoic	TeCF	Trees and grasses
42	17.0	1,400	Sedimentary	Upper Paleozoic	TeBMF	Grasses
59	19.0	350	Endogenous	Lower Paleozoic	MedV	Grasses
36	12.5	1,080	Sedimentary	Mesozoic	TeBMF	Trees
30	12.5	1,080	Sedimentary	Mesozoic	TeBMF	Trees
43	8.0	1,160	Endogenous	Lower Paleozoic	TeBMF	Grasses
11	14.2	1,115	Extrusive	Lower Paleozoic	TeBMF	Grasses
12	23.9	194	Endogenous	Precambrian	DrXS	Trees and grasses
41	11.7	1,207	Sedimentary	Lower Paleozoic	TeBMF	Trees
44	23.0	210	Endogenous	Precambrian	DrXS	Grasses
45	23.0	1,230	Sedimentary	Cenozoic	TrSMBF	Grasses
46	20.0	1,220	Sedimentary	Cenozoic	TrSGSS	Grasses
47	14.0	510	Sedimentary	Cenozoic	TeCF	Grasses
48	7.0	390	Sedimentary	Cenozoic	TeGSS	Grasses
49	19.0	880	Sedimentary	Lower Paleozoic	TeBMF	Grasses
58	13.2	1,036	Sedimentary	Cenozoic	TeBMF	Grasses
13	13.5	535	Sedimentary	Precambrian	TeBMF	Trees
38	12.1	685	Extrusive	Cenozoic	TeCF	Trees and grasses
60	13.8	1,113	Sedimentary	Lower Paleozoic	TeBMF	Trees
50	14.0	770	Sedimentary	Cenozoic	TeGSS	Grasses
10	9.0	350	Sedimentary	Upper Paleozoic	TeCF	Trees and grasses
14	17.4	1,278	Sedimentary	Upper Paleozoic	TrSMBF	Trees and grasses
51	20.0	110	Endogenous	Precambrian	DrXS	Grasses
52	14.0	240	Endogenous	Precambrian	DrXS	Grasses
53	11.0	940	Extrusive	Cenozoic	TeCF	Grasses
54	10.0	400	Sedimentary	Upper Paleozoic	TeCF	Grasses

(Continued)

(Continued)

Study	MAT (°C)	MAP(mm yr⁻¹)	Lithology	Stratigraphy	Ecoregion	Vegetation
4	12.6	1,052	Sedimentary	Lower Paleozoic	TeBMF	Trees
55	4.0	790	Undifferentiated	Precambrian	TeBMF	Grasses
35	14.5	1,075	Sedimentary	Cenozoic	TeBMF	Grasses
56	10.0	970	Sedimentary	Lower Paleozoic	TeBMF	Grasses
5	20.8	1,850	Sedimentary	Precambrian	TrSMBF	Trees
23	10.8	1,038	Sedimentary	Lower Paleozoic	TeBMF	Trees
25	21.0	647	Sedimentary	Lower Paleozoic	MedV	Trees and grasses
2	7.9	748	Sedimentary	Upper Paleozoic	TeBMF	Trees and grasses
63	14.5	622	Sedimentary	Lower Paleozoic	TeBMF	Trees and grasses
1	9.4	796	Sedimentary	Lower Paleozoic	TeBMF	Grasses
61	12.5	1,080	Sedimentary	Mesozoic	TeBMF	Trees and grasses
28	1.7	575	Endogenous	Upper Paleozoic	TeBMF	Trees and grasses
7	9.7	620	Sedimentary	Lower Paleozoic	TeBMF	Trees and grasses
33	10.0	400	Sedimentary	Upper Paleozoic	TeCF	Trees and grasses
26	21.0	647	Sedimentary	Lower Paleozoic	MedV	Trees and grasses
31	14.5	1,075	Sedimentary	Cenozoic	TeBMF	Trees and grasses
34	14.5	1,075	Sedimentary	Cenozoic	TeBMF	Trees and grasses
57	7.0	750	Undifferentiated	Precambrian	TeBMF	Grasses
29	20.3	1,425	Sedimentary	Cenozoic	TrSMBF	Trees and grasses
37	14.8	1,054	Sedimentary	Mesozoic	TeBMF	Trees and grasses
9	9.7	620	Sedimentary	Lower Paleozoic	TeBMF	Trees
17	10.7	992	Sedimentary	Lower Paleozoic	TeGSS	Trees and grasses
39	10.7	992	Sedimentary	Lower Paleozoic	TeGSS	Trees and grasses
62	5.8	650	Sedimentary	Upper Paleozoic	TeBMF	Trees and grasses
8	8.5	803	Sedimentary	Lower Paleozoic	TeBMF	Trees and grasses
27	10.8	665	Sedimentary	Lower Paleozoic	TeBMF	Trees and grasses
3	8.4	591	Sedimentary	Lower Paleozoic	TeBMF	Trees and grasses
40	8.3	825	Sedimentary	Mesozoic	TeBMF	Trees and grasses

(Continued)

(Continued)

Study	Age	Land use	Population (#)	Population Density (# km⁻²)	Soil Order
32	N/A	Mix	622,793	2,962	Ultisols
19	51–100	Mix	156,480	1,024	Mollisols
20	N/A	Mix	1,537,058	1,080	Mollisols
6	<20	Residential	96,207	95	Ultisols
15	20–50	Mix	394,072,000	434	Ultisols
18	N/A	Mix	3,648,475	15,046	Entisols
16	N/A	Transportation	7,188,000	6,544	Ultisols
24	<20	Mix	18,384,100	3,360	Inceptisols
22	20–50	Mix	7,188,000	6,544	Ultisols
21	N/A	Mix	603,500	5,249	Spodosols
42	<20	Residential	5,475,213	1,218	Ultisols
59	20–50	Parks and openlands	258,386	1,241	Alfisols
36	N/A	Transportation	8,491,079	10,430	Inceptisols
30	51–100	Forest	8,491,079	10,430	Inceptisols
43	<20	Residential	66,666	1,200	Spodosols
11	51–100	Residential	220,289	1,319	Entisols
12	N/A	Residential	1,537,058	1,080	Aridisols
41	20–50	Residential	83,322	1,341	Entisols
44	<20	Residential	1,537,058	1,080	Aridisols
45	20–50	Residential	262,372	898	Entisols
46	<20	Residential	2,239,558	1,352	Vertisols
47	<20	Residential	852,469	6,633	Ultisols
48	<20	Residential	62,845	936	Alfisols
49	20–50	Residential	1,281,047	1,358	Mollisols
58	>100	Parks and openlands	658,893	3,806	Ultisols
13	N/A	Mix	2,423,000	1,067	Inceptisols
38	51–100	Mix	24,767	1,341	Mollisols
60	N/A	Transportation	612,780	709	Ultisols
50	<20	Residential	388,413	927	Mollisols
10	51–100	Residential	663,862	1,515	Mollisols
14	20–50	Residential	60,258	355	Ultisols
51	<20	Residential	613,599	1,659	Aridisols
52	20–50	Residential	557,169	1,122	Aridisols
53	<20	Residential	668,342	2,800	Inceptisols
54	20–50	Residential	663,862	1,515	Alfisols
4	20–50	Residential	96,207	95	Entisols
55	20–50	Residential	86,238	492	Alfisols
35	20–50	Residential	622,793	2,962	Ultisols
56	20–50	Residential	26,540	618	Alfisols
5	20–50	Forest	104,300,000	586	Ultisols
23	<20	Residential	96,207	95	Ultisols
25	20–50	Parks and openlands	4,529,500	453	Alfisols
2	51–100	Forest	300,000	2,100	Alfisols
63	N/A	Mix	780,000	359	Inceptisols
1	20–50	Residential	465,700	3,889	Inceptisols
61	51–100	Mix	8,491,079	10,430	Entisols

(Continued)

(*Continued*)

Study	Age	Land use	Population (#)	Population Density (# km⁻²)	Soil Order
28	N/A	Forest	61,875	64	Alfisols
7	N/A	Mix	310,000	4,605	Inceptisols
33	20–50	Mix	663,862	1,515	Mollisols
26	51–100	Parks and openlands	4,529,500	453	Alfisols
31	N/A	Mix	622,793	2,962	Ultisols
34	51–100	Mix	622,793	2,962	Ultisols
57	<20	Residential	407,207	2,737	Alfisols
29	N/A	Mix	2,231	449	Spodosols
37	N/A	Mix	655,884	5,009	Inceptisols
9	N/A	Mix	310,000	4,605	Inceptisols
17	N/A	Residential	2,722,389	4,572	Mollisols
39	N/A	Mix	2,722,389	4,572	Mollisols
62	>100	Mix	12,197,596	4,581	Alfisols
8	N/A	Parks and openlands	2,600,000	302	Inceptisols
27	N/A	Mix	590,000	2,950	Inceptisols
3	N/A	Mix	203,431	1,100	Alfisols
40	>100	Mix	21,784	1,200	Entisols

Study	Bulk Density (g cm⁻³)	Clay (%)	Sand (%)	pH
32	1.2	17.0	53.0	5.2
19	1.3	22.5	58.5	7.8
20	N/A	N/A	N/A	N/A
6	N/A	N/A	N/A	N/A
15	N/A	N/A	N/A	N/A
18	1.4	N/A	N/A	6.6
16	1.7	6.7	81.1	8.7
24	N/A	N/A	N/A	N/A
22	1.3	N/A	N/A	6.4
21	N/A	N/A	N/A	N/A
42	N/A	N/A	N/A	N/A
59	N/A	N/A	N/A	N/A
36	N/A	N/A	N/A	N/A
30	0.9	74.0	10.1	4.5
43	N/A	N/A	N/A	N/A
11	1.1	N/A	N/A	N/A
12	1.7	7.3	74.0	8.3
41	N/A	N/A	N/A	N/A
44	N/A	N/A	N/A	N/A
45	N/A	N/A	N/A	N/A
46	N/A	N/A	N/A	N/A
47	N/A	N/A	N/A	N/A
48	N/A	N/A	N/A	N/A
49	N/A	N/A	N/A	N/A
58	1.7	18.9	48.3	6.4

(*Continued*)

(Continued)

Study	Bulk Density (g cm⁻³)	Clay (%)	Sand (%)	pH
13	N/A	N/A	N/A	N/A
38	1.5	16.6	23.0	6.9
60	1.2	23.4	10.9	6.4
50	N/A	N/A	N/A	N/A
10	1.5	6.5	63.2	7.7
14	1.5	44.2	27.0	N/A
51	N/A	N/A	N/A	N/A
52	N/A	N/A	N/A	N/A
53	N/A	N/A	N/A	N/A
54	N/A	N/A	N/A	N/A
4	1.2	N/A	N/A	5.0
55	N/A	N/A	N/A	N/A
35	1.2	10.0	65.0	5.0
56	N/A	N/A	N/A	N/A
5	1.3	N/A	N/A	N/A
23	N/A	N/A	N/A	N/A
25	1.3	N/A	N/A	6.1
2	N/A	11.7	70.2	7.1
63	1.3	N/A	N/A	N/A
1	1.5	4.0	55.2	7.7
61	1.7	N/A	N/A	N/A
28	1.4	75.0	8.0	5.8
7	N/A	N/A	N/A	N/A
33	N/A	N/A	N/A	N/A
26	1.4	N/A	N/A	N/A
31	N/A	N/A	N/A	N/A
34	N/A	N/A	N/A	N/A
57	N/A	N/A	N/A	N/A
29	1.2	N/A	N/A	5.4
37	0.9	N/A	N/A	N/A
9	N/A	N/A	N/A	N/A
17	N/A	N/A	N/A	N/A
39	1.2	N/A	N/A	6.6
62	N/A	N/A	N/A	N/A
8	N/A	N/A	N/A	N/A
27	N/A	N/A	N/A	N/A
3	N/A	7.8	74.3	7.1
40	1.3	7.0	69.3	7.1

REFERENCES

Beesley, L. 2012. Carbon storage and fluxes in existing and newly created urban soils. *Journal of Environmental Management* 104: 158–165.

Beyer, L., H-P. Blume, D-C. Elsner et al. 1995. Soil organic matter composition and microbial activity in urban soils. *Science of the Total Environment* 168 (3): 267–278.

Beyer, L., P. Kahle, H. Kretschmer et al. 2001. Soil organic matter composition of man-impacted urban sites in North Germany. *Journal of Plant Nutrition and Soil Science* 164 (4): 359–364.

Campbell, C. D., J. R. Seiler, P. E. Wiseman et al. 2014. Soil carbon dynamics in residential lawns converted from Appalachian mixed oak stands. *Forests* 5 (3): 425–438.

Chen, H., W. Zhang, F. S. Gilliam et al. 2013a. Changes in soil carbon sequestration in *Pinus massoniana* forests along an urban-to-rural gradient of southern China. *Biogeosciences* 10: 6609–6616.

Chen, Y., S. D. Day, A. F. Wick et al. 2013b. Changes in soil carbon pools and microbial biomass from urban land development and subsequent post-development soil rehabilitation. *Soil Biology and Biochemistry* 66: 38–44.

Edmondson, J. L., I. S. Jonathan Potter, E. Lopez-Capel et al. 2015. Black carbon contribution to organic carbon stocks in urban soil. *Environmental Science & Technology* 49 (14): 8339–8346.

Edmondson, J. L., Z. G. Davies, N. McHugh et al. 2012. Organic carbon hidden in urban ecosystems. *Scientific Reports* 2: 963.

Edmondson, J. L., Z. G. Davies, S. A. McCormack et al. 2014. Land-cover effects on soil organic carbon stocks in a European city. *Science of the Total Environment* 472: 444–453.

Golubiewski, N. E. 2006. Urbanization increases grassland carbon pools: Effects of landscaping in Colorado's front range. *Ecological Applications* 16 (2): 555–571.

Gough, C. M., and H. L. Elliott. 2012. Lawn soil carbon storage in abandoned residential properties: An examination of ecosystem structure and function following partial human-natural decoupling. *Journal of Environmental Management* 98: 155–162.

Green, D. M., and M. Oleksyszyn. 2002. Enzyme activities and carbon dioxide flux in a Sonoran Desert urban ecosystem. *Soil Science Society of America Journal* 66 (6): 2002–2008.

Hao, C., J. Smith, J. Zhang et al. 2013. Simulation of soil carbon changes due to land use change in urban areas in China. *Frontiers of Environmental Science & Engineering* 7 (2): 255–266.

Huyler, A., A. H. Chappelka, S. A. Prior et al. 2014. Drivers of soil carbon in residential 'pure lawns' in Auburn, Alabama. *Urban Ecosystems* 17 (1): 205–219.

Jia-Guo, J.I.A.O., Y.A.N.G. Lin-Zhang, W.U. Jun-Xi et al. 2010. Land use and soil organic carbon in China's village landscapes. *Pedosphere* 20(1): 1–14.

Jim, C. Y. 1998. Urban soil characteristics and limitations for landscape planting in Hong Kong. *Landscape and Urban Planning* 40 (4): 235–249.

Jo, H-K. 2002. Impacts of urban greenspace on offsetting carbon emissions for middle Korea. *Journal of Environmental Management* 64 (2): 115–126.

Jo, H-K., and G. E. McPherson. 1995. Carbon storage and flux in urban residential greenspace. *Journal of Environmental Management* 45 (2): 109–133.

Kaye, J. P., A. Majumdar, C. Gries et al. 2008. Hierarchical Bayesian scaling of soil properties across urban, agricultural, and desert ecosystems. *Ecological Applications* 18 (1): 132–145.

Kaye, J. P., R. L. McCulley, and I. C. Burke. 2005. Carbon fluxes, nitrogen cycling, and soil microbial communities in adjacent urban, native and agricultural ecosystems. *Global Change Biology* 11 (4): 575–587.

Kellett, R., A. Christen, N. C. Coops et al. 2013. A systems approach to carbon cycling and emissions modeling at an urban neighborhood scale. *Landscape and Urban Planning* 110: 48–58.

Kong, L., Z. Shi, and L. M. Chu. 2014. Carbon emission and sequestration of urban turfgrass systems in Hong Kong. *Science of the Total Environment* 473: 132–138.

Layman, R. M., S. D. Day, D. K. Mitchell et al. 2016. Below ground matters: Urban soil rehabilitation increases tree canopy and speeds establishment. *Urban Forestry and Urban Greening* 16: 25–35.

Liu, Y., C. Wang, W. Yue et al. 2013. Storage and density of soil organic carbon in urban topsoil of hilly cities: A case study of Chongqing Municipality of China. *Chinese Geographical Science* 23 (1): 26–34.

Livesley, S. J., A. Ossola, C. G. Threlfall et al. 2016. Soil carbon and carbon/nitrogen ratio change under tree canopy, tall grass, and turf grass areas of urban green space. *Journal of Environmental Quality* 45 (1): 215–223.

Livesley, S. J., B. J. Dougherty, A. J. Smith et al. 2010. Soil-atmosphere exchange of carbon dioxide, methane and nitrous oxide in urban garden systems: Impact of irrigation, fertiliser and mulch. *Urban Ecosystems* 13 (3): 273–293.

Lorenz, K., and E. Kandeler. 2005. Biochemical characterization of urban soil profiles from Stuttgart, Germany. *Soil Biology and Biochemistry* 37 (7): 1373–1385.

Meshcheryakov, P. V., E. V. Prokopovich, and I. N. Korkina. 2005. Transformation of ecological conditions of soil and humus substance formation in the urban environment. *Russian Journal of Ecology* 36 (1): 8–15.

Nagy, R. C., B. Graeme Lockaby, W. C. Zipperer et al. 2014. A comparison of carbon and nitrogen stocks among land uses/covers in coastal Florida. *Urban Ecosystems* 17 (1): 255–276.

Pouyat, R. V., I. D. Yesilonis, and D. J. Nowak. 2006. Carbon storage by urban soils in the United States. *Journal of Environmental Quality* 35 (4): 1566–1575.

Pouyat, R. V., I. D. Yesilonis, and N. E. Golubiewski. 2009. A comparison of soil organic carbon stocks between residential turf grass and native soil. *Urban Ecosystems* 12 (1): 45–62.

Pouyat, R. V., I. D. Yesilonis, J. Russell-Anelli et al. 2007. Soil chemical and physical properties that differentiate urban land-use and cover types. *Soil Science Society of America Journal* 71 (3): 1010–1019.

Pouyat, R. V., P. M. Groffman, I. D. Yesilonis et al. 2002. Soil carbon pools and fluxes in urban ecosystems. *Environmental Pollution* 116: S107–S118.

Raciti, S. M., L. R. Hutyra, and A. C. Finzi. 2012a. Depleted soil carbon and nitrogen pools beneath impervious surfaces. *Environmental Pollution* 164: 248–251.

Raciti, S. M., L. R. Hutyra, P. Rao et al. 2012b. Inconsistent definitions of "urban" result in different conclusions about the size of urban carbon and nitrogen stocks. *Ecological Applications* 22 (3): 1015–1035.

Raciti, S. M., P. M. Groffman, J. C. Jenkins et al. 2011. Accumulation of carbon and nitrogen in residential soils with different land-use histories. *Ecosystems* 14 (2): 287–297.

Scharenbroch, B. C., J. E. Lloyd, and J. L. Johnson-Maynard. 2005. Distinguishing urban soils with physical, chemical, and biological properties. *Pedobiologia* 49 (4): 283–296.

Scharenbroch, B. C., R. T. Fahey, and M. Bialecki. (unpublished). Soil organic carbon storage in the Chicagoland ecosystem.

Schleuß, U., Q. Wu, and H-P. Blume. 1998. Variability of soils in urban and periurban areas in Northern Germany. *Catena* 33 (3): 255–270.

Schmitt-Harsh, M., S. K. Mincey, M. Patterson, B. C. Fischer, and T. P. Evans. 2013. Private residential urban forest structure and carbon storage in a moderate-sized urban area in the Midwest, United States. *Urban Forestry and Urban Greening* 12 (4): 454–463.

Selhorst, A., and R. Lal. 2012. Effects of climate and soil properties on US home lawn soil organic carbon concentration and pool. *Environmental Management* 50 (6): 1177–1192.

Short, J. R., D. S. Fanning, M. S. McIntosh et al. 1986. Soils of the Mall in Washington, DC: I. Statistical summary of properties. *Soil Science Society of America Journal* 50 (3): 699–705.

Sun, Y., J. Ma, and C. Li. 2010. Content and densities of soil organic carbon in urban soil in different function districts of Kaifeng. *Journal of Geographical Sciences* 20 (1): 148–156.

Townsend-Small, A., and C. I. Czimczik. 2010. Carbon sequestration and greenhouse gas emissions in urban turf. *Geophysical Research Letters* 37(2). doi:10.1029/2009GL041675

Trammell, T. L. E, B. P. Schneid, and M. M. Carreiro. 2011. Forest soils adjacent to urban interstates: Soil physical and chemical properties, heavy metals, disturbance legacies, and relationships with woody vegetation. *Urban Ecosystems* 14 (4): 525–552.

Vasenev, V. I., J. J. Stoorvogel, and I. I. Vasenev. 2013. Urban soil organic carbon and its spatial heterogeneity in comparison with natural and agricultural areas in the Moscow region. *Catena* 107: 96–102.

Index

Printed in the United States
by Baker & Taylor Publisher Services